Construction Estimating Reference Data

By Ed Sarviel, A.S.P.E.

Craftsman Book Company

6058 Corte del Cedro P.O. Box 6500 Carlsbad, CA 92008

Table of Contents

Library of Congress Cataloging in Publication Data Sarviel, E., Construction estimating reference data. Includes index. 1. Building--Estimates. I. Title. TH435.S25 692'.5 80-22796
ISBN 0-910460-89-2 AACR1   ©1981 Craftsman Book Company   Seventh printing 1990

Introduction

Most construction estimators accumulate tables, man-hour records, and useful data during their entire career. Few ever organize and compile this valuable information into a form that could be used by other estimators. This manual is an attempt to assemble in a convenient form a career's worth of estimating reference data.

The content of the book is limited to estimating reference materials; no dollar costs and very little engineering data are included. If you need engineering, architectural or dollar cost data, many other publications are available. Nowhere in this book do we explain how to compile an estimate. Again, many other books explain estimating procedures.

We generally follow the Uniform Construction Index format to organize the book into sections. If you've used the 16 division UCI system before, you'll have no trouble turning right to the section that has the information you need. For those not familiar with the UCI system, an exhaustive index is provided at the back of this book.

The information you find here falls into four categories: Construction Descriptions, Conversion Tables, Material Tables, and Labor Tables.

Construction Descriptions
It's difficult or impossible to estimate the cost of any work you can't visualize. Consequently, many pages of text and illustrations are included in this book to familiarize you with the work involved. But a much larger volume would be needed to describe all construction work. In fact, a whole library of books would be required. The descriptions included here only highlight information that will be of particular interest when estimating the type of work covered.

Conversion Tables
Every estimator needs to make conversions: from tons to cubic yards; from linear feet to board feet; from bank to loose measure. There's no substitute for the exact conversion table you need, and this manual will most likely have what you're looking for.

Material Tables
Properties of materials are an important subject for every construction estimator: What types of materials are generally available, in what finishes, in what weights, in what sizes? A good, non-technical summary of key materials and how they are used will be found in many sections.

Labor Tables
Most of this book is devoted to man-hour estimating tables, and for good reason. The labor required is a key element in every construction estimate. Because there are so many labor tables in this book, and because labor is such an important part of every estimate, an in-depth explanation of the labor tables is appropriate.

Experienced construction estimators recognize that no two jobs are exactly alike. Labor productivity varies widely from job to job, even if the crew remains the same. Thus, judgement is an essential element in every construction estimate. And judgement will be required when using the labor tables in this manual.

The man-hour tables in this book are **not** based on "ideal" conditions. They assume conditions typical of what most contractors encounter on better planned and managed jobs. The labor productivity indicated in the tables will apply to the extent that these conditions apply to the job you are figuring. Where conditions differ, modification will be necessary.

Specifically, the man-hour tables are based on the following assumptions:

Size of the job is moderate, about what most contractors handling this type of work are accustomed to bidding.

The materials needed are readily available at a storage point relatively close to the point of installation. The materials are service grade and meet generally accepted standards for the type of use intended.

Layout and installation are relatively uncomplicated because the plans and specifications are adequate, access to the work is good, and work done by other trades was done according to the plans and is professional quality.

Labor productivity is fair to good. The crew is experienced in this type of work, motivated to complete the work as required, and is just large enough to get the job done using routine procedures.

Temperature and working conditions do not adversely affect progress of the job.

Tools and equipment appropriate for the job are available and used to best advantage during the course of construction.

Work is professional quality. However, exceptional work involving great detail, decorative materials, custom treatments or unique skills is not considered. Any defects or omissions are remedied before the crew leaves the job site.

Only new construction is involved. Repair, replacement and remodeling-type work often involves problems of limited access, matching of materials, working with non-standard sizes, patching, and control of the construction environment. The tables will be a useful guide to the extent that repair or remodeling work is similar to new construction.

Scope of the Work Described

The man-hour tables in this book will be a useful guide if you can visualize what work is included and what work is excluded from the tasks listed in each table. No man-hour estimate is useful if there is considerable doubt about what the figures actually cover. Most labor tables in this book have a footnote which should clarify the scope of work included. But it would be nearly impossible to describe in detail every element of each man-hour estimate.

It is safe to assume, however, that every task essential to performing the work has been included in the estimate. If the scope of work covered still is not clear, understand that estimates are for the "complete" job and include all of the associated work usually performed along with the named task unless noted otherwise.

But be aware that most tables include work by only one trade classification. This should help you define what is included and excluded from each table. For example, man-hours for installing a cabinet (by a carpenter) would not include the time to stain and seal the cabinet (by a painter).

Two categories of work are specifically included and excluded from the man-hour tables:

Non-productive labor is not included unless noted otherwise. On larger crews it is assumed that the supervisor works along with the crew when not actually directing the work.

Mobilization and demobilization on the site are included. Time usually spent unloading tools, materials and equipment, and preparing to do the work is included in the man-hour estimates. So is time spent reloading tools and equipment at job completion and cleanup of surplus material. Naturally, no travel time or delays off the site are included unless specifically noted.

How Accurate are the Tables?

The figures published here are the result of actual observations compiled, interpreted, and verified by professional estimators. This implies an exercise of judgement. And it should be clear to the user that this manual is the product of personal judgement. Crew productivity varies widely. Even the most well-informed, professional judgement can not guarantee that the figures here will apply to the job you are estimating.

In the aggregate the man-hour estimates in this manual will be accurate to within about 20% on most jobs where conditions are similar to the conditions outlined. On most of the remaining jobs the figures will be too high by 20% or more — estimating more man-hours than are actually required. This is intentional, as an estimate slightly too high is better than an estimate slightly too low. Most contractors would agree.

Let's look at an example. This book lists labor installing asphalt strip shingles at 1.5 man-hours per 100 square feet. A skilled shingle specialist working under ideal conditions on a larger job will be able to handle considerably more work than this. You'll hear claims of 200 or more square feet per man-hour. At that rate a two man crew would finish two 1600 square foot roofing jobs in a day. That's an excellent rate. But it's not a rate most estimators should use until they saw their crews produce results like that on several jobs.

Now look at the other extreme. Estimators who have figured asphalt shingles on commercial, better industrial or military jobs claim that 2 to 2.5 man-hours per 100 square feet is a reasonable figure. We don't doubt the validity of these estimates for that type of work.

Most jobs fall between these extremes. Many experienced roofing contractors would insist that their crews can average close to 100 square feet per man-hour. Thus a two man crew would finish that 1600 square foot job in one 8 hour day. That's 33% faster than the 1.5 man-hours listed on page 192 of this manual.

Again, we don't doubt the validity of these figures. But we recognize that they are based on specialized crews working under experienced supervisors and with exactly the right equipment.

To summarize, reasonable estimates for installing asphalt strip shingles may vary from .5 to 2.5 man-hours per square, with most experienced crews producing about 100 square feet per hour.

Why the difference? There are many reasons. The highest and lowest productivity rates vary from the conditions outlined above and probably don't include and exclude the same tasks. Many shinglers can put down 200 square feet of shingles in an hour. But every shingle job involves checking and cleaning the deck, moving tools and materials into place, laying out the job, placing felt and starter strips, some flashing work, making minor repairs and cleanup. All of this should be included in any realistic estimate and is included in the 1.5 hour estimate in this book.

Another difference is the type of work itself — even though the material applied fits the same description. Commercial jobs nearly always receive higher quality workmanship than the typical residential job. Every tradesman worthy of your payroll will give more care and attention to a highly visible roof on a commercial building covered with top quality shingles than a garage roof using strictly standard grade shingles.

Recognize also that an experienced crew working under the direct supervision of an owner-entrepreneur will out-produce less motivated tradesmen on nearly every job. And a crew working on a piecework basis will do its level best to get the job done by quitting time so a return trip isn't necessary the next day.

No single figure will cover all work and a 500 word essay would be needed to describe most common situations. In this book we have selected what we feel are reasonable first approximations for labor productivity. Use these figures when the productivity of the crew is unknown and the exact type of work is not specified. In the case of asphalt strip shingles, 1.5 man-hours was our choice. And that brings us to our final, and most important point.

Every man-hour estimate in this volume is a poor second choice when compared to the figures you develop yourself from the work your crews have handled. Your most reliable guide will always be your own cost records. Where you must supplement your experience with the reference data in this manual, we hope our judgement proves worthy of your trust.

1
general requirements section

Estimating Service

Date	Job No.	LOC.	For
Auth.	Owner	DWG.	Sheet_____of_____
Est. By	Hrs.	Act. Hrs.	Project

No.	Item and/or Event	Unit	Material Cost		Labor Cost		Extention
			Unit $	$	Unit $	$	$

section contents

Tips for Maximizing Productivity

Order materials correctly and issue specific delivery times and places. Make certain that the materials supplier understands the correct amount to be delivered to the proper site at the proper job at the proper time. This avoids delays and extra costs due to incomplete material orders or delivery errors. Keep materials lists up to date. Give your supplier instructions on how piles of materials are to be built up.

Create a flexible material handling system. Because building requires a wide range of materials, flexibility is very important. Most builders have to perform a wide variety of work. Develop systems capable of handling many different types of materials. If possible, anticipate the breakdown of one system and have an alternate ready to avoid delay.

Reduce all materials handling time to a minimum. Storing and moving materials from one site on the job to another can be avoided almost entirely with a good delivery system. Proper materials placement on delivery at the site reduces hand carrying and sorting. Move materials in quantity wherever possible. For example, deliver the lumber package for rough framing to a location predetermined by the supervisor or indicated on the drawings. Stack lumber in the sequence of use with the first material needed on top to eliminate the need for sorting and restacking.

Use the best method of material handling. Find and use the best way of handling the materials for each operation or phase of construction. Methods vary between operations due to differences in materials handled. The best method does not always mean carrying the largest possible quantity with each trip. For example, if a small amount of brick is required for a house, the brick should be packaged so that it can be carried by a two-wheeled handcart rather than a forklift.

Use packaged and unitized materials. This discourages pilferage and materials scattering and helps reduce the total cost of materials. It also tends to speed up production because workmen have no trouble finding the necessary materials when needed. For example, have lumber delivered in banded stacks, have block delivered in pallet-sized units, and use boxes for tools and equipment when possible.

Reduce interdependency of men and machines. Provide holding or storage facilities so workmen can continue their operations without being delayed by waiting for other men or machines. For example, when material is raised with a forklift, have a storage platform on the roof so the man below does not have to wait for the man above to deposit each piece and return to take the next piece.

Schedule for continuous flow of production and materials. Schedule production so there is minimum distance between jobsites. A diagram may help determine the best schedule. Certain crews must make several trips to each jobsite. The need for scheduling an even flow of materials and production should be obvious. Schedules must include time for both production and material delivery. For example, assume lumber to be used in the construction of two houses will be delivered to the jobsites in one truck. The material can be unloaded much faster and placed in the proper locations with a minimum of travel if the houses are adjacent to each other.

Use good housekeeping. Whenever possible, use unitized and packaged material to keep good material together. Be certain that material is not scattered around the jobsite and susceptible to damage by vehicles and weather. Separate waste material from usable material. This allows easier equipment movement around each site.

Make full use of equipment. Use the right kind of equipment for each job. Keep equipment operating near its maximum capacity to provide the largest possible benefit. But don't overextend the capacity of equipment. Make sure operators are familiar with each piece of equipment. Consider equipment use when scheduling all work involving equipment. This keeps equipment working on productive jobs and helps eliminate conflicts. For example, don't use a large capacity front-end loader for carrying lumber around a jobsite. Don't try to use a small loader to dig an entire basement. Remember that your goal is to keep every piece of equipment working all during every work day. No builder can do this, but keep it in mind when buying equipment. Don't buy equipment that is going to stand idle much of the time. Nearly anything you need only occasionally will be available at an equipment rental yard. Your success as a builder is not measured by the value of the equipment parked in your yard or on jobsites each night. Your success is in the profit you make on each job and at the end of each year.

Use multipurpose equipment. Equipment flexibility depends on the size of the operation and the type of construction done. In a large organization, specific uses of equipment can be justified. But builders with a smaller volume must be able to perform a variety of operations with each major piece of equipment. To keep equipment operating continuously, you need equipment that can perform several types of operations. For example, a front-end loader with backhoe can be used for small excavation work and for lifting and carrying materials. It can also be used for pulling tool wagons and equipment such as concrete mixers around the jobsite. A belt conveyor used for positioning gravel in the basement can also carry shingles to the roof.

Use simple low-cost jigs and equipment. Job fabricated jigs and simple tools can greatly reduce materials handling. For example, build a plywood ladder out of 2'' x 6'' lumber for moving plywood to the first and second floors. Make a simple chute out of plywood for positioning gravel in the basement for slab base.

Select equipment carefully. When considering the purchase of equipment, compare annual savings produced by the equipment with the annual cost of the equipment. In determining equipment cost, consider the initial price, capital costs, repair costs and operating costs.

Set up a maintenance program. Good maintenance extends equipment life and reduces equipment down time. If equipment is to operate properly when needed, it must be maintained.

Reduce worker fatigue. Good material handling reduces worker fatigue and allows operations to be performed better and faster. This is especially important when heavy, bulky or a large number of items must be moved. For example, transporting concrete block to the basement can be exhausting work. Many masonry jobs are 30% or more materials handling. Be especially alert to reduce double handling when heavy materials are involved. Try to get materials delivered close to the point of use and at the same height if possible. This reduces fatigue and can significantly improve production late in the day and during warm weather.

Increase safety awareness. Good materials handling methods reduce on-the-job injuries and cut lost time. Time lost due to injury normally costs more than a good materials handling program. For example, place a waste box on each job to avoid rehandling and stepping on discarded materials. This also highlights waste and scrap material accumulation by putting it where everybody can see it.

Don't let the work expand to fill the time available. Most tradesmen can drag out a task until it is completed just at quitting time. If your crews begin carrying tools back to the truck an hour before quitting time and spend the last 15 minutes waiting in the truck for four o'clock to arrive, you have a problem. If you don't do any other scheduling, schedule the last hour of each day. Make sure each lead man has a list of productive work that can be put off until some job is finished just before the end of the day. If nothing else, the crew can spend the time maintaining tools, cleaning up the jobsite or moving materials to where they will be needed the next day. Likewise, make it your policy that work begins at the beginning of the work day. Some employees arrive at work with the intention of spending the first half hour drinking coffee, eating a roll and discussing last night's T.V. programs or the morning news. An employee who needs a half hour to get ready to go to work should arrive at the job a half hour early.

Use the lowest paid workers to haul materials. If you look at most building tasks carefully, you will discover that about 40% of the work does not require highly skilled labor. Most crews should be one half laborers or apprentices, and the laborers or apprentices should do nearly all the materials handing. A single laborer can often supply materials to two craftsmen working in different trades. For example, if you have a one man finish carpentry crew and a one man painting crew on the job, a single laborer may be able to mix paint for the painter when he is not helping the carpenter carry doors and hardware. Every crew of two or more men on most jobs should include a laborer or trainee to handle materials.

Don't waste valuable materials. Good material handling policy requires that carpenters take the time to pick up and use every sound piece of lumber over 11 inches long. There are probably 500 places in a home where a 12 inch 2 x 4 or 2 x 6 can be used. At $500 per thousand board feet, a 2 x 6 twelve inches long is worth 50 cents. Even at a cost of $20 per hour, it is worth a minute of a carpenter's time to pick up a 12 inch 2 x 6. At $500 per thousand board feet, stud material costs considerably more than the labor required to install or repair it. All scrap material is not waste material. Educate supervisors on the differences: scrap is what is left over after some material has been installed; waste material has no value for its original intended use. Scrap materials are waste only if they can not be used economically somewhere else on the job. Separate scrap and waste and encourage crews to go to the scrap pile when they need bits and pieces to complete a job.

Instruct your staff. Show your entire staff the importance of the materials handling systems that you develop. Everyone should assist in putting these systems into use. Workmen should be trained in the correct methods for handling materials.

Encourage your supervisors to "microschedule." Some of the most effective planning on a residential job is done by lead men and foremen who have the appropriate materials available before the need arises. As each task is finished, the skilled craftsmen should be able to go directly to the next task. The tools and materials should be prepositioned to where they are needed and the lead man should have thought out just how the work will be done. This type of planning pleases many craftsmen and frees them to be nearly 100% productive. Once that task is progressing smoothly, the lead man should begin to think about preparing for the next task. This is scheduling on a very small and very personal scale. Many crew leaders do it instinctively. Every leader should "microschedule" for high productivity.

Use the smallest crew that can get the job done. There is a tendency to have more people available to do the work than are really needed. Job foremen often feel that they need more men or could get the job done better if they had a larger crew. This is especially true where the work is unpleasant or the working conditions difficult. Make it clear to your foremen that they are to use the smallest crew size that will get the job done on time. This means that for many jobs only one man will be assigned. Seldom will you need a seven man carpentry crew, for example.

There are several reasons for using the smallest crew possible. Delays occur on every job. There are many reasons why the work can't go forward temporarily. Many times, it is expedient to drag out the work until quitting time for the day. There are just about the same number of these delays on a given job when a small crew is working as when a large crew is working. But where there are more men, there are more manhours wasted during the delay. Unfortunately, your lead man, foreman or superintendent may not realize how much time is nonproductive. Look at your jobs through the eyes of an industrial engineer. Spot time lost due to unnecessary delay.

Building Material Weights per Volume

Asbestos 110-120 lbs. per C.F.
Brick
    Common 2½" x 4" x 8¼", 5.4 lbs. each; 2.7 tons per 1000
    Fire, standard 9" x 4½" x 2½", 7.0 lbs. each; 3.5 tons per 1000
    Hard 2¼" x 4¼" x 8½", 6.48 lbs. each; 3.24 tons per 1000
    Paving 2¼" x 4" x 8½", 6.75 lbs. each; 3.37 tons per 1000
    Paving brick 3¼" x 4" x 8½", 8.75 lbs. each; 4.37 tons per 1000
    Soft 2¼" x 4" x 8¼", 4.32 lbs. each; 2.6 tons per 1000
Cement bag 94 lbs. each; bbl. weighs 376 lbs.
Clay
    Dry 63-95 lbs. per C.F.; 1700-2295 lbs. per C.Y.
    Fire 130 lbs. per C.F.; 3510 lbs. per C.Y.
    Wet 120-140 lbs. per C.F.; 2970-3200 lbs. per C.Y.
Concrete 138 lbs. per C.F.; 3726 lbs. per C.Y.
    Cinder concrete 112 lbs. per C.F.
    Gravel and limestone concrete 150 lbs. per C.F.
    Trap-rock concrete 155 lbs. per C.F.
Crushed stone 100 lbs. per C.F.; 2700 lbs. per C.Y.
Gravel 95 lbs. per C.F.; 2565 lbs. per C.Y.
Hydrated lime 40 lbs. per C.F.
Mortar 103 lbs. per C.F.
Plaster of paris 98 lbs. per C.F.
Reinforced concrete 150 lbs. per C.F.
Sand
    Dry 97-117 lbs. per C.F; 2619-3159 lbs. per C.Y.
    Wet 120-140 lbs. per C.F.; 3240-3780 lbs. per C.Y.
Shingles, bundles 24" long, 20" wide, 10" high weighs 50 lbs. Approximately 250 per bundle.
Slag 1755-1890 lbs. per C.Y.; 65-70 lbs. per C.F.
Slag concrete 135 lbs. per C.F.
Stone riprap 65 lbs. per C.F.; 1755 lbs. per C.Y.

Live Load Allowances
Minimum Uniformly Distributed Live Loads

Occupancy or Use	Live Load in Lbs. Per S.F.
Apartments (see Residential)	
Assembly halls and other places of assembly	
Fixed seats	60
Movable seats	100
Balcony (exterior)	100
Bowling alleys, poolrooms, etc.	75
Corridors	
First floor	100
Other floors, same as occupancy served	
Dance halls, dining rooms and restaurants	100
Dwellings (see Residential)	
Garages (passenger cars)	100
Floors shall be designed to carry 150% of the maximum wheel load anywhere on the floor.	
Grandstands (see Reviewing stands)	
Gymnasiums, main floors and balconies	100
Hospitals	
Operating rooms	60
Private rooms	40
Wards	40
Hotels (see Residential)	
Libraries	
Reading rooms	60
Stack rooms	150
Manufacturing	125
Marquees	75
Office buildings	
Offices	80
Lobbies	100
Residential	
Multifamily houses	
Private apartments	40
Public rooms	100
Corridors	60

Live Load Allowances (Continued)
Miminum Uniformly Distributed Live Loads

Occupancy or Use	Live Load in Lbs. Per S.F.
Dwellings	
First floor	40
Second floor and habitable attics	30
Uninhabitable attics	20
Hotels	
Guest rooms	40
Public rooms	100
Corridors serving public rooms	100
Public corridors	60
Private corridors	40
Reviewing stands and bleachers	100
Schools	
Classrooms	40
Corridors	100
Sidewalks, vehicular driveways, and yards subject to trucking	250
Skating rinks	100
Stairs, fire escapes, and exitways	100
Storage warehouse	
Light	125
Heavy	250
Stores	
Retail	
First-floor, rooms	100
Upper floors	75
Wholesale	125
Theaters	
Aisles, corridors and lobbies	100
Orchestra floors	60
Balconies	60
Stage floors	150
Yards and terraces, pedestrian traffic	100

Dead Loads — Approximate Weights per Square Foot

Roof or Ceiling Type	Pounds
Roofs	
Asphalt, felt and gravel (3-5 ply built-up)	5 - 6½
Asphalt, felt and slag (3-5 ply built-up)	4½-5½
Composition 3-ply	1
Concrete, cinder (per inch thickness)	9
Concrete, nailing (per inch thickness)	8
Corrugated aluminum (.024″ thick)	½
Corrugated asbestos (¼″-3/8″ thick)	3 - 4½
Corrugated iron-steel (20-18 gauge)	2 - 3
Gypsum slab (per inch thickness)	8
Sheathing boards (1″ WP, Spr., Hmlk.)	2½ - 3
Shingles, asbestos	3 - 6
Ceilings	
Gypsum lath and plaster (3/8″ plus 1/2″ thick)	5½
Lath and ¾″ plaster	8
Suspended metal lath and plaster	10
Gypsum board (½″ thick)	2

Building Material Weights Per Square Foot

Block, creosoted wood, 3″	15.00
Boards, fiber insulating	
1″	1.50
¾″	1.10
½″	0.80
Ceiling, wood	
¾″	2.50
5/8″	1.80
½″	1.40
3/8″	1.10
Copper, sheet	1.00
Lead, sheet	4.00 to 8.00
Plywood	
¼″	0.70
5/16″	1.00
3/8″	1.10
Shingles	
Asphalt	2.00 to 3.00
Wood	2.50
Slate	10.00
Tile, plain	9.00 to 12.00
Tin, painted	1.00
Zinc, sheet	1.00 to 2.00

Use these figures when determining dead loads on floors and roofs.

Inch Fractions to Decimal Equivalents

1/32 = .03125	5/32 = .15625	9/32 = .28125	13/32 = .40625
1/16 = .0625	3/16 = .1875	5/16 = .3125	7/16 = .4375
3/32 = .09375	7/32 = .21875	11/32 = .34375	15/32 = .46875
1/8 = .125	1/4 = .250	3/8 = .375	1/2 = .500
17/32 = .53125	21/32 = .65625	25/32 = .78125	29/32 = .90625
9/16 = .5625	11/16 = .6875	13/16 = .8125	15/16 = .9375
19/32 = .59375	23/32 = .71875	27/32 = .84375	31/32 = .96875
5/8 = .625	3/4 = .750	7/8 = .875	1 = 1.0000

Area and Volume Conversions

Area of a square = length x breadth or height.

Area of a rectangle = length x breadth or height.

Area of a triangle = base x ½ altitude.

Area of parallelogram = base x altitude.

Area of trapezoid = altitude x ½ the sum of parallel sides.

Area of trapezium = divide into two triangles, total their areas.

Circumference of circle = diameter x 3.1416.

Circumference of circle = radius x 6.283185.

Diameter of circle = circumference x .3183.

Diameter of circle = square root of area x 1.12838.

Radius of a circle = circumference x .159155.

Area of a circle = half diameter x half circumference.

Area of a circle = square of diameter x .7854.

Area of a circle = square of circumference x .07958.

Area of a sector of circle = length of arc x ½ radius.

Area of a segment of circle = area of sector of equal radius minus area of a triangle, when the segment is less, and plus area of triangle, when segment is greater than the semi-circle.

Area of circular ring = sum of the diameter of the two circles x difference of the diameter of the two circles and that product x .7854.

Side of square that shall equal area of circle = diameter x .8862.

Side of square that shall equal area of circle = circumference x .2821.

Diameter of circle that shall contain area of a given square = side of square x 1.1284.

Side of inscribed equilateral triangle = diameter x .86.

Side of inscribed square = diameter x .7071.

Side of inscribed square = circumference x .225.

Area of ellipse = product of the two diameters x .7854.

Area of a parabola = base x 2/3 of altitude.

Area of a regular polygon = sum of its sides times perpendicular from its center to one of its sides divided by 2.

Surface of cylinder or prism = area of both ends plus length and times circumference.

Surface of sphere = diameter x circumference.

Solidity of sphere = surface x 1/6 diameter.

Solidity of sphere = cube of diameter x .5236.

Solidity of sphere = cube of radius x 4.1888.

Solidity of sphere = cube of circumference x .016887.

Diameter of sphere = cube root of solidity x 1.2407.

Diameter of sphere = square root of surface x .56419.

Circumference of sphere = square root of surface x 1.772454.

Circumference of sphere = cube root of solidity x 3.8978.

Contents of segment of sphere = (height squared plus three times the square of radius of base) x (height x .5236).

Contents of a sphere = diameter x .5236.

Side of inscribed cube of sphere = radius x 1.1547.

Side of inscribed cube of sphere = square root of diameter.

Surface of pyramid or cone = circumference of base x ½ of the slant height plus area of base.

Contents of pyramid or cone = area of base x ⅓ altitude.

Contents of frustum of pyramid or cone = sum of circumference at both ends x ½ slant height plus area of both ends.

Contents of frustum of pyramid or cone = multiply areas of two ends together and extract square root. Add to this root the two areas and x ⅓ altitude.

Contents of a wedge = area of base x ½ altitude.

Square Measure

1 square centimeter	0.1550 square inch
1 square decimeter	0.1076 square feet
1 square meter	1.196 square yard
1 acre	3.954 square rods
1 hectare	2.47 acres
1 square kilometer	0.386 square mile
1 square inch	6.452 square centimeters
1 square foot	9.2903 square decimeters
1 square yard	0.8361 square meter
1 square rod	0.259 acre
1 acre	0.4047 hectare
1 square mile	2.59 square kilometers
144 square inches	1 square foot
9 square feet	1 square yard
30¼ square yards	1 square rod
40 square rods	1 rood
4 roods	1 acre
640 acres	1 square mile

Square Tracts of Land

Acres	Length of One Side of Square Tract, L.F.	Area S.F.
1/10	66.0	4,356
1/8	73.8	5,445
1/6	85.2	7,260
1/4	104.4	10,890
1/3	120.5	14,520
1/2	147.6	21,780
3/4	180.8	32,670
1	208.7	43,560
1½	255.6	65,340
2	295.2	87,120
2½	330.0	108,900
3	361.5	130,680
5	466.7	217,800

Linear Conversions

1 centimeter	0.3937 inches
1 inch	2.54 centimeters
1 decimeter	3.937 inches or 0.328 foot
1 foot	3.048 decimeters
1 meter	39.37 inches or 1.0936 yards
1 yard	0.9144 meter
1 dekameter	1.9884 rods
1 rod	0.5029 dekameter
1 kilometer	0.62137 mile
1 mile	1.6093 kilometers

Volume Conversions

1 cubic centimeter	0.061 cubic inch
1 cubic inch	16.39 cubic centimeters
1 cubic decimeter	0.0353 cubic foot
1 cubic foot	28.317 cubic decimeters
1 cubic yard	0.7646 cubic meter
1 stere	0.2759 cord
1 cord	3.624 steres
1 liter	0.908 dry quarts or 1.0567 liquid quarts
1 dry quart	1.101 liters
1 liquid quart	.09463 liter
1 dekaliter	2.6417 gallons or 1.135 pecks
1 gallon	0.3785 dekaliter
1 peck	0.881 dekaliter
1 hektoliter	2.8375 bushels
1 bushel	0.3524 hektoliter

Length Conversion Tables for English to Metric Systems

Inches
Centimeters   Example: 2 inches = 5.08 cm
Feet
Meters
Yards

Meters Miles Kilometers	Kilometers Miles	Miles Kilometers	Meters Yards	Yards Meters	Meters Feet	Feet Meters	Centimeters Inches	Inches Centimeters
1	0.62	1.61	1.09	0.91	3.28	0.30	0.39	2.54
2	1.24	3.22	2.19	1.83	6.56	0.61	0.79	5.08
3	1.86	4.83	3.28	2.74	9.84	0.91	1.18	7.62
4	2.49	6.44	4.37	3.66	13.12	1.22	1.57	10.16
5	3.11	8.05	5.47	4.57	16.40	1.52	1.97	12.79
6	3.73	9.66	6.56	5.49	19.68	1.83	2.36	15.24
7	4.35	11.27	7.66	6.40	22.97	2.13	2.76	17.73
8	4.97	12.87	8.75	7.32	26.25	2.44	3.15	20.32
9	5.59	14.48	9.84	8.23	29.53	2.74	3.54	22.86
10	6.21	16.09	10.94	9.14	32.81	3.05	3.93	25.40
20	12.43	32.19	21.87	18.29	65.62	6.10	7.87	50.80
30	18.64	48.28	32.31	27.43	98.42	9.14	11.81	76.20
40	24.85	64.37	43.74	36.58	131.23	12.19	15.75	101.60
50	31.07	80.47	54.68	45.72	164.04	16.24	19.68	127.00
60	37.28	96.56	65.62	54.86	196.85	18.29	23.62	152.40
70	43.50	112.65	76.55	64.00	229.66	21.34	27.56	177.80
80	49.71	128.75	87.49	73.15	262.47	24.38	31.50	203.20
90	55.92	144.34	98.42	82.80	295.28	27.43	35.43	228.60
100	62.14	160.94	109.36	91.44	328.08	30.48	39.37	254.00

Basic Metric Length Relationships

One Unit (Below) Equals	Millimeters	Centimeters	Meters	Kilometers
Millimeter (mm)	1.	0.1	0.001	0.000,001
Centimeter (cm)	10.	1.	0.01	0.000,01
Meters	1,000.	100.	1.	0.001
Kilometer (km)	1,000,000.	100,000.	1,000.	1.

Weight[1] Conversion Tables for English to Metric Systems

Number	Metric Ton / Short Ton	Short Ton / Metric Ton	Kilograms / Pounds	Pounds / Kilograms	Grams / Ounces	Ounces / Grams
1	1.10	0.91	2.20	0.46	0.04	28.4
2	2.20	1.81	4.41	0.91	0.07	56.7
3	3.31	2.72	6.61	1.36	0.11	85.0
4	4.41	3.63	8.82	1.81	0.14	113.4
5	5.51	4.54	11.02	2.67	0.18	141.8
6	6.61	5.44	13.23	2.72	0.21	170.1
7	7.72	6.35	15.43	3.18	0.25	198.4
8	8.82	7.26	17.64	3.63	0.28	226.8
9	9.92	8.16	19.84	4.08	0.32	255.2
10	11.02	9.07	22.05	4.54	0.35	283.5
20	22.05	18.14	44.09	9.07	0.71	567.0
30	33.07	27.22	66.14	13.61	1.06	850.5
40	44.09	36.29	88.18	18.14	1.41	1134.0
50	55.12	45.36	110.23	22.68	1.76	1417.5
60	66.14	54.43	132.28	27.22	2.12	1701.0
70	77.16	63.50	154.32	31.75	2.47	1984.5
80	88.18	72.57	176.37	36.29	2.82	2268.0
90	99.21	81.65	188.42	40.82	3.17	2551.5
100	110.20	90.72	220.46	45.36	3.53	2835.0

Example: Convert 28 pounds to kilograms.
28 pounds = 20 pounds + 8 pounds
From the tables: 20 pounds = 9.07 kg and 8 pounds = 3.63 kilograms
Therefore, 28 pounds = 9.07 kg + 3.63 kg = 12.70 kg
[1]The weights used for the English system are avoirdupois (common) weights. The short ton is 2,000 pounds. The metric ton is 1,000 kilograms.

Volume Conversion Tables for English to Metric Systems

Cubic Meters Cubic Yards Cubic Feet	Cubic Feet To:		Cubic Yards To:		Cubic Meters To:	
	Cubic Yards	Cubic Meters	Cubic Feet	Cubic Meters	Cubic Feet	Cubic Yards
1	0.037	0.028	27.0	0.76	35.3	1.31
2	0.074	0.057	54.0	1.53	70.6	2.62
3	0.111	0.085	81.0	2.29	105.9	3.92
4	0.148	0.113	108.0	3.06	141.3	5.23
5	0.185	0.142	135.0	3.82	176.6	6.54
6	0.212	0.170	162.0	4.59	211.9	7.85
7	0.259	0.198	189.0	5.35	247.2	9.16
8	0.296	0.227	216.0	6.12	282.5	10.46
9	0.333	0.255	243.0	6.88	317.8	11.77
10	0.370	0.283	270.0	7.65	353.1	13.07
20	0.741	0.566	540.0	15.29	706.3	26.16
30	1.111	0.850	810.0	22.94	1059.4	39.24
40	1.481	1.133	1080.0	30.58	1412.6	52.82
50	1.852	1.416	1350.0	38.23	1765.7	65.40
60	2.222	1.700	1620.0	45.87	2118.9	78.48
70	2.592	1.982	1890.0	53.52	2472.0	91.56
80	2.962	2.265	2160.0	61.16	2825.2	104.63
90	3.333	2.548	2430.0	68.81	3178.3	117.71
100	3.703	2.832	2700.0	76.46	3531.4	130.79

Example: 3 cubic yards = 81.0 cubic feet.
Volume: The cubic meter is the only common dimension used for measuring the volume of solids in the metric system.

Conversion Factors

(°C. x 9/5) + 32 = °F.
(°F — 32) x 5/9 = °C.
Liter x 1.05671 = U.S. quarts
Quarts x .946333 = liters
Liters x 61.025 = cubic inches
Gallons x 231 = cubic inches
Kilograms x 2.2046 = pounds
Pounds x 453.59 = grams
Ounces (avdp) x 28.35 = grams
Kilowatts x 1.341 = horsepower
Horsepower x 746 = watts
1 atmosphere = 33.899 feet of water at 39 1°F
1 atmosphere = 760 mm. of mercury
1 atmosphere = 14.7 pounds per square inch
1 cubic foot water = 62.37 pounds @ 60°F
1 cubic inch water = 0.036 pounds @ 60°F
Cubic meters x 35.314 = cubic feet
Cubic feet x 0.02832 = cubic meters
Centistokes x density = centipoises
Pounds/gallon at 20°C. = specific gravity at 20/20°C. x 8.3216
1 centimeter = 0.3937 inches
1 inch = 2.540 centimeter

Cubic Measure

1,728 cubic inches	1 cubic foot
128 cubic feet	1 cord wood
27 cubic feet	1 cubic yard
40 cubic feet	1 ton shipping
2,150.42 cubic inches	1 standard bushel
268.8 cubic inches	1 standard gallon dry
231 cubic inches	1 standard gallon liquid
1 cubic foot	About 4/5 of a bushel
1 Perch	A mass 16½ feet long, 1 foot high and 1½ feet wide, containing 24-2/3 cubic feet.

Miscellaneous

3 inches	1 palm
4 inches	1 hand
6 inches	1 span
18 inches	1 cubit
21.8 inches	Bible cubit
2½ feet	1 military pace

Fractions of an Inch

Inch	1/16	1/8	3/16	1/4	5/12	3/8	7/16	1/2
Centimeters	0.16	0.32	0.48	0.64	0.79	0.95	1.11	1.27

Inch	9/16	5/8	11/16	3/4	13/16	7/8	15/16	1
Centimeters	1.43	1.59	1.75	1.91	2.06	2.22	2.38	2.54

Units of Centimeters

Centimeters	0.1	0.2	0.3	0.4	0.5	0.6	0.7	0.8	0.9	1.0
Inches	0.04	0.08	0.12	0.16	0.20	0.24	0.28	0.31	0.35	0.39

Surveyor's Measure

7.92 inches	1 link
25 links	1 rod
4 rods	1 chain
10 square chains or 160 square rods	1 acre
640 acres	1 square mile
36 square miles or 6 miles square	1 township

Weight Conversions

1 gram	0.03527 ounce
1 ounce	18.35 grams
1 kilogram	2.2046 pounds
1 pound	0.4536 kilogram
1 metric ton	0.98421 English ton
1 English ton	1.016 metric ton

Time Conversion Factors

Man-Hours to Minutes		Man-Days to Hours	
Fractional Man-Hours	Minutes Equivalent	Fractional Man-Days	Man-Hours Equivalent
.04	2.4	.1	48 min.
.05	3.0	.2	1 hr. 36 min.
.06	3.6	.3	2 hr. 24 min.
.07	4.2	.4	3 hr. 12 min.
.08	4.8	.5	4 hr.
.09	5.6	.6	4 hr. 48 min.
.10	6.0	.7	5 hr. 36 min.
.15	9.0	.8	6 hr. 24 min.
.20	12.0	.9	7 hr. 12 min.
.25	15.0		
.30	18.0		
.40	24.0		
.50	30.0		
.60	36.0		
.70	42.0		
.80	48.0		
.90	56.0		
1.00	60.0		

Basic Metric Weight Relationships

One Unit (Below) Equals	Grams	Kilograms	Metric Ton
Gram (gm)	1.	0.001	0.000,001
Kilogram (kg)	1,000.	1.	0.001
Metric ton	1,000,000.	1,000.	1.

2
site work section

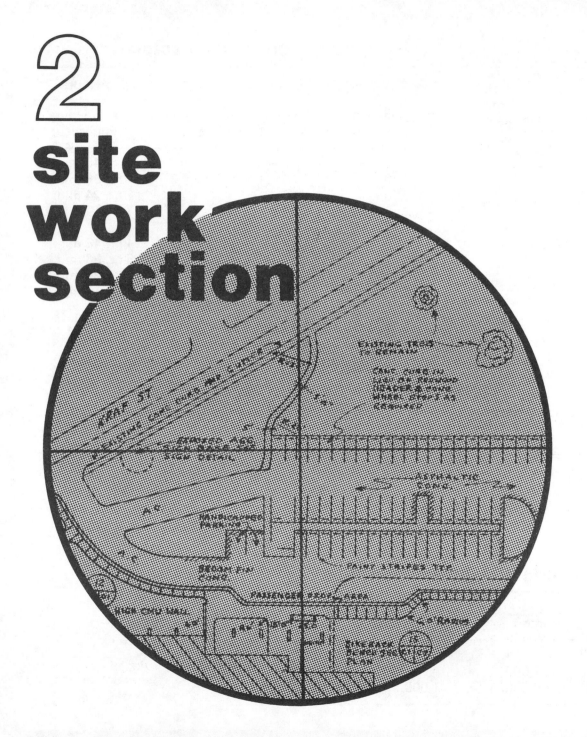

section contents

Sitework Definitions

Asphalt, Liquid Asphalt in a fluid or soft state which can not be tested with any normal penetration testing method.

Asphalt Paving The act of laying down a surface consisting of various sized aggregates coated and bound together with an asphalt cement supported on a base course of gravel, crushed stone, concrete or brick.

Asphalt Plant A facility where aggregate is mixed with liquid asphalt and heated to form an asphalt mix ready for use in the construction of paved surfaces.

Blasting Use of high explosives to move or fracture rock or earth to facilitate its removal from the work area.

Cement Asbestos Pipe Pipe made from a specialized cement with a high degree of strength and low degree of water permeability. Usually used for water distribution.

Chain Link Fencing A woven wire fabric formed by interlocking vertical strands to form a unit with a high degree of strength and flexibility.

Compaction To reduce any loose bulk such as earth, sand or crushed rock to a volume less than its original condition by tamping, rolling or watering down.

Concrete Pipe Pipe made from concrete and reinforced with steel reinforcing bars. Widely used in the construction of storm drains.

Corrugated Metal Pipe A metal pipe with grooves and ridges around the circumference, giving added strength and rigidity to metal pipe.

Cubic Yard The volume of material contained in a space three feet by three feet by three feet.

Demolition The planned and deliberate removal of an object by destruction. Demolition is usually done with hand tools to allow salvage of valuable materials. High explosives and machines such as wrecking balls and dozers are also used.

Earth Moving The removal of earth from one location to another with the use of machines or by hand labor.

Erosion Control The prevention of wear on the earth's surface by wind or water. Includes natural vegetation such as trees, shrubs and grasses and man-made objects such as riprap, fiber mats, chemicals, plastics, and wind screens.

Excavation Any man-made depression or cavity on the earth's surface resulting from removal of earth from one place to another.

Excavation By Hand The displacement of earth from one location to another with manpower and hand tools only.

Excavation Under Water The removal of mud, sand, rocks, gravel or clay from beneath the surface of water.

Flanged Steel Pipe Pipe made of steel with a circular rim at the end, usually with hole patterns for attaching bolts or studs; couples with a like pipe flange to form a water or gas-tight seal.

Grubbing To clear an area of trees, brush, shrubs, grasses and roots using hand tools and manpower only, or excavation equipment such as dozers, scrapers or tractors.

Lead Joint A hand formed joint, usually in a cast iron pipe, made by pouring molten lead into a void to form a water tight seal.

Pile Cap Placed over a pile or a group of piles to transmit heavy structure loads directly to the piles.

Pipe, Cast Iron Pipe which has been cast into a particular shape using molten cast iron.

Pipe Insulation Any material used to prevent the loss of heat or cold from pipe. Insulation around the pipe creates an air space between the pipe and the outside atmosphere.

P.V.C. Pipe Polyvinylchloride synthetic resin formed into a pipe.

Repose The maximum angle above the horizontal plane at which a given material will lie without movement.

Rock Crushing Plant A facility where large rocks are reduced in size using heavy equipment.

Scaffolding A temporary elevated platform and its supporting structure used to support workmen and equipment during the course of construction.

Sheet Piling A vertical structure constructed of individual piles to form a single line overlapping one another or interlocking. Provides a tight wall to resist lateral pressures from water, earth, mud or sand. Sheet piles are constructed primarily of concrete, steel or wood.

Shoring Posts made of heavy steel or timbers to support horizontal or vertical loads in excavations, forms or unstable structures.

Steel Pile A long structural element of various diameter sizes driven into the earth as a support for heavy structures.

Subbase A layer of material placed between the subgrade and basecourse in road construction.

Swell Factor The percentage of growth in volume of earth or rock after it has been excavated.

Swing Angle The degrees a piece of heavy earth moving equipment moves to deposit its load from the point of excavation.

Temporary Construction Buildings or other structures used during the course of construction and removed upon completion of the primary construction.

Thrust Block Concrete or other material placed under or around pipe to resist thrust forces caused by the movement of fluid or air.

Trench An excavation made by hand or by machine to permit installation of underground pipe or conduit.

Vitrified Clay Pipe Pipe which has been made from a clay product and then baked and glazed.

Welded Steel Pipe Pipe made of steel joined by the welding of a seam rather than the use of a mechanical joint.

Erecting and Operating an Asphalt Plant

Work Element	Unit	Man Hours Per Unit
Set up and dismantle plant	Each	420
Operation of asphalt plant	1,000 tons	60
Hauling asphalt to job	1,000 tons/mile	36

Figures are based on a 125 tons per hour batch plant.
Site preparation and concrete curing time not included in table.

Suggested crew:
Set up and dismantle plant: four equipment operators, two laborers, one electrician, one mechanic.
Asphalt plant operation: five equipment operators.
Maintenance (support): two laborers, one electrician, one mechanic.

Labor Operating a Rock Crushing Plant

Work Element	Unit	Man Hours Per Unit
Set up and dismantle	Each	240
Operating crushing plant	1,000 C.Y.	145
Stockpiling crushed material	1,000 C.Y	18
Hauling crushed material to job	1,000 C.Y./mile	20

The production figure is based on a 200 tons per hour plant operating at 50% of rated capacity crushing granite at 3,000 lbs. per cubic yard. For plants of other sizes, use 50% of the rated capacity.
Adjust production figures for the type of material being processed.

Suggested crew:
Set and dismantle plant: seven equipment operators, one mechanic, one electrician.
Operating crushing plant: three equipment operators.
Stockpiling crushed material: four equipment operators.
Maintenance (support): one steelworker, one equipment operator.

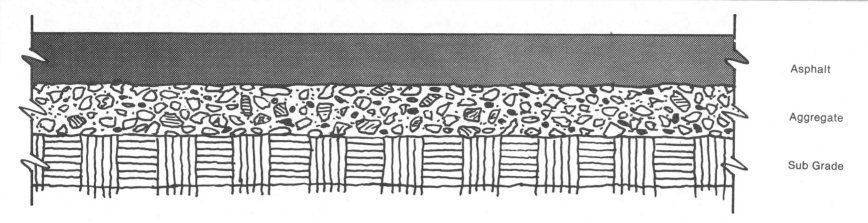

Asphalt

Aggregate

Sub Grade

Paving

All bituminous paving consists of a relatively thin wear surface built over a base course and subbase course, which rest on the compacted subgrade. The thickness of pavement includes all components of the pavement above the compacted subgrade. Thus, the subbase, base, and wearing surface are the structural components of the pavement. The load carrying capacity of the pavement results from the load distributing characteristics of the layered system.

Concrete paving may or may not have a base course between the pavement and the subgrade. The concrete exclusive of the base is referred to as the pavement. In some cases, Portland cement concrete is used as a base course for a flexible-type wearing surface. Rigid concrete pavement tends to distribute the load over a relatively wide area of bases. A major portion of the structural capacity is supplied by the slab itself. The major factor considered in the design of concrete pavement is the strength of the concrete. For this reason, minor variations in subgrade strength have little effect on the capacity of the pavement.

Components of Bituminous Paving

The aggregate in bituminous pavement may be crushed stone, gravel, sand or mineral filler. Regardless of the type of aggregate, it has three functions. First, it must transmit the load from the surface down to the base course. This happens because of the mechanical interlocking of the aggregate particles. Second, the aggregates must take the abrasive action of the traffic. If a wearing surface was binder alone, it would soon be worn away by the abrasive action of tires. The third function of the aggregate is to provide a non-skid surface. The other chief material in bituminous paving is the bituminous binder. There are two func-

tions for the binder. One, it binds the aggregate together, thus preventing the displacement of the aggregate. Two, the binder provides a waterproof cover for the base and keeps surface water from seeping into and weakening that material. The binder must be waterproof and have the ability to bind aggregate particles together. All bituminous materials possess these qualities. All bituminous materials are composed largely of bitumen, a black solid that gives the bituminous binders their black color, cementing ability, and waterproofing properties.

Cutback Asphalt

Asphalt must be applied in a liquid state. But asphalt is liquid in a natural state only at high temperatures. The equipment needed to heat asphalt to a liquid is not always available. In many instances, asphalt in fluid form is required for a smaller job where bringing in heating equipment is not practical. Asphalt is kept fluid even at low temperatures by combining the asphalt with any of the petroleum distillates called "cutterstock". The product of the combined materials is called "asphalt cutback."

The distillates used for cutterstock are normally gasoline, kerosene and diesel fuel (fuel oil). The important difference in the three cutterstocks is their evaporation rate. For example, asphalt cement plus gasoline will cure (harden) at the fastest rate; this mixture is known as "rapid curing cutback", abbreviated "RC".

Asphalt cement plus kerosene will cure at a slower rate, so we refer to this material as "medium curing cutback", abbreviated "MC".

Asphalt cement plus diesel fuel (fuel oil) will cure at the slowest rate, so this mixture is known as "slow curing cutback", abbreviated "SC". This material is often called "road oil".

Liquid Asphalt On Base

All cutbacks are liquids at room temperature and therefore are usually more convenient to use than asphalt cements. However, cutbacks are not without drawbacks. Probably their biggest disadvantage is that flammable materials are used for the thinners. At some temperature every cutback becomes a serious fire hazard. This is known as the "flash point". The flash point of a material is the lowest temperature at which the vapors from the material will ignite.

Pavement Components

Pavement work begins with a prepared subgrade. Then a prime coat of tar or cutback asphalt is applied to the soil. The prime coat penetrates and seals the voids in the surface of the base course, binds the particles together to form a tight tough surface and helps bind the subbase to the bituminous pavement. Pressure distributors are used in applying the prime material. Sufficient tar or asphalts should be used to seal the voids but not more than can be readily and completely absorbed. When excessive amounts are used, the surplus material softens the overlying pavement and can lead to failure. The quantity to be applied to a base course generally ranges from 0.1 to 0.5 gallons per square yard. The exact amount depends on the density or porosity of the base course.

Where a bituminous pavement is to be placed on an old concrete or bituminous pavement or on new bituminous courses, a heated tack coat of a soft grade of asphalt cement or tar, or an emulsified asphalt is applied to the cleaned surface just prior to laying the new course. This ensures a good bond between the two. The application should be made with a distributor at a rate of not over 0.2 gallons per square yard. The purpose of the tack coat is to provide the strongest possible bond between binder course and surface course. When asphaltic cement is used, it must be heated to comparatively high temperature and sprayed uniformly on to the surface to be "tacked". Little can be done to spread the material after it strikes the cold surface.

Emulsified asphalt can be used as a tack coat under some weather conditions. Emulsified asphalt has water mixed with the asphalt. The amount of water it contains should be taken into consideration in arriving at quantities required.

The seal coat is a type of surface treatment. It consists of a sprayed application of binder covered with a small aggregate and therefore does not add to the strength of a pavement structure. Properly designed and constructed bituminous pavements have a surface texture that does not require a seal coat to fill surface voids. Seal coats are used to treat old bituminous pavements which are dry, raveling, or beginning to hair crack. The normal seal coat is 0.1 to 0.2 gallons per square yard of hot cutback asphalt, a very soft grade of asphalt cement or tar followed by an application of 8 to 15 pounds per square yard of sand or other fine aggregate. The additional bitumen tends to liven the surface bitumen. The aggregate protects the bitumen from wear from traffic. Heavier application of bituminous materials should be used when treating very old pavements.

During construction it may be desirable to protect a completed layer or subbase or subgrade material from rain. This is important when a rather plastic material which would be sensitive to the additional moisture is used. The coating should consist of heated cutback asphalt and should be applied with a pressure distributor. The best type of asphalt for this coat depends on the climate and type of material being coated.

Labor For Subbase and Base

Work Element	Equipment	Unit	Machine-Hours Per Unit
Subbase:			
Scarify	D7E dozer with rippers	1000 S.Y.	3.33
Shape	CAT 12 grader	1000 S.Y.	2.50
Compaction, 6'' layers			
2 passes	Sheepsfoot roller	1000 S.Y.	0.19
4 passes	2—40 inch rollers	1000 S.Y.	0.39
6 passes	Abreast	1000 S.Y.	0.58
Compaction			
2 passes	9 tire pneumatic tire roller	1000 S.Y.	0.25
4 passes	9 tire pneumatic tire roller	1000 S.Y.	0.49
6 passes	9 tire pneumatic tire roller	1000 S.Y.	0.74
2 passes	2 axle, 5—8 ton tandem roller	1000 S.Y.	0.29
4 passes	2 axle, 5—8 ton tandem roller	1000 S.Y.	0.58
6 passes	2 axle, 5—8 ton tandem roller	1000 S.Y.	0.89
2 passes	50-ton pneumatic tire roller	1000 S.Y.	0.29
4 passes	50-ton pneumatic tire roller	1000 S.Y.	0.58
6 passes	50-ton pneumatic tire roller	1000 S.Y.	0.89
Base course:			
Spread material	D7E dozer	1000 C.Y.	3.33
Spread material	CAT 12 grader	1000 C.Y.	5
Shape surface	CAT 12 grader	1000 S.Y.	2.22
Compact gravel	Tandem roller, 8 ton	1000 S.Y.	1.33
Compact gravel	Rubber-tired roller, 10 passes	1000 S.Y.	2.50
Spread, sprinkle, and compact, man-hours per unit	Shovels, pneumatic tamps	C.Y.	2
Fine grade, sprinkle, and compact, man-hours per unit	Rake, shovel, pneumatic tamp	S.Y.	0.20

Suggested crew:
Machine work: two to four operators on equipment, two to six operators on hauling equipment, three to five men spreading, compacting, and fine grading.

Hand work: two to ten men spreading, compacting, and fine grading.

Volume of Material in Piles

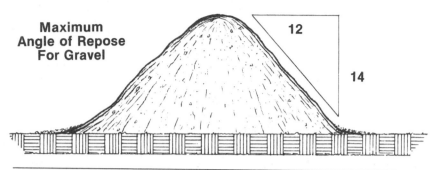

Maximum Angle of Repose For Gravel

Material	Ratio	Degrees
Dry sand	4½ in 12 to 7 in 12	20-30
Moist sand	7 in 12 to 12 in 12	30-45
Wet sand or dry earth	4½ in 12 to 12 in 12	20-45
Moist earth	5½ in 12 to 12 in 12	25-45
Wet earth	5½ in 12 to 7 in 12	25-30
Gravel	4½ in 12 to 14 in 12	20-50

Angle of Repose

Height Feet	20 Degrees	25 Degrees	30 Degrees	35 Degrees	45 Degrees
5	36	22	15	10	5
10	291	177	116	79	39
15	981	598	292	268	132
20	2,330	1,420	928	635	313
25	4,550	2,775	1,780	1,240	607
30	7,850	4,800	3,110	2,145	1,050
40	18,600	11,350	7,420	5,060	2,480
50	36,400	22,100	14,500	9,900	4,850
60	62,600	41,500	25,000	17,100	8,400
70	100,000	61,000	39,800	27,250	13,300
80	149,000	91,000	60,000	40,600	19,800
90	215,000	129,200	84,500	55,600	28,300
100	291,000	177,500	116,000	79,200	38,900

Cubic yards assuming the pile is free standing with a circular base.

Labor for Asphalt Paving

Work Element	Equipment	Unit	Machine-Hours Per Unit
Spread and finish asphalt concrete	Asphalt finisher	10000 S.Y.	4.0
Roll asphalt concrete	9 to 14 ton 3 wheel tandem	10000 S.Y.	3.2
	5 to 8 ton 2 wheel tandem	10000 S.Y.	3.4
	10 ton 3 wheel roller	10000 S.Y.	3.0
	9 tire pneumatic roller	10000 S.Y.	3.1
	50 ton pneumatic roller	10000 S.Y.	3.5
Spread aggregate			
4' width	Hopper type spreader	10000 S.Y.	4.2
6' width	Hopper type spreader	10000 S.Y.	3.3
8' width	Hopper type spreader	10000 S.Y.	2.5
Sweep base prior to spraying	30 dbhp tractor sweeper	10000 S.Y.	4.0

Recommended operating speed for asphalt finishers is about 50 f.p.m. Therefore, 3.5 hours is the minimum time required for a finisher to cover 10,000 yards. For paving thicknesses of ¾ inch or less, use 3.5 hours.

For double-lane roads using one paver, additional time will be required if hot joint construction is required.

Asphalt paving construction includes manpower for heating asphalt, marking pavement edges, brooming, priming, spreading and finishing asphaltic concrete, rolling asphaltic concrete, applying seal coat, applying tack coat, loading and hauling chips or gravel, spreading and rolling chips or gravel, and brooming chips or gravel. The time required to spread asphalt concrete with an asphalt finisher and to roll this material is important in only a very few cases. A roller can only compact material as fast as it is spread, and an asphalt finisher can only spread asphalt concrete as fast as it is produced from an asphalt plant.

Suggested crew: Asphalt finisher: six men; Rollers: one operator; Spreaders: one operator and two men.

Labor Applying Liquid Asphalt — Tank Emptying Time

Application Rate (Gal/S.Y.)	Speed (f.p.m.)	Spray bar length with ⅛'' nozzle		
		12'	18'	24'
.05	1300	7	5	3.5
.1	900	7	5	3.5
.2	450	7	5	3.5
.3	300	7	5	3.5
.5	180	7	5	3.5
1.0	90	7	5	3.5

Use these figures for applying prime, tack or seal coats with an 800 gallon truck mounted distributor. The application rate depends on the truck speed. The time required varies only with changes in bar lengths. To determine how much area is covered with one 800 gallon tank, multiply the speed times the spray bar length times the time required to empty for the area covered.

The most time-consuming part of this operation is travel time to and from the refill point and the time required to reload (up to ½ hour per tank).

For larger trucks or different application rates, adjust the table accordingly.

Labor For Concrete Paving

Work Element	Equipment	Unit	Hours Per Unit
Place and remove formwork		100 L.F.	5.5 man-hours
Reinforcing mesh and dowels		100 L.F.	2.0 man-hours
Reinforcing steel and dowels		Ton	35.0 man-hours
Place ready mix		C.Y.	0.4 man-hour
Spread			
First gear	Self-propelled	1000 S.Y.	1.0 machine-hour
Second gear	concrete	1000 S.Y.	0.7 machine-hour
Third gear	spreader	1000 S.Y.	0.6 machine-hour
Finish, by hand		1000 S.Y.	324 man-hours
Finish, by machine			
First gear	Traverse concrete finisher	1000 S.Y.	1.0 machine-hour
Second gear	Traverse concrete finisher	1000 S.Y.	0.8 machine-hour
Third gear	Traverse concrete finisher	1000 S.Y.	0.7 machine-hour
Fourth gear	Traverse concrete finisher	1000 S.Y.	0.5 machine-hour
Place premolded expansion joint		1000 L.F.	15 man-hours
Cut joints	Concrete joint saws	1000 L.F.	20 man-hours
Place joint sealer		1000 L.F.	12 man-hours
Cure and clean-up		1000 S.F.	15 man-hours

Includes placing forms, placing reinforcement, placing dowels, placing, finishing and curing concrete, removing and cleaning forms, cutting or forming joints, pouring joint sealer and installing expansion joints.

Suggested crew: two to three men forming, four men reinforcing, six to eight men mixing and placing, six men finishing, four men sawing and sealing joints.

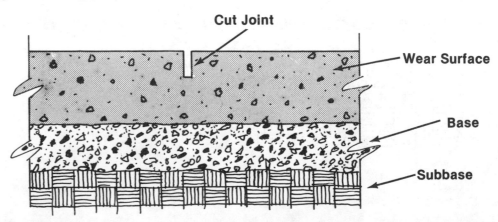

Cut Joint

Wear Surface

Base

Subbase

Labor For Curbs & Walks

Work Element	Unit	Man-Hours Per Unit
Concrete curbs		
Formwork integral with paving	100 L.F.	10.5
Formwork separate from paving	100 L.F.	22.5
Combined curb and gutter form	100 L.F.	25.5
Placing reinforcing	Ton	35.0
Mix and place finish top	100 S.F.	5.5
Cure and clean-up	1000 S.F.	1.0
Concrete walks		
Formwork	100 L.F.	4.5
Mix and place (hand)	C.Y.	4.5
Mix and place (machine mixer)	C.Y.	3.0
Finish	100 S.F.	5.0
Place ready mix	C.Y.	1.0
Cure and clean-up	1000 S.F.	1.0
Asphalt walks		
Formwork	100 L.F.	4.5
Prime coat	100 S.F.	0.7
Spread asphalt	100 S.F.	2.9
Roll asphalt	100 S.F.	0.6

Suggested crews:

Curbs: three to four men forming, two men reinforcing, four men mixing and placing, two men general labor.

Walks: three men forming, six to seven men mixing and placing, two to three men finishing and general labor.

Asphalt walks: three men forming, five men placing, one man rolling.

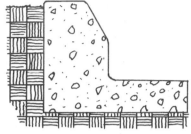

Integral Curb

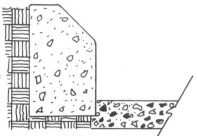

Separate Curb

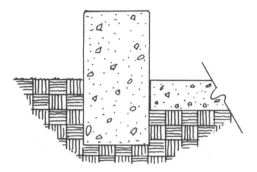

Straight Curb and Separate Gutter

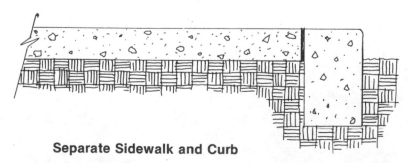

Separate Sidewalk and Curb

Straight Curb and Integral Gutter

Special Soil Types

Hard material is usually defined as "solid rock, firmly cemented unstratified masses or conglomerate deposits possessing the characteristics of solid rock that is not ordinarily removed without systematic drilling and blasting, and any boulder, masonry, or concrete except pavement, exceeding one-half cubic yard in volume."

The need for blasting is usually clear. However, some types of unstratified or cemented material are questionable. This material can often be removed by heavy equipment or rippers rather than blasting and should not be considered as hard material.

Measure solid rock volumes very accurately as the work progresses to ensure that charges to the owner can be computed exactly. Where large quantities of rock are to be removed, a topographic survey may be required before rock excavation is started. Establish the extent and character of underlying rock by probing and boring. Remove the soil overburden if necessary to determine the nature of the rock strata below.

The term **unsuitable** in the field of soils engineering is difficult to define except by understanding the soil type and structure being constructed. A satisfactory subgrade soil will have good stability, good bearing value, low shrinkage or swell, and good drainage capabilities. The soil type, its location in the subgrade strata, and its moisture content will determine its degree of suitability. Unsuitable materials include rich or fatty clays, high moisture content silts, unstable fine grain sands, organic peat, and mulch. Even when an adequate soils investigation has been performed, unsuitable soil may be found in isolated pockets or strips across a generally suitable soil subgrade area. Soil borings taken during the foundation investigation may indicate general areas where unsuitable soil is likely to be found.

Boring legs and soil classification may be shown on the drawings, along with ground water level. All unsuitable material should be removed as indicated. However, the borings may not show the exact extent and depth of the material to be removed. Keep a close check on the excavation and have measurements taken if unsuitable material is not shown on the drawings. When in doubt about foundation conditions when excavation is completed, request assistance from a soil engineer.

Unsuitable material should be deposited in a designated area and should not be used in embankments. However, if the

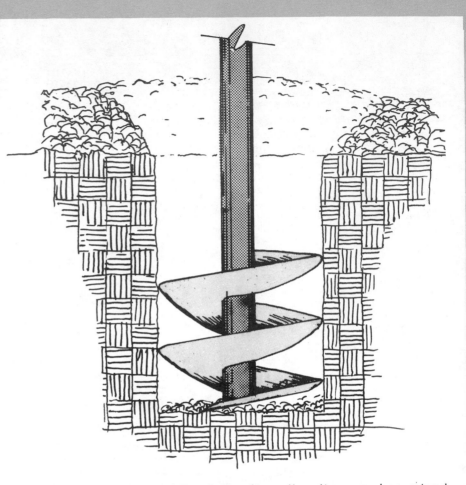

material is peat, specifications often allow its use when mixed with topsoil and applied to areas which will be seeded.

Topsoil is usually required on all finished grading around buildings to promote growth of lawn, trees and shrubbery. Where existing topsoil is available on the site, the specifications probably require that it be removed to a certain depth and stockpiled for future use. Topsoil is an expensive item in many areas so it is important that this material be saved rather than mixed with subsoil and used in embankments. As previously mentioned, peat makes an excellent mulch when mixed with topsoil and should be salvaged when needed.

Clearing

All existing trees and shrubbery that are to remain on the site should be marked prior to starting clearing operations. Trees to remain within the area or in a material stock pile area should be protected from accidental damage. Barricades or wooden planks strapped around their trunks will usually prevent most damage. Any tree which is to remain and is scarred or damaged while clearing goes on should be painted with a tree wound dressing.

In areas where burning is permissible, local regulations probably require that fire guards with special fire equipment be on site during burning. Usually fires must be extinguished or guards posted in the area at the end of the day.

Specifications usually require removal of stumps and matted roots to not less than 18" below finished grade. Depressions made by grubbing should be filled with suitable material and compacted to the density of the surrounding soil prior to starting embankment.

In embankment areas where the subgrade or slope elevation is more than 5 feet above the natural ground surface, all trees, existing stumps and large roots under 18" in diameter usually can be close cut or removed, at the contractor's option. However, where a structure is to be built or subdrainage trenches are to be excavated, unsuitable material must be removed. If hillsides are to be terraced, complete grubbing and removal is usually required.

When permanent survey monuments or bench marks are within the construction area, preservation or relocation is always required. It is important that these monuments be preserved since extensive survey work may be required to replace them. The plans or specifications should cover relocation of monuments that may be destroyed.

Borrow and waste areas should be located and a truck route established to the construction site. The plans should indicate the location of these areas. Disposal sites, routes, and local requirements are important cost considerations. Investigate and verify any restrictions on street closures prior to starting earthmoving.

Trenching Problems

Pipe bedding is often required when the trench bottom will not make a good bedding for the pipe. A layer of muck, peat, rock or soft clay 12" or less below invert grade must be excavated and replaced with granular material. High water table or quicksand can be corrected with well points. When there is unsuitable material well below the level of the pipe, it may be necessary to support the pipe on piles.

Trenching in industrial areas is usually difficult because there is no place to leave excavated material. In areas that can not be closed to traffic, it may be necessary to haul the earth away and bring it back later when pipe installation is complete. Fire regulations may require that a roadway be kept open at all times for fire fighting equipment.

Specifications usually require de-watering of trenches by either pumping, draining, or well points. Contractors usually are allowed to select their own method of removing water. The water level must be kept below the trench bottom when laying pipe and until backfill is placed above the top of the pipe.

Ground water is often a problem when excavating for water and sewer lines. Rising water makes the pipe bed unsuitable, creates difficult working conditions and is sure to result in a poor installation. It may be necessary on some jobs to excavate deeper and backfill with crushed rock or other suitable granular material.

Site investigations often do not locate underground obstructions such as abandoned pipe lines, cables, old foundations, logs, etc. When these obstructions are encountered during excavation, a change order should be required to cover the additional cost for their removal. It is appropriate to stop work and require authorization before going forward. Before starting excavation of a building site, verify the location of all existing underground utility lines shown on the drawings. The location or elevation of these lines may be incorrectly shown. Some very costly change orders have occurred after building foundations were completed. It is also important at this time to verify the invert elevations of sewer lines and the size and depth of water lines so there will be no problem in connecting to these lines in the future. A change in building location or elevation may be preferable to relocation of a major utility line.

Contractors can be held liable for damage to telephone or utility lines which are indicated on the plans or are located by the owner prior to construction. Therefore, it may be wise to do some hand excavation to locate telephone and power cables, water mains and other utility lines which cross the trench. Failure to do this may result in a major interruption of service or injury to personnel.

Labor Clearing and Grubbing by Hand

Work Element	Unit	Man-Hours Per Unit
Light clearing with axes, brushhooks and hatchets	100 S.Y.	2.5
Medium clearing with axes and chain saws	100 S.Y.	5.0
Cutting trees, removing branches, cutting into short lengths with chain saws and axes, by tree diameter		
8" to 12"	Each	2 to 3
13" to 18"	Each	3 to 4
19" to 24"	Each	5 to 6
25" to 36"	Each	6 to 8
Removing stumps with picks, shovel, and axes, by tree diameter		
8" to 12"	Each	8
13" to 18"	Each	10
19" to 24"	Each	12
25" to 36"	Each	15
Removing stumps with dozer, by tree diameter		
6" to 10"	Each	1.6
11" to 14"	Each	2.1
15" to 18"	Each	2.6
19" to 24"	Each	3.1
25" to 30"	Each	3.3
Removing stumps by blasting and pulling with a tractor, by tree diameter		
6" to 10"	Each	1.1
11" to 14"	Each	1.4
15" to 18"	Each	1.6
19" to 24"	Each	2.0
25" to 30"	Each	2.5
Piling and burning brush	100 S.Y.	.8

Suggested crew for clearing operations is one foreman, four men with brushhooks, axes and chain saws. Crew for removing stumps, cutting trees and blasting is three men. Labor includes loading on a truck and hauling 5 miles to a dump site.

Light Clearing   **Medium Clearing**   **Cutting Trees**

Labor Clearing and Grubbing by Machine

Work Element	Unit	Machine-Hours Per Unit
Light clearing with D7 dozer	1000 S.Y.	.75
Medium clearing with D7 dozer	1000 S.Y.	1.80
Removing tree stumps with D7 dozer		
8" - 12" diameter	10 stumps	1 to 2.5
13" - 24" diameter	10 stumps	3 to 5
25" - 36" diameter	10 stumps	5 to 10

The figures assume reasonably good conditions, good traction and no steep slopes. The time given is machine time for one operator. A typical crew would be one operator, one foreman, and two to five men with chain saws and axes cutting and trimming. Total man-hours per unit would thus be four to seven times the machine-hours shown.

Cubic Yards Per 100 Linear Feet of Trench

Depth	12"	14"	16"	18"	20"	22"	24"	30"	36"	42"	48"
12"	3.7	4.3	4.9	5.6	6.2	6.2	7.4	9.3	11.1	13.0	14.8
14"	4.2	5.0	5.8	6.5	7.2	7.9	8.6	10.8	12.9	15.1	17.3
16"	4.9	5.8	6.6	7.4	8.2	9.0	9.9	12.4	14.8	17.4	19.8
18"	5.6	6.5	7.4	8.3	9.3	10.2	11.1	13.9	16.7	19.4	22.2
20"	6.2	7.2	8.2	9.3	10.3	11.3	12.3	15.4	18.5	21.5	24.6
22"	6.8	7.9	9.0	10.2	11.3	12.4	13.6	17.0	20.4	23.6	27.2
2'- 0"	7.4	8.6	9.9	11.1	12.3	13.6	14.8	18.5	22.2	26.0	29.6
2'- 6"	9.4	10.9	12.5	13.9	15.4	17.0	18.5	23.2	27.8	32.4	37.0
3'- 0"	11.2	13.0	14.9	16.7	18.5	20.4	22.2	27.8	33.3	38.9	44.5
3'- 6"	13.0	15.1	17.3	19.4	21.6	23.8	25.9	32.4	38.9	45.4	52.0
4'- 0"	14.8	17.2	19.7	22.2	24.7	27.2	29.6	37.0	44.5	52.0	59.2
4'- 6"	16.6	19.3	22.1	24.8	27.6	30.6	33.3	41.6	50.0	58.4	66.7
5'- 0"	18.5	21.5	24.6	27.7	30.8	34.0	37.0	46.3	55.5	64.9	74.1
5'- 6"	20.4	23.8	27.2	30.5	34.0	37.3	40.7	51.0	61.1	71.3	81.6
6'- 0"	22.2	25.8	29.6	33.3	37.1	40.7	44.4	55.5	66.7	77.9	89.0
6'- 6"	24.0	27.8	31.8	36.1	40.0	44.2	48.1	60.2	72.2	84.2	96.5
7'- 0"	25.9	30.1	34.4	38.9	43.1	47.6	51.9	64.8	77.8	90.8	103.8
7'- 6"	27.8	32.3	36.9	41.6	46.2	51.0	55.6	69.5	83.4	97.3	111.4
8'- 0"	29.6	34.4	39.2	44.5	49.4	54.5	59.2	74.1	88.9	102.0	118.6

For shallow trenches (up to two feet deep), use the width of the footing.
For trenches two to four feet deep, use the width of the footing or 1'-6", whichever is greater.
For trenches four to six feet deep, use a width of at least two feet.

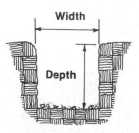

Labor for Trench Excavation — Machine

Work Element	Unit	Machine-Hours Per Unit
3/4 CY backhoe		
Medium gravel	100 CY	7.6
Light clay	100 CY	8.0
Heavy clay	100 CY	9.7
Hardpan	100 CY	10.5
1 CY backhoe		
Medium gravel	100 CY	6.0
Light gravel	100 CY	6.3
Heavy clay	100 CY	6.6
Hardpan	100 CY	7.0
Ladder type trencher, 24" trench		
Light to medium soil	100 CY	1.9
Sand or gravel	100 CY	2.2
Medium clay	100 CY	2.4
Ladder type trencher, 18" trench		
Light or medium soil	100 CY	2.3
Sand or gravel	100 CY	2.7
Medium clay	100 CY	2.9
Ladder type trencher, 12" trench		
Light or medium soil	100 CY	2.6
Sand or gravel	100 CY	3.0
Medium clay	100 CY	3.3
Fine grade trench bottom	10 SY	0.9
Trim trench banks	10 SY	0.5
Hand work around obstructions	CY	3.0
Backfilling trench, no compaction		
Hand work, man-hours	100 CY	85.0
Dozer	100 CY	1.9
Backhoe	100 CY	2.5
Compacting backfill to 95% of maximum density		
Pneumatic tamper, man-hours	100 CY	30.0
Vibratory compactor	100 CY	10.5

These figures assume that excavated material is deposited adjacent to the trench. Loading a truck with the backhoe will slow production 10 to 20%. **Suggested Crew:** Straight line (utilities/piping): one operator on trencher, two men cleaning and fine grading trench bottom. Short line (footings/foundations): one operator on trencher, two to five men finishing ends and squaring corners. Backfilling: one operator on dozer, one operator on compactor (optional).

Labor for Rock Excavation

Work Element	Unit	Man-Hours Per Unit
Drilling for rock bolts	10 LF	.6
Drilling for pre-splitting	10 LF	.5
Splitting rock, per SF of face area		
Large open areas	100 SF	1.6
Restricted areas or small jobs	100 SF	3.1
Surface blast & load on truck		
Bulk, over 500 CY	CY	.35
In trenches	CY	1.10
In pier holes	CY	1.50
Drill rock, blast & load on truck, deep hole method		
Bulk, over 500 CY	CY	.35
Less than 500 CY	CY	.40
In trenches	CY	1.35
In pier holes	CY	1.50
Load exposed boulders on truck		
Drill, blast, & load	CY	.50
Load only, up to 2 CY each	CY	.25
Break up rock with air hammer, no loading included		
Very dense rock	CY	2.2
Medium density rock	CY	1.6
Soft rock	CY	.7
Level and clean rock area for concrete		
For slab	SY	.15
For footings	SY	.25
For trenches	SY	.22
For piers	SY	.21

These figures assume an urban area without noise restrictions. Blasting costs will be lower where no structures are adjacent to the job site and up to 60% higher in residential areas where a noise ordinance is enforced.

Labor Blasting and Quarrying

Work Element	Unit	Man-Hours Per Unit
Stripping overburden:		
Strip and cast	1,000 C.Y.	23
Strip and load	1,000 C.Y.	21
Haul to spoil area	1,000 C.Y./Mi.	18
Spread spoil pile	1,000 C.Y.	8
Drilling and blasting:		
Drill holes	1,000 L.F.	50
Load and shoot holes	1,000 Each	200
Load quarried material	1,000 C.Y.	45

Use these figures for large scale quarry operations.

Notes for Blasting and Quarrying

The time required for drilling and blasting will vary with the type of equipment and the type of explosive used.
Figures are based on the use of an airtrack for drilling and stick dynamite for blasting.
Adjust figures accordingly for estimates using other types of drilling equipment or explosives.

Suggested crew:
Stripping and casting: one operator on dragline or dozer, one helper.
Stripping and hauling: one operator on frontend loader, one operator on dozer, two to six operators on dump trucks, two helpers.
Spreading soil: one operator on dozer.
Drilling: two to ten men operating compressor and drilling.
Blasting: two to three operators loading holes.
Handling quarried material: one man loading trucks as required.

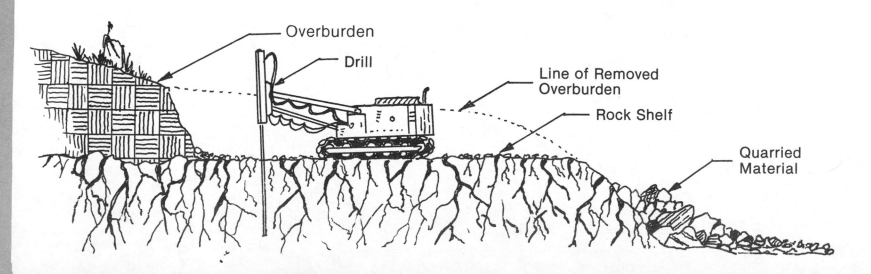

Overburden

Drill

Line of Removed Overburden

Rock Shelf

Quarried Material

Material Weights and Swell Factors

Material	Per C.Y. (Loose)	Per C.Y. (In Place)	Percent of Swell	Swell Factor
Cement, portland	2450	2950	20	.83
Clay, natural red	2700	3500	30	.77
Clay and gravel, dry	2300	3100	34	.74
Clay and gravel, wet	2600	3500	34	.72
Concrete	2650	3700	40	.72
Concrete, wet mix	3600	3600	40	.72
Earth, dry loam	2300	2850	25	.81
Earth, wet loam	2750	3400	24	.81
Granite	2800	4560	65	.60
Gravel, ¼ to 2 in. dry	2850	3200	12	.89
Gravel, ¼ to 2 in. wet	3200	3600	13	.89
Laterite	3900	5200	33	.75
Limestone, blasted	2500	4250	69	.59
Limestone, crushed	2700	4500	67	.60
Limestone, marble	2700	4550	69	.59
Mud, dry	2100	2550	21	.82
Mud, wet	2650	3200	21	.83
Sand, dry	2750	3100	13	.89
Sand, wet	3150	3600	14	.88
Sandstone, shot	2700	4250	58	.64
Shale, riprap	2100	2800	33	.75
Slate	3600	4700	30	.77
Coral, class No. 2, soft	1760	2900	65	.61
Coral, class No. 1, hard	2030	2900	67	.60

Percent of swell times the bank (in place) cubic yards equals the loose cubic yards to be moved.

Swell factor times the loose cubic yards equals bank cubic yards being moved.

Excavation Factors

Depth	Cubic Yards per Square Foot
2"	.00617
4"	.01235
6"	.01852
8"	.02469
10"	.03086
1'-0"	.03704
1'-6"	.05555
2'-0"	.07407
2'-6"	.09259
3'-0"	.11111
3'-6"	.13333
4'-0"	.14815
4'-6"	.16667
5'-0"	.18519
5'-6"	.20370
6'-0"	.22222
6'-6"	.24074
7'-0"	.25926
7'-6"	.27777
8'-0"	.29630
8'-6"	.31481
9'-0"	.33333
9'-6"	.35185
10'-0"	.37037

Multiply the excavation factor by the area in square feet to find the cubic yards of soil to be excavated.
Example: Assume an excavation 24 ft. x 30 ft. and 6 ft. deep. 24 x 30 = 720. In the table the 6 ft. depth has a factor of .22222 (the number of cu. yd. in an excavation 1 ft. square and 6 ft. deep).

720 x .22222 = 160 Cu. Yds.

Swing Angle Conversion Table

Equipment Type	Swing Angle in Degrees							
	30	45	60	75	90	120	150	180
Backhoe	0.76	0.84	0.90	0.95	1.00	1.10	1.21	1.30
Clamshell, dragline	0.76	0.84	0.90	0.95	1.00	1.10	1.21	1.30
Shovel	---	0.80	0.86	0.93	1.00	1.14	1.27	1.41

The rate of excavation to be used times the swing angle factor equals the new rate in hours.

Earthmoving tables in this section assume the excavation equipment makes a 90 degree swing to deposit excavated soil. For other swing angles, modify production figures as given in this table.

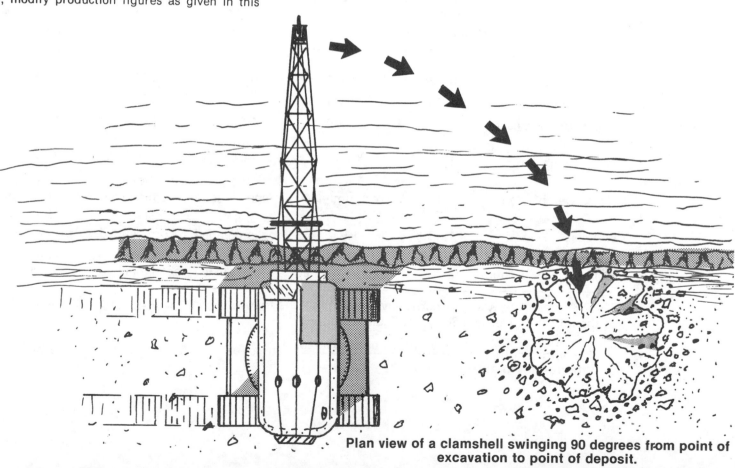

Plan view of a clamshell swinging 90 degrees from point of excavation to point of deposit.

Conversions for Soil Types

Soil Type	Efficiency Factors by Machine Type				Hand Efficiency Factors
	¾ C.Y. Power Shovel	¾ C.Y. Dragline	¾ C.Y. Clamshell	2½ C.Y. Loader	Manual
Loose sand—clay or moist loam	0.82	0.81	0.74	0.77	0.65
Sand—gravel	0.88	0.84	0.83	0.83	0.75
Good common earth	1.00	1.00	1.00	1.00	1.00
Hard, tough clay	1.23	1.17	Not recommended	1.26	1.29
Rock, well blasted	1.43	Not recommended	Not recommended	Not recommended	Not recommended
Wet clay	1.93	1.91	Not recommended	Not recommended	Not recommended
Rock, poorly blasted	2.70	Not recommended	Not recommended	Not recommended	Not recommended

Unless noted otherwise, the earthmoving tables on the following pages assume "good common earth" as the norm. For other soil types, figure as follows:

Hours required = $\dfrac{\text{quantity to be excavated}}{\text{unit}}$ **times** unit hours required **times** the efficiency factor listed above.

Labor for Underwater Excavation

Work Element	Equipment	Unit	Machine-Hours Per Unit
Machine work:[1]			
Operate dredge	Hydraulic dredge	1000 C.Y.	10
Underwater excavation	Clamshell — ¾ yard	1000 C.Y.	25
	Dragline — ¾ yard	1000 C.Y.	17
Spoil disposal: (Assumes 6 C.Y. per truck, 4 round trips per hour)	5-ton dump	1000 C.Y.	42
Spoil disposal: Barge	Barge with clamshell[2]	1000 C.Y.	50
	Barge with dragline[2]	1000 C.Y.	33.3
Install and remove discharge lines (hand work in man-hours)	Hydraulic dredge	100 L.F.	10

[1] Typical crew: one crew leader, four to seven men installing and removing discharge lines, three to five men per shift operating dredge (usually on a two or three shift basis), one man with equipment building dike with three men installing drain pipes through dike. For dragline or clamshell excavation, one operator and signalman on dragline, two to five trucks with operators hauling spoil, one man to direct loading of spoil on barge, two barges (one loading and one unloading), two men and one bulldozer unloading barge at disposal area, one tugboat and crew (usually three to five men).

[2] Assumes rotation of two barges per machine with short haul distance.

Dragline excavating in shallow water

Labor for Dewatering

Work Elements	Unit	Man-Hours Per Unit
Sump hole construction		
Unlined hole	10 CF	0.25
15" diameter hole, metal liner	10 LF	1.6
24" diameter hole, metal liner	10 LF	2.1
Wood lined hole, SF of contact area	10 SF	0.8
Construct gravity run-off system[1]		
Trench drain	10 CY	0.7
6 inch pipe	10 LF	0.6
8 inch pipe	10 LF	1.1
Gravel drain	10 CY	1.4
Well point system[2]		
Install & remove pump & accessories	Ea	11.0
Install & remove header pipe	LF	1.5
Install & remove wellpoints	100 Ea	8.8
Jet in wellpoints (if needed)	100 LF	1.8
Drill in wellpoints (if needed)	100 LF	2.8
Pea gravel surround for points	CF	0.1

[1] Includes installation, excavation and removal.
[2] Assumes use of a 6 inch diesel pump and 2 inch wellpoints.

Labor for Bulk or Bank Excavation

Work Element	Unit	Machine-Hours Per Unit
1 CY backhoe		
Sand or gravel	100 CY	6.4
Light clay	100 CY	7.3
Heavy clay	100 CY	8.7
Hardpan	100 CY	9.0
1½ CY backhoe		
Sand or gravel	100 CY	5.5
Light clay	100 CY	6.0
Heavy clay	100 CY	7.9
Hardpan	100 CY	8.8
1½ CY crawler mounted shovel		
Sand or gravel	100 CY	3.7
Light clay	100 CY	4.1
Heavy clay	100 CY	4.5
Hardpan	100 CY	5.0

Labor for Bulk or Bank Excavation (continued)

Work Element	Unit	Machine-Hours Per Unit
2 CY crawler mounted shovel		
Sand or gravel	100 CY	2.1
Light clay	100 CY	2.5
Heavy clay	100 CY	2.8
Hardpan	100 CY	3.5
1½ CY rubber tired front end loader		
Sand or gravel	100 CY	2.7
Light clay	100 CY	3.2
Heavy clay	100 CY	3.7
Hardpan	100 CY	4.4
2½ CY rubber tired front end loader		
Sand or gravel	100 CY	1.8
Light clay	100 CY	2.2
Heavy clay	100 CY	2.4
Hardpan	100 CY	3.5
1 CY tracked front end loader		
Sand or gravel	100 CY	2.9
Light clay	100 CY	3.7
Heavy clay	100 CY	3.9
Hardpan	100 CY	4.1
2 CY tracked front end loader		
Sand or gravel	100 CY	2.4
Light clay	100 CY	2.7
Heavy clay	100 CY	2.9
Hardpan	100 CY	3.3
2½ CY tracked front end loader		
Sand or clay	100 CY	1.5
Light clay	100 CY	1.8
Heavy clay	100 CY	2.2
3 CY tracked front end loader		
Sand or gravel	100 CY	1.1
Light clay	100 CY	1.3
Heavy clay	100 CY	1.5

These figures assume a selection of equipment appropriate for the work being done. The work includes digging and piling on site or loading on trucks. Use this table for bulk or bank excavation only.

Labor for Earthmoving

Work Element	Typical Equipment	Unit	Machine-Hours Per Unit
Excavate and load into trucks	Power shovel — ¾ yard[1]	1000 C.Y.	9
	Dragline — ¾ yard[1]	1000 C.Y.	12
	Clamshell — ¾ yard[1]	1000 C.Y.	17
One-way distance:			
50 feet	Scoop — 2½ yards	1000 C.Y.	3.5
100 feet	Scoop — 2½ yards	1000 C.Y.	5.2
200 feet	Scoop — 2½ yards	1000 C.Y.	7.1
300 feet	Scoop — 2½ yards	1000 C.Y.	10.4
Haul in trucks: (6 C.Y. per truck, approximately 4 cycles per hour)	5-ton dump[2]	1000 C.Y.	42
Excavate, load, haul (round trip) cycle time:			
3 minutes	18 yard scraper	1000 C.Y.	3.5
5 minutes	18 yard scraper	1000 C.Y.	5.6
7 minutes	18 yard scraper	1000 C.Y.	7.8
10 minutes	18 yard scraper	1000 C.Y.	11.1
Strip topsoil/stockpile soil, shallow excavation	D7 dozer	1000 C.Y.	37
Spread fill	D7 dozer	1000 C.Y.	16
Sprinkle	1000-gallon distributor	1000 S.Y.	0.04
Compact			
2 passes	Sheepsfoot roller	1000 S.Y.	0.19
4 passes	113 inches wide	1000 S.Y.	0.39
6 passes	(2 rollers)	1000 S.Y.	0.58

[1]See the Swing Angle Conversion Table for swing angles other than 90 degrees.

[2]To use for other than four trips per hour, calculate cycle time and use appropriate proportion.

Example: If cycle time of 15 minutes (4 cycles per hour) takes 42 equipment hours to haul 1,000 C.Y., then cycle time of 20 minutes (3 cycles per hour) will take 42 x ⁴⁄₃ = 56 equipment hours.

Number of dump trucks will vary with the length of haul.

More than one endloader and additional trucks may be advantageous if the pit site is large enough to work and job priorities warrant.

These operations will require a dozer part or full time in pit or cut area and a grader part or full time on haul roads.

Suggested crew: Dump operations — one operator on end loader, five operators on dump trucks; one operator on water truck; one pit and one fill boss.

Scraper operations — two or more operators on scrapers; one or more operators on "push cats;" one operator on compactor; one operator on water truck; one cut and one fill boss. Clamshell/Dragline operations: two operators on clamshell; two operators on dragline.

Labor for Ditch and Swale Excavation

Work Element	Unit	Machine-Hours Per Unit
Excavate ditch and spread adjacent to ditch, including shaping ditch[1]		
Sand or gravel	100 CY	1.8
Medium clay	100 CY	2.2
Heavy clay	100 CY	2.7
Trim banks[1]		
Large job	100 SY	0.2
Small job	100 SY	0.3
Bulk swale excavation, including casting only[2]		
Large job	100 CY	8.2
Small job	100 CY	9.2

[1]Based on use of a small dozer or a gradall.
[2]The production rate for a power shovel will be about the same but the material excavated can be stacked or loaded on trucks with no loss of production.

Labor Excavating Piers — Machine

Work Element	Unit	Man-Hours Per Unit
3/8 CY backhoe		
Medium gravel	10 CY	6.1
Light clay	10 CY	7.0
Heavy clay	10 CY	8.0
Hardpan	10 CY	9.3
3/4 CY backhoe		
Medium gravel	10 CY	4.7
Light clay	10 CY	5.3
Heavy clay	10 CY	6.4
Hardpan	10 CY	6.9
1 CY backhoe		
Medium gravel	10 CY	4.0
Light clay	10 CY	4.3
Hardpan	10 CY	5.0
Backfilling against pier, no compaction		
Hand work, man-hours	CY	0.9
Backhoe	10 CY	0.3

Labor Excavating Piers — Machine (continued)

Work Element	Unit	Man-Hours Per Unit
Compacting backfill to 95% of maximum density		
Hand-tamping, man-hours	CY	0.5
Vibratory compactor	CY	0.1

These figures assume that the work is in scattered locations and that excavated soil is piled adjacent to the pit. Loading a truck with the backhoe will slow production 10% or more.

Labor for Trench or Pier Excavation — By Hand

Work Element	Unit	Man-Hours Per Unit
Up to 2 feet deep		
Normal soil	CY	.9
Sand or gravel	CY	1.2
Medium clay	CY	1.1
Heavy clay	CY	1.4
Loose rock	CY	2.2
Over 2 feet to 6 feet deep		
Normal soil	CY	1.3
Sand or gravel	CY	1.5
Medium clay	CY	1.4
Heavy clay	CY	1.8
Loose rock	CY	2.6
Fine grade excavation bottom	10SY	.9
Trim excavation banks	10SY	.5
Hand work around obstructions	CY	3.0
Backfilling, no compaction	CY	.9
Compacting backfill to 95% of maximum density		
Hand tamping	CY	.5

Excavation includes only piling soil adjacent to the trench or pit. Add for loading on trucks or spreading soil.

Backfill

Backfill is required around building foundations, back of retaining walls and around utility lines in trenches. Backfill should not be placed against foundations or retaining walls until the concrete has reached its design strength, usually 28 days. Premature backfill could result in cracking or failure of the concrete. Backfilling too soon makes it possible to damage a retaining wall by excessive compaction from heavy equipment or by puddling. These conditions can result in a loading higher than those assumed in the design of the wall.

Backfill should be brought up evenly on each side of a wall whenever possible. Slope the fill to drain away from the wall. Brace the inside of a basement before backfill is placed outside if the first floor is not in place. Backfill should also be placed in layers and compacted similar to embankment. Excavated material may be used for backfill provided it is suitable and can be compacted to the required density. Unsuitable material removed during excavation should be wasted and replaced with granular fill.

Backfill in pipe trenches requires special care to prevent future settlement or damage to the pipe. Backfill should be placed in 6 inch compacted layers to a depth of one foot over the top of the pipe. The remainder of the trench is backfilled in well-compacted one-foot layers. This procedure provides adequate support to the sides of the pipe.

When backfilling around or immediately on top of the pipe, be sure that no rock falls on the pipe or bears against it. Rock falling on the pipe could cause cracks which may not show up as a leak until some time after the job is compacted.

Unsatisfactory fill materials include peat, silts and very fine sands. However, any material which is too wet or too soft to provide a stable foundation for structures, utility lines or pavement may be classed as unsatisfactory.

Select backfill material usually consists of sand, stone, gravel or stone screenings, uniformly graded and free from excess soft or unsound particles or other objectionable material.

ASTM C 136 requires that select backfill conform to the following graduation limits:

Passing 3/8 in. sieve - 100%
Passing No. 4 sieve - 85-100%
Passing No. 100 sieve - 0-10%
Passing No. 200 sieve - 0-3%

Select backfill is often specified for backfill around or above utility lines such as water, sewer, gas and electrical ducts or cables.

Compaction to the specified density is usually required for all fill and backfill. It is also necessary in some cases to compact soil in cut sections. More compaction is needed under floor slabs, foundations and pavement than in embankments. A higher degree of compaction is required when only slight settlement can be tolerated.

Compaction Factors

Material (Compacted)	Multiplier
Sand	1.17
Loam	1.39
Clay	1.59
Rock (blasted)	1.15
Coral	1.15

Cubic volume in place times the multiplier equals the cubic yards of loose material needed.

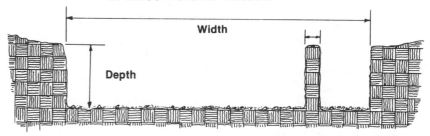

Labor Backfilling

Work Element	Unit	Machine-Hours Per Unit
Spread borrow material from dumped piles with motor grader, minimum compaction		
Over 2,000 SY	100 CY	1.1
Small area	100 CY	1.7
Small areas around building spreading by machine and by hand	100 CY	3.0
Spread dumped fill or gravel with motor grader in open area, no compaction		
6 inch layers	100 SY	0.5
12 inch layers	100 SY	0.3
Backfill trench, no compaction		
By hand, man-hours	100 CY	85.0
With small tracked dozer	100 CY	1.9
With small front end loader	100 CY	1.6
Backfill against foundation, no compaction		
Dumped & placed by machine	100 CY	10.0
Filled by loader, 300' haul	100 CY	1.8
Compaction to 95% maximum density		
Confined area, hand tamp	CY	0.5
Vibrator in open area	CY	0.1
Vibrator roller	100 CY	3.7
Pneumatic tired roller	100 CY	3.7
Sheepsfoot roller	100 CY	3.7

These figures assume equipment appropriate for the type of work performed.

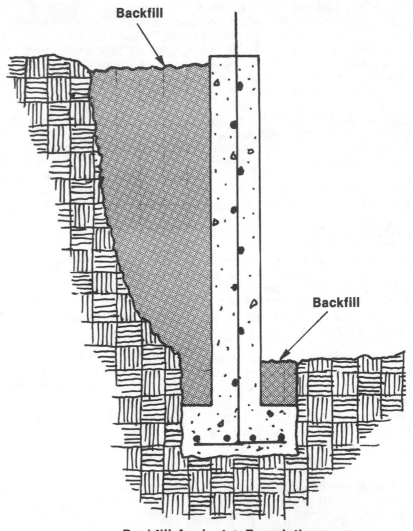

Backfill Against A Foundation

Compaction

Compaction can be done with various types of rollers, vibrating or hand-operated compactors. Compaction equipment can be classified under the general types and purposes as follows: Steel-wheeled rollers compact from the surface downward with a uniform, direct weight application.

Rubber-tired rollers compact from the surface with a non-uniform kneading action, with direct weight application. Some lateral movement results from the effect of the pneumatic tires.

Grid rollers compact from the surface downward with partial surface contact, some lateral effect, and consolidation effort.

Vibratory rollers combine direct surface compaction effort with a penetrating consolidation effort.

Vibratory pans are the most basic consolidation equipment. They are used for local and confined application such as around steel forms or foundation walls.

Soil types, moisture content, and characteristics of the subgrade will determine the most effective type of compaction equipment. The project specifications may limit the choice of equipment. However, it is usually the contractor's option to use any type of compactor that he selects. Regardless of the type of compaction equipment used, you must obtain the specified compacted density.

Optimum moisture conditions must be maintained within set limits during compaction operations. The contractor must either dry the soil by exposing it to the air or add water to each layer to reach the ideal moisture content necessary for optimum compaction.

Borrow is often allowed from designated areas on the building site. Approval to use material from the designated pit does not mean that all material found in the deposit will be acceptable. It is the contractor's responsibility to select material that meets specifications or request approval of another borrow pit site if it does not. All clearing and grubbing of the borrow site must be done by the contractor. When excavation is completed, leave the borrow pit uniformly graded to drain and with edges neatly trimmed and protected from erosion.

Fill

Before any fill is placed on the building site, the area must be cleared and grubbed. The existing ground surface may also require leveling and compaction to prevent settlement of the fill. Be especially careful to compact any new utility trenches which may have been constructed in the fill area.

Embankments for support of buildings, structures, roads and parking areas are built up from successive layers of fill material compacted to the required density. Placing fill in lifts of a greater thickness than specified makes it difficult to reach the necessary compaction.

It is important to see that fill materials meet any specification requirements. No stumps, roots, trees, brush or unsuitable material should be allowed in the embankment. Material of this type would eventually result in settlement of the embankment. Rock is perfectly acceptable as fill. If rock is placed in a shallow fill where soil or fine grained granular overlayment is intended, the voids between the rock should be filled. In deep fills, dock, groin, and pier construction, rock interlocking and consolidation is necessary. But the voids will not need to be filled.

Rock should not be placed in locations where underground utility lines or piling may be placed in the future.

When embankments are constructed on hillsides or placed on the sides of existing embankments, the original ground or embankment is usually terraced to prevent sliding. This should be covered by a detail on the drawings or in the specifications.

Drainage channels adjacent to fill slopes will often need lining with turf, rock, asphalt or concrete to prevent erosion. The drawings usually provide for protection. If field conditions indicate the need for a lining where not shown on the drawings, the contractor should call this to the attention of the owner.

Labor for Erosion Control

Work Element	Unit	Man-Hours Per Unit
Machine work:		
Sloping shoulders, banks, and ditches	1000 S.Y.	11
Hauling riprap or rubble	1000 C.Y./mile	30
Placing riprap or rubble	1000 C.Y.	60
Hydro-Mulch	1000 S.Y.	.5
Hand work:		
Sloping shoulders, banks, and ditches	1000 S.Y.	70
Placing riprap or rubble	S.Y.	1.0
Planting grass sprigs and roots or sod	1000 S.Y.	48

Grass planting may involve plowing, harrowing, fertilizing, digging sprigs, disking, seeding, and watering or placing sod. Therefore, the man-day figure may have to be adjusted to meet specific job requirements according to the hand work required. Hydro-Mulch crew requires at least one experienced operator. Figure is based on a 1,000 gallon working capacity machine. Figure also includes charging time.

Suggested crew:
Sloping shoulder, banks and ditches; one or two operators on equipment; two to eight men with hand tools.
Riprap: one to twenty men hauling and placing stone.
Grass: three operators on Hydro-Mulch machine, one to twenty men with hand tools.

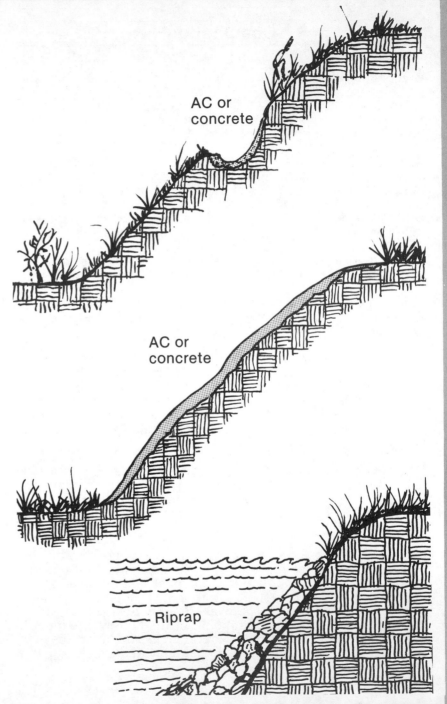

AC or concrete

AC or concrete

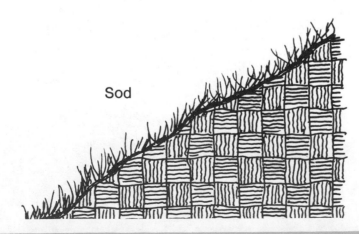

Sod

Riprap

Driving Piles

The safe load that can be placed on a pile is usually computed with a formula which considers the energy of the blow of the hammer and the penetration of the pile from that blow. Some formulas in common use also consider the weight of the pile in relation to the weight of the hammer.

These formulas are by no means an exact index of actual supporting power of the pile. They are a guide and may be fairly reliable where previous experience with load tests or actual structures has shown that they can be used with confidence. In the majority of cases, the pile-driving formula is merely a guide which indicates when it is safe to stop driving.

When the specifications direct that piles be driven to a bearing value as determined by a given formula, that formula must be followed. If driving conditions make the formula results seem unreliable, the project engineer may have to be consulted.

Some engineers use a "wave equation" for pile foundation analysis to increase the accuracy of bearing capacity calculations. The basic equation describes the travel of an impulse or wave down a pile as it is being driven.

Piles driven through soft material into a very hard substratum generally develop their bearing capacity by end bearing. Most pile-bearing capacity formulas are not reliable under such conditions. Piles usually are driven until they can not be driven further in such cases. Thus the formula becomes irrelevant.

Most piles develop their carrying capacity by the friction of the soil against the surface of the pile. Pile formulas apply best to these conditions. There is a general relationship between penetration per blow and frictional value per unit of surface area, which the formulas attempt to evaluate. You should understand that there are many variables that distort the reliability of all such formulas. In some soils piles drive easily but, if permitted to stand for a few days, or even a few hours, may on redriving develop much greater resistance. In these cases, the actual bearing capacity also increases greatly after a rest.

Occasionally a pile will lose more bearing capacity after a period of rest and can be driven more easily. Failure of foundations may result if special attention is not given to driving under these conditions.

Soil mechanics is also used to estimate the safe capacity of friction piles. The shearing strength of the soils is determined by laboratory tests on undisturbed samples. The safe frictional load per square foot for each principal stratum is determined by applying a factor of safety to the ultimate values obtained.

Load Capacities
A group of closely spaced piles does not have as much load-bearing capacity as the same number of piles driven at wide distances apart. This is because the cones or bulbs of pressure transmitted to the subsoil by each pile overlap. The combined pressure overloads the weaker strata and causes settlement. The plans and specifications make allowance for this deficiency. This is why it is important that piles be spaced as indicated, that the outer rows be battered if so specified, and that piles already driven not be pushed up when other piles are driven.

Pile load tests can be performed in various ways:

a. A reaction applied to the head of a pile by direct load on a platform.

b. By jacking down against a sufficient load placed over the platform or the load on a structure.

c. By jacking down against a cross beam held down by uplift piles at each end.

d. With a relatively small variable or movable load on the end of a long cantilever beam pivoting over the test pile and held down by two or three uplift reaction piles on the end of the short back arm of the cantilever. Platform load for a direct load test may be pig iron, rails, or any material that will not absorb rain water.

Types of Drivers
The five basic types of pile rigs in general use are skid drivers, floating pile drivers, railway pile drivers, rotating locomotive or truck cranes with hanging leads, and vibratory pile drivers.

Skid rigs are used when a large number of piles are to be driven in a relatively small area. They are also used for trestles and pier construction when the work can be done progressively with the driver supported on the part of the structure already installed. Skid rigs may be equipped with pendulum or pivoted leads for driving batter piles. Skid rigs for ordinary work usually are of timber construction and consist of sills, rigid tower leads, backstrap, and bracing. They may be equipped with rollers for ease of movement. The stream boiler and hoist are mounted at the back end of the sills. Two lines are used: one carries the hammer and the other hoists the pile into position. Skid rigs for driving precast or cast-in-place concrete piles and for other heavy work are frequently of steel construction.

Floating pile drivers are similar to land drivers except that the rig is mounted in a fixed position on a barge. The leads are frequently braced laterally to minimize side whip. The boiler and hoist are mounted separately at the stern of the barge.

Several different types of **railway pile drivers** have been perfected for use on railroads when piles have to be driven between passage of trains. They are mounted on standard-gage trucks, are generally self-propelled, and have full rotation on turntables. Leads usually are of the swinging type to permit driving batter piles.

Most contractors now prefer to use **truck crane drivers** with swinging or rigid leads for driving piles. These rigs are fast and flexible. Swinging leads are more difficult to control than rigid tower leads because swinging leads rest on the pile and may buckle out of line as the cable is slacked.

Vibratory pile drivers use the centrifugal force of a rapidly rotating eccentric weight resting on top of the pile. This causes very rapid vibrations which "shake" the pile into the ground. Another type of vibratory hammer which has seen limited use is the sonic hammer. It operates on the same principle as the vibratory hammer. However, the sonic hammer operates at much higher frequencies than the vibratory hammer.

Types of Piles

Wood Piles

Wood piles should not have large or loose knots, shakes or splits, or sharp, short, or reverse bends. Crooks should not exceed half the diameter of the pile at the middle of the bend. In short bends, the distance from the center line of the pile to a line stretched from the centers above and below the bend should not exceed 4 percent of the bend length or 2½ inches. A straight line between centers of butt and tip should fall within the pile. The taper should be uniform from butt to tip. Friction piles should be debarked

Before driving, butts of wood piles should be squared or chamfered to fit the pile cap. The tips are pointed or squared. It was once common practice to trim tips to 4 inches in diameter, but points may cause piles to drive out of line. Squared tips are now considered preferable. Tips may be trimmed slightly to fit in short pipe sleeves to prevent brooming in hard materials.

Salt treated wood piles are used where the wood may be painted or the odor or color of creosote would be objectionable. A variety of chemicals is used for salt treatment: zinc chloride, chromated zinc chloride, zinc metal-arsenate, and copper sulfate - sodium dichromate solution. The chemicals usually are in waterborne solutions.

Oilborne preservatives are also used to preserve wood piles. Such organic materials as pentachlorophenol and copper napthanate are used. The treating process is adapted to standard methods. As a temporary expedient, these preservatives are sometimes brushed on the surface.

The processes of creosoting lumber and timber are generally referred to as "full cell" or "empty cell," depending on the manner in which the material is treated. Records are maintained during the process, giving the amount of creosote introduced and retained in the wood, the temperatures reached during the various stages. and the air pressure and vacuum introduced.

Creosote is a distillate of coal tar. The composition of creosote varies with the properties of the coal from which it was produced and the method of distillation. Refined coal tar is often added to creosote to obtain a product known as creosote-coal-tar solution.

Wood piles are sometimes either impregnated or temporarily surfaced with a chemical that keeps the wood from supporting combustion. Many chemical components are used, including chromated zinc chloride, sodium borate, and various ammonium salts.

Steel Piles

Steel H bearing piles penetrate hard soils with a minimum of effort and in minimum time. In some cases steel H piles may be the only type which can be driven to the required depth without jetting or coring. Steel H piles can be driven on close centers or even close to existing structures. They can also carry heavy loads. A 14-inch 138 pound H pile 138 feet long has been tested to carry 450 tons. They have been driven 6 or 8 feet in mica schist, 22 feet in cemented sand and gravel, and 10 to 12 feet in hardpan. H piles resist lateral forces well because of their inherent bending strength.

H pile caps may consist of flat plates, pieces of rolled beams, or bars extending through web-and-flange holes. Research tests indicate that caps are unnecessary.

Riveted, bolted, or welded splices can be made on H piles. Connections are furnished by the mill. Splice connections should be made on inner face of the H pile to avoid loss of lateral ground support due to enlarged soil holes. A full butt weld made after scarfing one section (and possibly using small backing plates or single field welded web-and-flange plates) is enough if piles are in firm soil clusters or splice elevations are staggered, so long as only moderate driving force is needed. Splices should develop one-third of the full strength of the section if the splices are all at the same elevation and if driving is hard, even if laterally supported for their full length. Splices in long piles not braced laterally should develop full strength.

Box Piles

Box piles are composed of pieces of steel sheeting or beams. They can support large lateral loads where a bank of sliding rock has to be retained, for example. Box piles are also used in deep water or where large loads are to be carried on piles through poor material. Box piles can be driven into soft rock. They are not usually filled with concrete, but can be cleaned out and filled down to any desired distance for strength and protection of the interior against corrosion.

Precast Concrete Piles

Precast concrete piles are used to carry heavy loads through soft material to firmer strata. They can be reinforced to resist bending and uplift, and are also used where wood piles might deteriorate or decay. The largest size used is 30 inches in diameter and slightly over 100 feet long.

Judging the exact length needed may be difficult. Cutting or extending concrete piles is more difficult. Concrete pile splices can be designed if lengths are definitely known. An epoxy bonding compound splice is used in combination with dowels or post tensioning if lengths are not predetermined.

Precast piles may have a central core hole to save weight, and may be circular, octagonal, or square. The lower few feet are often tapered, although flat ends appear to be the best.

Prestressed concrete piles resist spalling during driving because the concrete is compressed by the steel reinforcing. In ordinary piles, concrete spall is caused by recoil during the driving.

Square piles up to about 50 times longer than they are thick can be handled by a single-point pickup. Piles 60 times longer than they are thick require two-point pickups. Piles 20 inches square and smaller are usually solid. 24 inch piles usually have 12 inch diameter cored holes. Octagonal 20 inch piles 132 feet long, with 11 inch diameter cored holes, have been driven through 30 feet of riprap to reach 60 ton design loads.

Cylinder type prestressed concrete piles are well adapted to soil or water conditions requiring very long piles of high carrying capacity. The piles are made up of a series of hollow-spun concrete sections 16 feet long, and are reinforced with longitudinal and spiral steel. Most of these piles are 14 to 30 inches in diameter. The walls are about 4 inches thick and have longitudinal holes. Sections are assembled after curing. High strength steel wires are threaded through the holes and then tensioned by hydraulic jacks and locked in place. After the wire holes are grouted and allowed to set, the pile is picked up as a

unit and driven. The weight is less than for solid concrete piles, and it can be driven more easily. Piles 36 inches in diameter and 192 feet long have been driven.

Cast-in-Place Piles

Cast-in-place uncased concrete piles can be used where soil or water will not fall into the hole after withdrawal of the shell before pouring concrete. Cast-in-place piles can not be used where driving for adjacent piles will damage the green concrete. Uncased piles need no casting yard, no storage space, avoid high freight costs, and do not require cutting off or splicing.

Compressed concrete piles are satisfactory if the soil type will permit placing concrete under pressure. Where the soil has little resistance, a temporary casing and coring are driven together. After replacing the core with concrete and withdrawing the casing, the weight of about 7 tons of hammer and plunger is left on the concrete. This is known as a straightshaft pile. Lengths up to 60 feet long with diameters of 14 to 24 inches have been made. To form a mushroom base, deposit concrete on the lower end of the casing, replace the core on it and draw the casing up to meet the core. Then redrive casing and core through the concrete, remove the core, fill the casing, replace the core on the concrete, and pull the casing over the core.

Compressed concrete pedestal piles are formed by driving a casing and core, removing the core, dropping a charge of concrete, pulling up the casing 18 inches to 4 feet with pressure of the core and hammer on the concrete, ramming out the concrete into a pedestal, removing the core, filling the case with concrete, and withdrawing the casing with the weight of the core and hammer on the concrete. Usual diameters are 14 or 16 inches. The result is the equivalent of a spread footing. A bearing stratum of limited thickness can be reached economically.

Cast-in-place concrete piles are used where support is required for the sides of the hole while concreting, or where the soil is difficult to compress and might be forced into the shaft by adjacent driving. A temporary outer casing is driven, and a permanent light shell inserted with reinforcing cages used if bending, shear, or uplift occurs or if the piles project above side support. Diameters range from 12 to 20 inches with the length limit approximately 75 feet. A core and casing are driven together, the core is removed, and a shell inserted to serve as a container for the concrete to protect it while setting. The core is set in the case on top of the shell. The hammer is dogged, so the position of the core is fixed. The lower collar is drawn toward the upper collar to withdraw the casing while the shell is held in place. Shells are thin corrugated or plain metal, or may be laminated fiber.

Casing concrete pedestal piles may be used where a soft soil overlies a stratum that may be developed into an adequate bearing stratum if the load can be spread somewhat. This type is best suited for use on irregular or sloping rock surfaces to secure a grip and avoid eccentric forces or sliding. The core and casing are driven together, the core removed, a charge of concrete put in the bottom core and casing driven down through the concrete. The core is then removed, a corrugated shell placed in the casing, the core placed on the shell, the hammer dogged so that the position of the core becomes fixed, the casing withdrawn, and the shell fully concreted. These cased piles are from 12 to 20 inches in diameter and up to 75 feet long. Reinforcing may be used.

Pipe Piles

Pipe piles may be open-end or solid-point. When supporting a load carried to depths of more than 30 feet and supported by end bearing or friction, pipe piles are most economical. Solid point pipe piles carrying loads ot 65 to 75 tons, with a diameter of 10¾ inches and lengths of over 100 feet are not uncommon. Solid-point piles are useful for under-pinning or for foundations which support additional loads in or under buildings where headroom is low.

Seamless steel pipe piles range in diameter from 6 to 24 inches. Spiral-welded pipe piles range from 8 to 36 inches, but closed-end pipe piles generally do not exceed 18 inches. Shell thicknesses for seamless pipe piles range 5/16 to 5/8 inch and from 7/64 to 5/8 inch for spiral-welded pipe. Closed-end pipe piles are jacked, driven or jetted, using cast-steel drive sleeves between sections or with the sections welded together. Yield points ranging from 24,000 to 45,000 p.s.i. are obtained when the piles are filled with concrete after driving.

Driving Hammers

Six principal types of pilehammers are in general use. They are drop hammers, single-acting steam hammers, double-acting steam hammers, differential-acting hammers, diesel hammers, and vibratory hammers.

Drop hammers are steel castings weighing from about 2,000 to 5,000 pounds. They are equipped with guide flanges to fit the leads and a removable pin for attachment to the hammer line. This pin is sometimes designed so that the grips attaching it to the hammer line can be automatically tripped after the hammer is hoisted. This permits free fall from a regulated height. Usually the line is attached permanently and the hammer is allowed to drop by releasing the hoist clutch and overhauling the hammer line as it falls. Drop hammers have been largely replaced by steam hammers.

Single-acting steam hammers consist of a frame and a hammer. The hammer is raised by steam pressure and drops by gravity. The force of the blow depends on the length of the stroke and the weight of the hammer. The stroke force varies with the steam pressure and the adjustment of the valves.

Double-acting steam hammers differ from the single-acting type in that steam pressure is used both to raise the moving part and to accelerate its drop. The force and frequency of the blows is governed by the steam pressure. Double-acting hammers have relatively light moving parts as compared with single-acting hammers but develop comparable force per blow. They operate at higher speeds and use shorter strokes.

Diesel hammers have been used to a considerable extent in Europe. The hammers form a self-contained unit, including fuel tank and injectors, and are lighter and more portable than steam hammers. Diesels operate more slowly than double-acting hammers. The harder the driving, the longer the strokes. This causes difficulty in rating energy and in operation when driving is soft.

Differential-acting hammers use steam or air to both raise and drive the ram. Operating speeds are greater than the single-acting hammers.

No single hammer is best for all classes of work. When driving heavy piles, or driving piles into dense strata, a heavy blow with a heavy ram, fairly short stroke, and low velocity is best. This reduces the impact and shattering effect of a high-velocity, lightweight ram striking a heavy mass.

When driving light or average weight piles or casings in materials of average consistency, rapid blows from a double-acting hammer keep the pile in motion and reduce driving resistance.

Differential-acting hammers combine some of the advantages of both single and double-acting hammers. Ram weights fall in the same range as for single-acting hammers, while the number of blows per minute approaches that of double-acting hammers.

Driving the Pile

Damage to the pile is common when driving thin pipe or monotube shells. A wider spacing of piles or the use of a lighter hammer will reduce pile damage. If a large hammer is crushing the shell, stop driving tor a few minutes to allow the soil to readjust itself. This will often allow driving to the desired depth without damage.

Time required for driving is an important factor in large jobs and may influence equipment selection. A drop hammer may strike few blows a minute, but the energy of each blow is greater. On the other hand, the energy of the drop hammer may require a drop limitation in firm strata to avoid overstress. This saving in time may or may not be significant.

The accuracy with which piles must be located varies with the character of the work. Extreme accuracy in positioning is essential, for example, in a large mat foundation supported on piles at relatively wide centers. Accuracy is important in closely spaced footing clusters, and particularly important in bridge bent piers, where the upper part of the pile is exposed and any misalignment will be apparent. Be sure that piles are positioned with the accuracy required by the circumstances.

When close tolerances are essential, the specifications may require the use of templates to assure proper centering. Templates must be strong enough to withstand the abuse they receive.

Piles can never be driven absolutely vertical and true in position. Even in ideal conditions the center of a pile head will deviate a certain amount from the required location. The lower depth will nearly always vary from the required vertical or batter line. Still, every precaution should be taken to maintain the piles in position. The general procedure for determining the pile alignment and elevations is as follows:

(1) Measure the elevation at the top of the pile immediately after driving. Check the final elevations after the adjacent piles are driven or at the completion of all pile driving. If point-bearing piles are uplifted, they should be redriven. A small amount of uplift in friction piles is not harmful. (2) Check the location of all piles after the adjacent piles are driven or at the completion of all pile driving. In ordinary soil conditions a 3" tolerance is considered reasonable. Piles which are driven at greater variation may throw a greater load to some piles in the group. In such cases, the pile reaction and cap design must be checked.

(3) Inspect the pile shaft for vertical alignment. For cast-in-place piles, the general practice is to lower an electric light into the shaft before placing concrete. If the light can not be seen from the top, the pile is rejected. In the case of heavily loaded piles, such crude procedure may not be enough. Measurement should be made with special instruments.

(4) A pile is considered defective if it is damaged by driving or if it is driven out of position, is bent, or is bowed along its length.

Avoid damage to fresh concrete in a cast-in-place pile when driving adjacent piles. Delay concrete work until all piles within a certain radius are driven. The radius depends on the soil condition, the length and size of pile, and the pile spacing.

A defective pile may be withdrawn and replaced by another pile. It may be left in place and another pile be driven adjacent to it. Sometimes the damaged part of the pile can be removed and a new length of pile spliced in. Unless the remaining portion of the pile is absolutely intact, this method is not desirable.

Pile Accessories

Pile accessories in common use include rings, caps, shoes, splices and extension leads:

(1) Timber piles are usually protected against splitting under the impact of the hammer with steel or iron pile rings. Generally 3" by ½" stock is driven over the chamfered head of the pile with a few light blows of the hammer.

(2) All types of piles can be protected against damage with a pile cap. This is a steel casing, fitted with side guides to engage the leads. It is designed with a lower recess into which the head of the pile fits and an upper recess in which an expendable cushion block is placed. The latter absorbs the hammer impact and requires replacement when crushed or set on fire. Cushion blocks usually are of hardwood or of rope or fiber mats.

(3) The points of timber piles are sometimes protected with metal shoes when the piles are to be driven into hard strata. They should be used only when permitted by the specifications.

(4) When it is necessary to splice two piles together, the joint is reinforced by either hardwood fishplates bolted through each pile or by steel pipe sleeves secured to the two pile sections by lag screws. Details of the splice should be approved by the engineer unless they are shown on the contract drawings.

(5) When it is necessary to drive piles to cutoffs considerably below the base of the piledriver, extension leads are fitted to the fixed leads. The hammer then travels in the extension leads, which are lowered progessively as the pile is driven.

Labor Driving Precast Concrete Piles

Work Element	Unit	Machine-Hours Per Unit
Pile, complete		
20-foot	Each	0.5
40-foot	Each	1.5
60-foot	Each	2.5
80-foot	Each	3.5
100-foot	Each	5.0

Pile driving includes the work of assembling leads and hammer, preparing equipment for driving, squaring and trimming pile butts, moving the driver into place, placing pile in leads, and driving the pile. **Typical crew: one supervisor, eight men.**

Labor Driving Wood Piles

Work Element	Unit	Machine-Hours Per Unit
25-foot pile		
Preparation	Each	1.5
Drive	Each	0.5
50-foot pile		
Preparation	Each	2.0
Drive	Each	1.5
75-foot pile		
Preparation	Each	2.5
Drive	Each	3.0
Rigging leads and hammer (2-3 men)	Each	6.0
Cut pile at required level	Each	0.2
Dismantle leads and hammer	Each	6.0
Lash piles to form dolphin	Each	1.5

Pile driving includes the work of assembling leads and hammer, preparing equipment for driving, sharpening pile tips, installing steel tips on the piles, squaring and trimming pile butts, moving the drive into place, placing pile in leads, driving pile, and cutting pile to required grade. No time is included for delivery of material to the jobsite.

The type of driving equipment will have a considerable effect on the time required for this work. Whether a steam, diesel, or drop hammer is used to drive piling affects the time required for a given unit of work considerably more than it affects the number of men required for the cperation. The estimator should know what equipment is to be used in performing this work.

Suggested crew: One leader, six men; ten men when placing dolphins. If an additional crane is required to support construction, increase driving figures by 15 percent.

Labor Driving Steel Piles

Work Element	Unit	Machine-Hours Per Unit
25-foot pile		
Preparation	Each	1.5
Drive	Each	0.8
50-foot pile		
Preparation	Each	2.0
Drive	Each	2.3
75-foot pile		
Preparation	Each	2.5
Drive	Each	4.5
Rigging leads and hammer (2-3 men)	Each	6.0
Cut pile at required level	Each	0.3
Dismantle leads and hammer	Each	6.0

Pile driving includes the work of assembling leads and hammer, preparing equipment for driving, sharpening pile tips, squaring and trimming pile butts, cutting holes to facilitate handling, moving the driver into place, placing pile in leads, driving pile, and cutting pile to required grade. No time is included for delivery of material to the jobsite.

These figures are for preliminary estimates only. The many variables involved in this type of work require on-site determinations for accurate estimates. Variables of prime importance are: design, soil, equipment and method used, tides, access to the site, currents, and material storage. For concrete filled, fluted hollow steel piling and pipe piling, use these steel bearing pile figures. If an additional crane is required to support construction, increase driving figures by 15 percent.

Suggested crew: one leader, six men.

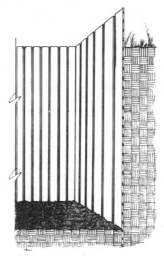

Labor Placing Sheet Piling

Work Element	Unit	Man-Hours Per Unit
Wood (20 feet deep)		
Preparation	1000 S.F.	4.0
Drive	1000 S.F.	35.0
Bracing	1000 S.F.	20.0
Cutting	1000 S.F.	1.5
Steel (30 feet deep)		
Preparation	1000 S.F.	6.0
Drive	1000 S.F.	50.0
Bracing	1000 S.F.	30.0
Cutting	1000 S.F.	2.0
Concrete (30 feet deep)		
Preparation	1000 S.F.	35.0
Drive	1000 S.F.	75.0
Bracing (steel)	1000 S.F.	30.0
Cutting	1000 S.F.	4.0
Install deadman and tieback	Each	24.0

Includes preparation of leads and equipment for driving, preparation of pile for driving, placing pile in leads, driving pile, cutting and bracing pile, and installing deadmen and tiebacks.

Suggested crew: one leader, six men.

Labor Bracing and Capping Piles

Work Element	Unit	Man-Hours Per Unit
Bracing		
Horizontal	Each brace	1.0
Diagonal	Each brace	0.8
Capping		
Wood	1000 L.F.	100
Steel	1000 L.F.	150
Concrete	1000 L.F.	200

The table assumes 4" x 10" x 4' bracing members and includes cutting bracing, drilling, handling into place, and fastening.

Suggested crew: one leader, six men.

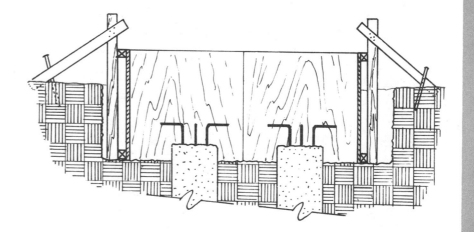

Labor Extracting Piling

Work Element	Unit	Man-Hours Per Unit
Pile removal		
Piles with extractor	Each	1.5
Sheet piling with extractor	1000 S.F.	25.0
Sheet piling with crane	1000 S.F.	20.0
Cut pile below water level	Each	1.0
Pile disposal	Each	0.5

Includes rigging equipment, extracting and handling piling.
Suggested crew: one supervisor, four men.

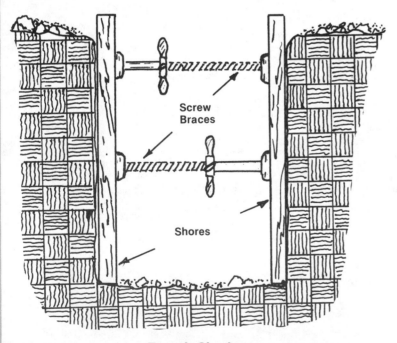

Trench Shoring

Labor for Trench Shoring

Work Element	Unit	Man-Hours Per Unit
Open wood sheeting and bracing		
Trench to 6' wide	100 S.F.	5.5
Over 6' to 10' wide	100 S.F.	6.8
Over 10' to 16' wide	100 S.F.	8.0
Closed 2" wood sheeting and bracing		
Trench to 6' wide	100 S.F.	11.2
Over 6' to 10' wide	100 S.F.	13.0
Over 10' to 16' wide	100 S.F.	13.6
Pits, one side only, braced	100 S.F.	16.8
Open bank, 4' high retainers braced on shelves	100 S.F.	21.8

Unit is 100 square feet of trench face on one side. These figures include bracing to a depth of 15 feet. Add 6 man-hours per 100 square feet for trenches 15 to 22 feet deep.

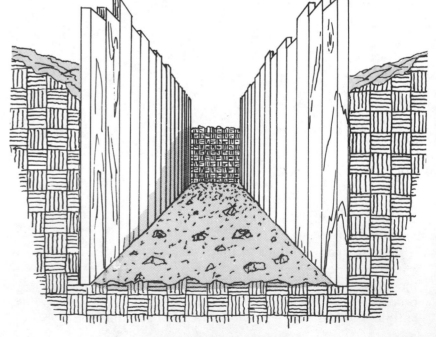

Heavy Timber Sheeting

Piping

Some plans give trench elevations of the pipe invert, the lowest point on the interior of the pipe. Others indicate only the minimum depth below the surface. In any case, the trench must be deep enough to provide cover, and deep enough to receive bedding if specified. Bedding with a minimum thickness of four inches is usually required to receive the pipe. Bedding rock has a maximum size of one-half inch with 50 percent or less passing a No. 22 sieve. To support the lower part of the pipe, a 14 inch outside diameter pipe would be bedded not less than two inches into the material. This would support the pipe over about 70 percent of its diameter or 10 inches.

When the pipe is laid on undisturbed soil, the trench is under-excavated. The trench bottom is shaped and prepared by hand for a smooth, even, uniform bearing for the full length of the barrel. Bell holes are dug at each joint to allow proper construction. If tunneling is permitted under surface obstructions, care must be taken to shore the roof where necessary.

Begin trench excavation at the lower end of the line and proceed upgrade to protect the work from possible flooding.

Laying Pipe

Pipe should be laid to the prescribed line and grade and should be firmly and uniformly bedded. It is important that the pipe not rest on hard spots, ledge rock or soft spots where settlement may occur. Each length of pipe must be sound and clean and handled carefully to prevent damage. Joints should be made as specified for the type of pipe. The pipe should not be moved after the joint is made (except in case of welded joints).

Pipe usually must be laid within 0.02 foot (¼") above or below the specified grade for gravity lines. The best method of alignment uses 1" x 6" batter boards placed across the trench, firmly bedded, and anchored by stakes at either side. A short upright of ¾" x 2" stock is aligned by transit so one edge is exactly vertical and on line. The upright is nailed to the batter board. Avoid errors by using the same edge (preferably the right edge) as the work progresses upgrade. A nail is then brought to true grade on this edge, at a given number of even feet above correct invert grade.

Stretch strong string or fishline between the nails and pull the line taut against the edge of the upright. Two nails 1 foot apart will be required where the depth of cut makes a change in vertical offset necessary. The batter boards should be spaced 25 feet on centers. If spacing is greater than that, make some correction for sag in the line.

For a sound joint, keep jointing surfaces clean. Moisture in a hot-poured joint may cause explosion and injury. Jointing should be done according to specifications or the manufacturer's recommendations. Joint gauges or pipe markings are used to check joints, particularly those with a groove in the spigot end for the ring gasket.

Backfill pipe trenches very carefully, paying particular attention to the materials and methods of compaction around, and immediately above the pipe. Specifications usually require that selected fine excavation or imported material be used for this portion of the backfill to 1 foot above the pipe. This material should be carefully tamped in layers. Excessive compaction may crack or break an unyielding pipe, or force a flexible pipe out of round.

Once backfill is one foot above the top of the pipe, the balance of the trench can usually be backfilled with the run of excavated materials and compacted. The surface is mounded to allow for settlement if the upper trench was not compacted.

Testing Pipe

After a section of pipeline is in place and before the trench has been completely backfilled, it is usually tested under pressure. The section of pipe is hydrostatically tested under pressure for a period of one hour. Following any necessary repairs, the section of pipe should be subjected to a leakage test for a period of two hours.

The standard American Water Works Association test requirements stipulate an allowable loss of water based on the pipe diameter, the number of joints in the section under test, the average test pressure, and time. Be careful to remove all air from the section of line under hydrostatic pressure or leakage testing. Under certain conditions trapped air may create a hazardous situation.

In areas where it is extremely hot, testing should be done when there is no direct sun light on the exposed sections of pipe.

The pipe is partially backfilled leaving the points exposed and with trust blocking in place. Water is pumped into a valved-off section of pipe until the required pressure is reached. The additional water that must be pumped in to maintain this pressure during the specified time will indicate the water loss. Most contractors perform a preliminary test for their own information before conducting the actual test for record purposes. Don't complete backfilling until the test has been successfully completed.

Types of Pipe

Water distribution piping may be cast-iron, concrete, asbestos-cement, plastic, or steel. Cast-iron pipe with a cement lining and asphalt coating is common. Checking cast-iron pipe for leaks with a hammer is poor practice. The cement deadens the ring. Swabbing the exterior with kerosene is a better way to disclose cracks. Pipe installed in corrosive soil must be carefully selected. Some soils may require cathodic protection for ferrous materials. Other soils may attack Portland cement pipe. Plastic pipe is more inert but is not always able to withstand the trench loading or the hydraulic pressure. Concrete pipe is reinforced for water lines. Steel pipe is used for water lines above grade. Asbestos cement pipe is made up to 36 inch diameter, cast-iron up to 48 inch diameter, and concrete or steel up to very large diameters. Galvanized steel pipe, copper or plastic pipe may be used in the smaller sizes for water service and irrigation systems.

Specific Uses

High points in water lines should be avoided as much as possible. Where they occur, check the specifications for vacuum and relief valves required. Be sure the water lines are flushed and sterilized as specified. Backflow preventers are always used where there is a chance for contamination.

Storm drains are most commonly concrete, reinforced concrete, or corrugated metal pipe. In the smaller sizes, asbestos cement or other materials might be used. Joints may be bell and spigot, tongue and groove, or couplings, as specified, using mortar, mastic or gaskets. Pipe with elliptical reinforcing must be installed carefully.

Subdrains are used to drain off ground water or leakage. Sometimes they are used in a drain field to dispose of an effluent into the ground. Common materials include vitrified clay, concrete, asbestos cement, corrugated metal, and bitumenized fiber. The pipe is frequently perforated along the lower portion to permit entry or exit of water. Joints are not sealed. Plain end butted joints are usually covered over the top with waterproof strips. A fine screen may also be used over joints to keep sand and fines out of the pipe. The backfill around the subdrain should be carefully graded and may consist of two envelopes of different gradings. This is to keep fines from migrating into and clogging the drain.

Sewer Piping

Sewers usually are laid for gravity flow, but may include lift stations if required. Some lift stations discharge through a relatively long force main at low or medium pressure. Safety precautions are necessary when working where any sewage has been present. The atmosphere may be explosive, poisonous, or lacking the oxygen required to support life. Check the distance separating sewers from water lines. Sewers should be below water lines if they are within 10 feet horizontally. Special provisions are taken at crossings.

Materials used in gravity sewers are usually vitrified clay with bell and spigot compression joints. There has been a growing use of plastic pipe in the smaller sizes, with either bell and spigot and rubber-type gasket joints, or plastic-welded coupling joints. Force mains use either cement-asbestos pipe, (usually epoxy lined), or cast-iron cement-lined pipe, or plastic pipe where diameter is small.

Gravity sewer lines can not use cement pipe because they are not full and thus are subject to corrosion. Hydrogen sulfide gas which is released from the sewage would attack the cement. Hydrogen sulfide does not attack vitrified clay or plastic pipe.

Sewers passing through soil that is above the water table are subject to leakage of sewage into the soil. A test for exfiltration, if required by the specifications, is started after all pipes connected to the section to be tested are plugged. The entire section is filled with water to a specified level and the drop in the water level during a specified time indicates the leakage. Exfiltration tests should be made when backfill has been compacted to the level of the top of the pipe.

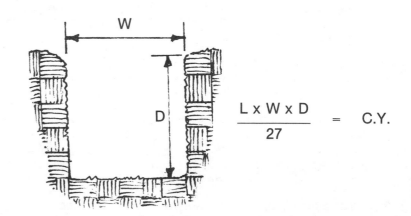

$$\frac{L \times W \times D}{27} = C.Y.$$

Labor Installing Concrete Culvert Pipe

Work Element	Unit	Man-Hours Per Unit
Concrete pipe, unreinforced, bell and spigot		
6" diameter	100 L.F.	9.0
8" diameter	100 L.F.	10.9
10" diameter	100 L.F.	15.2
12" diameter	100 L.F.	17.8
15" diameter	100 L.F.	23.0
18" diameter	100 L.F.	39.0
21" diameter	100 L.F.	43.0
24" diameter	100 L.F.	58.0
Concrete pipe, reinforced, bell and spigot		
12" diameter	100 L.F.	19.2
15" diameter	100 L.F.	24.8
18" diameter	100 L.F.	42.0
24" diameter	100 L.F.	62.0
30" diameter	100 L.F.	78.0
36" diameter	100 L.F.	83.0
42" diameter	100 L.F.	103.0
48" diameter	100 L.F.	119.0
60" diameter	100 L.F.	124.0
72" diameter	100 L.F.	207.0
84" diameter	100 L.F.	312.0
96" diameter	100 L.F.	495.0

Time does not include trenching, layout, backfilling or compaction above the top of the pipe. Work includes handling, placing, caulking, grouting and bedding pipe. If spedi-seal type joints are used, reduce the time estimate by 20%.
Suggested Crew: crane operations — 1 operator and 1 signalman; installing operations — 3 to 7 laborers depending on pipe size

Labor Installing Porous Concrete Underdrain

Work Element	Unit	Man-Hours Per Unit
Porous concrete pipe, standard strength		
4" diameter	100 L.F.	12.0
6" diameter	100 L.F.	13.1
8" diameter	100 L.F.	15.3
10" diameter	100 L.F.	17.6
12" diameter	100 L.F.	19.6
15" diameter	100 L.F.	23.4
18" diameter	100 L.F.	27.3
21" diameter	100 L.F.	35.8
24" diameter	100 L.F.	44.8
Concrete extra heavy strength		
6" diameter	100 L.F.	13.8
8" diameter	100 L.F.	15.5
10" diameter	100 L.F.	17.0
12" diameter	100 L.F.	19.5
15" diameter	100 L.F.	23.0
18" diameter	100 L.F.	27.4
Concrete manhole or catch basin connection	Each	4.0

Time does not include trenching, layout, backfilling or compaction above the top of the pipe. Time does include move on and off site, unloading, placing, repair and cleanup.
Suggested Crew: 4 or 5 laborers. For larger diameter pipe, crew size will vary with lifting requirements.

Diameter

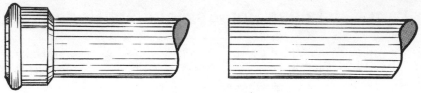

Cast Iron Pipe

Labor For Cast Iron Pipe

Work Element	Unit	Man-Hours Per Unit
Install pipe		
3" and smaller	100 L.F.	14.8
4" to 6"	100 L.F.	19.2
8" to 12"	100 L.F.	30.0
14" to 18"	100 L.F.	35.0
20" to 24"	100 L.F.	40.0
30" to 36"	100 L.F.	80.0
Install fire hydrant	Each	4.8
Install air release valves including risers and connections	Each	2.0
Install line valves and fittings:		
3" and smaller	Each	1.2
4" to 6"	Each	0.8
8" to 12"	Each	0.8
20" to 24"	Each	0.8
30" to 36"	Each	1.6
Install cast iron drain line (hot poured joint):		
4" to 6"	100 L.F.	10.8
8" to 12"	100 L.F.	16.0
Test system	100 L.F.	8.0

Includes unloading, placing and joint make-up.
Lifting equipment will be required on larger sizes. No excavation is included.
Suggested crew: two to six men, depending on pipe size and job scope.

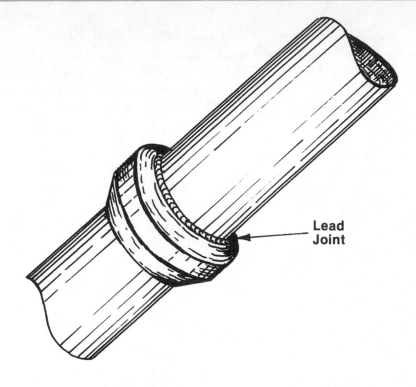

Lead Joint

Material for Lead Joints in Cast Iron Bell and Spigot Soil and Waste Lines

Nominal Pipe Size (Inches)	Materials Required Per Joint	
	Oakum, Lbs.	Lead, Lbs.
2	0.4	2
3	0.6	3
4	0.8	4
5	1.0	5
6	1.2	6
8	1.6	8
10	2.0	10
12	2.4	12

Labor for Asbestos Cement Pipe and Fittings

Work Element	Unit	Man-Hours Per Unit
Install pipe		
4" to 6"	100 L.F.	6.0
8" to 12"	100 L.F.	8.0
14" to 18"	100 L.F.	10.0
Install pipe fittings		
4" to 6"	Each	0.3
8" to 12"	Each	0.4
14" to 18"	Each	0.6
Concrete manhole or catch basin connections	Each	4.0
Test	100 L.F.	4.0

Time includes unloading, placement and joint make-up only.
Suggested Crew: 3 to 5 men, depending on pipe size.

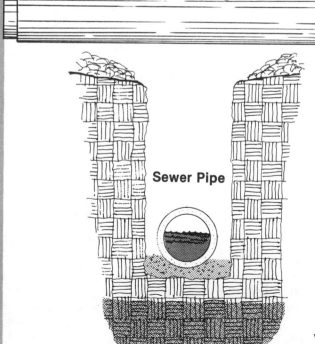

Diameter

Sewer Pipe

Water Table

Labor Installing Corrugated Metal Culvert Pipe

Work Element	Unit	Man-Hours Per Unit
Standard galvanized culvert pipe		
12" to 24"	100 L.F.	24
26" to 45"	100 L.F.	37
48" to 72"	100 L.F.	54
Bolted galvanized culvert pipe		
12" to 24"	100 L.F.	36
26" to 45"	100 L.F.	49
48" to 72"	100 L.F.	68

Work includes unloading, fine grading, placing, caulking, and installing joint clamps.

Installation of bolted pipe includes bolting together sections, unloading, fine grading, and placement.

Estimates for bolted pipe assume sections are bolted into desired lengths in a prefabrication yard.

When installing culverts over 48" in diameter, it is recommended that cross bridging be used to prevent culverts from being bent or twisted during hauling or installation. Cross bridging can easily be removed upon completion of backfilling and compaction.

Suggested crew is three to seven laborers depending on size of culvert to be installed.

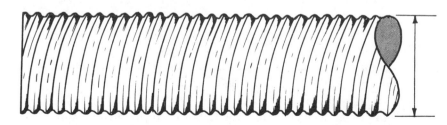

Welded Steel Pipe

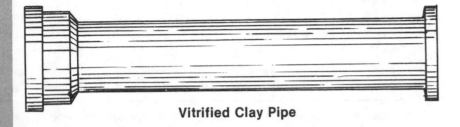

Vitrified Clay Pipe

Labor for Vitrified Clay Pipe and Fittings

Work Element	Unit	Man-Hours Per Unit
4″ to 6″	Joint[1]	0.25
8″	Joint[2]	0.30
10″	Joint	0.50
12″	Joint	0.60
15″	Joint	0.95
18″	Joint	1.1
21″	Joint	1.25
24″	Joint	1.4
30″	Joint	2.00
36″	Joint	3.5

[1]Section length is 2′-6″.
[2]Section length is 3′ for 8″ to 36″ diameter.
This table is based on a crew of four to six men for all pipe up to and including 21″. For larger pipe a crane and operator are needed.
Labor includes unloading, placing, grouting or caulking, gaskets and testing. Figure the time required for making the tie to an existing manhole or catch basin at six man-hours.

Labor for Welded Steel Pipe

Work Element	Unit	Man-Hours Per Unit
Install schedule 40 pipe, by oxyacetylene welding, butt weld; positions include horizontal, vertical and overhead		
1″	Joint	.9
1¼″	Joint	1.02
1½″	Joint	1.12
2″	Joint	1.15
2½″	Joint	1.43
3″	Joint	1.67
3½″	Joint	1.87
4″	Joint	2.13
5″	Joint	2.74
6″	Joint	3.73
8″	Joint	4.91
10″	Joint	6.72
Install schedule 40 pipe, by metallic arc welding, butt welds; positions include horizontal, vertical, and overhead		
1″	Joint	.9
1¼″	Joint	1.1
1½″	Joint	1.15
2″	Joint	1.39
2½″	Joint	1.76
3″	Joint	2.07
3½″	Joint	2.30
4″	Joint	2.65
5″	Joint	3.34
6″	Joint	4.33
8″	Joint	5.9
10″	Joint	8.12

The time for installation of the pipe includes unloading, erecting and aligning pipe in hangers, cutting and beveling one end of pipe, and welding pipe. Add 6% of total man-hours for testing. For schedule 80 pipe multiply man-hours by 1.6.
Suggested crew: one pipefitter, one welder, and one helper.

Labor for Pipe Insulation and Lagging

Work Element	Unit	Man-Hours Per Unit
Install magnesia covering:		
1½″ and smaller	100 L.F.	14
2″ to 3″	100 L.F.	16
3½″ to 4″	100 L.F.	18
5″ to 6″	100 L.F.	22
Install fiberglass with metal covering		
1½″ and smaller	100 L.F.	14
2″ to 3″	100 L.F.	16
3½″ to 4″	100 L.F.	18
5″ to 6″	100 L.F.	22
8″ to 10″	100 L.F.	24
Install molded cork covering:		
1½″ and smaller	100 L.F.	15
2″ x 3″	100 L.F.	18
3½″ to 4″	100 L.F.	24

Work includes the following items: mudding fitting and valves, installing metal lagging and waterproofing valves and fittings. Suggested crew: two to eight men, depending on pipe size and job scope.

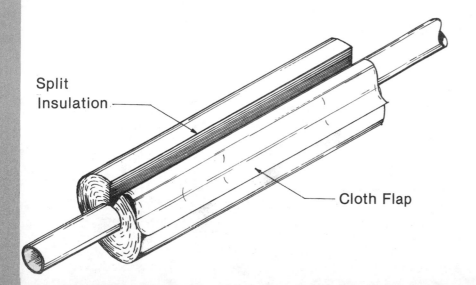

Split Insulation

Cloth Flap

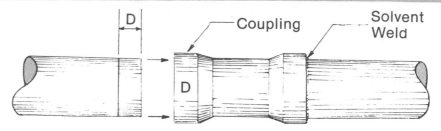

Coupling — Solvent Weld — D

Labor for PVC Pipe, Solvent Welded

Work Element	Unit	Man-Hours Per Unit
Install pipe		
½″	100 L.F.	1.1
¾″	100 L.F.	2.2
1″	100 L.F.	3.2
1¼″	100 L.F.	4.5
1½″	100 L.F.	5.4
2″	100 L.F.	5.5
3″	100 L.F.	6.0
4″	100 L.F.	8.0
Install couplings		
½″	Per 10	1.2
1″	Per 10	2.0
2″	Per 10	3.6
3″	Per 10	6.0
Install elbows		
½″	Per 10	1.2
1″	Per 10	2.0
2″	Per 10	4.0
3″	Per 10	6.0
4″	Per 10	10.0
Install tees		
½″	Per 10	1.6
1″	Per 10	2.4
2″	Per 10	6.0
3″	Per 10	8.0
4″	Per 10	10.0

Includes cleaning, applying solvent and drying time, installation of hangers and supports. PVC solvent will not work with CPVC. Each must have its own solvent cement. **Suggested crew:** two to four men per crew.

Labor for Threaded and Flanged Steel Pipe

Work Element	Unit	Man-Hours Per Unit
Install threaded pipe, schedule 40		
½'' — ¾''	Joint	1.00
1''	Joint	1.16
1¼''	Joint	1.24
1½''	Joint	1.36
2''	Joint	1.60
2½''	Joint	2.00
3''	Joint	2.40
3½''	Joint	2.80
4''	Joint	3.20
5''	Joint	4.40
6''	Joint	5.60
8''	Joint	7.60
Install schedule 40 pipe, flange fittings		
2''	Joint	1.08
2½''	Joint	1.30
3''	Joint	1.40
3½''	Joint	1.50
4''	Joint	2.12
5''	Joint	2.70
6''	Joint	3.30
8''	Joint	4.50
10''	Joint	7.00
12''	Joint	8.20

Threaded Pipe

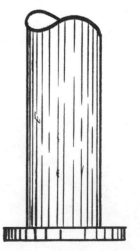

Flanged Pipe

This table is based on the fabrication and installation of pipe per joint. The job operations taken into account are making fittings service tight, installing, handling materials and tools, and threading one end per joint. Add 6% of total man-hours for testing.

Crew size is one pipefitter and one helper for pipe under four inches. For four inches and over one pipefitter and two helpers are used.

For extra-heavy (schedule 80) pipe and screwed fittings, multiply man-hours by 2.

For schedule 120 pipe and screwed fittings, multiply man-hours by 3.

Labor Installing Thrust Blocks, Fittings and Valves for Threaded and Flanged Steel Pipe

Work Element	Unit	Man-Hours Per Unit
Install thrust block		
12" and smaller	Each	6.4
14" and larger	Each	9.6
Install flanged fittings (schedule 40)		
2"	Each	0.8
2½"	Each	0.88
3"	Each	0.96
3½"	Each	1.06
4"	Each	1.6
5"	Each	1.94
6"	Each	2.2
8"	Each	2.68
10"	Each	2.72
12"	Each	5.8
Install threaded and flanged valves (schedule 40)		
2"	Each	0.6
2½"	Each	0.72
3"	Each	0.82
3½"	Each	1.00
4"	Each	1.1
5"	Each	1.4
6"	Each	1.9
8"	Each	3.1
10"	Each	5.0
12"	Each	8.0

This table is based on the average time required to unload, erect, align, and make up a service tight joint. All material and tool handling time is included. Crew size is one pipefitter and one helper. For schedule 80 pipe, multiply man-hours by 2. For schedule 120 pipe multiply man-hours by 3. Allow two hours installing air release valves.

The installation of thrust blocks includes bracing, forming, reinforcing, placing concrete, and stripping forms.

PVC Solvent Requirements

Size Fittings	Pint			Quart		
	No. of Joints	No. of Coupling or Elbows	No. of Tees	No. of Joints	No. of Coupling or Elbows	No. of Tees
½"	350	175	115	700	350	230
¾"	200	100	65	400	200	130
1"	150	75	50	300	150	100
1¼"	110	55	35	220	110	70
1½"	80	40	25	160	60	50
2"	45	22	15	90	45	30
3"	35	17	12	70	35	25
4"	25	12	2	50	25	15
6"	16	8	5	32	16	10
8"	10	5	3	20	10	6

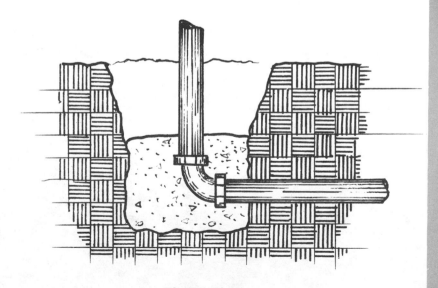

Thrust Block

Labor for Cast-in-Place Concrete Culverts

Work Element	Unit	Man-Hours Per Unit
Plywood forms and bracing	100 S.F.	26.7
Set reinforcing bars and tie in place	Ton	31.5
Place ready-mix concrete	C.Y.	2.2
Finish	100 S.F.	3.6
Cure and clean-up	100 S.F.	0.1
Finish with carborundum stone	100 S.F.	3.2

Suggested crew:
Two carpenters, one helper for forms.
One crew leader, four men erecting and stripping forms, three men placing reinforcing, six to eight men placing, spreading, and vibrating concrete, two to three men finishing, one to two men clean-up.

Labor Setting Trench Covers

Work Element	Unit	Man-Hours Per Unit
Angle iron frames		
1½″ x 1½″, 1.8 lbs./L.F.	100 L.F.	2.1
2″ x 2″, 3.2 lbs./L.F.	100 L.F.	3.9
2½″ x 2½″, 4.6 lbs./L.F.	100 L.F.	5.6
Steel cover plates		
6 lbs./S.F.	100 S.F.	2.9
8 lbs./S.F.	100 S.F.	3.8
11 lbs./S.F.	100 S.F.	5.3

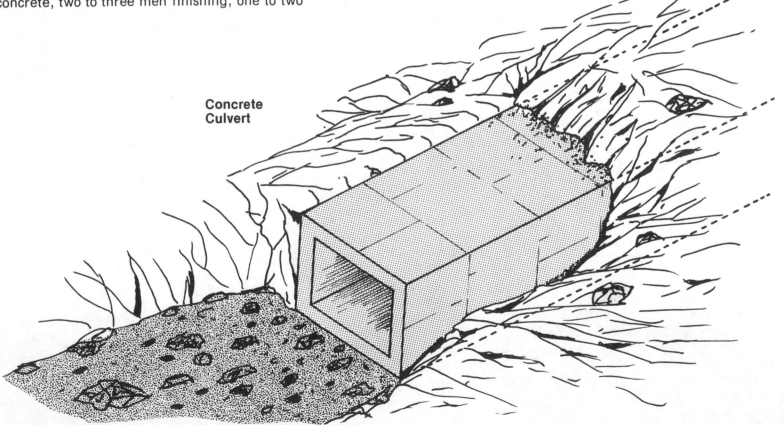

Concrete
Culvert

Labor Forming Transformer Vaults, Manholes, Catch Basins. and Valve Boxes

Work Element	Unit	Man-Hours Per Unit
Forms, plywood sheathing (contact area)	100 S.F.	26.7
Set reinforcing bars and tie in place	Ton	31.5
Place manhole frame and cover	Each	3.0
Place catch basin grate	Each	2.0
Place ready-mix concrete	C.Y.	2.7
Finish	100 S.F.	3.6
Cure and clean-up	Each	1.0
Average for all sizes of all units	C.Y.	15.0

Forming crew: two carpenters, one helper.
Placing crew: one crew leader, three to five men setting and removing forms, four to seven men reinforcing, three to five men placing concrete.

2 Sitework

Labor Installing Manholes and Catch Basins

Work Element	Unit	Man-Hours Per Unit
Brick manholes, 4' diameters		
Up to 6' deep	Each	39.0
7' to 9' deep	Each	49.0
10' to 12' deep	Each	77.0
Precast concrete manholes		
5' deep	Each	13.4
Add for each 1' over 5'	Each	.8
Brick catch basins		
2' deep	Each	30.0
4' deep	Each	37.0
6' deep	each	46.0
Precast concrete manholes		
2' x 3' x 3' deep	each	8.3
2' x 3' x 4' deep	each	11.3
Precast septic tanks		
300 gallon	Each	48.0
500 gallon	Each	60.0
750 gallon	Each	72.0
1000 gallon	Each	90.0
3500 gallon	Each	110.0
5000 gallon	Each	120.0

Includes handling materials into place, installation in a prepared location, cleanup and repairs as needed but no excavation or connections.

Labor Erecting Retaining Walls

Work Element	Unit	Man-Hours Per Unit
Rubble stone	C.Y.	8.0
Precast cribbing, S.F. of face	100 S.F.	9.5
Timber cribbing, S.F. of face	100 S.F.	20.0
Concrete, complete	C.Y.	11.0
Flagstone, laid dry	C.F.	.7
Flagstone, mortared	C.F.	.9

Labor includes handling materials into place, erection in a prepared location, cleanup and repairs as needed. No excavation included.

Labor Installing Street and Security Lighting

Work Element	Unit	Man-Hours Per Unit
Install foundations for metal standards[1]	Each	4.0
Install metal light standards (30')[2]		
Aluminum standard	Each	8.5
Steel standard	Each	10.5
Install wood street light pole[3]	Each	6.5
Install wood floodlight pole (2 floods)	Each	10.0
String one conductor for series lighting[4]	100 L.F.	1.5
Connect streetlight floodlight to power[5]	Each	4.5
Install lighting transformer	Each	9.0
Install constant current regulator and control devices for street lighting		
Installed in vault	Each	6.5
Installed on pole	Each	8.0

[1] Electrician work includes approximately ten feet of 2'' rigid steel conduit and pull box.
[2] Assembly and wiring performed on the ground.
[3] If light is to be added to existing pole, increase time by 10%.
[4] Work is approximately the same as over-head power construction. Use 30% of primary conductors man-hours for series circuits. Secondary installation of floodlights is the same as secondary conductors and service drops.
[5] Work does not include installation of power source.
Suggested crew:
Install foundation for metal standards: one electrician
Install metal light standards: two electricians, one equipment operator with crane.
Install streetlight pole: two electricians
String one conductor for series lighting: two electricians
Connect streetlight floodlight to power: two electricians.

Underground Electrical Distribution

Underground electrical distribution consists of cables installed in one or more conduits which are buried in a trench with manholes provided at intervals.

A simpler type of underground distribution system involves direct burying of cables in trenches. The draw-in type of underground distribution system is more suitable to cable maintenance than the direct burial distribution systems because repairs can be made without digging up the cable.

Cable replacements on underground systems are costly. Materials used in underground distribution systems should be carefully selected and properly installed. The basic materials used in underground distribution systems are power cables, duct, manholes, handholes and vaults.

The materials used for ducting include the following:

Fiber conduit is a tube made of wood pulp fiber treated and impregnated with a bituminous compound. There are two types of fiber conduit, thin wall and heavy wall. Thin-walled fiber conduit is usually encased in concrete. Heavy-wall fiber conduit is suitable for direct earth installation or can be encased in concrete.

Plastic conduit is available in two types, thin-wall Type I and heavy wall Type II. Plastic conduit suitable for underground distribution systems normally is extruded from plastics with acrylonitrile copolymer resin or cellulose acetate butyrate. Plastic duct made from either of the materials is suitable for direct burial in earth. Thin-wall plastic conduit, Type I, is not suitable for direct burial and requires concrete encasement.

Vitrified clay conduit is similar to sewer pipe. This is a durable conduit and is quite satisfactory for underground distribution systems.

Asbestos cement conduit is a tube formed around a polished steel mandrel in layers from a slurry of asbestos fibers and cement. The conduit is cured under steam, resulting in a hard and durable tube. There are two types of asbestos cement conduit, thin wall and heavy wall. Thin-wall asbestos cement requires concrete encasement.

Unless indicated or specified otherwise, the kind of conduit used should not be mixed in any one duct bank and should not be smaller than 4 inches inside diameter unless specified otherwise.

Conduits should be separated by a minimum concrete thickness of 2 inches, except that light and power conduits should be separated from control, signal and telephone conduits by a minimum of 3 inches. The top of the concrete envelope should be at least 18 inches below grade. Under roads and pavement it should be at least 24 inches below grade and under railroad tracks, not less than 36 inches below grade.

Conduits installed in the ground or beneath floors slabs should be encased in concrete. Unless specified otherwise, the top of the concrete envelope can be directly under the floor slab.

Multiple-type vitrified clay conduits are usually used for telephone, signal and control circuits.

Duct lines should have a continuous slope downward toward manholes and away from buildings. The pitch should be at least 3 inches in 100 feet. Changes in direction of runs exceeding a total of 10 degrees, either vertical or horizontal, use long sweeping bends with a minimum radius of 25 feet. Manufactured bends can be used at ends of short runs of 100 feet or less when close to the end of the run.

Long sweep bends may be made up of one or more curved or straight sections or combinations of straight and curved sections. Conduits should terminate in end bells where duct lines enter manholes or handholes.

Conduit joints should be staggered a minimum of 6 inches apart, the joints of conduits should be staggered by rows and layers to provide a duct line with maximum strength. Duct spacing should be maintained by the use of spacers constructed from precast concrete, high impact polystyrene or steel.

Seal duct entrances into manholes or handholes to prevent water from entering the manholes from the junction of duct and the manhole. Pull wires are left in the duct for pulling in the wire rope which in turn is used to pull in the cable.

Rigid metallic conduit is suitable for underground distribution systems. Steel conduits generally are used for building service installations and riser laterals. Rigid metal conduit installed underground must be zinc coated and encased in concrete.

Manholes, handholes and underground vaults are constructed of reinforced concrete. The strength of the concrete and type and size of the reinforcing bars are determined by the project specifications. They are built so that construction joints between the wall and base are absent or are adequately sealed.

Underground cables which come out of the ground and connect to utilization equipment often require a change from one cable type to another. Potheads are used to terminate cables.

Potheads should be provided for all terminations of single and multiconductor power and lighting cables for service above 600 volts and when specified on project plans for service at 600 volts and below. The pothead consists of a porcelain insulator, top cable conductor connector, aerial lug, metal body and supporting bracket.

Pothead type of termination can be used on any nonleaded or leaded cable. Pothead terminations use high dielectric strength filling compounds which permit considerable reduction in the required length of termination. Stress relief cones are used in terminating and splicing a shielded cable to relieve the concentration of voltage stress at the shielding tape termination.

Duct plugs are used during construction of partially completed ductlines to provide protection against the entrance of debris. Duct plugs should be used on all spare conduits.

Conduit installed without concrete encasement must not touch rocks or sharp edged materials. The top of the conduit should have a minimum slope of 3 inches in each 100 feet away from buildings and toward manholes and drainage points. Maintain at least a 3-inch clearance between the conduit and each side of the trench.

Labor Installing Underground Electrical Service

Work Element	Unit	Man-Hours Per Unit
Plastic conduit, Type 1		
2″	100 L.F.	4.0
3″	100 L.F.	4.4
3½″	100 L.F.	5.0
4″	100 L.F.	5.6
5″	100 L.F.	6.7
6″	100 L.F.	8.9
Plastic conduit, Type II direct burial		
1½″	100 L.F.	4.4
2″	100 L.F.	4.8
3″	100 L.F.	5.6
3½″	100 L.F.	6.6
4″	100 L.F.	7.3
5″	100 L.F.	8.0
6″	100 L.F.	9.8
Fibre conduit, Type I concrete encasement		
2″	100 L.F.	4.7
3″	100 L.F.	5.2
3½″	100 L.F.	5.6
4″	100 L.F.	5.9
4½″	100 L.F.	6.1
5″	100 L.F.	6.4
6″	100 L.F.	6.7
Fibre conduit, Type II direct burial		
2″	100 L.F.	7.7
3″	100 L.F.	8.8
3½″	100 L.F.	10.0
4″	100 L.F.	11.0
Transite conduit, Type I		
2″	100 L.F.	7.6
3″	100 L.F.	8.8
3½″	100 L.F.	9.8
4″	100 L.F.	11.2
4½″	100 L.F.	11.8
5″	100 L.F.	14.3
6″	100 L.F.	16.1

Labor Installing Underground Electrical Service (continued)

Work Element	Unit	Man-Hours Per Unit
Transite, Type II		
2″	100 L.F.	7.9
3″	100 L.F.	9.0
4″	100 L.F.	13.3
5″	100 L.F.	14.2
6″	100 L.F.	15.0
Galvanized conduit		
¾″	100 L.F.	4.0
1″	100 L.F.	5.0
1¼″	100 L.F.	6.0
2″	100 L.F.	8.0
2½″	100 L.F.	12.3
3″	100 L.F.	14.2
3½″	100 L.F.	16.3
4″	100 L.F.	18.0
5″	100 L.F.	25.2
6″	100 L.F.	35.0

Time does not include trenching, concrete encasement, or backfilling. Time does include move on and off site, cleanup and repairing, unloading and stacking.
Suggested Crew: 1 electrician and 1 laborer

Labor Splicing Cable in Manholes

Work Element	Unit	Man-Hours Per Unit
600 volt cable	Each	1.0
5,000 volt cable	Each	4.0
15,000 volt cable	Each	6.0

Time includes move-on and off-site and clean-up as needed.
Suggested Crew: 1 electrician

Labor Installing Wood Fencing

Work Element	Unit	Man-Hours Per Unit
Picket fence, pickets 1½" O.C., posts at 6'		
2 rail, 3' high	10 L.F.	1.5
3 rail, 5' high	10 L.F.	1.8
3' wide gate	Each	.9
Post and rail fence, posts at 8'		
2 rail, 3' high	10 L.F.	.9
3 rail, 6' high	10 L.F.	1.4
3' wide gate	Each	.8
Board fence, 1" x 6" boards on posts at 8'		
2 rail, 4' high	10 L.F.	1.8
3 rail, 8' high	10 L.F.	2.3
3' wide gate	Each	1.3
Snow fence, 4' high on steel posts at 5'	10 L.F.	.7

Labor includes handling materials into place, digging post holes, setting posts in concrete, erecting fencing, cleanup and repairs as needed.

Labor Installing Temporary Fencing

Work Element	Unit	Man-Hours Per Unit
½" plywood on 2" x 4" frame	10 S.F.	.3
Guard rail and wood fence around opening in pavement	10 L.F.	.7
Two 4" x 8" timber rails on wood posts at 4'	10 L.F.	2.4
Chain link on 10' posts at 12', 2' bury		
8' fence on 2" steel posts	10 L.F.	2.0
8' fence on concrete filled 4" steel posts	10 L.F.	2.4
End sections	Each	.8

Labor includes handling materials, excavation, erection, cleanup and repairs as needed.

Labor Installing Mesh & Chain Link Fence

Work Element	Unit	Man-Hours Per Unit
9 gage chain link, 2" posts at 10'		
6' high	10 L.F.	2.0
5' high	10 L.F.	1.8
4' high	10 L.F.	1.5
End or corner posts	Each	1.0
Gate, L.F. of opening	L.F.	.2
Barbed wire, 3 strand top, 1 side	10 L.F.	.4
Barbed wire, 3 strand top, 2 sides	10 L.F.	.7
Stucco net mesh fence with posts at 5'		
1" mesh, 4' high	10 L.F.	.6
2" mesh, 6' high	10 L.F.	.7
12 gage galvanized steel mesh (2" x 4") with posts at 6'		
3' high	10 L.F.	.7
5' high	10 L.F.	.8

Labor includes handling materials into place, digging post holes, setting in concrete, erecting fence, cleanup and repairs as needed.

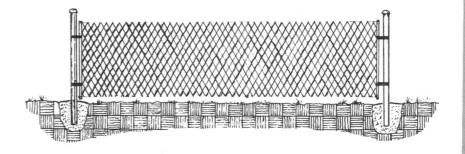

Labor Installing Athletic Equipment

Work Element	Unit	Man-Hours Per Unit
Baseball fields		
Diamond layout, sodded	100 S.Y.	32.0
Wire mesh backstop, typical	Each	40.0
Dugout, concrete block with bench, floor, roof and steps, S.F. of floor	10 S.F.	3.4
Football fields		
Goalposts	Each	26.0
Steel frame bleachers with plastic benches and wood walkways, demountable, per seat space	Each	1.5
Sodded field with drainage and markings	100 S.Y.	47.0
Complete synthetic turf field with pad, base, curbing, drainage, markings and posts	100 S.Y.	180.0
Basketball basket and metal backstop on pole	Each	8.4
Tennis courts		
Clay courts	100 S.Y.	36.0
Grass courts	100 S.Y.	22.0
Synthetic surfaced courts	100 S.Y.	45.0
Tennis nets	Each	3.0
Track and field equipment		
Cinder running track	100 S.Y.	57.0
Synthetic track with pad	100 S.Y.	25.0
Jumping pit	Each	16.0
Playground equipment		
4 seat swings	Each	5.5
10 seat swings	Each	10.6
Sliding boards, 10' high	Each	7.5
Monkey bars, 14' long	Each	6.0
Carousels, 10' diameter	Each	9.8
See-saw, 4 boards	Each	9.0
Lacrosse or hockey goals	Each	13.0
Soccer goals	Each	21.0
Football & soccer goal combo	Each	42.0

Use these figures for installation of college or public park grade athletic equipment installed according to manufacturer's recommendations. Labor includes handling materials into place, installation in a prepared location, cleanup and repairs as needed.

Labor for Landscaping

Work Element	Unit	Man-Hours Per Unit
Seed at 300 lbs. per acre & fertilize	Acre	22.0
Seed at 4.5 lbs. per 100 S.Y.	100 S.Y.	1.4
Fine grade and seed including lime and fertilizer	100 S.Y.	2.0
Fertilizer applied at 3 lbs. per 100 S.Y.	100 S.Y.	.2
Grass seed at 7 lbs. per 100 S.Y.	100 S.Y.	.3
Hydraulic seeding for large areas	100 S.Y.	.7
Sodding on level ground	100 S.Y.	13.0
Sodding on slopes	100 S.Y.	26.0
Spread and grade topsoil	100 C.Y.	34.0
Tree planting, including hole preparation and staking		
6' tree, 5 gallon	Each	.8
8' tree, 15 gallon	Each	1.5
10' tree, boxed	Each	2.5
Shrub planting, 1 gallon container	10 Each	2.0
Ground cover, most types	100 Each	1.4

Labor includes handling materials into place, setting or spreading in prepared locations, and cleanup.

Labor for Railroad Trackwork

Work Element	Unit	Man-Hours Per Unit
Excavate for trackwork, typical	100 C.Y.	4.0
Fill and compact, on site material	100 C.Y.	6.0
Fill and compact, 5 mile haul	100 C.Y.	8.5
Trim banks	100 S.Y.	1.6
Place ballast	100 C.Y.	11.0
Track, including ties and rails		
90 lb. rail	100 L.F.	105.0
100 lb. rail	100 L.F.	125.0
Siding bumper	Each	16.0
Switch assembly	Each	39.0
Turnout	Each	130.0

Use these figures for small jobs such as short sidings. Time includes handling materials into place, setting, cleanup and repair as needed.

Labor for Highway Guard Rail

Work Element	Unit	Man-Hours Per Unit
Cable type with three ¾" cables and posts at 5'		
Railroad tie posts	10 L.F.	1.7
6" x 10" wood posts	10 L.F.	1.6
Steel I beam posts	10 L.F.	1.9
Remove cable guide posts & cable	10 L.F.	1.2
Reset cable guide post & cable	10 L.F.	1.5
Steel beam type with 10 gage beam and posts at 10'		
Wood posts	10 L.F.	2.3
Railroad tie posts	10 L.F.	2.3
Steel I beam posts	10 L.F.	2.4
Mall barrier brackets	10 L.F.	2.3
Tubular steel bridge type railing		
2 rail, set in concrete	10 L.F.	4.1
4 rail, set in concrete	10 L.F.	6.3
Tubular aluminum bridge type railing		
2 rail, set in concrete	10 L.F.	3.2
4 rail, set in concrete	10 L.F.	4.4

Labor includes handling materials into place, excavation, setting and aligning, cleanup and repairs as needed.

Labor for Marine Structures

Work Element	Unit	Man-Hours Per Unit
Jetties and boat slips, per S.F. of solid docking surface		
Wood, on 50' timber piles	S.F.	2.6
Floating type	S.F.	1.8
Breakwaters		
3" x 10" treated wood	100 S.F.	18.0
6" thick precast concrete	100 S.F.	29.0
Steel, 19 lbs/S.F.	100 S.F.	40.0

Includes moving materials into place, installation in a prepared location, cleanup and repairs as needed.

Labor Erecting and Dismantling Scaffolding, Runways, and Ramps

Work Element	Unit	Man-Hours Per Unit
Erect and dismantle tubular scaffold (including planks and leveling)	1,000 S.F. of wall surface	20
Construct runways and ramps	1,000 S.F.	40
Place and remove runways and ramps	L.F.	.12

The first tier requires more time due to leveling and alignment procedures.
Suggested crew for scaffolding erection is three laborers.
Increase crew size for multiple tiers.

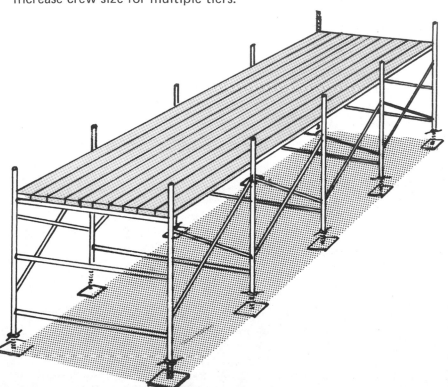

Labor for Demolition and Removal

Work Element	Unit	Man-Hours Per Unit
Concrete foundations	C.Y.	2.5
Concrete slabs on grade, no reinforcing	C.Y.	2.0
Concrete slabs on grade, with wire mesh reinforcing	C.Y.	2.5
Ceilings, plaster	100 S.F.	1.5
Ceilings, suspended acoustic	100 S.F.	1.5
Ceilings, acoustic tile, stapled or cemented	100 S.F.	1.5
Doors and frames, 3' x 7', wood	Each	.5
Doors and frames, 3' x 7', steel	Each	1.0
Flooring, ceramic or quarry tile	100 S.F.	3.0
Flooring, resilient tile	100 S.F.	1.0
Flooring, wood subfloor	100 S.F.	3.5
Wallboard, gypsum	100 S.F.	1.5
Wallboard, plywood	100 S.F.	1.5
Roofing, corrugated metal	100 S.F.	1.5
Roofing, builtup 5 ply	100 S.F.	2.0
Windows, metal, remove only	100 S.F.	5.8
Windows, wood, remove only	100 S.F.	6.5

Work includes removal of item and stacking or piling on site for removal at ground level.

Second floor or upper story work includes dumping into rubbish chutes.

For disposal up to five miles, use one man-hour per cubic yard for rubbish and rubble.

Concrete demolition is based on using pneumatic tools with crew of two tool operators and three laborers.

No allowance for salvage of materials (cleaning, pulling nails, etc.) is included.

The crew sizes for various operations will be dictated by safety, weight, or bulk of materials handled.

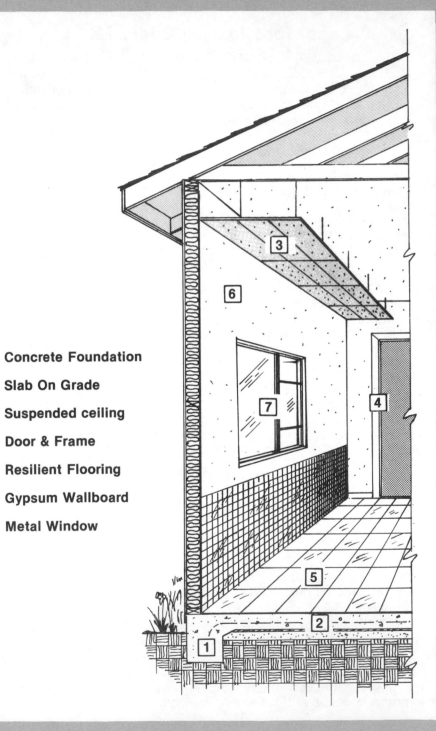

1 Concrete Foundation

2 Slab On Grade

3 Suspended ceiling

4 Door & Frame

5 Resilient Flooring

6 Gypsum Wallboard

7 Metal Window

Labor for Building Demolition

Work Element	Unit	Man-Hours Per Unit
Reinforced concrete buildings, per cubic yard of concrete[1]		
Blasting	CY	2.1
Pneumatic tools	CY	3.3
Headache ball	CY	1.4
Drilling and rock jack	CY	3.0
Structural steel building frames, per cubic yard of frame[2]		
Blasting	CY	2.5
Headache ball & torch	CY	2.6
Pneumatic tools and torch	CY	5.2
Entire buildings, per cubic yard of enclosed space		
Masonry	CY	.21
Wood frame	CY	.17
Steel and concrete	CY	.25
Reinforced concrete	CY	.21
Fireproofed steel	CY	.23

[1][2] work includes demolition of the structural system and building skin only and hauling debris 5 miles to a disposal site.
[1] Deduct 33% for non-reinforced concrete
[2] Deduct 33% if the frame has spray-on or other fireproofing.
No salvage of materials is assumed.

Labor for Partition Wall Demolition

Work Element	Unit	Man-Hours Per Unit
Concrete block partitions		
4" thick	100 SF	4.5
8" thick	100 SF	6.8
12" thick	100 SF	9.4
Brick masonry partitions		
4" thick	100 SF	6.1
8" thick	100 SF	8.1
12" thick	100 SF	13.5
Cast in place concrete partitions		
6" thick	100 SF	6.2
8" thick	100 SF	8.6
12" thick	100 SF	11.7
Plastered gypsum or terracotta partitions		
4" thick	100 SF	6.8
8" thick	100 SF	10.5
10" thick	100 SF	12.5
Metal or wood stud partitions with drywall		
4" thick	100 SF	1.2
6" thick	100 SF	2.5
8" thick	100 SF	3.7
Metal or wood stud partitions with lath & plaster		
4" thick	100 SF	1.6
6" thick	100 SF	3.3
8" thick	100 SF	4.8

Labor includes construction of a rubbish chute to a truck or trailer below and haul 5 miles to a disposal site. The figures are based on alteration type work where the building structural and mechanical systems are not demolished. Increase labor hours for work above 10 stories.

Labor For Demolition of Site Improvements

Work Element	Unit	Man-Hours Per Unit
Bituminous paving and base course		
3" thick	100 SY	8.0
Concrete paving and base course		
To 6" thick, with wire mesh	100 SY	10.5
To 6" thick, with rebar	100 SY	16.0
Airport runway, heavy reinforcing	100 CY	9.0
4" sidewalk	SY	.13
Curbing	100 LF	5.0
Granite curbing		
For disposal	100 LF	6.0
For reuse, stacked on site	100 LF	12.5
Masonry catch basins, 4' diameter		
5' deep	each	2.6
8' deep	each	4.0
10' deep	each	4.8
Concrete catch basins, 4' diameter		
5' deep	each	3.1
8' deep	each	4.8
10' deep	each	5.8
Precast catch basins, 4' diameter		
5' deep	each	2.2
8' deep	each	3.4
10' deep	each	4.0
Manhole removal and backfill, 4' diameter, concrete or masonry		
5' deep	each	2.3
10' deep	each	7.8
20' deep	each	16.0
25' deep	each	21.0
Railroad trackage		
Salvage rail for scrap	LF	.1
Salvage rail for reuse	LF	.17
For reuse, stacked on site	100 LF	12.5
Masonry catch basins, 4' diameter		
5' deep	each	2.6
8' deep	each	4.0
10' deep	each	4.8

Labor for Demolition of Site Improvements continued

Work Element	Unit	Man-Hours Per Unit
Concrete catch basins, 4' diameter		
5' deep	each	3.1
8' deep	each	4.8
10' deep	each	5.8
Precast catch basins, 4' diameter		
5' deep	each	2.2
8' deep	each	3.4
10' deep	each	4.0
Manhole removal and backfill, 4' diameter, concrete or masonry		
5' deep	each	2.3
10' deep	each	7.8
20' deep	each	16.0
25' deep	each	21.0
Railroad trackage		
Salvage rail for scrap	LF	.1
Salvage rail for reuse	LF	.17
Remove stone ballast	CY	.12
Remove track ties	each	.18
Chain link fencing		
Remove salvage and store for reuse on same job	100 SF	1.0
Remove and dispose	100 SF	.6

Labor includes breakout with equipment and load on a truck and 5 mile haul to disposal site. These figures are based on small jobs in urban areas.

3

concrete section

section contents

Concrete

Concrete is composed of cement, water and aggregate. The chemical process known as hydration transforms the cement and water combination into another material which binds the aggregate to form a solid mass. When freshly mixed with water, concrete can be formed into a strong, dense mass.

The strength, durability and workability of concrete depend on the combination of ingredients used in the mix. Concrete mix is controlled by the mixing plant to meet job requirements. Where a precise mix is required, the specifications will require that samples of the mix be taken from the truck and tested in a laboratory. Where quality of the concrete must be controlled, ready-mix concrete is usually used. However good quality concrete can be mixed at the job site in small batches. But the risk of improper mixing and the high cost of labor on site make use of pre-mixed concrete the best choice for most jobs.

Concrete is a highly plastic material with both large and small particles. The rock and sand tend to separate from the cement and water in handling. This can severely reduce the strength of the mass as it hardens. Even good quality concrete can be damaged by poor handling.

Well-graded aggregates of the largest practicable maximum size and highest specific gravity produce the heaviest concrete. Strength and durability increase generally as the weight increases for concrete made from specific materials. You can use 144 pounds per cubic foot as the unit weight of normal weight fresh concrete in computations for yield and cement content.

Workability of concrete is an important consideration. A workable mix has a uniform character and a plastic consistency. The workability required depends on the intended use of the concrete and the physical spacing of the forms and reinforcing steel. A slump test is the usual measure of workability. Smooth, well-graded aggregates, entrained air, and certain admixtures increase workability.

Two mixes having the same slump may have different workabilities if one shows less tendency to segregate during handling. A change in graduation or fineness of the sand often affects workability.

Water-cement ratio is the ratio of water (exclusive of water absorbed by the aggregates) to the amount of cement in a concrete or mortar mixture. It is a measure of the quality of the hardened cement paste binder. Strength, impermeability and many other characteristics of concrete are improved by lowering the water-cement ratio is less workable. But high workability is not required in uses such as foundations where strength is more important.

Compressive strength is the single most important factor controlling abrasion resistance of a given concrete. This resistance increases with the increase in compressive strength. The wearing resistance of concrete can be increased by delaying floating, troweling, proper curing and by surface treatment with liquid chemicals, or with special compounds.

Permeability of concrete is the ability of concrete to permit the flow of water through it. It is a factor which permits corrosion of reinforcing steel. Low water-cement ratios tend to produce less permeable concrete provided well-graded aggregates are used. Certain admixtures decrease permeability. Low slump concrete given proper vibration supplemented by spading next to the forms reduces segregation and decreases permeability.

Freezing and thawing can damage concrete. Frost damage under usual winter exposures is caused by water expanding as it freezes in the concrete pores. Avoid frost damage by using **air entrainment**, durable aggregate, and a low water-cement ratio concrete.

Admixtures are specified for concrete mixtures to improve workability, reduce segregation, entrain air, accelerate setting, obtain a specified strength at lower cement content, or to increase the slump of a given mixture without an increase in water content. They are essential when placing concrete with a pump or tremie.

Accelerator is commonly used in cold weather and has become known as an antifreeze solution. This is a misconception. The use of an accelerator doesn't reduce the need for protective cover, heat or other protection. An accelerator should not be used in hot weather. The use of calcium chloride as an accelerator tends to decrease the concrete strength and can corrode reinforcing and any other embedded steel materials.

Set retarders are specified to offset the accelerating effects of high temperature. They help the concrete remain plastic longer and thus eliminate the development of cold joints if placing requires successive lifts. The quantity of retarder can vary during placing, provided the admixture is not a combined retarder and air retaining one. Retarders are not ordinarily effective in controlling false set.

Air entrainment is the artificial introduction of finely divided air bubbles which produces a resistance to severe frost action and to salt. Air entrained mix is more cohesive and workable than normal concrete and has less tendency to segregate. Air entrainment also reduces the permeability of concrete since the air bubbles reduce capillary action. Too much air will reduce strength and abrasion resistance.

Portland Cement

There are 5 types of Portland cement:

Type I cement is used in general concrete construction when special properties of the other types are not required, when there is no exposure to sulfate in the soil or ground water, and where the hydration of the cement will not cause an objectionable rise in temperature.

Type II cement is used where moderate sulfate attack may occur. It produces less "heat of hydration."

Type III cement is used where rapid strength development of concrete is essential.

Type IV cement generates less heat than the other types and at a lesser rate. It has greater resistance to sulphate solutions.

Type V cement is used in concrete which will be in contact with soils and ground waters containing sulfate in concentrations which would deteriorate the concrete. Type V cement is nearly ideal for most applications.

Often cement carries a suffix letter "A" after the type. This indicates that an air-entraining agent has been added at the mill. Use of cement with the air entrainment agent already added is not recommended because there is no control over the admixture during mixing.

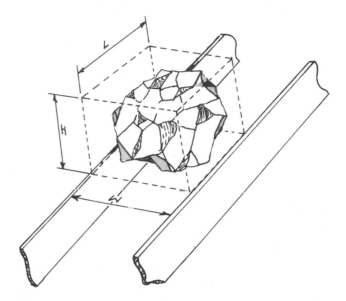

Grizzley screen used to grade aggregate

Aggregate

Natural gravels and crushed rock are the two main types of coarse aggregates. Both types should be clean, uncoated, properly shaped, strong, and durable. Gravel aggregate is inexpensive and generally available and produces a workable mix requiring less cement and sand. Air-entraining agents are frequently used with crushed rock aggregate to give added workability. Blast furnace slag may used when available and permitted by the specifications.

Soft, porous, or light rocks generally make inferior aggregates. Gravel deposits in river beds produce superior aggregate.

Aggregate sizes are usually determined by a series of grizzley screens positioned at a fixed angle. The screens remove and sort out various rock sizes. As a rule a grizzley is made up of heavy tough steel bars arranged in parallel with fixed spaces between each bar. Some grizzlies are made up to form squares or open end boxes with a fixed dimension. Other screens are round holes in a steel plate which allow the smaller rocks to pass through.

Sand is the main type of fine aggregate. It is mined in natural deposits. Sand from any source may require water washing to remove organic material such as leaves or peat.

The following aggregates are used in lightweight concrete:

Cinders are the residue from high-temperature combustion of coal or coke.

Expanded slag aggregates are produced by treating blast-furnace slag with water.

Expanded shale and clay are made by heating prepared materials to the fusion point where they become soft and expand because of trapped gas.

Several natural aggregates are usable for low density concrete. They are pumice, scoria, volcanic cinders, tuff and diatomite rocks. Diatomite is the only one of these which is not of volcanic origin.

Lightweight concrete is often used for elevated slabs because it weighs only 90 to 110 pounds per cubic foot. It can also be used for structural purposes. The lightness of the concrete is due to the lightweight aggregate used. The weight of the concrete is also influenced by cement, water and air content. In some cases an admixture is added, generally to provide entrained air and occasionally to modify setting time or reduce water content.

3 Concrete

The term "all-light-weight" indicates concrete in which both the coarse and fine aggregates are lightweight. The term "sand-lightweight" indicates concrete with coarse lightweight aggregate but fine material of natural sand.

Low density concrete (lightweight nonstructural concrete 50 pounds per cubic foot or less) is of two types: (1) aggregate type made predominantly with low density mineral aggregates such as expanded perlite or vermiculite; (2) cellular type made by forming a cement matrix around air voids which are generated by preformed foams or special foaming agents. This concrete is generally used for fills, thermal insulation and to provide stiffness for metal decking. An expansion or a contraction joint is usually recommended at the junction of all roof projections because large temperature changes will cause major expansion of the deck.

Grout is used for sealing precast pipe joints, in seating machinery and structural steel members on foundations, or repairing defective or deteriorated concrete.

Grout is made up of various combinations of cement, water, sand, and possibly gravel. Grout must readily and completely fill the space to be grouted and harden without shrinking. Ordinary plastic and fluid mortars are unsatisfactory because they settle and leave a layer of water at the top surface and shrink as the mortar dries.

A **slump test** is a method of evaluating the consistency of concrete. Several representative samples of the concrete are taken from the mixer or truck. These samples are brought to an open ended cone shaped mold. The sample is first mixed then placed in the moistened mold in three equal volumed layers. After each layer is rodded, the top layer is leveled and the form is removed. The slump is the difference between the height of mold and height of the concrete which has "slumped" when the form was removed.

False set is the loss of concrete consistency shortly after mixing. It is indicated by the loss of slump between the mixer and the forms. If it becomes severe, the concrete is nearly impossible to place.

Plastic **shrinkage cracks** in concrete are caused by too rapid loss of moisture during the period before curing begins. They appear when the rate of drying is more rapid than the movement of bleeding water to the surface.

Vibration consolidates concrete in the forms into a uniformly dense mix. Vibration is especially helpful when placing low slump concrete where voids would reduce the wall strength.

When vibrating concrete, avoid over-vibration which can damage the forms. When over-vibration occurs, slump should be decreased. Vibration or revibration is acceptable if the running vibrator can sink under its own weight. Both compressive strength and bond with reinforcing is increased by late revibration. Any separation due to settlement shrinkage will be reconsolidated. Internal vibrators, however, should not penetrate partially hardened layers on exposed work since a wavy line may appear on the surface. Eliminate air bubble holes on vertical surfaces by using up to 100 percent more vibration than is necessary.

Materials for Concrete

Sacks of Cement	Gallons of Water Per Sack of Cement	Weights of Saturated Surface-Dry Aggregate Per Sack of Cement, Lbs.		28-Day Compressive Strength P.S.I.
		Sand	Gravel	
4.9	8	280	380	2,250
5.9	7	240	330	2,750
6.0	6½	210	310	3,000
6.5	6	190	280	3,300
7.2	5½	160	260	3,700
8.0	5	140	230	4,250

Maximum Size of Aggregate Recommended

Minimum Dimension of Section (Inches)	Maximum Size of Aggregate* in Inches, for:			
	Reinforced Walls, Beams, and Columns	Unreinforced Walls	Heavily Reinforced Slabs	Lightly Reinforced or Unreinforced Slabs
2½— 5	½— ¾	¾	¾—1	¾—1½
6 —11	¾—1½	1½	1½	1½—3
12 —29	1½—3	3	1½—3	3
30 or more	1½—3	6	1½—3	3 —6

*Based on square openings.

Labor and Materials for Excavating and Pouring Footings, Foundations and Grade Beams

Size	Material			Labor	
	C.F. Concrete Per L.F.	C.F. Concrete Per 100 L.F.	C.Y. Concrete Per 100 L.F.	Man-Hours Per 100 L.F. for Hand Excavation	Man-Hours Per C.Y. for Concrete Placement
6 x 12	0.50	50.00	1.9	3.8	1.1
8 x 12	0.67	66.67	2.5	5.0	1.1
8 x 16	0.89	88.89	3.3	6.4	1.1
8 x 18	1.00	100.00	3.7	7.2	1.1
10 x 12	0.83	83.33	3.1	6.1	1.0
10 x 16	1.11	111.11	4.1	8.1	1.0
10 x 18	1.25	125.00	4.6	9.1	1.0
12 x 12	1.00	100.00	3.7	7.2	1.0
12 x 16	1.33	133.33	4.9	9.8	1.0
12 x 20	1.67	166.67	6.1	12.1	0.9
12 x 24	2.00	200.00	7.4	15.8	0.9

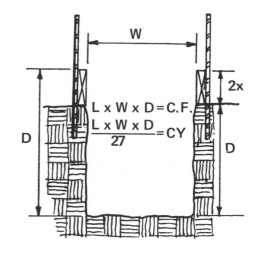

$$L \times W \times D = C.F.$$
$$\frac{L \times W \times D}{27} = CY$$

Trench and form

Reduce excavation hours by ¼ for sand or loam. Increase excavation hours by ¼ for heavy clay soil.

Placement labor is based on ready-mix concrete direct from chute. Add 50% if concrete is pumped into place. Add 60% if concrete is placed with a crane and bucket. Add two hours per cubic yard if concrete is wheeled 40 feet into place and an additional 25% for each additional 40 feet.

Excavation includes loosening and one throw from trench only.

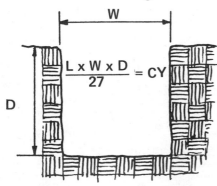

$$\frac{L \times W \times D}{27} = CY$$

Trench

Wood form for footing

Labor Forming Footings

Work Element	Unit	Man-Hours Per Unit
Wall Footings		
1 use	100 S.F.	10.1
3 uses	100 S.F.	7.4
5 uses	100 S.F.	6.8
Column Footings		
1 use	100 S.F.	12.0
3 uses	100 S.F.	8.7
5 uses	100 S.F.	8.0
Grade beam or tie beam footings		
1 use	100 S.F.	8.8
3 uses	100 S.F.	6.5
5 uses	100 S.F.	6.0
Pile cap forms		
1 use	100 S.F.	17.0
3 uses	100 S.F.	13.6
5 uses	100 S.F.	11.0

Labor is man-hours for each use of plywood forms. Unit is 100 square feet of contact area. These figures include fabrication, erection, stripping, and repairing forms for multiple uses. **Suggested crew:** 2 carpenters, 1 helper. For steel forms, use the labor hours for 5 uses less 10%.

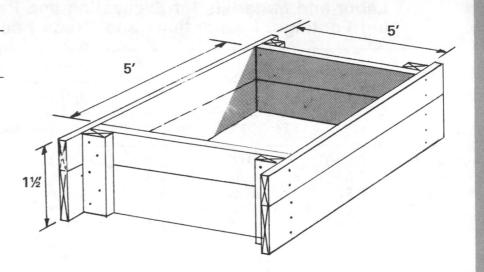

Column footing

Wall footing

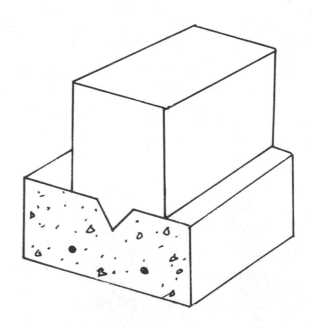

Wall on footing with key joint

Square Feet of Concrete From
1 Cubic Yard of Concrete

Thickness Inches	S.F.	Thickness Inches	S.F.	Thickness Inches	S.F.	Thickness Inches	S.F.
1	324	4	81	7	46	10	32
1¼	259	4¼	76	7¼	44	10¼	31
1½	216	4½	72	7½	43	10½	31
1¾	185	4¾	68	7¾	42	10¾	30
2	162	5	65	8	40	11	29½
2¼	144	5¼	62	8¼	39	11¼	29
2½	130	5½	59	8½	38	11½	28
2¾	118	5¾	56	8¾	37	11¾	27½
3	108	6	54	9	36	12	27
3¼	100	6¼	52	9¼	35	12¼	26½
3½	93	6½	50	9½	34	12½	26
3¾	86	6¾	48	9¾	33	12¾	25½

Thickened edges must be calculated separately

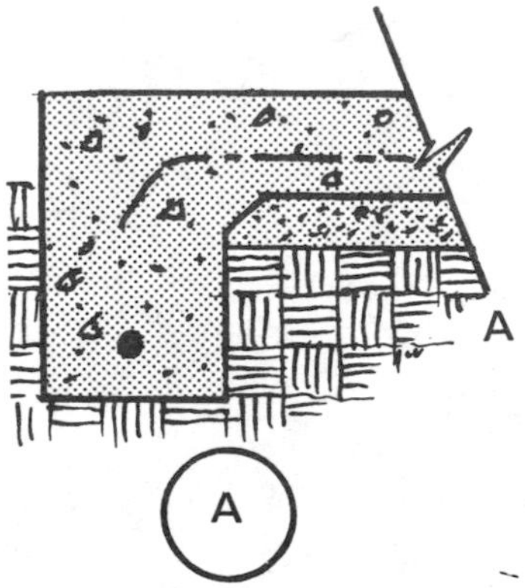

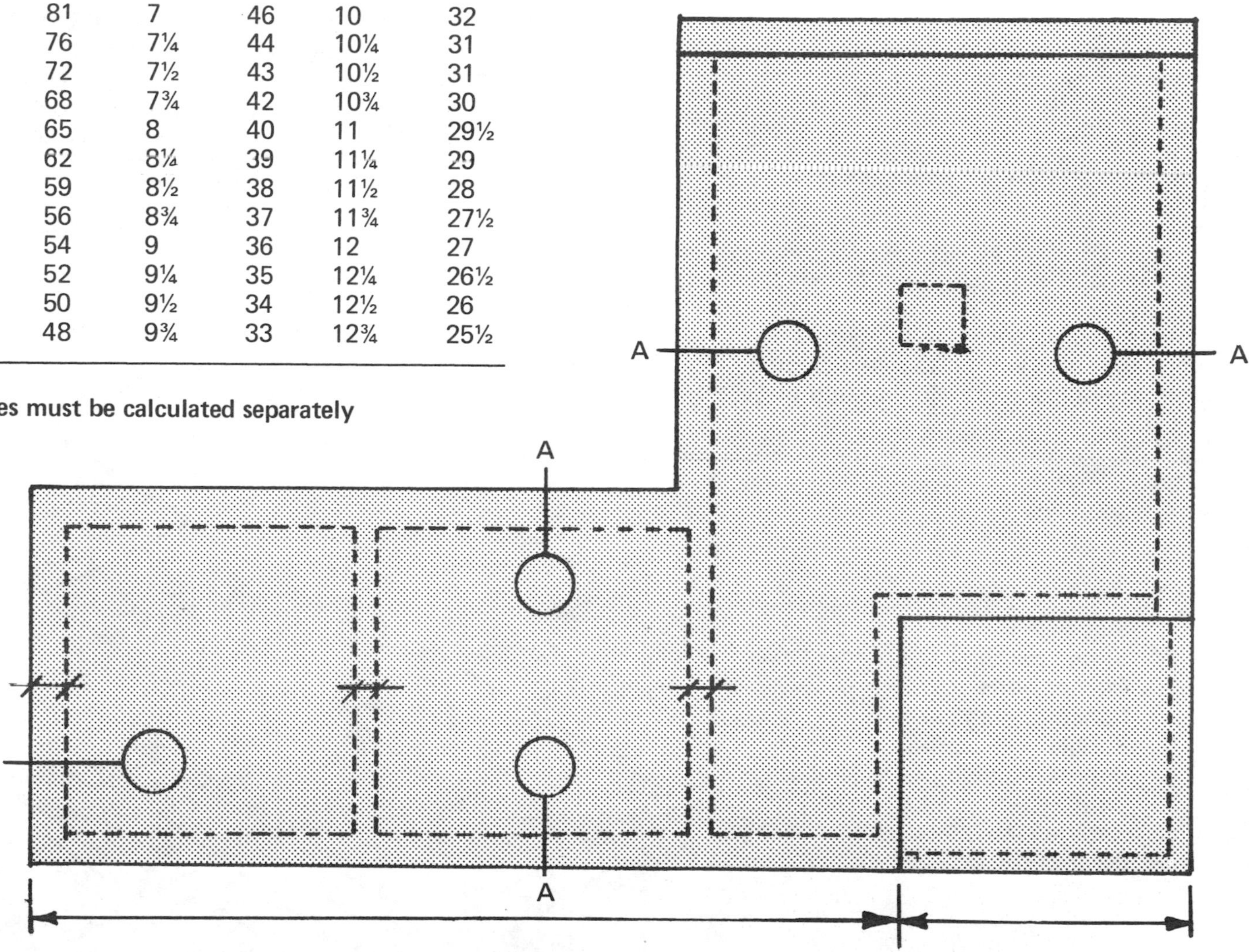

Calculating quantity of concrete for slabs

Labor and Materials for Slabs on Grade

	Material per Square Foot		Labor	
Thickness	C.F. of Concrete	S.F. per C.Y. of Concrete	Man-Hours Per 100 L.F., Forms and Screeds	Man-Hours Placing Concrete per 100 S.F.
2″	0.167	162	Averages	Aver-
3″	0.25	108	22 linear	ages
4″	0.333	81	feet	2
5″	0.417	65	per hour	hours
6″	0.50	54		

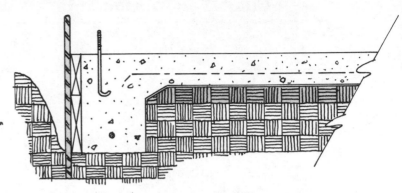

Typical slab cross section

Placement includes finishing with topping. If topping is omitted, deduct 1.2 hours.

Placement labor is based on ready-mix concrete direct from chute. Add ½ hour per cubic yard if concrete is pumped into place and two hours if concrete is wheeled up to 40 feet into place.

1. Sidewalk
2. Stoop, porch
3. Anchor bolts
4. Concrete slab
5. Rough-in plumbing
6. Depressed slab
7. Curb
8. Garage slab
9. Driveway

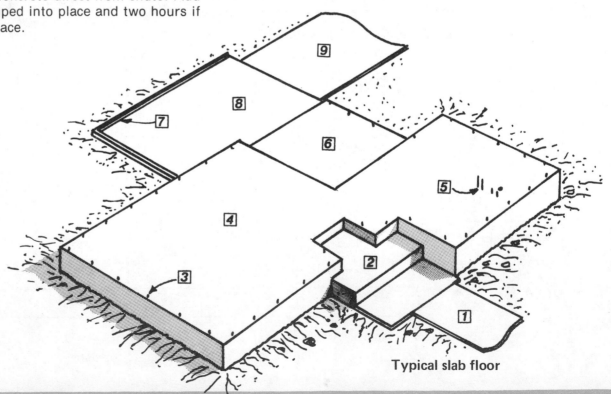

Typical slab floor

Labor for Concrete Slabs on Grade

Work Element	Unit	Man-Hours Per Unit
Fine grading	100 S.F.	1.0
Edge forms	100 L.F.	4.5
Set screeds	100 L.F.	1.0
Place wire mesh	100 S.F.	0.6
Pour concrete, 4''	100 S.F.	0.8
Trowel finishing	100 S.F.	1.6
Float finish only	100 S.F.	0.9
Strip forms	100 L.F.	1.2

Suggested Crew: For setting forms is two carpenters and one laborer

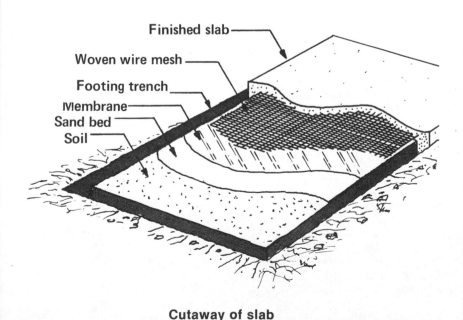

Finished slab

Woven wire mesh

Footing trench

Membrane
Sand bed
Soil

Cutaway of slab

Placing Concrete

Concrete should be placed in the forms as soon as possible, in no case more than 45 minutes after mixing. It should not be placed in large quantities at a given point and allowed to run or to be worked over a long distance in the form. This practice results in segregation because the mortar tends to flow out ahead of the coarser material. In general, concrete should be placed in horizontal layers of uniform thickness, each layer thoroughly compacted before the next is placed. Layers should be 6'' to 12'' thick for reinforced members and up to 18'' thick for mass work. The thickness depends on the width between forms, the amount of reinforcing and the requirement that each layer be placed before the previous one sets.

The concrete should not be allowed to drop freely more than 3 or 4 feet. In thin sections, drop chutes of rubber or metal should be used. In narrow wall forms, the metal drop chutes may be made rectangular to fit between reinforcing steel. Drop chutes should be provided in several lengths or should be in sections which can be hooked together so that the length can be adjusted as concreting progresses.

To avoid cracking due to settlement, concrete in columns and walls should be allowed to stand for at least two hours before concrete is placed in the slabs, beams or girders they are to support. Haunches and column capitals are considered part of the floor or roof and should be placed at the same time with them.

In slab construction, placing of the concrete should be started at the far end of the work so each batch will be dumped against previously placed concrete, not away from it. The concrete should not be dumped in separate piles and the piles then leveled and worked together.

The order of placing concrete is also important. In walls, the first batches should be placed at either end of the section; the placing should then progress toward the center. This method can also be followed in placing beams and girders. In large open areas, the first batches should be placed around the outsides. But always be careful to prevent water from collecting at the ends and corners of forms and along form faces.

Occasionally the work has to stop long enough for the concrete to begin hardening. The top surface should be roughened just before it hardens to remove laitance or scum and provide a good bonding surface for the next layer of concrete. Just before resuming concreting, the roughened surface should be cleaned and brushed with a cement-water paste of a thick, creamy consistency. This paste is applied in a thick brush coat just a few feet ahead of the concreting operation so that it does not have a chance to dry. If there isn't a good bond between different layers of concrete, there is danger of seams developing that will cause leakage. This precaution is very important wherever the concrete must be watertight.

3 Concrete

Labor Forming Slabs on Grade

Work Element	Unit	Man-Hours Per Unit
Edge forms for slabs		
Up to 6 inches high	100 L.F.	4.5
Over 6 to 12 inches	100 L.F.	5.2
Over 12 to 24 inches	100 L.F.	6.8
Over 24 to 36 inches	100 L.F.	8.5
Keyed joint forms		
Up to 6 inches high	100 L.F.	6.8
Over 6 to 12 inches high	100 L.F.	7.5
Pad and base edge forms		
1 use	100 S.F.	6.8
3 uses	100 S.F.	6.2
5 uses	100 S.F.	5.7
Keyway forms, 4 uses		
2 x 4	100 L.F.	1.5
2 x 6	100 L.F.	1.6

Unit is per 100 linear feet or 100 square feet of contact area. Does not include excavation, backfill or trench clean-up.
Suggested Crew: 1 carpenter and 1 helper

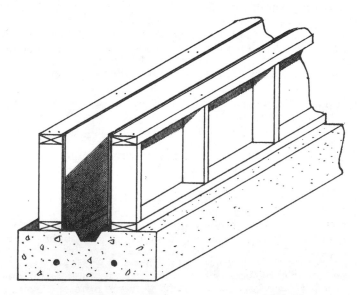

Concrete foundation on footing wall with key

Labor Placing Concrete In Forms

Work Element	Unit	Man-Hours Per Unit
Wall footings	C.Y.	.55
Column footings	C.Y.	.82
Pile caps	C.Y.	.91
Tie beams	C.Y.	.86
Grade beams	C.Y.	.86
Walls below grade	C.Y.	.90
Columns, piers, pilasters	C.Y.	.95
Slabs on grade to 6" thick	C.Y.	.65
Slabs on grade 6" to 12" thick	C.Y.	.54
Slabs on grade 12" to 24" thick	C.Y.	.44
Slabs on grade over 24" thick	C.Y.	.38
Steps on grade	C.Y.	.90
Pit slabs	C.Y.	.74
Pit walls	C.Y.	.90
Entrance platforms on grade	C.Y.	.66
Pads and equipment bases	C.Y.	.69
Slab edge curbs	C.Y.	.86
Columns	C.Y.	1.15
Walls above grade	C.Y.	.95
Buttresses	C.Y.	1.00
Beams	C.Y.	1.00
Suspended flat slabs	C.Y.	.82
Suspended beams and slabs	C.Y.	.86
Suspended pan formed slabs	C.Y.	.90
Stairways, reinforced	C.Y.	1.00
Concrete filled metal pan stairs and platforms	C.Y.	1.20
Concrete filled metal decks	C.Y.	.95
Concrete beam encasement	C.Y.	1.00
Concrete column encasement	C.Y.	1.15

These figures assume that all concrete placed at ground level is truck-accessible. Concrete placement above or below ground level includes hoisting time to 100 feet. For over 100 feet, add 0.2 hours per cubic yard. For hoisting over 200 feet, add 0.4 hours per cubic yard. If concrete is pumped into place, deduct 0.1 man-hours per cubic yard.
Suggested Crew: 3 laborers and 1 foreman

Labor and Materials for 100 S.F. of Foundation Wall Forms

Work Element	B.F. per S.F. of Forms	Make and Place Forms			Removing Forms	
		S.F. in 8 Hours	Carpenter Hours	Labor Hours	S.F. in 8 Hours	Labor Hours
Foundation wall forms						
4' to 6'	2	190-210	4.0	2.0	640	1.3
7' to 8'	2	165-190	4.5	2.3	550	1.5
9' to 10'	2.5	150-160	5.3	2.5	450	1.8
11' to 12'	3	135-150	5.5	2.8	425	1.9
Retaining walls						
16' to 20'	3.5	105-115	7.3	3.5	325	2.5

Conversions for Reinforcing Bars

Bar Number	Lbs. Per L.F.	Diameter in Inches	Cross-Sectional Area, Inches	Perimeter in Inches	L.F. Per Ton
3	.376	0.375	0.11	1.178	5,319
4	.668	0.500	0.20	1.571	2,994
5	1.043	0.625	0.31	1.963	1,918
6	1.502	0.750	0.44	2.356	1,332
7	2.044	0.875	0.60	2.749	978
8	2.670	1.000	0.79	3.142	749
9	3.400	1.128	1.00	3.544	588
10	4.303	1.270	1.27	3.990	465
11	5.313	1.410	1.56	4.430	376
14	7.650	1.693	2.25	5.320	261
18	13.600	2.257	4.00	7.090	147

The nominal dimensions of a deformed bar are equivalent to those of a plain round bar having the same weight per foot as the deformed bar.

Bar numbers are based on the number of eighths of an inch included in the nominal diameter of the bars.

Labor and Materials for Foundation Walls

Wall Thickness	Concrete Per 100 S.F. of Wall		Forming Man-Hours Per 100 S.F.			Man-Hours per C.Y. for Concrete Placement
	C.F. Required	C.Y. Required	Place to 4'	Place 4'-8'	Remove	
4"	33.3	1.24	4.7	7.13	1.5	
6"	50.0	1.85	4.7	7.75	to	
8"	66.7	2.47	5.0	7.75	3.0	Averages
10"	83.3	3.09	5.0	7.90	(Varies	1.0
12"	100.0	3.70	5.0	7.90	with	Hours
					Height)	

Placement labor is based on ready-mix concrete direct from chute. No bracing is included.

Labor Placing Reinforcing Steel

Work Element	Unit	Man-Hours Per Unit
Reinforcing bars		
Footings and foundations	Ton	22
Slabs on grade	Ton	19
Structural slabs	Ton	18
Columns and beams	Ton	28
Walls	Ton	24
Cast-in-place culverts, manholes and catch basins	Ton	25
Structural frames	Ton	35
Lintels, sills and coping	Ton	24
Apexes	Ton	31
Concrete paving	Ton	25
Concrete curbs	Ton	24
Precast roof and wall panels	Ton	25
Welded wire fabric		
Slabs on grade	1000 S.F.	6.0
Concrete paving	1000 S.F.	6.0
Footings	1000 S.F.	7.0
Precast roof panels	1000 S.F.	6.0
Gunite	1000 S.F.	9.0
Head walls and culvert end walls	1000 S.F.	7.0

Placement of reinforcing steel includes handling into place, tying, supporting, and any cutting necessary at the site such as cutting around imbedded items or cutting stock lengths of straight bars to fit slab dimensions.

Man-hours estimates are based on all reinforcing steel being shop fabricated (cut to length and bent ready to place in the structure).

If reinforcing steel is to be welded in place, add 50% to the time.

Labor for wire mesh reinforcing includes handling into place, cutting to fit, tying at overlaps, and pulling up into position during placement of concrete.

Suggested Crew: 2 to 6 reinforcing ironworkers

Labor Fabricating Reinforcing Steel

Work Element	Unit	Man-Hours Per Unit
Light bending (less than 40 bends per 100 linear feet)		
No. 4 bars and smaller	Ton	23
No. 5 to No. 7 bars	Ton	20
No. 8 to No. 9 bars	Ton	19
Heavy bending (40 or more bends per 100 linear feet)		
No. 4 bars and smaller	Ton	42
No. 5 to No. 7 bars	Ton	36
No. 8 to No. 9 bars	Ton	35
Light cutting (10 or less cuts per 100 linear feet)	Ton	7
Heavy cutting (over 10 cuts per 100 linear feet)	Ton	14

Reinforcing steel fabrication includes cutting, bundling, tagging, storing, loading, and hauling to the job site. Assembly and tying into mats and beams are also included. Man-hour bending estimates are based on hand bending. If bending machines are used, subtract 30 percent.

Suggested Crew: 2 to 8 reinforcing ironworkers, depending on amount of reinforcing steel to be cut or bent.

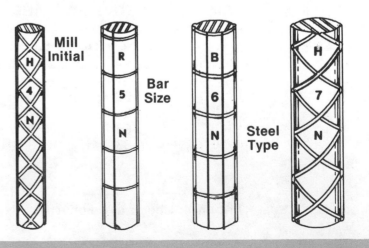

Labor for Miscellaneous Concrete Items

Work Element	Unit	Man-Hours Per Unit
Place anchor bolts	100 Each	4.0
Pickup and brace inserts	100 Each	3.2
Curb angles	100 L.F.	8.0
Ceiling inserts	100 Each	6.5
Dovetail anchor slots	L.F.	0.8
Steel base plates 12" x 14" x ½"	Each	0.4
Waterstops (PVC dumbells or copper)	100 L.F.	2.0
Vapor barriers	1000 S.F.	1.0
Floor hardeners (magnesium fluosilicate) 1 coat	1000 S.F.	6.0
Silicone waterproofing	1000 S.F.	5.4
Acid wash (walls)	1000 S.F.	7.5
Non-shrink grout under steel plate 1"	S.F.	0.4
Concrete cutting		
2" deep, cured	100 L.F.	3.5
4" deep, cured	100 L.F.	6.0
green, 2" deep	100 L.F.	2.5
green, 4" deep	100 L.F.	4.0
Concrete core drilling		
Slab, vertical, 3" diameter, 6" thick	Each	0.8
Wall, horizontal, 4" diameter, 8" thick	Each	1.2

Table is for installation only and does not include fabrication time.

For architectural concrete wall surfaces that must be patched, honed, and sack rubbed, use eight man-hours per 100 S.F.

Concrete sawing is based on 400 L.F. per inch blade life, which reduces in proportion to depth.

Example: 4" depth — 500 L.F.
3" depth — 1,000 L.F.
2" depth — 1,500 L.F.

Core drilling labor will increase if slabs or walls are heavily reinforced (¾" to 1½" rebar) or if holes are widely spaced.

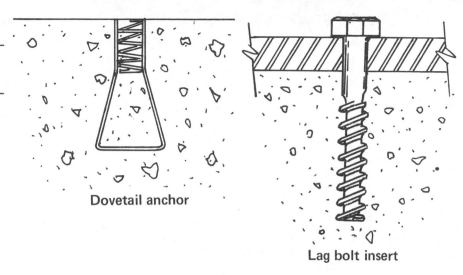

Dovetail anchor

Lag bolt insert

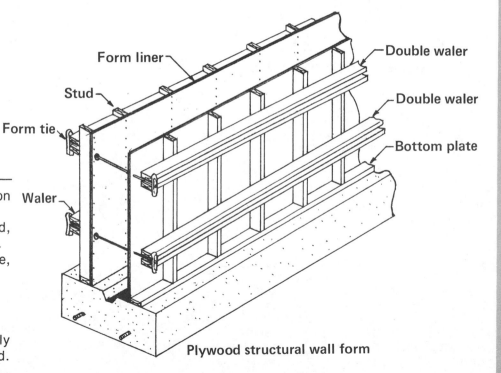

Form liner

Double waler

Stud

Double waler

Form tie

Bottom plate

Waler

Plywood structural wall form

Labor for Encasement Forms

Work Element	Unit	Man-Hours Per Unit
Encasing steel beams		
1 use	100 S.F.	27.0
3 uses	100 S.F.	15.0
5 uses	100 S.F.	14.5
Encasing steel columns		
1 use	100 S.F.	27.0
3 uses	100 S.F.	15.0
5 uses	100 S.F.	14.5

Labor is man-hours per use for plywood forms. Unit is 100 square feet of contact area. These figures include fabricating, erecting, stripping, repairing and cleaning for multiple uses. Allow 0.5 hours extra for round steel columns. **Suggested crew:** 1 carpenter and 1 helper.

Labor Forming Structural Beams

Work Element	Unit	Man-Hours Per Unit
Beam bottoms		
1 use	100 S.F.	19.6
3 uses	100 S.F.	17.1
5 uses	100 S.F.	16.1
Beam sides		
1 use	100 S.F.	17.4
3 uses	100 S.F.	14.8
5 uses	100 S.F.	14.4

Labor is man-hours for each use of plywood forms. Unit is 100 square feet of contact area. These figures include fabricating, erecting, shoring, stripping, repairing and cleaning beam forms for multiple uses. Beam bottom and side dimension is assumed to be 12 to 16 inches. For dimensions of 18 inches or more, deduct 2 hours per 100 S.F. **Suggested crew:** 1 carpenter and 1 helper.

Labor Forming Structural Walls Above Grade

Work Element	Unit	Man-Hours Per Unit
Walls to 10 feet high		
1 use	100 S.F.	13.4
3 uses	100 S.F.	12.5
5 uses	100 S.F.	12.0
Walls over 10 to 14 feet		
1 use	100 S.F.	14.6
3 uses	100 S.F.	13.0
5 uses	100 S.F.	12.5
Walls over 14 to 18 feet		
1 use	100 S.F.	15.9
3 uses	100 S.F.	14.0
5 uses	100 S.F.	13.0
Walls over 18 to 22 feet		
1 use	100 S.F.	16.9
3 uses	100 S.F.	14.5
5 uses	100 S.F.	14.0
Buttress forms		
1 use	100 S.F.	13.8
3 uses	100 S.F.	12.7
5 uses	100 S.F.	12.6

Labor is man-hours for each use of plywood forms. Unit is 100 square feet of contact area. These figures include assembling, erecting, bracing, stripping, repairing and cleaning forms for multiple uses. **Suggested crew:** Walls from 10 feet to 14 feet, 2 carpenters and 2 helpers; walls 14 feet to 18 feet, 2 carpenters and 3 helpers; walls 18 feet to 22 feet, 3 carpenters and 3 helpers. Figures apply to buttress forms up to 12 feet high. Larger buttress forms or buttress forms with wings take more time because they must be reinforced to prevent damage when being removed.

Structural Slabs

There are several types of suspended floor systems in common use. Each has advantages and disadvantages that make selection of one system over the others appropriate for a particular application.

One-way solid slabs have a uniform depth, with no filler material, and with the main reinforcement in one direction only. These slabs were the earliest floor systems to be used. They are suitable and economical for spans from 6 to 12 feet. For light loads, these spans may be increased to 14 to 15 feet.

Solid slabs make a stiff and rigid floor capable of withstanding vibration and shock from machinery. They are used where heavy concentrated loads must be supported. One-way solid slabs can not be used on long spans because of their high dead weight. Also, they require more beams than other systems and this may complicate the floor layout below. The slab weight may be reduced by the use of lightweight aggregate. This reduction usually means that the sizes of columns and footings can be reduced. Lightweight aggregate also improves heat insulation.

If forms are made of pressed wood, fiber boards or similar material, plaster or other surface treatment other than paint can be omitted.

Ordinary electric conduits and pipes about the same size can generally be run in the slab. Mechanical equipment may require space in either a suspended ceiling or an extra fill between the slab and the finished floor.

One-way ribbed slabs with block filler: use hollow filler blocks of lightweight concrete or clay tile laid in rows in the bottom of the slab. The dead load is considerably less than a solid slab of equal load carrying capacity although the depth of the slab is increased. The width of the concrete joists that separate the rows of filler and encase the reinforcement may be any desired width to meet the strength requirements. It is customary to include a solid concrete top of two inches or more over the blocks. This provides a space for concealing small pipes and conduits and adds considerable strength. The filler blocks improve sound and heat insulation of the slab.

Ribbed slabs are good for floors of medium span with light to intermediate loading. They cannot carry heavy concentrated loads as well as solid slabs and are less economical when the spans are increased.

To be sure of having straight joists, the filler blocks must be set accurately in line. The blocks are very porous and must be thoroughly sprinkled to prevent absorption of the water in the concrete, particularly in warm weather.

The clay or gypsum tiles provide a surface which serves as a plaster base for the ceiling under the slab. The ends of the end tiles may be closed by a thin slab made for that purpose. They may also be left open.

One-way ribbed slabs with metal pans are the lightest type of concrete floors. They use metal pan fillers between concrete joists. There are several kinds of these pans on the market. The steel pans may be made of heavy material which will stand being removed after the concrete has set and thus can be used several times. They may also be made of thinner material designed to be left in place. If removable forms are used, metal lath is fastened to the underside of the joists to serve as a base for plaster for the ceiling below. If the pans are to be left in place, metal lath is laid over the forms before the pans are set. The metal lath is wired to the reinforcing bars in the joists. In one kind of steel-pan, the lath is fastened before the pan is placed. Sheet-metal closers or end caps are made for the ends of the end pans.

Both permanent and temporary pans are furnished with either straight or tapered closed ends. Tapered pans are best on long spans as the width of the joists is increased at points where the width is needed, near the supports where stresses are the highest. The usual width of metal pans is 20", although 30" pans and other sizes can be used. The 20" wide pan with a joist 4" wide gives a 24" joist spacing. The depths vary from 6", 8", 10" up to 14". They are usually corrugated to increase their stiffness and are lapped one or one-half corrugations or more if necessary to provide the exact lengths required. The pans are supported on forms called **falsework** placed under the joists.

Metal pan ribbed floors are particularly suitable for light loads, even on quite long spans. While the weight is light, the thickness of the floor makes it very rigid. Formwork is merely a board under each rib or joist. The spaces under the pans are left open.

These floors are not well suited to support concentrated loads as the topping between joists is somewhat thin (2½" to 3"). Be careful to reinforce the topping across construction joints to prevent these joints from opening.

Removable pans may be used several times. To stand moving, they must be constructed of heavier metal than the pans which are left in place. The cost of moving and the greater cost of placing the lath must be considered. Special types of steel pans can be used for the two-way type of construction.

Both the block-filled and pan-formed slab may be built with the one-way or two-way system of reinforcement. The two-way is usually more rigid. It is made by using square or rectangular tiles or pans which are separated on all four sides to form a series of small T-beams at right angles to each other.

Two-way solid slabs are reinforced in two directions and carry the floor load both ways to surrounding beams. Modern building codes allow use of improved designs which make this type of floor economical and well suited to support intermediate and heavy loads on spans up to about 30 feet.

The beams run in two directions to columns, producing excellent horizontal bracing. Wind and earthquake forces as well as vibrating machinery loads are effectively resisted. Other advantages of two-way slab are the simplicity of formwork and ease of erection. If lightweight aggregate is used, the dead load may be reduced about 30 percent.

Flat slab floors are concrete slabs reinforced in two or more directions and without beams connecting into columns. They are especially economical for heavy loads and are efficient for light loads. These floors have a high degree of rigidity. The absence of beams offers better lighting, ventilation and arrangement of mechanical equipment if used in factory buildings. It is customary to have supporting columns with flared heads or capitals. In some cases, it is economical to use panels of additional thickness at each column. These are called **drop panels.**

Flat slabs make rigid, efficient floors for such heavy and concentrated loads as are found in warehouses, loft buildings and industrial buildings with heavy or vibrating machinery. Panels with a medium span and nearly square are preferable. The slab must be made continuous over two or more spans. Story heights may be reduced because of the greater clear ceiling height.

Labor Forming Suspended Slabs

Work Element	Unit	Man-Hours Per Unit
Closed deck forms[1]		
1 use	100 S.F.	8.4
2 uses	100 S.F.	8.1
3 uses	100 S.F.	8.0
4 uses	100 S.F.	7.9
6 uses	100 S.F.	7.6
Open deck forms for pans or domes[1]		
1 use	100 S.F.	6.1
2 uses	100 S.F.	6.0
3 uses	100 S.F.	5.9
4 uses	100 S.F.	5.9
6 uses	100 S.F.	5.6
Metal pan forms[2]		
20 inch pans	100 S.F.	6.5
30 inch pans	100 S.F.	8.0
Metal dome forms[2]		
19 or 20 inch domes	100 S.F.	4.0
30 inch domes	100 S.F.	6.0
Fiberglass pan forms[3]		
20 inch pans	100 S.F.	1.3
30 inch pans	100 S.F.	1.9
Fiberglass dome forms[3]		
19 or 20 inch domes	100 S.F.	1.4
30 inch domes	100 S.F.	2.1
Screeds for slabs[4]		
4 uses or more	100 L.F.	4.0

Labor is man-hours per use. Unit is 100 square feet of horizontal area or 100 linear feet.

[1] Includes erecting, shoring, reshoring, and removal. Figures are based on 8 to 15 foot floor-to-floor height. Add .8 hour per 100 S.F. for 16 to 20 foot floor-to-floor height and 2.3 hours per 100 S.F. for over 20 to 35 foot floor-to-floor height. For more than 6 uses on the same project, deduct 5% for each use over 6 up to 11 uses.
[2] Includes handling, placing, stripping, oiling and cleaning.
[3] Includes handling, placing, stripping and cleaning.
[4] For screed 4 inches to 12 inches high.
Suggested Crew: For metal pan forms, 3 carpenters and 2 helpers; fiberglass pan forms, 2 carpenters and 2 helpers

Labor For Suspended Arch Forms

Work Element	Unit	Man-Hours Per Unit
8 foot height		
1 use	100 S.F.	13.7
3 uses	100 S.F.	12.1
5 uses	100 S.F.	10.9
15 foot height		
1 use	100 S.F.	14.0
3 uses	100 S.F.	12.3
5 uses	100 S.F.	11.6
20 foot height		
1 use	100 S.F.	15.1
3 uses	100 S.F.	13.4
5 uses	100 S.F.	12.7
35 foot height		
1 use	100 S.F.	19.1
3 uses	100 S.F.	17.5
5 uses	100 S.F.	17.1
Stairway arch forms		
1 use	100 S.F.	25.0
3 uses	100 S.F.	22.2
5 uses	100 S.F.	21.6
Arch slab edge forms		
6 inches high	100 L.F.	5.1
12 inches high	100 L.F.	5.8
18 inches high	100 L.F.	8.4

Labor is man-hours for each use of plywood forms. Unit is 100 square feet or linear feet of contact area. These figures include fabricating, erecting, adjustable shoring and reshoring, stripping, repairing and cleaning for multiple uses. Prefabricated steel truss supported flying arch forms require 15 to 25% fewer man-hours.
Suggested Crew: 3 carpenters and 2 helpers. Contact area is measured from the flat area of the rake. Labor includes forming the stairs, edge risers and rake.

Labor for Fiber Column and Void Forms

Work Element	Unit	Man-Hours Per Unit
Tube column forms, 1 use		
8 inch diameter	100 L.F.	16.1
12 inch diameter	100 L.F.	16.8
16 inch diameter	100 L.F.	16.9
20 inch diameter	100 L.F.	17.9
24 inch diameter	100 L.F.	20.4
30 inch diameter	100 L.F.	22.6
36 inch diameter	100 L.F.	26.4
Round fiber void forms, 1 use		
6 inch diameter	100 L.F.	6.6
10 inch diameter	100 L.F.	7.4
12 inch diameter	100 L.F.	7.5
15 inch diameter	100 L.F.	7.7
18 inch diameter	100 L.F.	8.0
20 inch diameter	100 L.F.	8.1
24 inch diameter	100 L.F.	8.2
26 inch diameter	100 L.F.	8.4
Lined forms for duct use		
6 inch diameter	100 L.F.	6.7
10 inch diameter	100 L.F.	7.4
12 inch diameter	100 L.F.	7.5
15 inch diameter	100 L.F.	7.6
18 inch diameter	100 L.F.	8.0
20 inch diameter	100 L.F.	8.1
24 inch diameter	100 L.F.	8.2
26 inch diameter	100 L.F.	8.3
Square fiber and styrofoam column forms		
12 inch square, 1 use	100 L.F.	20.7
12 inch square, 2 uses	100 L.F.	10.9

These figures include setting, aligning, and stripping.
Suggested Crew: 3 carpenters and 1 helper

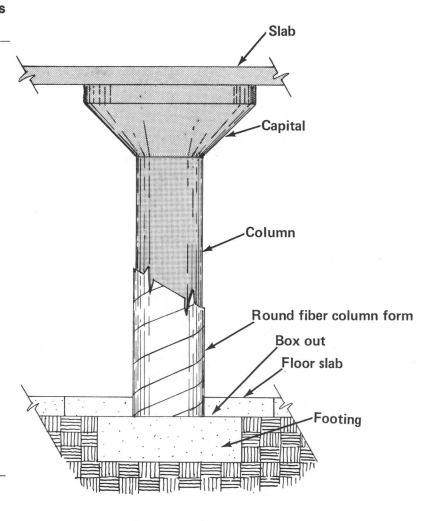

Tube column form

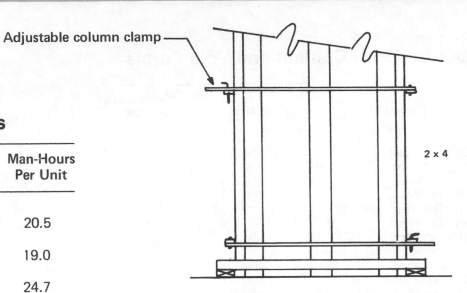

Plywood column form (Side View)

Labor For Concrete Columns

Work Element	Contact Area	Man-Hours Per Unit
Formwork		
Column forms, plywood sheathing	100 S.F.	20.5
Inside beam and girder forms with shoring	100 S.F.	19.0
Spandrel beam or lintel forms with shoring	100 S.F.	24.7
Reinforcing		
Set bars and stirrups	Ton	26.5
Set bars and stirrups and tie in place	Ton	29.5
Finish (carborundum stone)	100 S.F.	2.8
Cure and clean-up	100 S.F.	.10
Patch tie holes	100 S.F.	1.2

Labor required for forming includes fabrication, handling into place, erection, and oiling; installing form ties, tie wire, struts, chamfer strips, screed guides, bracing, and shoring; erecting runways and scaffolds; checking forms during placement of concrete; and stripping.

Forming crew: two carpenters, one helper.
Finishing crew: one finisher, one laborer.

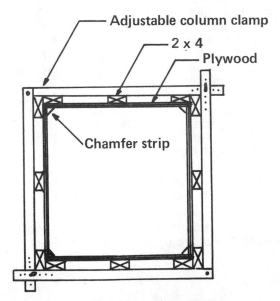

Plywood column form (Plan View)

Labor For Round Steel Column Forms

Work Element	Unit	Man-Hours Per Unit
16 inch diameter		
1 use	100 L.F.	27
3 uses	100 L.F.	25
5 uses	100 L.F.	24
20 inch diameter		
1 use	100 L.F.	39
3 uses	100 L.F.	37
5 uses	100 L.F.	36
24 inch diameter		
1 use	100 L.F.	43
3 uses	100 L.F.	39
5 uses	100 L.F.	37

These figures include assembly, erection, bracing, stripping and cleaning. Does not include lifting operator or riggers. **Suggested crew:** 2 carpenters and 1 helper. Allow 10 additional hours for operator, riggers and oiler to place columns 24 inches or more.

Labor Installing Poured Gypsum Decks

Work Element	Unit	Man-Hours Per Unit
Placing board and deck supports, bulb tees	100 S.F.	1.2
Placing form board for lightweight decking	100 S.F.	1.4
Placing class A gypsum fill		
2" thick	100 S.F.	1.1
2½" thick	100 S.F.	1.2
3" thick	100 S.F.	1.3
3½" thick	100 S.F.	1.4
4" thick	100 S.F.	1.5

No reinforcing steel is included in these figures.
Suggested Crew: 3 laborers and 1 foreman

Labor Installing Expansion Joints

Work Element	Unit	Man-Hours Per Unit
Premoulded expansion joints		
½ x 6 inches	100 L.F.	2.1
½ x 8 inches	100 L.F.	2.9
½ x 12 inches	100 L.F.	3.9
¾ x 6 inches	100 L.F.	2.8
¾ x 8 inches	100 L.F.	3.7
¾ x 12 inches	100 L.F.	5.2
1 x 6 inches	100 L.F.	2.8
1 x 8 inches	100 L.F.	2.9
1 x 12 inches	100 L.F.	3.9
Poured asphalt expansion joints		
½ x 1 inch	100 L.F.	0.6
½ x 2 inches	100 L.F.	1.1
Poured rubberized expansion joints		
½ x ½ inch	100 L.F.	0.7
½ x 1 inch	100 L.F.	0.9
½ x 2 inches	100 L.F.	0.9
½ x 4 inches	100 L.F.	1.4
Poured 2-part polysulphide joints		
½ x 1½ inch	100 L.F.	1.5
½ x 1 inch	100 L.F.	1.1
1 x 2 inches	100 L.F.	3.4
1 x 4 inches	100 L.F.	3.4

Labor for poured joints includes a backer for the joint if required. **Suggested crew:** 1 carpenter.

Prestressed Concrete

Prestressed concrete uses high tensile strength steel cable to produce compression in areas of the concrete where tension would normally occur. Pre-compressing those areas before loads are applied counteracts the normal stress in the concrete and reduces the quantity of steel required.

The plans, specifications and the prestressed concrete shop drawings will show bed layouts, cable tensioning data, and sequences for stressing and detensioning (releasing). Requirements for testing and sampling materials will be as stated in the specifications.

Tensioning the steel cables can be done before or after concrete is poured, or both before and after. Pre-tensioning is done by placing the concrete around tendons which have been held and stressed to controlled values. This method is common in factory assembled members such as beams and slabs. Post-tensioning will usually be done on the job site where tendons are drawn and anchored after the concrete is in place.

Prestressed concrete construction has several advantages:

1. Cracking is reduced. Pretension in the reinforcing compresses the concrete, especially in the lower faces of beams and slabs. Normal "temperature" and flexural cracks are less evident.

2. Prestressed concrete is more resistant to freezing and thawing.

3. Structural weight can be reduced.

Pre-tensioning

The steel cable tendons are tensioned before the concrete is placed and are not detensioned until the concrete has developed the strength specified by the design requirements. Strands may be tensioned individually or in groups.

Pre-tensioned precast pieces are very common. They may be planks, channel slabs, beams and girders, pilings, pipes, panels, and in general those items which can be mass-produced of a size and weight not exceeding over-the-road limits. On-site precasting can be done but is not usually economical.

Post-tensioning

This method uses tendons drawn through ducts or voids formed in the pre-cast member. The strands are stressed after the concrete has developed a specified strength. Post-tensioning is common in greater dimension applications such as continuous-pour slabs or girders, tilt-up slabs, bridge structures, smoke stacks, tanks, roof shells, paving and runways.

Combination of Pre-tensioning and Post-tensioning

Elements of the structure may be reinforced by pre-tensioned strands. After erection additional tendons can be tensioned to complete the structure. This is common in structural frames and multi-span girders.

In post-tensioning, the tendons are drawn and stressed through ducts or voids which are then grouted to bond the strands. The concrete must have been cured to the strength required, the tensioning undertaken in sequence, and then elongation and gauging done just as in the pre-tensioning procedures.

Initial tensioning takes up slack in the strands. Elongation measurements, gauging and final tensioning can then be done. Friction losses as predicted by the engineer must be accounted for in determining prestress. Jacking (simultaneously) from each end of strands, especially with draped strands, will reduce friction losses. Uniform forces can be found by checking for equal pressure on both jacks. Grouting should be done within 2 days after tensioning. Ducts should be blown with compressed air and protected until grouted. When using compressed air, be sure there is no oil injected into the system.

The grout must be carefully mixed using precise measurements of the components. Then screen the grout to remove lumps before use. The voids should be flushed with water and blown dry again before grouting. Grout is pumped at moderate pressure to evacuate the air through the opposite end or through a vent. The vents are then closed and the grout pressure raised to at least 75 psi. This reading should be held for at least ¼ minute. The grouting vent is then closed.

Steel side forms and steel or concrete bottom forms are better than less durable forms. Cardboard tubes or other approved material may be used to produce voids. Forms must be supported, anchored, and braced to withstand vibration and sturdy enough to hold alignment and shape. They must be clean, smooth, tight, leak-free and possibly chamfered. (Chamfered edges and bearing ends reduce edge chipping and cracking that can occur with sharp-cornered, brittle shapes.) Internal forms for voids must be held in alignment and shape. Forms should be oiled or waxed to break easily from the concrete. It is extremely important that the tendons be kept free of this material or, in fact, from any foreign matter. The strands can be cleaned with solvent if they become contaminated.

Bowing or chamfering may cause minute temperature cracking to occur in the top compression face of simple spans. Under load this cracking may disappear. No real weakness occurs where this involves the compressive face. Cracks in areas that may be closed by compression are not important. Surface defects such as "bug holes" caused by air bubbles in the forms usually are repaired by patching. Patch steam-cured concrete, which is white in appearance, with white Portland cement. All repairs should be made before release of strands.

Honeycomb cracking is a more serious defect and is the result of incomplete vibration. Members with honeycomb cracks that expose the tendons should not be used.

Labor Installing Precast Structural Beams

Work Element	Unit	Man-Hours Per Unit
Simple rectangular beams, edge beams		
Spans of 20 to 25 feet		
1.5 to 2.0 S.F. cross section	C.F.	.25
2.0 + to 2.5 S.F. cross section	C.F.	.20
2.5 + to 3.0 S.F. cross section	C.F.	.16
Spans over 25 to 30 feet		
2.0 to 2.5 S.F. cross section	C.F.	.18
2.5 + to 3.0 S.F. cross section	C.F.	.15
3.0 + to 3.5 S.F. cross section	C.F.	.13
Spans over 30 to 35 feet		
2.0 to 2.5 S.F. cross section	C.F.	.18
2.5 + to 3.0 S.F. cross section	C.F.	.18
3.0 + to 3.5 S.F. cross section	C.F.	.12
Spans over 35 feet to 40 feet		
2.5 to 3.0 S.F. cross section	C.F.	.12
3.0 + to 3.5 S.F. cross section	C.F.	.10
3.5 + to 4.0 S.F. cross section	C.F.	.09
4.0 + to 4.5 S.F. cross section	C.F.	.09
Inverted tee beams, interior beams[1]		
Spans from 20 to 25 feet		
1.5 to 2.0 S.F. cross section	C.F.	.26
2.0 + to 2.5 S.F. cross section	C.F.	.21
2.5 + to 3.0 S.F. cross section	C.F.	.18
Spans over 25 to 30 feet		
2.0 to 2.5 S.F. cross section	C.F.	.20
2.5 + to 3.0 S.F. cross section	C.F.	.16
3.0 + to 3.5 S.F. cross section	C.F.	.14
3.5 + to 4.0 S.F. cross section	C.F.	.13
Spans over 30 to 35 feet		
2.0 to 2.5 S.F. cross section	C.F.	.19
2.5 + to 3.0 S.F. cross section	C.F.	.16
3.0 + to 3.5 S.F. cross section	C.F.	.13
3.5 + to 4.0 S.F. cross section	C.F.	.12
Spans over 35 to 40 feet		
2.5 to 3.0 S.F. cross section	C.F.	.14
3.0 + to 4.0 S.F. cross section	C.F.	.12
4.0 + to 4.5 S.F. cross section	C.F.	.10

These figures include hoisting to 100 feet. For hoisting over 100 feet to 200 feet, add 10%. For hoisting over 200 feet, add 25%.
[1]Use the stem width for determining the cross section area.
Suggested Crew: 1 foreman, 2 structural steel workers, 1 crane operator

Labor Installing Flat Precast Wall Panels

Work Element	Unit	Man-Hours Per Unit
3" thick	200 S.F.	17.8
4" thick	100 S.F.	18.5
5" thick	100 S.F.	19.2
6" thick	100 S.F.	19.3

Use these figures for applying custom precast architectural panels up to 100 S.F. per panel. Larger panels will require 1 to 3 fewer man-hours per panel.
Suggested Crew: 3 rigger, 1 crane operator and 1 foreman

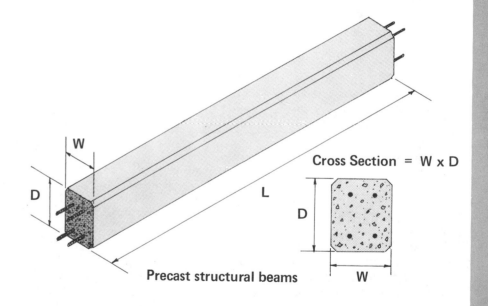

Cross Section = W x D

Precast structural beams

Labor Installing Cementitious Unit Decking

Work Element	Unit	Man-Hours Per Unit
Cement fiber T&G plank		
1" thick	100 S.F.	1.5
1½" thick	100 S.F.	1.5
2" thick	100 S.F.	1.6
2½" thick	100 S.F.	1.6
3" thick	100 S.F.	1.7
3½" thick	100 S.F.	1.7
4" thick	100 S.F.	1.7
Install bulb tees, sub-purlins and grout	100 S.F.	1.7

Suggested Crew: 1 foreman, 3 laborers, 1 crane operator

Labor Erecting Precast Rectangular Columns

Work Element	Unit	Man-Hours Per Unit
8 to 12 feet high	Each	10.0
12 + to 16 feet high	Each	11.3
16 + to 20 feet high	Each	13.0
20 + to 24 feet high	Each	15.6
Add per column base plate connection	Each	3.2
Add per column to column connection	Each	2.3
Add per column to beam connection	Each	1.9

These figures include hoisting to 100 feet. For hoist over 100 200 feet, add 10%. For hoisting over 200 feet, add 25%.
Suggested Crew: 3 riggers, 1 crane operator and 1 foreman

Labor Installing Concrete Roof Decks

Work Element	Unit	Man-Hours Per Unit
Lightweight channel slab		
2¾" thick	100 S.F.	2.8
3½" thick	100 S.F.	3.1
3¾" thick	100 S.F.	3.2
4¾" thick	100 S.F.	3.8
Gypsum plank		
2" thick	100 S.F.	1.9
3" thick	100 S.F.	3.2
Lightweight concrete plank[1]		
2" thick	100 S.F.	2.2
2½" thick	100 S.F.	2.4
3" thick	100 S.F.	2.5
3½" thick	100 S.F.	2.7
4" thick	100 S.F.	2.9

No reinforcing, forming or finishing is included in these figures.
[1]Sloping roof decks take longer. For over 4 in 12 slope, add 0.2 man-hours per 100 feet. For over 6 in 12 slope add 1 man-hour per 100 square feet.
Suggested Crew: 1 foreman, 3 laborers, 1 crane operator

Labor Installing Precast Prestressed Concrete

Work Element	Unit	Man-Hours Per Unit
Single tees		
8' wide, 28" deep	100 S.F.	11.0
8' wide, 32" deep	100 S.F.	11.7
8' wide, 36" deep	100 S.F.	12.0
8' wide, 48" deep	100 S.F.	14.2
6' wide, 28" deep	100 S.F.	13.2
6' wide, 32" deep	100 S.F.	14.0
6' wide, 36" deep	100 S.F.	15.0
6' wide, 48" deep	100 S.F.	16.8
Double tees		
8' wide, 12" to 16" deep	100 S.F.	4.8
8' wide, 20" deep	100 S.F.	4.7
8' wide, 24" deep	100 S.F.	4.9
8' wide, 28" deep	100 S.F.	4.8
8' wide, 32" to 36" deep	100 S.F.	5.2
4' wide, 12" to 20" deep	100 S.F.	6.4
4' wide, 24" deep	100 S.F.	6.6
4' wide, 28" deep	100 S.F.	6.8
4' wide, 32" deep	100 S.F.	7.1
4' wide, 36" deep	100 S.F.	6.9
Longspan channels		
6" leg, 1¼" web, 14' span	100 S.F.	4.8
10" leg, 1¼" web, 22' span	100 S.F.	5.5
12" leg, 1¼" web, 30' span	100 S.F.	6.2
Hollowcore floor planks		
4" thick	100 S.F.	3.4
6" thick	100 S.F.	3.3
8" thick	100 S.F.	3.7
10" thick	100 S.F.	4.4
12" thick	100 S.F.	4.8

Figures include hoisting to 100 feet. For hoisting over 100 feet to 200 feet, add 10%. For hoisting over 200 feet, add 25%. No topping included in these installation times.
Suggested Crew: 1 foreman, 1 crane operator, 3 laborers, 1 structural steel worker

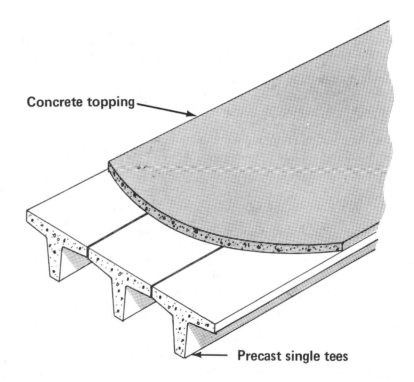

Single tees with concrete topping

Finishing Concrete

Finishing should be done in the following order: spreading, consolidating, screeding, darbying, bull floating, edging, floating and troweling. **Screeding** is straight edging or striking off the surface of the concrete to a predetermined grade. **Darbying** eliminates ridges and fills surface voids. **Bull floating** is done for the same reason as darbying. These two operations are never done on the same surface. Bull floating is usually done on large areas because the long handled tool does not allow work in small areas. **Edging** is done on horizontal surfaces to form a radius on the slab edge.

Floating helps embed large aggregate just beneath the surface, removes slight imperfections, humps, and voids, produces a level plane surface, compacts the concrete, and consolidates mortar at the surface to prepare for other finishing operations. **Troweling** produces a smooth, hard surface. Frequently, additional troweling is done to compact fines at the surface. **Brooming** helps produce a nonslip surface. **Jointing** forms a predetermined path for slab cracking. When floor slabs are tilted, jointing is ordinarily unnecessary.

If work is stopped for a period of time and concrete sets, a bonded construction joint is required. Be careful when preparing the joining surfaces. Rough and clean the surface, exposing and then coating the aggregates with a film of mortar of the same strength.

Tool joints are made with a deep bit that cuts at least one-fifth the slab thickness. The cut forms a point weakness along which the slab will crack when it contracts. Joints can also be made with a concrete saw. Floating is best when done by hand with a metal float. Wood floats tend to stick to the surface. Power floating done with a floating - trowel machine (rather than disc type) must have float shoes attached.

Screeding must be complete before any excess moisture or bleeding water is present on the surface. Either darbying or bullfloating immediately follows screeding. After slight stiffening of the concrete (when concrete can sustain a foot pressure with only a ¼" inch indentation), finishing can continue.

Floating is finished when the water sheen disappears and concrete will support a man. Troweling is the last finishing operation. In hot weather, trowel only the minimum necessary to create the desired surface. Otherwise, delay in applying water retaining materials will result in plastic shrinking, cracking, low-surface strength and dusting. Any finishing or troweling tool which wets the surface is creating a surface which will result in dusting under use.

The specification will state the type of finish needed on exposed floor slabs. The type of finish depends on the expected use of the floor. Bleed water should be removed only by dragging with a rubber hose over the surface and not by dusting with dry cement to absorb water. Rakes should not be used to spread concrete because they usually cause segregation. Screeding by magnesium straightedges or power strikeoffs are recommended.

Don't use lumber or roller screeds unless care is used to avoid high and low spots. Vibrating screeds must be moved forward rapidly to avoid raising too much mortar to the surface in normal weight concrete. In the case of lightweight concrete, a vibrating screed tends to raise too much coarse aggregate to the surface.

Darby normal concrete with a low handled wood darby. For lightweight concrete, magnesium darbies are recommended because they are less apt to "tear" the surface. High handled darbies make it difficult to get proper leverage. This also applies to tools for bull floating.

Curing Concrete

Concrete curing materials delay evaporation of water to ensure complete hydration of concrete. Incomplete hydration results in reduced compressive strength. Curing compounds are spray applied. Hand spraying requires care to make sure that the film is uniform and complete. Application is required while the surface of the concrete is still damp but not wet with free water.

Curing can be done by placing polyethylene waterproof sheeting or other plastic membranes over the surface. This is ideal for curing floors to be covered with tile since the membrane does not leave a film. Membrane should not, however, be used for curing colored floors because it can cause uneven moisture distribution over the top surface, leaving a blotchy surface appearance.

Wet burlap kept continuously moist can be used to slow drying. On small jobs where discoloration of the slab is not critical, moist hay, straw, earth or sand can be laid over the surface.

Colored mortars are usually made by adding a coloring agent to the cement. To assure batches of uniform color, the exact amount of color must be thoroughly and evenly mixed. Finishing operations, including the direction of the final finishing stroke, should be uniform over the entire colored area.

There are two forms of **concrete sealers.** One is a curing-sealer compound or membrane curing compound. The other is only a sealer. They are used to provide an abrasion resistant floor. Plain sealer is ordinarily used to improve abrasive resistance of an old floor of poor quality. If it is used on a new floor, moist curing is specified because a curing compound would hinder penetration of the sealer.

Joint sealants prevent passage of gases, liquids or other undesirable substances into or through the joints. There are two groups of joint sealant: field-molded and preformed. The first are applied in a liquid or semiliquid form. They are formed into the required shape within the mold provided at the joint. Preformed sealants are functionally preshaped, usually at the manufacturers' plant.

Labor Applying Concrete Toppings

Work Element	Unit	Man-Hours Per Unit
Integral topping ¼″ to 1″ thick	100 S.F.	1.8
Granolithic topping		
½″ thick	100 S.F.	2.2
1″ thick	100 S.F.	2.4
1½″ thick	100 S.F.	2.7
2″ thick	100 S.F.	3.4

Includes a hard trowel surface finish.
Suggested Crew: 1 laborer and 1 finisher

Labor Finishing Concrete Ceilings

Work Element	Unit	Man-Hours Per Unit
Rubbed and filled, two passes	100 S.F.	2.6
Sanded and filled, two passes	100 S.F.	3.9
Light sandblast, per pass	100 S.F.	5.5
Heavy sandblast, per pass	100 S.F.	7.0
Brushhammer, lightly	100 S.F.	5.5
Exposed aggregate, acid wash	100 S.F.	6.8

Suggested Crew: 1 finisher

Labor Finishing Concrete Slabs

Work Element	Unit	Man-Hours Per Unit
Hand screed	100 S.F.	.54
Vibratory screed	100 S.F.	.25
Wood float finish	100 S.F.	1.05
Machine trowel, normal	100 S.F.	.55
Machine trowel, very dense	100 S.F.	.70
Steel trowel, one pass by hand	100 S.F.	1.30
Steel trowel, two passes by hand	100 S.F.	1.85
Broom finish	100 S.F.	1.05
Apply concrete hardener	100 S.F.	1.00
Exposed aggregate, wash process	100 S.F.	4.60

Suggested Crew: 1 foreman and 2 finishers

Labor Finishing Concrete Walls

Work Element	Unit	Man-Hours Per Unit
Break form ties, plug holes and patch	100 S.F.	1.8
Rubbed and filled, two passes	100 S.F.	2.1
Sanded and filled, two passes	100 S.F.	3.8
Etched with acid	100 S.F.	1.5
Light sandblast, per pass	100 S.F.	4.4
Heavy sandblast, per pass	100 S.F.	6.2
Brushhammer, lightly	100 S.F.	5.2
Exposed aggregate, acid wash	100 S.F.	3.5

Suggested Crew: 1 finisher

4
masonry section

section contents

Masonry Materials

Masonry in common use includes concrete block, solid clay masonry (brick) and hollow clay masonry (hollow tile). All units are known by their "nominal" size, the dimension of the unit plus the assumed join on one side and bottom. The actual length of a unit whose nominal length is 12 inches would be 11½ inches if the unit were designed to be laid with a ½ inch mortar joint, or 11-58 inch if the unit were designed to be laid with a ⅜ inch joint.

The standard joint thickness is subtracted from each of the three nominal dimensions of a masonry unit to obtain the standard or specified dimensions of the unit. The actual dimensions of a single unit can vary from the standard or specified dimensions by not more than an established tolerance. Unit joints for "slump" block, however, vary with the type of units being used.

Concrete Block
Concrete block units are manufactured in Grades N and S. Grade N is used in exterior walls below and above grade that may or may not be exposed to moisture penetration or weather and for interior walls and back-up. Grade S block is limited to use above grade in exterior walls with weather protective coatings and in walls not exposed to the weather.

Clay Masonry
The brick is made from clay just as it comes from the pits or banks. Brick not exposed to weather may be grade NW when used in walls, but should be grade SW when used in floors.

Structural clay load-bearing wall tile is manufactured in Grade LBX and LB. Grade LBX is suitable for general use in masonry construction and adapted for use in masonry exposed to weathering, provided it is burned to the normal maturity of the clay. It may also be considered suitable for the direct application of stucco.

Grade LB is suitable for general use where not exposed to frost, or for use in exposed masonry where protected with a facing of 3 inches or more of stone, brick, terra cotta, or other masonry.

Mortar for Masonry
Masonry units are set on a mortar bed which creates a bond with adjoining units and permits adjustment of the level of each unit. Each block, brick or tile must be placed correctly so that the mortar will create a strong, weathertight wall. Mortar for reinforced masonry structures should be either Type M or Type S with minimum compressive strengths of 2,500 psi and 1,800 psi respectively. Mortar proportions are listed in Table 4-10.

Cement used in mortars may be Portland cement or masonry cement. Portland cement called for in the specifications will generally be Type I or Type II. Type III is used when high early strength is required.

Masonry cements are sometimes specified for mortar though they produce weaker mortar than equal amounts of Portland cement. The advantage of masonry cement is that the setting time is generally shorter. There are two types of masonry cement though Type II is generally used.

Quicklime is used to prepare lime putty. Quicklime should be slaked according to the specifications. Any active unslaked particles left in the lime after slaking may result in later "popping", disintegration of the mortar, and efflorescence. Quicklime slaked at the site should not be used unless specifically allowed by the specifications.

Hydrated lime is quicklime that has been slaked at the factory. Dry hydrated lime is added to the mix along with the cement. Lime putty is a mixture of lime and water. Specifications may require that lime be added to the mix in this form. Lime putty should be mixed and aged the specified length of time before it is used. Both cement and slaked lime must be protected from wet weather and should never be stored directly on the ground.

Commercial mortar additives are used to improve workability, plasticity and strength. Some additives that improve strength and adhesion may be very useful in thin reinforced walls and can greatly increase flexural strength.

Grout for Masonry
Masonry grout should be thin enough to be poured or pumped without segregation. Any excess water in the grout is absorbed by the masonry units. The moist masonry helps cure the cement grout and aids in the strength gain.

Grout should be proportioned by volume measurements. For example: Fine grout should be one part Portland cement, 2¼ to 3 parts sand to which may be added 1/10 part hydrated lime or lime putty. Fine grout is used in narrow grout spaces.

Coarse grout contains pea gravel and is used for hollow units and 2 inch or more grout spaces. It should be one part Portland cement, 2 to 3 parts sand, and not more than 2 parts pea gravel, to which may be added 1/10 part hydrated lime or lime putty. Finely ground pozzolans may be used in lieu of lime. This will counteract the free lime in the cement and thus reduce the possibility of efflorescence.

Water used in mixing mortar or grout should be free of oils, acids, alkalis, salts, organic materials, or other substances that may harm mortar, grout, or steel.

Masonry Reinforcing
The reinforcing steel generally used in masonry structures is Intermediate Grade ASTM A615 - Grade 40. However, where there are very heavy loads as in a high rise load-bearing wall building or highly loaded masonry columns, a 615 - Grade 40 high-strength steel might be used. Maximum size of reinforcing steel in concrete block masonry is:

Block Width	Horizontal	Vertical
4 inch	No. 5	None
6 inch	No. 8	No. 3
8 inch	No. 11	No. 7
12 inch	No. 11	No. 7

Prefabricated joint reinforcing used in horizontal masonry joints can be considered part of the minimum required reinforcing steel. It may be used as structural reinforcing to resist lateral forces and increase the strength of the wall. Joint reinforcing in 9-gauge, 8-gauge $\frac{3}{16}$ inch, $\frac{1}{4}$ and $\frac{5}{16}$ inch diameter wire is fabricated from high-strength cold-drawn ASTM A82 wire.

Laying Concrete Block

Concrete block should be clean and free of laitance, loose aggregate, grease or anything that will prevent the mortar from getting a proper bond. Surfaces should be level and at the correct grade so that the first bed joint does not exceed $\frac{3}{4}$" in height.

The first course on the foundation should have all webs and face shells set in a mortar bed for full bearing. The mortar, however, should project as little as possible into the cells because the cells will be grouted later. This guarantees that the grout will have the maximum direct bearing on the slab or foundation.

Bolts, anchors, and other inserts which attach adjoining construction to the wall should be embedded in mortar at the face shell and solidly grouted for the entire remaining embedment in the wall. Where possible, they should be wired to the bar reinforcement to keep them from moving during puddling of the grout.

Masonry walls can either abut and be doweled to existing walls or be separated. This detail should be shown clearly in the plans.

Roof flashing should not penetrate the mortar joints more than one-half inch. Metal door frames are set and braced in place before the masonry walls are erected. They should be anchored and solidly grouted in place as the wall is constructed.

Joints for Masonry

V-shaped and concave joints are recommended because they shed water. Joint tooling requires pressure which compresses the mortar and created a tight bond between the mortar and the unit. This also minimizes water penetration.

Flush or cut joints can be used where the finished surface will be painted, waterproofed, or plastered. Raked, struck, beaded, weathered, and squash (or weeping) joints increase chances of water leaks because they open up the body of the mortar and draw it away from the masonry units. This forms capillary areas and small ledges which collect water.

Brick Masonry

Brickwork should be started at the least conspicuous corner or wall. Horizontal surfaces which are to receive brickwork must be clean and damp to ensure a good bond between mortar, grout and the brick. All laitance must be removed. Roughness in itself is no guarantee of a good bond.

The corners or **leads** are built up first, but be careful that these leads are not too high, not over 4 feet. Each course in a corner lead must be grouted as laid, using a brick or other means to form a dam for the grout. All head joints must be solidly filled with mortar when laid.

Mortar must not fall into the grout space as the brick are laid. Keep the grout side of the mortar joint back about $\frac{1}{2}$" from the inside edge of the brick so that the mortar will spread only to the edges of the brick in the grout space without squeezing out.

Another method of keeping mortar out of the grout spaces is to trowel upward in the grout space over the mortar away from the grout space. This forms a beveled bed. The mortar will be squeezed only to the grout side edges without spilling. However, very light furrowing or riffling the bed joint with the point of the trowel helps create a solid bed joint.

Where the mortar squeezes out into the grout space 5/8" or less, it is best to leave it alone. The grout will bond around the mortar when poured and puddled. When mortar does drop into the grout space, as is bound to happen occasionally, it is best to leave it until the grouting is done. Puddling the grout will solidly mix the mortar droppings into the grout. Of course, this only applies where the grouting follows the bricklaying within a few minutes or so.

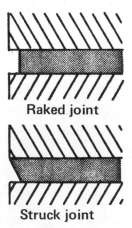

Raked joint

Struck joint

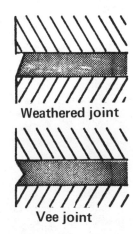

Weathered joint

Vee joint

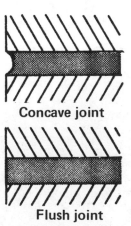

Concave joint

Flush joint

The height of any grout lift should not exceed approximately 4'. The grout is puddled or vibrated in place with a 1" x 2" wood pole or a flexible cable vibrator.

The vertical cells must be in vertical alignment to maintain a continuous unobstructed cell area of not less than 2" x 3".

The principal advantage to the "Low-Lift" grouting method is that cleanouts or inspection openings are not required. Inspection can be made from the top of the 4 foot high wall. Some building codes allow walls up to 8 feet to be grouted from the top of the wall without cleanouts being required. In this case a minimum of mortar is permitted to drop in the cells.

High Lift Grouting

In high-lift grouting the walls are laid up full height before grouting. Cleanout holes should measure not less than 2" x 3" and should be at the bottom of all cells containing reinforcement.

Mortar projections exceeding the vertical height of the mortar joint and mortar droppings must be cleaned out of the grout cells and off the reinforcing steel prior to grouting. Some masons wash out mortar droppings and joint over-hangs twice a day, or more often on hot days. Other methods are preferred since the cells should not be wet prior to grouting. Compressed air or a rod or stick can be used to clean the cells.

Grout should not be poured until the mortar has set a sufficient time to withstand the pressure of the grout. Three days is about the minimum time.

All reinforcing, bolts, other embedded items, and cleanout closures should be securely in place before the grouting is started.

Unless otherwise specified, the grout should have an 8" to 11" slump. Grout should be consolidated at the time of pouring by a flexible cable vibrator or a 1" x 2" wood pole. Grout should be reconsolidated later, before it loses its plasticity.

Grout should be poured in not more than 4' lifts. Wait 15 to 60 minutes for settlement and absorption of excess water, and then pour another 4' lift. Reconsolidation of the previous pour and consolidation of the succeeding pour may be done in the same operation. The full height of any section of wall should be completed in one day. Ungrouted walls should be braced during construction to prevent damage by wind or other forces.

When the temperature is below 40 degrees, protect the grout and completed work from freezing.

The beveled bed joint can best be made as seen here. It is very important for the bricklayer not to spread so much mortar on the joint that when the brick are shoved in place a "fin" protrudes from the brick into the grout space. Furrowed joints, made by plowing the center of the bed joint with the tip of the trowel, are poor practice.

An excess of mortar on the bed joint will result in fins which extend into the grout space. These fins interfere with proper grout flow. However, they should not be cut off, since all too often they fall into the grout space below. This is why the bed should be beveled.

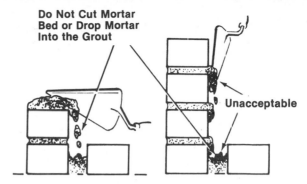

Do Not Cut Mortar Bed or Drop Mortar Into the Grout — Unacceptable

Head joints must be full either with grout or mortar. Grout will flow into a half-filled head joint if it is larger than ½" in thickness and the grout is puddled properly. A head joint buttered as in sketch (a) will not be filled when shoved, and will not later fill with grout. Head joints buttered as in sketch (b) will not be filled when shoved, and will not fill with grout unless they are larger than ½". Head joint buttered completely or ¾" full as in sketch (c) are good under almost any condition.

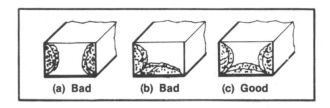

(a) Bad   (b) Bad   (c) Good

Proportions for Mortar

Mortar Type	Minimum Compressive Strength at 28 Days (p.s.i.)	Portland Cement	Hydrated Limes or Lime Putty(1)		Masonry Cements	Damp Loose Aggregate
			Min	Max		
M	2500	1	--	¼	--	Not less than 2¼ and not more than 3 times the sum of the volumes
		1	--	--	1	
S	1800	1	¼	½	--	
		½	--		--	
N	750	1	½	1¼	--	
		--	--	--	1	
O	350	1	1¼	2½	--	

(1)When plastic or waterproof cement is used, hydrated lime or putty may be added but not in excess of one-tenth the volume of cement.

S.F. of Wall per 1000 Common Brick

Width of Joint	Thickness of Wall			
	4"	8"	12"	16"
1/4"	143	71.6	47.7	35.8
3/8"	153	76.4	50.9	38.2
1/2"	162	81.2	54.1	40.6
5/8"	172	86.1	57.4	43.1
3/4"	182	91.2	60.8	45.6

Mortar Color for 1000 Brick

Description	Pounds or Quarts	
Black, standard (dark grey mortar)	125	80
Black, double strength (dense black)	100	70
Buff	100	75
Chocolate, double strength	100	55
Green	100	50
Red	125	60

Assumes brick are laid with ½-inch joints, running or common bond.

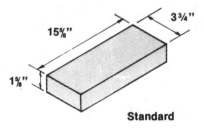

Standard

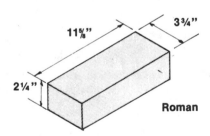

Roman

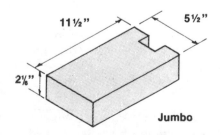

Jumbo

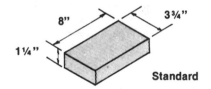

Standard

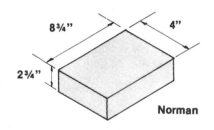

Norman

Brick Sizes and Shapes

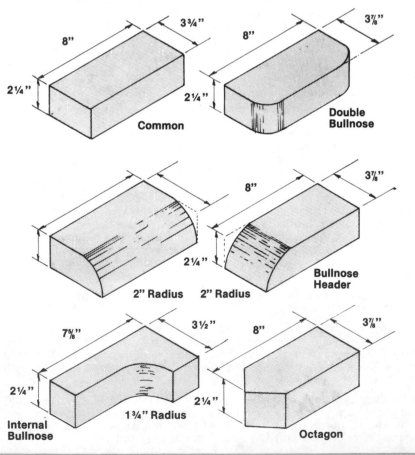

Common

Double Bullnose

2" Radius   2" Radius

Bullnose Header

Internal Bullnose

1¾" Radius

Octagon

Labor and Materials for 100 S.F. of Solid Exterior Basement Walls

Thickness of Wall and Type of Mortar	Number of Brick	C. F. Mortar	Sacks Cement	Sacks Hydrated Lime	C.Y. Sand	Bricklayer Man-Hours	Laborer Man-Hours
Lime mortar							
8" wall	1271	20	--	5.8	.7	11.7	10.0
12" wall	1926	32	--	9.3	1.2	16.3	15.4
16" wall	2580	44	--	12.9	1.6	19.5	20.8
Cement lime mortar							
8" wall	1271	20	2.6	2.9	.7	11.7	10.0
12" wall	1926	32	4.1	4.6	1.2	16.3	15.4
16" wall	2580	44	5.7	6.4	1.6	19.5	20.8
Cement mortar	1271	20	6.6	1.14	.8	13.1	10.0
8" wall	1271	20	6.6	1.14	.8	13.1	10.0
12" wall	1926	32	10.6	1.85	1.3	18.2	15.4
16" wall	2580	44	14.5	2.55	1.8	21.8	20.8

This table is based on common brick with joints ½" thick and all joints in the outside 4" thickness filled with mortar. The remaining brick are laid on a full bed of mortar but with brick touching end to end. Vertical space between each 4" thickness is left open. Every fifth course is a header course. Laborer time includes mortar mixing.

Labor and Materials for 1000 Common Brick in Solid Exterior Basement Walls

Work Element	S.F. Per 1000 Brick	C. F. Mortar	Sacks Cement	Sacks Hydrated Lime	C.Y. Sand	Bricklayer Man-Hours	Laborer Man-Hours
Cement-lime mortar							
8" wall	75	15.3	1.97	2.2	0.57	9.4	8.0
12" wall	51	16.3	2.10	2.4	0.60	8.5	8.0
16" wall	38	16.8	2.17	2.5	0.62	7.5	8.0
Cement mortar							
8" wall	78	15.3	5.1	0.95	0.57	10.5	8.0
12" wall	51	16.3	5.4	0.97	0.60	9.5	8.0
16" wall	38	16.8	5.6	0.98	0.62	8.4	8.0

This table assumes that the exterior 4" thickness of wall is laid with all joints filled. The remaining brick is laid on a full bed or mortar with brick touching end to end. Vertical space between each 4" thickness is filled with mortar, with every fifth course a header. Laborer time includes mixing and handling mortar. Joints are ½" thick.

Labor and Materials for 1000 Common Brick Laid Solid in Common Bond

Work Element	C.F. Mortar	Sack Cement	Sacks Hydrated Lime	C.Y. Sand	Bricklayer Man-Hours	Laborer Man-Hours
Lime mortar						
8" wall	10.6	--	3.0	0.39	10.5	9.0
12" wall	10.1	--	2.9	0.37	8.5	8.0
16" wall	9.9	--	2.9	0.37	7.5	8.0
Cement-lime mortar						
8" wall	10.6	1.36	1.54	0.39	10.5	9.0
12" wall	10.1	1.30	1.46	0.37	8.5	8.0
16" wall	9.9	1.28	1.44	0.37	7.5	8.0
Cement mortar						
8" wall	10.6	3.50	.61	0.39	11.5	8.0
12" wall	10.1	3.33	.59	0.37	10.0	8.0
16" wall	9.9	3.27	.57	0.37	8.4	8.0

This table assumes that the exterior 4'' thickness of wall is laid with all joints filled. The remaining brick is laid on a full bed of mortar with brick touching end to end. Vertical space between each 4'' thickness is filled with mortar, with every fifth course a header. The laborers' time includes mixing and handling of mortar. Joints are ½'' thick.

Common Bond

Labor and Materials for 1000 Common Brick in Solid Exterior Walls Laid in Flemish or English Cross Bonds

Work Element	S.F. Per 1000 Brick	C.F. Mortar	Sacks Cement	Sacks of Hydrated Lime	C.Y. Sand	Bricklayer Man-Hours	Laborer Man-Hours
Lime mortar							
8″ wall	78	15.9	--	4.7	0.59	12.6	9.4
12″ wall	51	13.5	--	4.0	0.50	10.5	8.2
16″ wall	38	12.3	--	3.6	0.46	9.5	8.2
Cement-lime mortar							
8″ wall	78	15.9	2.05	2.31	0.59	12.6	9.4
12″ wall	51	13.5	1.74	1.95	0.50	10.5	8.2
16″ wall	38	12.3	1.59	1.78	0.46	9.5	8.2
Cement mortar							
8″ wall	78	15.9	5.3	0.94	0.59	13.8	9.4
12″ wall	51	13.5	4.5	0.80	0.50	12.0	8.2
16″ wall	38	12.3	4.1	0.72	0.46	10.1	8.2

This table is based on joints ½″ thick. Brick in the outside 8″ thickness are laid with ½″ joints with as many of the vertical joints left open as possible. Remaining brick in 12″ and 16″ walls are laid on a full mortar bed with brick touching end to end and vertical space between 4″ thicknesses left open. Laborer time includes mixing mortar.

Flemish Bond

Labor and Materials for 100 S.F. of Common Brick Laid in Flemish, English or English Cross Bond

Work Element	Number of Brick	C.F. Mortar	Sacks Cement	Sacks of Hydrated Lime	C.Y. Sand	Bricklayers Man-Hours	Laborer Man-Hours
Lime mortar							
8" wall	1271	20	--	5.8	.7	15.8	11.8
12" wall	1926	26	--	7.5	.95	20.3	15.8
16" wall	2580	32	--	9.3	1.2	24.7	21.3
Cement-lime mortar							
8" wall	1271	20	2.6	2.9	.7	15.8	11.8
12" wall	1926	26	3.4	3.8	.95	20.3	15.8
16" wall	2580	32	4.1	4.6	1.2	24.7	21.3
Cement mortar							
8" wall	1271	20	6.6	1.13	.8	17.3	11.8
12" wall	1926	26	8.6	1.49	1.0	23.2	15.8
16" wall	2580	32	10.6	2.55	1.3	26.3	21.8

Based on joints ½" thick.

Laborers' time includes mixing of mortar.

Brick in the outside 8" thickness are laid with ½" joints with as many as possible of the vertical joints left open. Remaining brick in 12" and 16" walls are laid on a full mortar bed with brick touching end to end and vertical space between 4" thicknesses left open.

Number of 8" x 2¼" x 3¾" Brick for 100 S.F. of Wall

S.F. of Wall	Running Face Brick	Running Backing Brick In 8" Wall	Running Backing Brick In 12" Wall	Common Header Course Every 7th Course Face Brick	8" Wall	12" Wall	English and English Cross* Full Headers Every 6th Course Face Brick	8" Wall	12" Wall	Flemish Full Headers Every 5th Course Face Brick	8" Wall	12" Wall	Double Headers Alternating with Stretchers Every 5th Course Face Brick	8" Wall	12" Wall
1	6.16	6.16	12.32	7.04	5.28	11.44	7.19	5.13	11.29	6.57	5.75	11.91	6.78	5.54	11.70
5	31	31	62	36	27	58	36	26	57	33	29	60	34	28	59
10	62	62	124	71	53	115	72	52	113	66	58	120	68	56	117
20	124	124	248	141	106	229	144	103	226	132	115	239	136	111	234
30	185	185	390	212	159	344	216	154	339	198	173	358	204	167	351
40	247	247	494	282	212	458	288	206	452	263	230	477	272	222	468
50	308	308	616	352	264	572	360	257	565	329	288	596	339	277	585
60	370	370	740	423	317	687	432	308	675	395	345	715	407	333	702
70	432	432	864	493	370	801	504	360	791	460	403	834	475	388	819
80	493	493	986	564	423	916	576	411	904	526	460	953	543	444	936
90	555	555	1110	634	476	1030	648	462	1017	592	518	1072	611	499	1053
100	616	616	1232	704	528	1144	719	513	1129	657	575	1191	678	554	1170
200	1232	1232	2464	1408	1056	2288	1438	1026	2258	1314	1150	2382	1356	1108	2340
300	1848	1848	3696	2110	1584	3432	2157	1539	3387	1971	1725	3573	2034	1662	3510
400	2464	2464	4928	2816	2112	4576	2876	2052	4516	2628	2300	4764	2712	2216	4680
500	3080	3080	6160	3520	2640	5720	3595	2565	5645	3285	2875	5955	3390	2770	5850
600	3696	3696	7392	4224	3168	6864	4314	3078	6774	3942	3450	7146	4068	3324	7020
700	4312	4312	8624	4928	3696	8010	5033	3591	7903	4599	4025	8337	4746	3878	8190
800	4928	4928	9856	5632	4224	9152	5752	4104	9032	5256	4600	9528	5424	4432	9360
900	5544	5544	11088	6336	4752	10296	6471	4617	10161	5913	5175	10719	6102	4986	10530
1000	6160	6160	12320	7040	5280	11440	7190	5130	11290	6570	5750	11910	6780	5540	11700
2000	12320	12320	24640	14080	10560	22880	14380	10260	22580	13140	11500	23820	13560	11080	23400

*The quantities in this column also apply to common brick with headers in every sixth course.

For other than ½" joints, the following percentages must be added to or subtracted from the figures in the table.

Add: for 1/8" joint, 21%; for ¼" joint, 14%; for 3/8" joint, 7%.

Subtract: for 5/8" joint, 5%; for ¾" joint, 10%; for 7/8" joint, 15%; for 1" joint 20%.

Number of Nominal 2⅔" x 4" x 8" Modular
Brick in Common Bond per S.F. of Wall

Square Feet Wall Area	Nominal 4" Wall			Nominal 8" Wall			Nominal 12" Wall		
	No. of Brick	C.F. Mortar		No. of Brick	C.F. Mortar		No. of Brick	C.F. Mortar	
		3/8" Joint	1/2" Joint		3/8" Joint	1/2" Joint		3/8" Joint	1/2" Joint
1	6.75	.07	.08	13.5	.16	21	20.25	.26	.33
10	67.5	.66	.83	135	1.62	2.08	202.5	2.59	3.33
20	135	1.31	1.67	270	3.25	4.17	405.0	5.18	6.67
30	202.5	1.97	2.50	405	4.87	6.25	607.5	7.78	10.01
40	270	2.62	3.34	540	6.50	8.34	810.0	10.37	13.34
50	337.5	3.28	4.17	675	8.12	10.42	1012.5	12.96	16.68
60	405	3.93	5.00	810	9.74	12.51	1215.0	15.55	20.01
70	472.5	4.59	5.84	945	11.37	14.59	1417.5	18.15	23.35
80	540	5.25	6.67	1080	12.99	16.68	1620.0	20.74	26.68
90	607.5	5.90	7.51	1215	14.62	18.76	1822.5	23.33	30.02
100	675	6.56	8.34	1350	16.24	20.85	2025.0	25.92	33.35
200	1350	13.12	16.68	2700	32.48	41.69	4050	51.85	66.71
300	2025	19.67	25.02	4050	48.72	62.54	6075	77.77	100.06
400	2700	26.23	33.36	5400	64.96	83.38	8100	103.70	133.41
500	3375	32.79	41.70	6750	81.20	104.23	10125	129.62	166.76
600	4050	39.35	50.04	8100	97.45	125.08	12150	155.54	200.12
700	4725	45.91	58.38	9450	113.69	145.92	14175	181.47	233.47
800	5400	52.46	66.72	10800	129.93	166.77	16200	207.39	266.82
900	6075	59.02	75.06	12150	146.17	187.61	18225	233.32	300.18
1000	6750	65.58	83.40	13500	162.41	208.46	20250	259.24	333.43

Labor and Materials for Face Brick Veneer

Mortar Joint	Wall Thickness	Material				Labor	
		Brick		Wall Ties	Mortar	For 100 S.F. of Wall	
		Per 100 S.F. of Wall	S.F. Per 1000 Brick	Per 100 S.F.	C.F. Per 100 S.F.	Mason	Laborer
1/4	4″	698	143	100	4.48		
3/8	4″	655	153	93	6.56	6½	5
1/2	4″	616	162	88	8.34	hours	hours
5/8	4″	581	172	83	10.52	average	average
3/4	4″	549	182	78	12.60		

Mortar required assumes 20% waste for all head and bed joints.
Brick are assumed to measure 8″ x 2¼″ x 3¾″. No waste is included for brick.

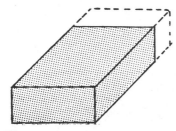

Three quarter

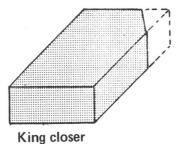

King closer

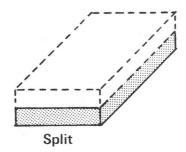

Split

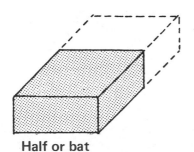

Half or bat

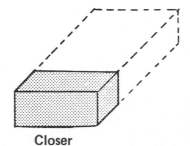

Closer

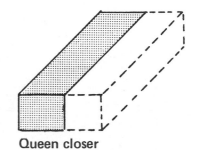

Queen closer

Number of Modular Face Brick For One S.F. of Wall

Size of Brick, Inches	Number of Courses	Height of Each Course	Thickness of Joint	Bricks for One S.F.
2-1/4 x 3-5/8 x 7-5/8	3″	2-2/3″	5/12″	6.75
2-5/16 x 3-5/8 x 7-5/8	3″	2-2/3″	3/8″	6.75
2-1/4 x 3-5/8 x 11-5/8	3 to 8″	2-2/3″	5/12″	4.50
3-5/8 x 3-5/8 x 7-5/8	2 to 8″	4″	3/8″	4.50
2-5/8 x 3-5/8 x 11-5/8	4 to 12″	3″	3/8″	4.00
3-5/8 x 3-5/8 x 11-5/8	2 to 8″	4″	3/8″	3.00

Brick Required in Piers

Size Pier	Face Brick	Common Brick
Brick laid flat, per 1' of height		
8″ x 8″	9	--
8″ x 12″	13½	--
12″ x 12″	18	2
12″ x 16″	24	4
16″ x 16″	28	8
Brick on edge, per 1' of height		
11″ x 11″	12	8

Note: Assumes common brick and ½ inch joints.

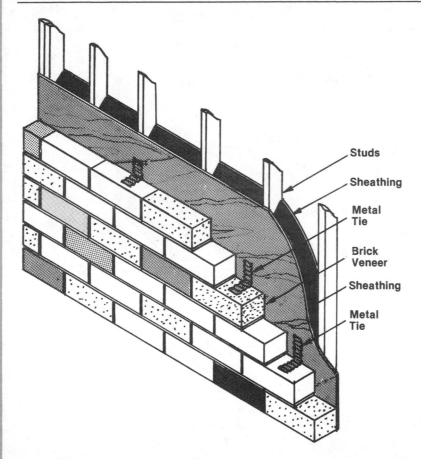

Studs

Sheathing

Metal Tie

Brick Veneer

Sheathing

Metal Tie

Brick Veneer

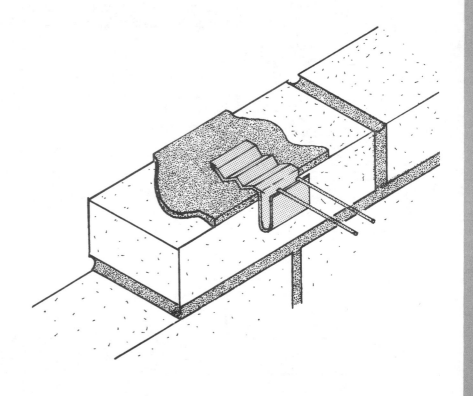

Furring Clip Set in Mortar

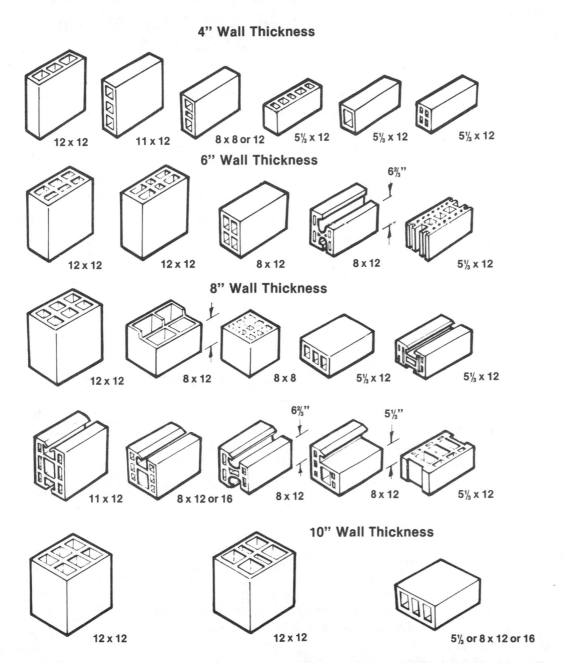

4" Wall Thickness

12 x 12   11 x 12   8 x 8 or 12   5⅓ x 12   5⅓ x 12   5⅓ x 12

6" Wall Thickness

6⅔"

12 x 12   12 x 12   8 x 12   8 x 12   5⅓ x 12

8" Wall Thickness

12 x 12   8 x 12   8 x 8   5⅓ x 12   5⅓ x 12

6⅔"   5⅓"

11 x 12   8 x 12 or 16   8 x 12   8 x 12   5⅓ x 12

10" Wall Thickness

12 x 12   12 x 12   5⅓ or 8 x 12 or 16

12" Wall Thickness

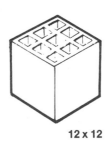

12 x 12

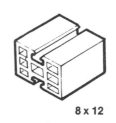

8 x 12

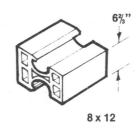

6⅔"

8 x 12

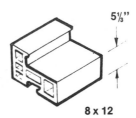

5⅓"

8 x 12

Common Sizes and Shapes for Clay Tile

Block and Tile per 100 S.F. of wall

Size of Units	Wall Thickness	Unit Per 100 S.F.	C.F. of Mortar
8″ x 12″ x 16″ block	12″	110	3.25
8″ x 10″ x 16″ block	10″	110	3.25
8″ x 8″ x 16″ block	8″	110	3.25
5″ x 8″ x 12″ tile	8″	220	5.00
3½″ x 8″ x 12″ tile	8″	300	6.00
5″ x 6″ x 12″ tile	6″	220	4.00
3½″ x 6″ x 12″ tile	6″	300	5.50
5″ x 4″ x 12″ tile	4″	220	4.50
3½″ x 4″ x 12″ tile	4″	300	5.50

These quantities are for load bearing units. The table does not include corner or jamb blocks or lintels. Mortar quantities are based on a ⅜ joint plus 25% for waste.

Lightweight Partition Block for 100 S.F. of Wall

Size of Unit	Wall Thickness	Units Per 100 S.F.	C.F. of Mortar
8″ x 6″ x 16″	6″	110	3.25
9″ x 4″ x 18″	4″	87	3.25
12″ x 4″ x 12″	4″	100	3.25
8″ x 4″ x 16″	4″	110	3.25
8″ x 4″ x 12″	4″	146	4.00
9″ x 3″ x 18″	3″	87	2.50
12″ x 3″ x 12″	3″	100	2.50
8″ x 3″ x 16″	3″	110	2.75
8″ x 3″ x 12″	3″	146	3.50

Labor Placing Hollow Clay Backing Tile

Work Element	Unit	Man-Hours Per Unit
Load bearing (12″ x 12″)		
4″ thick	100 S.F.	9.90
6″ thick	100 S.F.	10.90
8″ thick	100 S.F.	12.60
10″ thick	100 S.F.	13.50
12″ thick	100 S.F.	14.80
Non load bearing (12″ x 12″)		
2″ thick	100 S.F.	7.70
3″ thick	100 S.F.	8.30
4″ thick	100 S.F.	9.30
6″ thick	100 S.F.	9.80
8″ thick	100 S.F.	12.00

Time includes set-up, clean-up, striking where required, placing wall ties, truss reinforcing, grouting, steel alignment.
Suggested Crew: Small jobs, 1 mason, 1 helper

Labor Placing Structural Facing Tile

Work Element	Unit	Man-Hours Per Unit
4″ x 12″, 2″ & 4″ thick	100 S.F.	20
4″ x 12″, 6″ & 8″ thick	100 S.F.	25
5″ x 8″, 2″ & 4″ thick	100 S.F.	23
5″ x 8″, 6″ & 8″ thick	100 S.F.	29
5″ x 12″, 6″ & 4″ thick	100 S.F.	19

Production is affected by type and scope of work. Large open areas go faster than small rooms with many corners. If a crew arrives at a natural stopping point late in the day, hours are lost because the next phase is not started until the following day.
Suggested Crew: 2 bricklayers and 1 helper

Labor Placing Structural Glazed Tile

Work Element	Unit	Man-Hours Per Unit
6T Series (5⅓" x 12") glazed one side only		
2" thick	100 S.F.	18.60
4" thick	100 S.F.	20.40
6" thick	100 S.F.	22.20
8" thick	100 S.F.	23.10
Glazed two sides		
4" thick	100 S.F.	19.70
Series 8W hollow (8" x 16")		
2" thick	100 S.F.	15.50
4" thick	100 S.F.	17.60
6" thick	100 S.F.	18.70
8" thick	100 S.F.	20.50

Time includes set-up, striking and pointing both sides or as required, clean-up and washdown, steel alignment, grouting and repairs if needed. Suggested crew: small jobs, 1 mason, 1 helper.

Labor Setting Glazed Wall Coping

Size of Coping	L.F. Set in 8 Hours	Man-Hours for 100 L.F.	
		Mason	Labor
9"	190-200	4	4
12"	150-175	5	5
18"	125-150	6	6

Steel Shapes for Lintels

Size of Angle	Maximum Opening	Size of Angle	Maximum Opening
3½" x 3½" x 1/4"	8'-3"	6' x 3½" x 3/4"	17'- 2"
4" x 4" x 1/4"	9'-3"	6' x 4" x 1"	17'-11"
4" x 4" x 3/4"	12'-7"	7' x 4" x 3/8"	14'- 8"
5" x 3½" x 5/16"	11'-1"	7' x 4" x 1"	19'-10"
5" x 3½" x 3/4"	14'-6"	8' x 4" x 7/16"	16'- 9"
6" x 3½" x 3/8"	13'-4"	8' x 4" x 1"	21'- 7"
6" x 4" x 3/8"	13'-4"		

This chart shows the opening in a four inch masonry wall which may be spanned with various sized angles. An eight or twelve inch wall will require two and three such lintels respectively.

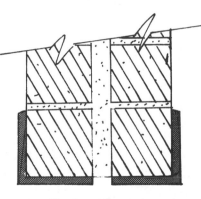

Exposed L Lintel

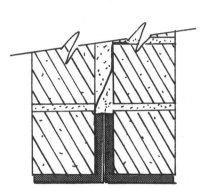

Channel

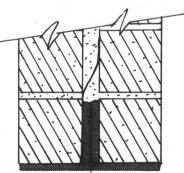

Welded Chair

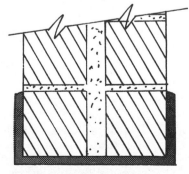

Exposed Channel

Types of Lintels

Number of 16" Block per Course

Width in Feet

Length in Feet	6	8	10	12	14	16	18	20	22	24	26	28	30	32	34	36	38	40
8	19	22	25	28	31	34	37	40	43	46	49	52	55	58	61	64	67	70
10	22	25	28	31	34	37	40	43	46	49	52	55	58	61	64	67	70	73
12	25	28	31	34	37	40	43	46	49	52	55	58	61	64	67	70	73	76
14	28	31	34	37	40	43	46	49	52	55	58	61	64	67	70	73	76	79
16	31	34	37	40	43	46	49	52	55	58	61	64	67	70	73	76	79	82
18	34	37	40	43	46	49	52	55	58	61	64	67	70	73	76	79	82	85
20	37	40	43	46	49	52	55	58	61	64	67	70	73	76	79	82	85	88
22	40	43	46	49	52	55	58	61	64	67	70	73	76	79	82	85	88	91
24	43	46	49	52	55	58	61	64	67	70	73	76	79	82	85	88	91	94
26	46	49	52	55	58	61	64	67	70	73	76	79	82	85	88	91	94	97
28	49	52	55	58	61	64	67	70	73	76	79	82	85	88	91	94	97	100
30	52	55	58	61	64	67	70	73	76	79	82	85	88	91	94	97	100	103
32	55	58	61	64	67	70	73	76	79	82	85	88	91	94	97	100	103	106
34	58	61	64	67	70	73	76	79	82	85	88	91	94	97	100	103	106	109
36	61	64	67	70	73	76	79	82	85	88	91	94	97	100	103	106	109	112
38	64	67	70	73	76	79	82	85	88	91	94	97	100	103	106	109	112	115
40	67	70	73	76	79	82	85	88	91	94	97	100	103	106	109	112	115	118
42	70	73	76	79	82	85	88	91	94	97	100	103	106	109	112	115	118	121
44	73	76	79	82	85	88	91	94	97	100	103	106	109	112	115	118	121	124
46	76	79	82	85	88	91	94	97	100	103	106	109	112	115	118	121	124	127
48	79	82	85	88	91	94	97	100	103	106	109	112	115	118	121	124	127	130
50	82	85	88	91	94	97	100	103	106	109	112	115	118	121	124	127	130	133
52	85	88	91	94	97	100	103	106	109	112	115	118	121	124	127	130	133	136
54	88	91	94	97	100	103	106	109	112	115	118	121	124	127	130	133	136	139
56	91	94	97	100	103	106	109	112	115	118	121	124	127	130	133	136	139	142
58	94	97	100	103	106	109	112	115	118	121	124	127	130	133	136	139	142	145
60	97	100	103	106	109	112	115	118	121	124	127	130	133	136	139	142	145	148

To find the number of block for any wall, always use outside measurements. A basement 22 feet by 32 feet, for example, would require 79 blocks for one course all around. Multiply 79 by the number of courses needed. Thus a ten-course basement would require a total of 790 blocks for the solid wall, from which deductions should be made for windows and doors. If any dimension is an odd number use the nearest smaller size listed in the table. For example, for a 22 foot by 31 foot enclosure use 22 feet by 30 feet and add ½ block per row.

Concrete Masonry Height by Courses

Height of Unit	7-1/2" 7-5/8"	7-5/8" 7-3/4"	7-3/4" 7-7/8"	7-7/8" 8"
Joint Thickness	1/2" 3/8"	1/2" 3/8"	1/2" 3/8"	1/2" 3/8"
No. of Courses				
1	8"	8-1/8"	* 8-1/4"	8-3/8"
2	1'- 4"	1'- 4-1/4"	1'- 4-1/2"	1'- 4-3/4"
3	2'- 0"	2'- 0-3/8"	2'- 0-3/4"	2'- 1-1/8"
4	2'- 8"	2'- 8-1/2"	2'- 9"	2'- 9-1/2"
5	3'- 4"	3'- 4-5/8"	3'- 5-1/4"	3'- 5-7/8"
6	4'- 0"	4'- 0-3/4"	4'- 1-1/2"	4'- 2-1/4"
7	4'- 8"	4'- 8-7/8"	4'- 9-3/4"	4'-10-5/8"
8	5'- 4"	5'- 5"	5'- 6"	5'- 7"
9	6'- 0"	6'- 1-1/8"	6'- 2-1/4"	6'- 3-3/8"
10	6'- 8"	6'- 9-1/4"	6'-10-1/2"	6'-11-3/4"
11	7'- 4"	7'- 5-3/8"	7'- 6-3/4"	7'- 8-1/8"
12	8'- 0"	8'- 1-1/2"	8'- 3"	8'- 4-1/2"
13	8'- 8"	8'- 9-5/8"	8'-11-1/4"	9'- 0-7/8"
14	9'- 4"	9'- 5-3/4"	9'- 7-1/2"	9'- 9-1/4"
15	10'- 0"	10'- 1-7/8"	10'- 3-3/4"	10'- 5-5/8"
16	10'- 8"	10'-10"	11'- 0"	11'- 2"
17	11'- 4"	11'- 6-1/8"	11'- 8-1/4"	11'-10-3/8"
18	12'- 0"	12'- 2-1/4"	12'- 4-1/2"	12'- 6-3/4"
19	12'- 8"	12'-10-3/8"	13'- 0-3/4"	13'- 3-1/8"
20	13'- 4"	13'- 6-1/2"	13'- 9"	13'-11-1/2"
25	16'- 8"	16'-11-1/8"	17'- 2-1/4"	17'- 5-3/8"
30	20'- 0"	20'- 3-3/4"	20'- 7-1/2"	20'-11-1/4"
35	23'- 4"	23'- 8-3/8"	24'- 0-3/4"	24'- 5-1/8"
40	26'- 8"	27'- 1"	27'- 6"	27'-11"
45	30'- 0"	30'- 5-5/8"	30'-11-1/4"	31'- 4-7/8"
50	33'- 4"	33'-10-1/4"	34'- 4-1/2"	34'-10-3/4"

*Dimensions in this column are also heights of standard concrete masonry backup course for standard face brick laid flat with ½-inch mortar joints and headers every sixth course.

Concrete Masonry Length by Stretchers

Number of Stretchers	Length of Concrete Masonry Wall		
	Thickness of Mortar in Head Joint		
	1/4"	3/8"	1/2"
1	1'- 3-3/4"	1'- 3-3/4"	1'- 3-3/4"
1½	1'-11-3/4"	1'-11-7/8"	2'- 0"
2	2'- 7-3/4"	2'- 7-7/8"	2'- 8"
2½	* 3'- 4"	3'- 4"	3'- 4-1/4"
3	4'- 0"	4'- 0"	4'- 0-1/4"
3½	4'- 8"	4'- 8-1/8"	4'- 8-1/2"
4	5'- 4"	5'- 4-1/8"	5'- 4-1/2"
4½	6'- 0"	6'- 0-1/4"	6'- 0-3/4"
5	6'- 8"	6'- 8-1/4"	6'- 8-3/4"
5½	7'- 4"	7'- 4-3/8"	7'- 5"
6	8'- 0"	8'- 0-3/8"	8'- 1"
6½	8'- 8"	8'- 8-1/2"	8'- 9-1/4"
7	9'- 4"	9'- 4-1/2"	9'- 5-1/4"
7½	10'- 0"	10'- 0-5/8"	10'- 1-1/2"
8	10'- 8"	10'- 8-5/8"	10'- 9-1/2"
8½	11'- 4"	11'- 4-3/4"	11'- 5-3/4"
9	12'- 0"	12'- 0-3/4"	12'- 1-3/4"
9½	12'- 8"	12'- 8-7/8"	12'-10"
10	13'- 4"	13'- 4-7/8"	13'- 6"
10½	14'- 0"	14'- 1"	14'- 2-1/4"
11	14'- 8"	14'- 9"	14'-10-1/4"
11½	15'- 4"	15'- 5-1/8"	15'- 6-1/2"
12	16'- 0"	16'- 1-1/8"	16'- 2-1/2"
12½	16'- 8"	16'- 9-1/4"	16'-10-3/4"
13	17'- 4"	17'- 5-1/4"	17'- 6-3/4"
13½	18'- 0"	18'- 1-3/8"	18'- 3"
14	18'- 8"	18'- 9-3/8"	18'-11"
14½	19'- 4"	19'- 5-1/2"	19'- 7-1/4"
15	20'- 0"	20'- 1-1/2"	20'- 3-1/4"
20	26'- 8"	26'-10-1/8"	27'- 0-1/2"
30	40'- 0"	40'- 3-3/8"	40'- 7"
35	46'- 8"	47'- 0"	47'- 4-1/4"
40	53'- 4"	53'- 8-5/8"	54'- 1-1/2"
45	60'- 0"	60'- 5-1/4"	60'-10-3/4"
50	66'- 8"	67'- 1-7/8"	67'- 8"

Cement Required to Lay 1,000 Concrete Block

Unit Size	Wall Thickness	C.F. Mortar	Sacks Cement	C.Y. Sand
8 x 8 x 16	8"	52.25	16.5	1.82
4 x 8 x 16	4"	32.3	10.2	1.12
8 x 12 x 16	12"	66.5	21.0	2.31
8 x 5 x 12	8"	26.0	8.2	.90

This table assumes ½" joints and a 1 to 3 mix of cement to sand.

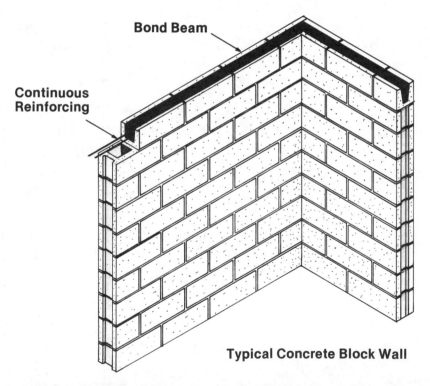

Bond Beam

Continuous Reinforcing

Typical Concrete Block Wall

Use the table at left for standard units 15¾ inches long and half units 7¾ inches long with ¼-inch, ⅜-inch and ½-inch head joints. Length is measured from outside edge to outside edge of units.

*By increasing the thickness of only two of the head joints by ⅛-inch each, the wall lengths from this point down in this column are multiples of 8 inches as shown.

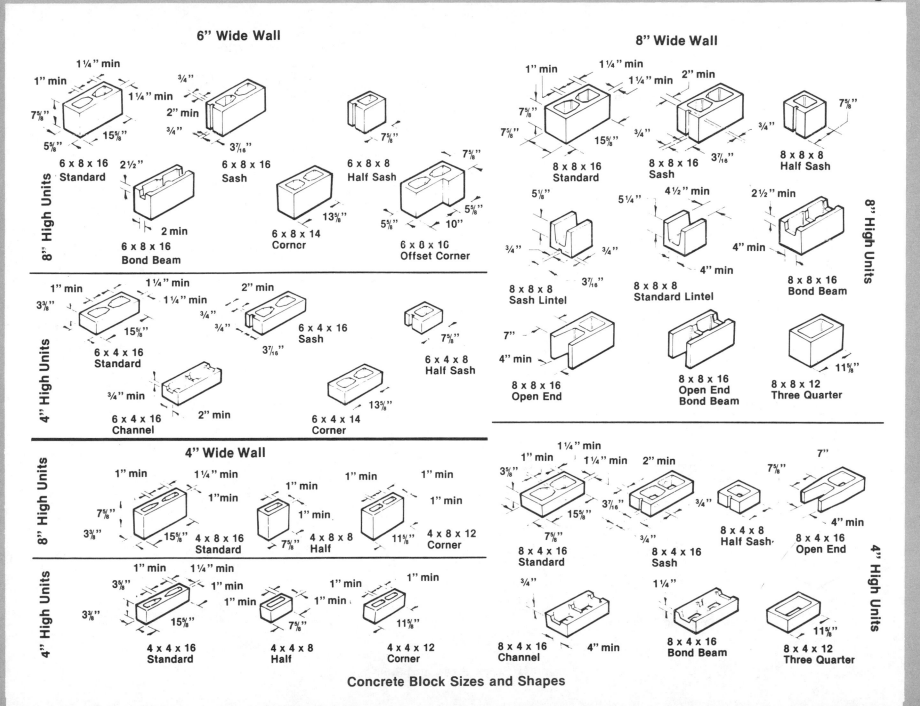

Concrete Block Sizes and Shapes

6" Wide Wall

8" High Units

6 x 8 x 16 Standard

6 x 8 x 16 Sash

6 x 8 x 8 Half Sash

6 x 8 x 16 Bond Beam

6 x 8 x 14 Corner

6 x 8 x 16 Offset Corner

4" High Units

6 x 4 x 16 Standard

6 x 4 x 16 Sash

6 x 4 x 8 Half Sash

6 x 4 x 16 Channel

6 x 4 x 14 Corner

4" Wide Wall

8" High Units

4 x 8 x 16 Standard

4 x 8 x 8 Half

4 x 8 x 12 Corner

4" High Units

4 x 4 x 16 Standard

4 x 4 x 8 Half

4 x 4 x 12 Corner

8" Wide Wall

8" High Units

8 x 8 x 16 Standard

8 x 8 x 16 Sash

8 x 8 x 8 Half Sash

8 x 8 x 8 Sash Lintel

8 x 8 x 8 Standard Lintel

8 x 8 x 16 Bond Beam

8 x 8 x 16 Open End

8 x 8 x 16 Open End Bond Beam

8 x 8 x 12 Three Quarter

4" High Units

8 x 4 x 16 Standard

8 x 4 x 16 Sash

8 x 4 x 8 Half Sash

8 x 4 x 16 Open End

8 x 4 x 16 Channel

8 x 4 x 16 Bond Beam

8 x 4 x 12 Three Quarter

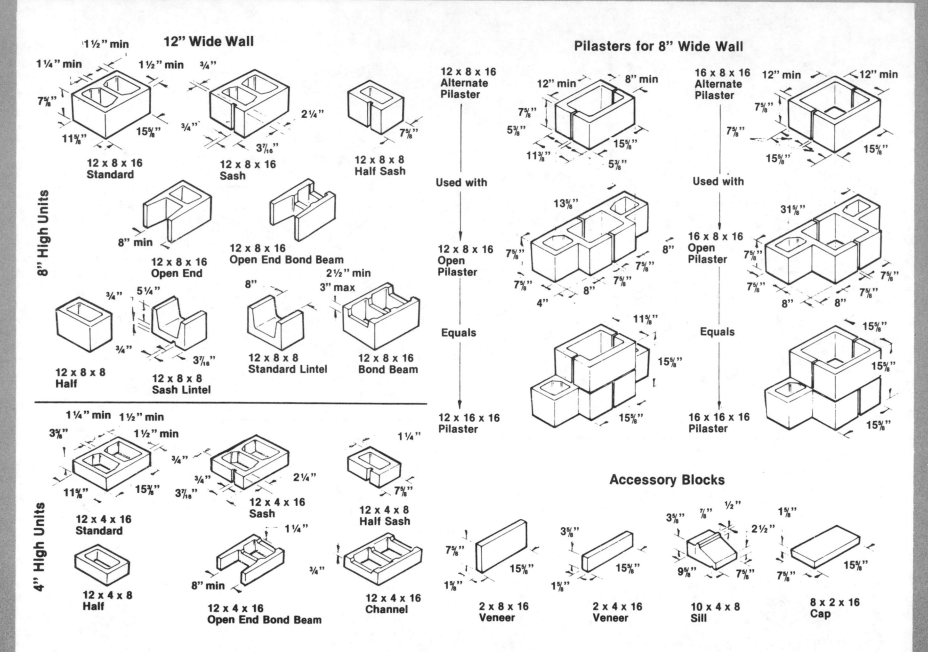

Concrete Block Sizes and Shapes

Labor Placing Concrete Block

Work Element	Unit	Man-Hours Per Unit
Concrete block, lightweight		
4" block	100 S.F.	10.50
6" block	100 S.F.	11.70
8" block	100 S.F.	12.80
10" block	100 S.F.	15.00
12" block	100 S.F.	17.90
Concrete block, hollow, standard weight		
4" block	100 S.F.	11.00
6" block	100 S.F.	12.00
8" block	100 S.F.	13.00
10" block	100 S.F.	15.00
12" block	100 S.F.	18.10

Time includes set-up, clean-up, joint striking one side only, cutting, pointing, steel alignment and grout.
Suggested Crew: Small jobs, 1 mason, 1 helper

Labor Placing Back-up Block

Work Element	Unit	Man-Hours Per Unit
Lightweight block		
2" wall furring soap	100 S.F.	7.30
3" wall furring soap	100 S.F.	7.50
4" wall furring or block	100 S.F.	8.40
6" back-up	100 S.F.	9.60
8" back-up	100 S.F.	10.80
10" back-up	100 S.F.	12.60
12" back-up	100 S.F.	13.60
Concrete block, solid load bearing type		
2" wall furring soap	100 S.F.	7.40
3" wall furring soap	100 S.F.	7.70
4" wall furring or block	100 S.F.	9.60
6" back-up	100 S.F.	10.60
8" back-up	100 S.F.	12.00
10" back-up (semi-solid)	100 S.F.	14.80

Time includes set-up, clean-up, joint striking, cutting, and pointing, steel alignment and grout. Suggested crew: small jobs, 1 mason, 1 helper.

Labor Setting Split Faced Block

Work Element	Unit	Man-Hours Per Unit
Exposed finished wall surface 8" x 16" hollow by size listed		
4" thick	100 S.F.	12.10
6" thick	100 S.F.	13.00
8" thick	100 S.F.	14.70

Time includes set-up, steel alignment, striking, pointing, grouting and repairs if needed.
Suggested Crew: Small jobs, 1 mason, 1 helper

 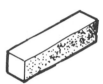

Split Faced Block

Labor Placing Glazed Block

Work Element	Unit	Man-Hours Per Unit
Hollow waylite 8 x 16 glazed		
2" block	100 S.F.	11.30
4" block	100 S.F.	13.30
6" block	100 S.F.	14.40
8" block	100 S.F.	15.70
10" block	100 S.F.	18.00
12" block	100 S.F.	27.70
Glazed two faces		
4" block	100 S.F.	13.30
6" block	100 S.F.	14.50
8" block	100 S.F.	15.70
Glazed base for above — coved		
2 inches high	L.F.	9.0
4 inches high	L.F.	9.70

Time includes set-up, clean-up, striking both sides, pointing both sides, grouting, steel alignment. Suggested crew: small jobs, 1 mason, 1 helper.

Labor Placing Clay Pavers

Work Element	Unit	Man-Hours Per Unit
Brick on mortared concrete bed		
Common bond	100 S.F.	9.70
Herringbone pattern	100 S.F.	21.70
Basket-weave pattern	100 S.F.	26.60
Brick steps (8" wide tread)	100 S.F.	33.60
Brick steps (12" wide tread)	100 S.F.	37.00

Time includes set-up, place, grout, strike and point as needed, brush clean and washdown. Suggested crew: small jobs, 1 mason, 1 helper.

Labor Placing Terracotta

Work Element	Unit	ManpHours Per Unit
Coping		
For 8" wall	100 S.F.	19.70
For 12" wall	100 S.F.	20.30
Terracotta all patterns and sizes	100 S.F.	9.60

Time includes set-up, place and grout, all coping sizes, all pattern terracotta included, place and grout on mortar bed. Suggested crew: small jobs, 1 mason, 1 helper.

Labor Placing Flue Lining

Work Element	Unit	Man-Hours Per Unit
Flue lining square or rectangular		
8" x 8" lining	L.F.	.14
8" x 12" lining	L.F.	.13
12" x 12" lining	L.F.	.16
Round flue lining		
24" diameter lining	L.F.	.25
18" diameter lining	L.F.	.24

Time includes all mortar joints. For heights over 16', add 10 to 15%. Suggested crew: 1 mason and 1 helper.

Labor Setting Marble

Work Element	Unit	Man-Hours Per Unit
Facing panels		
1" thick	S.F.	.41
1½" thick	S.F.	.50
2¼" thick	S.F.	.59
Column bases, 1" thick		
4" high	L.F.	.57
6" high	L.F.	.70
Columns		
Plain faced, solid	C.F.	1.71
Fluted, solid carved	C.F.	1.74
Flooring		
Floor tile mortar set*	S.F.	.24
Miscellaneous marble items		
Shower receptors, built-up	Ea.	5.50
Stair treads, 2" x 12"	L.F.	.43
Thresholds, 1½" x 3"	L.F.	.18
Window sills, 2" x 6"	L.F.	.25
Toilet stalls, without doors	Ea.	3.3

Time includes drilling and placing of shields, brackets, nuts, bolts and gaskets.
Suggested Crew: 1 foreman, 1 mason and 2 helpers for most jobs.
*Up to 1" thick ground face.

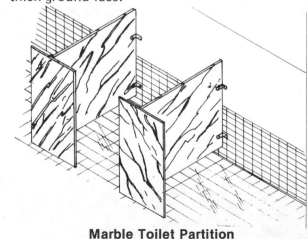

Marble Toilet Partition

Labor and Materials for Glass Block

Requirements for 100 S.F. of Wall	Size of Block in Inches		
	5¾ x 5¾ x 3-7/8	7¾ x 7¾ x 3-7/8	11¾ x 11¾ x 3-7/8
Number of block for 100 S.F. of wall	400	225	100
Cubic feet of mortar required	5	3.6	2.33
Mason time, hours	16 to 18	12 to 14	9 to 10
Labor time, hours	8 to 9	6 to 7	4½ to 5
Labor time for scaffolding, hours	2	2	2
Mason ramming oakum and caulking, hours	1½	1½	1½
Mason cleaning blocks, hours	2½	2½	2½
Expansion strips, 3/8″ x 4-1/8″, L.F.	30	30	30
Wall ties, L.F.	44	44	95
Oakum joints, L.F.	60	60	60
Caulking joints, L.F.	70	70	70
Asphalt emulsion, 40′-0″ x 4½″	½ pt.	½ pt.	½ pt.

¼″ joints are assumed.

Mortar for Glass Block

For One C.F. of Mortar	Mix by Volume	
	1:¼:3	1:1:6
Portland cement	0.3 bag	0.16 bag
Hydrated lime	0.06 bag	0.13 bag
Plastering sand	0.92 C.F.	1.0 C.F.
Waterproofing	0.3 quart	0.2 quart

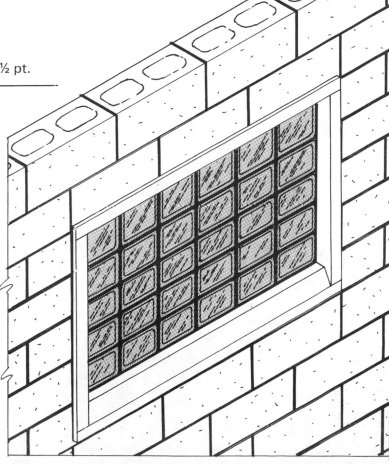

Glass Block in a Concrete Block Wall

Labor Setting Rubble Stone Walls

Work Element	Unit	Man-Hours Per Unit
Rough stone — rubble		
Set in mortar up to 18" thick	C.F.	1.01
Dry set up to 18" thick	C.F.	.69
Ashlar veneer (random size) square cut		
4" thick	S.F.	.43
6" thick	S.F.	.51

Estimated man-hours assume stone has been selected and dumped on site. Does not include placing or cutting of reinforcing steel. Time does include the placement of most wall ties.
Suggested Crew: 1 mason, 1 helper, small jobs. For jobs over 500 S.F. of surface wall area, 2 masons, 1 helper.

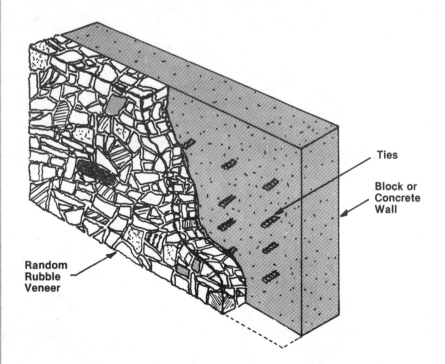

Ties

Block or Concrete Wall

Random Rubble Veneer

Labor Laying Stone Walkways

Work Element	Unit	Man-Hours Per Unit
Random limestone		
1" thick on concrete bed	10 S.F.	1.5
4" thick on sand bed	10 S.F.	1.9
1" thick irregular stone in concrete with mortar joints	10 S.F.	2.1
Snapped random rectangular flagstone		
1½" thick on concrete bed	10 S.F.	2.4
2" thick on sand bed	10 S.F.	2.2
3" thick on sand bed	10 S.F.	2.3
Slate, natural cleft, irregular		
¾" to 1¼" thick on sand bed	10 S.F.	1.8
Granite blocks		
4" x 4" x 2" thick, sand bed	10 S.F.	1.9
4" x 4" x 1" thick, concrete bed	10 S.F.	2.2
Granite pavers, 1" thick regular pieces		
3 S.F. each, in concrete	10 S.F.	1.8
4 to 6 S.F. each, in concrete	10 S.F.	1.6
7 to 10 S.F. each, in concrete	10 S.F.	2.1
Granite pavers, 3" thick regular pieces, on sand		
Up to 3 S.F. each	10 S.F.	1.8
Over 3 S.F. each	10 S.F.	1.3

Time includes handling materials, mixing concrete, preparing bed, setting stone, cleanup and repairs as needed, but no excavation.
Suggested Crew: 1 mason and 1 helper

5

metals
section

section contents

Structural Steel

Structural steel used in building construction is presently manufactured in approximately thirteen different grades. Most steel structures use A36 steel which has a yield strength of 36,000 pounds per square inch. This steel is ductile and can be either welded, bolted, or cold-worked. ASTM A36 steel has replaced A7 and A373 steels as an all-purpose carbon grade steel.

Where lighter members are required and local buckling or high deflection are not objectionable, the high-strength steels are an economical choice. Most of the high-alloy steels are weldable except steel manufactured under ASTM Specification A440.

Welding electrodes are furnished either bare or coated in grades to match the parent metal and weld process to be used. The American Welding Society standard AWS D1.1 lists the processes and types of rods used for each weld.

High-strength bolts for structural applications are furnished in two grades: ASTM 325 and ASTM 490. ASTM 490 bolts offer one-third greater shear capacity in friction-type connects. All high-strength bolts require high-strength nuts and hardened washers. High-strength bolts can be visually distinguished from common mild-steel bolts by three raised radial lines on the bolt head 120 degrees apart. The nuts are identified by three circumferential lines 120 degrees apart. Hardened washers have no identification markings, but can usually be identified by a scratch test.

Assembly of Structural Steel
Most structural steel is assembled with either bolted or welded connections. Both shop welding and field bolting can be used. Improvements in field-welding methods allow structural engineers to use field-welded connections for major structures rather than high-strength bolts. Where field-welding is permitted for major connections and splices, strict compliance with project specifications is essential.

The practice of the majority of the structural engineers in far west states is to detail all major connections in the project plans and specifications. The fabricator merely prepares his shop drawings from the project plans and bids the structural steel accordingly. On the east coast and in the midwest, many designers do not fully detail major connections. For example, on a truss design, a stress diagram with members adequately sized will be shown on the plans. The contractor must then hire a structural engineer to furnish design calculations and details. This may cause delays and differences in interpreting the intent of the design. Where project plans and specifications use contractor-furnished design, a meeting between the owner's engineer, the contractor and the design engineer should be held immediately afterward to detail the structural steel.

Labor for Structural Steel Fabrication

Work Element	Unit	Man-Hours Per Unit
Structural frames	Ton	10
Columns	Ton	10
Girders	Ton	10
Beams	Ton	6
Trusses	Ton	5
Purlins, girts and struts	Ton	8
Frames for openings	Ton	1.8
Cutting (gas)	100 L.F.	7
Welding (arc)	100 L.F.	7
Stairs	Ton	18
Platforms	Ton	22

Fabrication of structural steel includes cutting, riveting, burning, drilling, milling, fitting, assembling, welding, bolting, storing, loading, and hauling to the job site.

Man-hour units are based on bolted connections. If sections are to be welded add 25% for welded joint preparation.

For multiple pass welds, use the total linear feet of all weld passes.

Suggested Crew: two to six steel workers depending on the weight and length of the materials.

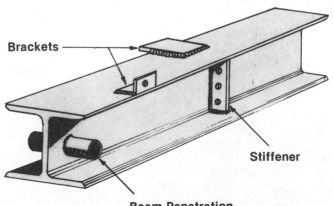

Brackets

Stiffener

Beam Penetration

Labor for Structural Steel Erection

Work Element	Unit	Man-Hours Per Unit
Handling (unloading steel from truck to ground location at erection site)	Ton	1.5
Erection of steel (erect, bolt, and plumb only)[1]		
Foundation work	Ton	5
Columns and struts	Ton	7
Beams and channels	Ton	5
Plate girders	Ton	5
Crane rails	Ton	5
Knee braces	Ton	9
Floor plates	Ton	7
Fittings. bolts, rods, and anchor plates	Ton	3
Grits, angles, angle braces, purlins	Ton	7
Skylight frames and curbs	Ton	10
Monitor frames	Ton	12
Dormers	Ton	10
Door frames	Ton	12
Roof trusses	Ton	9
Transmission towers	Ton	16 to 30
Light steel trestles	Ton	12 to 24
Steel mill buildings	Ton	4 to 12
Steel frame multistoried buildings	Ton	3 to 10
Temporary bolting[2] (3 to 10 bolts/ton)	100 bolts	6
Bolting, high-strength (15 to 30 high-strength bolts/ton)	100 bolts	5
Riveting, air driven[3]		
On ground, easy work	100 rivets	8
Trusses	100 rivets	10
Steel office buildings	100 rivets	12
Steel mill buildings	100 rivets	12
Light trestles and towers	100 rivets	18
Riveting, hand driven		
Easy work	100 rivets	14
Difficult work	100 rivets	21
Welding[4] (5-10 feet of ¼ weld/ton)	100 L.F.	22

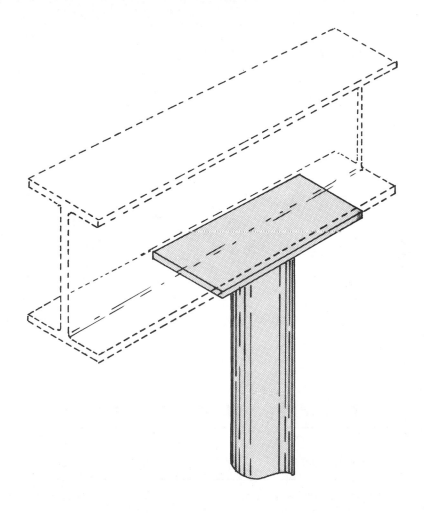

Notes for Structural Steel Erection

[1] For steel erection in this table, the crew consists of one foreman, one crane operator, and four ironworkers. Crew size can vary considerably with each job.

[2] For bolting, the crew in this table consists of four bolters and one helper.

[3] The riveting crew size in this table is five men: one helper, one catcher, two riveters, and one helper to handle air compressor and hoses.

[4] Welding crew is two welders and one helper.

Labor for Miscellaneous Steel Items

Work Element	Unit	Man-Hours Per Unit
Drawn steel tie rods		
5/8" and 3/4"	Ton	54.2
7/8" and 1"	Ton	49.6
Over 1"	Ton	45.1
Column base plates		
Up to 150 lbs. each	Ton	34.0
Over 150 lbs. each	Ton	24.0
Miscellaneous items		
Plain steel angles and tee section	Ton	49.6
Plain flat steel	Ton	63.2
Diamond plate	Ton	67.7
Structural pipe and tubes	Ton	58.9

Figures include fabrication, handling and erection.

Labor Painting Erected Structural Steel

Work Element	Unit	Man-Hours Per Unit
Field touch-up of shop coat	Ton	0.5
Field touch-up of galvanized coating	Ton	1.9
Field coats		
Red lead primer	Ton	1.4
Bitumastic coatings	Ton	1.7
Epoxy coatings	Ton	2.8
Sandblasting	Ton	2.1

Assumes steel sections less than 60 pounds per linear foot. Deduct 30% for heavier sections.
Suggested Crew: 1 ironworker, 1 painter, 1 operator and 1 laborer.

Labor For Specialized Painting on Steel and Iron

Work Element	Unit	Man-Hours Per Unit
On structural steel		
Field coat, red lead	Ton	1.5
Finish coat, enamel	Ton	1.7
On miscellaneous iron work		
Exposed metal, 2 coats of enamel	100 S.F.	1.7
Exterior pipe rail (2 coats)	100 L.F.	2.1
Interior pipe rail (2 coats)	100 L.F.	2.1
Steel stairs (2 coats)	100 S.F	2.2
Stair handrail (2 coats)	100 L.F.	1.2
Grating with frame (2 coats)	100 S.F.	2.3
Ladder (2 coats)	100 L.F.	2.9
Metal trim (2 coats)	100 L.F.	3.5

Time includes move-on and off-site, metal prep, surface protection, brush painting, clean-up and touch-up as required.
Suggested Crew: 1 painter

Labor Installing Construction Castings

Work Element	Unit	Man-Hours Per Unit
Miscellaneous castings		
Light sections up to 150 lbs.	100 lbs.	1.8
Heavy sections over 150 lbs.	100 lbs.	1.0
Steel angle guard corner protection, ¼" thick		
2" x 2"	10 L.F.	0.6
2" x 3"	10 L.F.	0.7
3" x 3"	10 L.F.	0.7
3" x 4"	10 L.F.	0.8
4" x 4"	10 L.F.	0.9
Cast-iron wheel guards		
3' high	Each	1.5
5' high	Each	3.2
Ventilation boxes, 4" deep		
5" x 8"	Each	0.3
5" x 16"	Each	0.4
8" x 16"	Each	0.5

If anchor shields are drilled into concrete, add 0.1 man-hours for each shield.

Labor Installing Expansion Joints

Work Element	Unit	Man-Hours Per Unit
1" floor joint	10 L.F.	2.1
2" floor joint	10 L.F.	2.4
1" ceiling or wall joint	10 L.F.	2.2
2" ceiling or wall joint	10 L.F.	2.4

Use these figures for installing expansion joints with metal covers.

Labor Installing Structural Aluminum

Work Element	Unit	Man-Hours Per Unit
Aluminum, rolled or plate shapes		
Up to 1 ton	100 lbs.	12.2
1 to 5 tons	100 lbs.	12.2
5 to 10 tons	100 lbs.	12.0
Over 10 tons	100 lbs.	11.8
Aluminum extrusions		
Under 5 tons	100 lbs.	7.7
Over 5 tons	100 lbs.	7.3

Time includes all erection, drilling, fitting and field welding as required.
Suggested Crew: 2 ironworkers, 1 welder, 1 crane operator

Labor Placing Bird and Squirrel Screens and Miscellaneous Wrought Iron

Work Element	Unit	Man-Hours Per Unit
Squirrel and bird screen (galvanized)		
8" x 8"	Each	.26
13" x 13"	Each	.59
24" x 24"	Each	1.61
Corbelling iron		
All sizes	Each	.20
Andiron brackets		
All sizes	Each	.20

Time includes drilling, bending, placing shields and special brackets as required. Time does not include field welding if required.
Suggested Crew: 1 mason

Open-Web Joists

Open-web steel joists are used primarily to support floor and roof systems. They make long spans possible and provide open space between the top and bottom chord for conduit, piping and ventilation ducts. Open-web steel joists can be adapted for either wood, steel or concrete floor and roof systems. Note Figure 5-1. Where a wood floor is planned, the steel joists can be furnished with a wood nailer factory installed in the top chord of the joist.

Joists are attached to support beams or walls by several methods. See Figure 5-2. The most common is the welded anchor. The joist beam anchor should not be used for roofs since it does not have any resistance to uplift wind loads.

All joists require bridging to adequately distribute concentrated loads and to restrain the top and bottom chords from lateral movement, as shown in Figure 5-3.

All bridging and bridging anchors must be completely installed before any construction loads are placed on the joists (except the weight of the men needed to install the bridging). Bridging must support the top chords against lateral movement during construction and must hold the steel joists in an approximately vertical plane. The ends of all bridging lines terminating at walls or beams must be anchored at top and bottom chords. Welds for attachment of bridging must not damage the joist members. Hoisting cables must not be released too soon. Install one bridging line nearest mid-span for spans up to 60'-0'' and two bridging lines nearest the third points of the span for spans over 60'-0''. Roof joists can be furnished with either parallel chords, single pitched top chords or double pitched top chords.

Joists 120 feet and longer will require field splicing. Field-splice plates and auxiliary chord members designed for high-strength bolt connections are specially engineered and fabricated to meet specific project conditions. Separate splice members are fitted, matched and pre-drilled in position in the fabricating plant, and are match-marked to assure exact fit and proper camber when erected in the field.

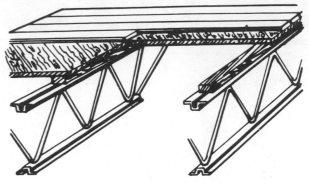

Wood Floor

Concrete Slab

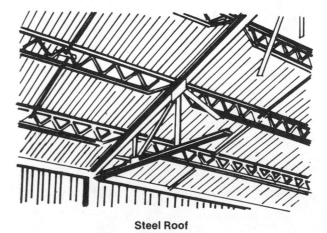

Steel Roof

Figure 5-1

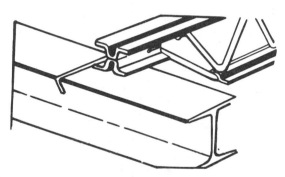

Welded Anchor

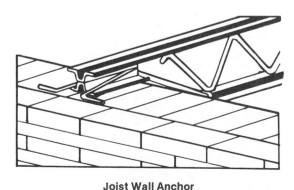

Joist Beam Anchor

Joist Wall Anchor

Figure 5-2

Labor Erecting Open Web Steel Joists

Work Element	Unit	Man-Hours Per Unit
H or J series joists (36,000 P.S.I.)	Ton	15.5
H series joists (50,000 P.S.I.)	Ton	16.5
LH or LJ series joists (36,000 P.S.I.)	Ton	15.0
LH series joists (50,000 P.S.I.)	Ton	16.1
Add for cross bracing, per 100 lbs of bracing	100 lbs	3.2

Figures include lifting to 50 feet. For lifting over 50 feet to 100 feet, add 1.3% man-hour per ton. For lifting over 100 feet to 200 feet, add one man-hour per ton. For lifting over 200 feet to 400 feet, add two man-hours per ton. For lifting over 400 feet, add 3 man-hours per ton.

Time includes placing, bolting or welding, primer touch-up and cross bracing if required.

Suggested Crew: 1 foreman, 2 riggers, 2 ironworkers, 1 crane operator, 1 supervisor. Jobs with less than 50 tons will require more time per ton.

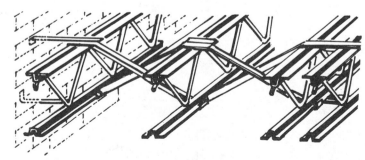

Cross Bracing
Figure 5-3

Metal Decking

Metal decking is light-gage cold-formed steel normally rolled from flat 14 to 26 gage material. The most commonly used metal decking is rolled 1½ inches deep with ribs at approximately six inches on center. The surface is usually galvanized and welded to supporting structural members. Material is rolled in deeper sections for long spans up to approximately 8 inches deep. The width of coverage varies from 12 inches to 30 inches. Lengths can be furnished up to 42 feet. Deck units are composed of a single-fluted sheet or a combination of a fluted and flat sheet welded together by resistance welds.

Light-gage metal decking can be used in combination with concrete as forming or as a composite section. When used as forms for concrete or as a composite steel and concrete deck, the system is designed and performs as a concrete system rather than a metal deck system. Critical connections are made with reinforcing steel and dowels from concrete deck to concrete or masonry shear and bearing walls. Metal decking without concrete requires very careful quality control during construction. Welding becomes critical and care must be taken to assure that all welding conforms to the plans and specifications. Figure 5-4 shows typical floor and roof systems using metal decking and concrete in combination.

Metal deck diaphragms are furnished in two general types. Decks having shear transfer elements directly attached to the framing members are Type "A" diaphragms. Figures 5-5 and 5-6 show typical Type "A" diaphragm applications. Decks with an elevated plane of shear transfer are known as Type "B" diaphragms as shown in Figure 5-7. This type of diaphragm has only welded seam attachments.

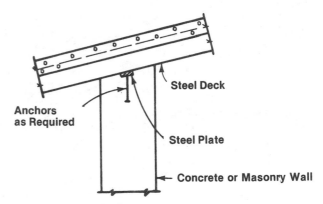

Floor with Direct Diaphragm Connection to Wall

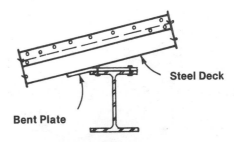

Roof with Diaphragm Connection Through Deck to Wall

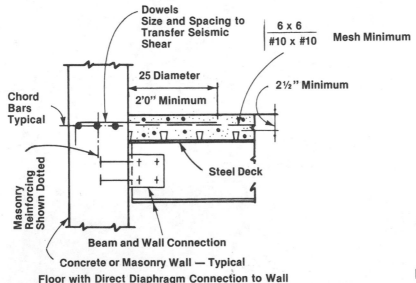

Floor with Direct Diaphragm Connection to Wall

Roof with Diaphragm Connection Through Deck to Framing

Figure 5-4

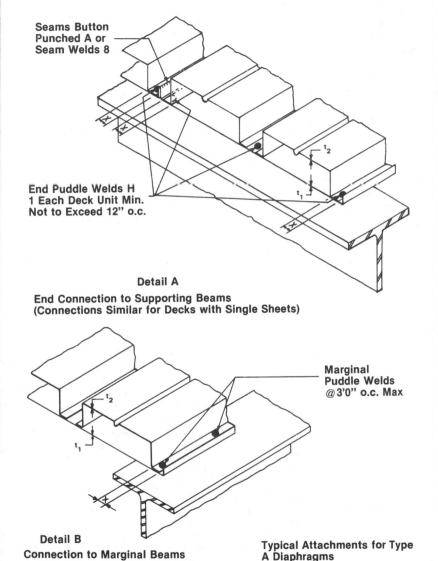

Seams Button Punched A or Seam Welds 8

End Puddle Welds H
1 Each Deck Unit Min.
Not to Exceed 12" o.c.

t_2

t_1

Detail A
End Connection to Supporting Beams
(Connections Similar for Decks with Single Sheets)

t_2

t_1

Marginal
Puddle Welds
@ 3'0" o.c. Max

Detail B
Connection to Marginal Beams

Typical Attachments for Type
A Diaphragms

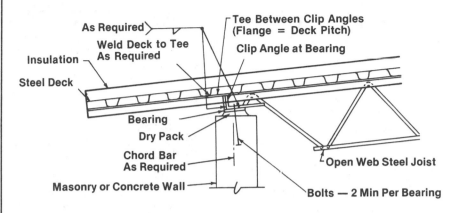

As Required

Weld Deck to Tee
As Required

Insulation

Steel Deck

Tee Between Clip Angles
(Flange = Deck Pitch)

Clip Angle at Bearing

Bearing

Dry Pack

Chord Bar
As Required

Masonry or Concrete Wall

Open Web Steel Joist

Bolts — 2 Min Per Bearing

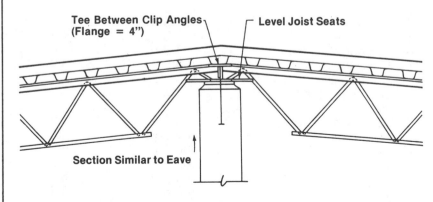

Tee Between Clip Angles
(Flange = 4")

Level Joist Seats

Section Similar to Eave

Typical Attachments for Type A Diaphragms
Figure 5-5

Steel Deck Type A Diaphragms
Figure 5-6

Labor Erecting Corrugated Metal Decking

Work Element	Unit	Man-Hours Per Unit
Corruform		
Standard 3' to 6' span	100 S.F.	1.0
Heavy duty 5' to 10' span	100 S.F.	2.1
Tufcor steel deck, 8' span	100 S.F.	1.8
Cofar steel deck		
20 gauge	100 S.F.	2.4
22 gauge	100 S.F.	2.3
24 gauge	100 S.F.	2.0
Arch-cor standard gauge steel deck	100 S.F.	2.4
Ribbed steel deck, 1½" rib		
16 gauge	100 S.F.	2.4
18 gauge	100 S.F.	1.9
20 gauge	100 S.F.	1.4
22 gauge	100 S.F.	1.1
Ribbed steel deck, 3" rib		
14 gauge	100 S.F.	3.6
16 gauge	100 S.F.	3.1
18 gauge	100 S.F.	3.0
20 gauge	100 S.F.	2.4
Flat and cellular steel decking, 1½" deep[1]		
12 - 12 gauge	100 S.F.	4.9
14 - 14 gauge	100 S.F.	4.6
16 - 14 gauge	100 S.F.	4.2
16 - 16 gauge	100 S.F.	3.6
16 - 18 gauge	100 S.F.	3.2
18 - 18 gauge	100 S.F.	3.1
Flat and cellular steel decking, 4½" deep[2]		
12 - 12 gauge	100 S.F.	7.2
14 - 14 gauge	100 S.F.	6.6
16 - 14 gauge	100 S.F.	6.1
16 - 16 gauge	100 S.F.	6.0
16 - 18 gauge	100 S.F.	5.4
18 - 18 gauge	100 S.F	4.8

Erecting Metal Decking (continued)

Work Element	Unit	Man-Hours Per Unit
High rib steel decking, 4½" deep		
12 gauge	100 S.F.	8.4
14 gauge	100 S.F.	7.2
16 gauge	100 S.F.	6.1
18 gauge	100 S.F.	4.9
20 gauge	100 S.F.	3.6
Holorib steel decking		
16 gauge	100 S.F.	6.1
18 gauge	100 S.F.	4.9
20 gauge	100 S.F.	3.6
22 gauge	100 S.F.	3.0

Installation is by the welded washer method on heights up to 50 feet. Over 50 feet to 100 feet, add ⅓ man-hour per ton for lifting. Over 100 feet to 200 feet, add one man-hour per ton. Over 200 feet to 400 feet, add 2 man-hours per ton. Over 400 feet, add 3 man-hours per ton.

[1] For 3 inch deep deck add 1 man-hour per 100 square feet.
[2] For 6 inch deep deck add 1.3 man-hours per 100 square feet.

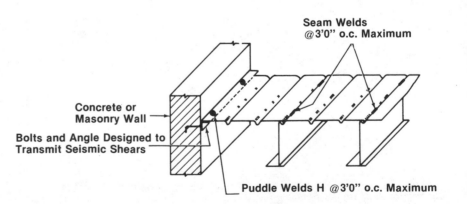

Typical Attachment to Frame for Type B Diaphragms
Figure 5-7

Labor Fabricating and Erecting Steel Stairs

Work Element	Unit	Man-Hours Per Unit
Metal grating tread stairs		
Less than 44" wide	Riser	.91
Over 45" to 56" wide	Riser	.96
Over 57" to 66" wide	Riser	1.05
Metal grating treads with space bar		
Less than 44" wide	Riser	.91
Over 45" to 56" wide	Riser	.96
Over 57" to 66" wide	Riser	1.05
Metal plate tread stairs		
Less than 44" wide	Riser	.86
Over 45" to 56" wide	Riser	.92
Over 57" to 66" wide	Riser	.96
Abrasive steel tread, self-supporting stairs		
Less than 44" wide	Riser	.86
Over 45" to 56" wide	Riser	.89
Over 57" to 66" wide	Riser	.96
Abrasive cross hatched iron tread stairs		
Less than 44" wide	Riser	.99
Over 45" to 56" wide	Riser	1.00
Over 57" to 66" wide	Riser	1.09
Abrasive aluminum tread stairs		
Less than 44" wide	Riser	.79
Over 45" to 56" wide	Riser	.84
Over 57" to 66" wide	Riser	.91
Landings		
To 44" x 44"	Each	2.70
Over 45" x 45" to 56" x 56"	Each	2.90
Over 57" x 57" to 66" x 66"	Each	3.15

Figures assume open riser stairs are field assembled and installed from factory cut materials. Includes setting channel iron stringers but no handrails or newel posts.

Labor Installing Steel Stairways

Work Element	Unit	Man-Hours Per Unit
Field assembled stairways[1]		
36" metal pan stairs and rail	Per Riser	.7
42" metal pan stairs and rail	Per Riser	.9
48" metal pan stairs and rail	Per Riser	1.1
54" metal pan stairs and rail	Per Riser	1.3
Pre-erected stairway, handrail and landing, 13' floor to floor height[2]		
7'4" width	Floor	10.5
7'10" width	Floor	11.0

[1] Time includes field assembly of factory cut materials and erection.
[2] Includes field erection only.

Labor Installing Floor Gratings

Work Element	Unit	Man-Hours Per Unit
Riveted aluminum grating, with bearing bars[1]		
¾" x ⅛", 1.7 lbs./S.F.	100 S.F.	6.1
1" x ⅛", 1.9 lbs./S.F.	100 S.F.	6.5
1¼" x ⅛", 2.3 lbs./S.F.	100 S.F.	8.4
1¼" x 3/16", 3.3 lbs./S.F.	100 S.F.	9.0
1½" x ⅛", 2.7 lbs./S.F.	100 S.F.	9.1
1½" x 3/16", 3.8 lbs./S.F.	100 S.F.	11.4
Added costs for gratings		
Straight cutting	100 L.F.	6.0
Circular cutting	100 L.F.	9.0
Straight bending	100 L.F.	12.0
Circular bending	100 L.F.	18.0
Straight toe plates	100 L.F.	2.0
Curved toe plates	100 L.F.	2.4

[1] Add 15% for welded gratings.

Labor Installing Steel Stair Rails

Work Element	Unit	Man-Hours Per Unit
Field assembled stairways[1]		
36" metal pan stairs and rail	Per Riser	.7
42" metal pan stairs and rail	Per Riser	.9
48" metal pan stairs and rail	Per Riser	1.1
54" metal pan stairs and rail	Per Riser	1.3
Landings	100 S.F.	16.0
Pre-erected stairway, handrail and landing, 13' floor to floor height[2]		
7'-4" width	Floor	10.5
7'-10" width	Floor	11.0

[1] Time includes field assembly of factory cut materials and erection.
[2] Includes field erection only.

Labor Installing Fire Escapes

Work Element	Unit	Man-Hours Per Unit
Balcony sections	L.F.	1.3
Stairs for balcony sections	Floor	1.2
Ground level movable sections	Vertical Foot	1.7
Ground level drop ladder	Linear Foot	1.3

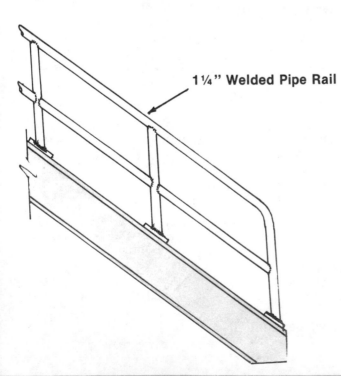

1¼" Welded Pipe Rail

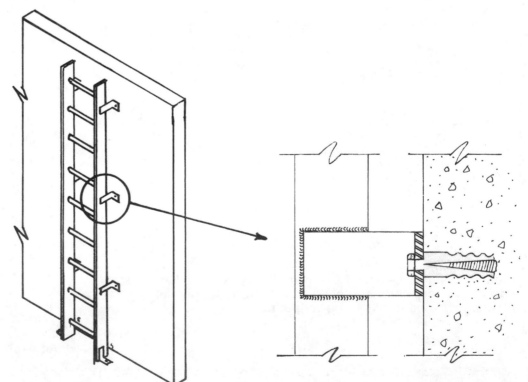

6

carpentry section

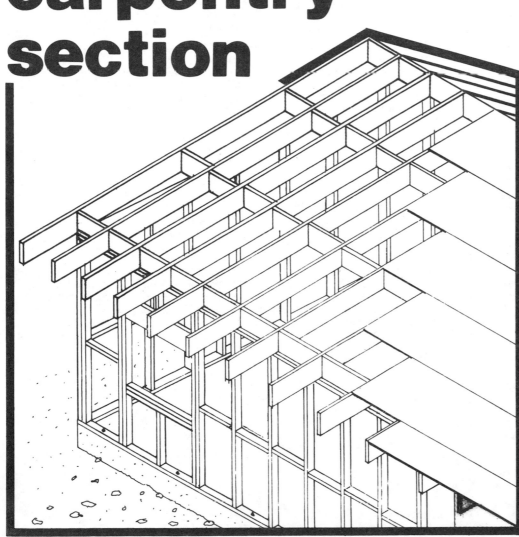

section contents

Rough Carpentry

Rough carpentry deals primarily with wood framing, sheathing, subflooring, and decking, normally that part of a building in which the wood members are not exposed to view.

Structural lumber and plywood for rough carpentry is graded according to rules set up for each species by various associations, inspection bureaus, and government agencies. Some of these groups both define the rules under which a species will be graded and maintain inspection at the mill by certified inspectors. These inspectors grade stamp each piece or certify grades by lot. Some grade stamps or certifications not only identify the grade, but also the species, mill number, intended use, and whether the piece was dry to within a specified moisture content when surfaced.

Framing or structural lumber is graded according to how well it will perform an intended use. The prime consideration is defects affecting strength. Finish lumber for siding, paneling, etc., is graded on the basis of appearance. Items intended to be clear finished will require a higher grade than those intended to be painted with opaque paint.

Finish grade lumber is manufactured in 2 inches or less nominal thickness. When structural or framing members are to be exposed, particularly when they are over 2 inches or more thick, they should be graded for strength and selected for appearance. Laminated members are graded on the basis of both strength and appearance. They may also be graded for strength and encased in appearance-grade lumber.

Generally any piece that carries a high-appearance grade will also be as strong as any piece for the particular species involved. Appearance-grade structural lumber may also command a premium price. Most appearance grades are judged from one side only, however, and serious defects on the reverse side might make it unsuitable as a structural member.

Milling Lumber
Most framing lumber is surfaced on 4 sides (S4S), but it might be unsurfaced, or surfaced on one side and one edge (S1S1E) or other combinations. Surfacing of framing lumber is not so much for appearance as it is for uniformity in width and thickness. and ease of handling.

Rough sawn lumber is produced in the sawmill, from logs. It is generally "sticked and stacked" (stacked with wood slat spacers) to air dry, or kiln dried before it is milled into its final form or dimension. If it is surfaced green, shrinkage will take place after it has been milled. Since moisture content will not be uniform, neither will the finished dimensions.

Resawn lumber refers to surfaced lumber that has been split, generally by bandsaw into roughly equal thicknesses, or saw texturing one surface of a milled piece. Resawn lumber will not generally have the thickness uniformity of S1S lumber, which is surfaced on one side after rough sawing. Rough and resawn lumber may be specified bandsawed or circular sawed, depending on the texture desired.

Lumber Characteristics
Vertical grain refers to the relationship of the annual growth rings to the flat side or face of a wood member. Wood cut with the growth rings at 45 to 90 degrees to the long dimension face (side) is said to be quartersawed or vertical grain. Wood sawed with the growth rings at an angle of 0 to 45 degrees to the side of the piece is said to be plainsawed or flat grain. Vertical grain should not be confused with straight grain, a term used to indicate that the wood fibers (or growth rings) are parallel to the long dimension of the piece.

Trees grow rapidly in the spring. As the season progresses the growth rate shows until it virtually stops in the fall. Spring growth is characterized by soft light fibers, and summer growth by hard, dense fibers. Closely spaced annual growth rings indicate slow growth, typical of dense virgin forests where trees must compete for sunlight and nutrients. Lack of sunlight among the lower limbs promotes "natural pruning". Wood from these trees is strong, dense and relatively knot free. Widely spaced annual growth rings indicate rapid growth, typical of second growth stands and managed tree farms. Wood from these trees is less dense, less strong, and tends to have more knots.

Wood shrinks as it loses moisture and swells as it absorbs moisture. This change in volume is most pronounced in the direction of the annual growth rings, less in the direction of the rays (across the rings), and very little in the direction parallel to the grain. Initial shrinkage takes place when the natural free water is removed from the cell cavities and intercellular spaces of the wood by air or kiln drying.

Air dried lumber has many advantages over green lumber: reduced weight and shrinkage, less checking and warping, increased strength and nail holding power, decrease in the attack of various fungi and insects, and improvement of the ability to hold paint and receive preservatives. Advantages of kiln drying over air drying are greater reduction in weight, control of moisture content to any desired value, reduction in drying time, killing of any fungi or insects, setting the resins in resinous wood, and reduced loss of quality during seasoning.

Using Lumber

Few pieces of long length lumber are chalkline straight. Since a certain amount of set is expected to take place in horizontal members, the crown side of a cambered piece should always be turned up. If the reverse side, however, has a structural defect, such as a large knot, especially if it is on the edge in the middle third of the piece, it should not be used for this purpose. Pronounced camber, bow, twist, or other distorted shapes should not be used where reasonably straight members are required.

Sheathing and decking often serve more than one function. They are both a base on which roofing, flooring, plastering or other veneers are built and serve as diaphragms to give the building rigidity. Plywood is commonly used, but lumber is still used occasionally. Decking may be tongue-and-grooved. Sheathing may be impregnated fiberboard or specially treated gypsum board. When lumber is used as a diaphragm, it is applied diagonally to the framing members. When tongue-and-groove lumber decking is used, an occasional butt joint between supports is acceptable. Tongue-and-groove plywood 2-4-1 decking is designed to span up to two feet unsupported at the long dimension (parallel to the face veneer grain) edges. All other plywood should be supported on all edges.

Base plates (also called sills, sill plates, or mudsills) are secured to concrete foundations under bearing walls. It should be foundation grade (dense heart) redwood or pressure treated structural lumber of other species. The base plate should have continuous level bearing at the proper elevation. Single shimming or mortar bedding will often be required. Sills are bolted to the concrete slab or tops of concrete foundation walls with anchor bolts cast into the concrete. These are generally ½ inch bolts, spaced 4 feet on center, and within about 8 inches of butted end joints.

Treated Sills

Sill plates, excluding formwork, is where rough carpentry for the building begins. If the sill plates are not level and straight, if they do not have continuous solid bearing, and if corners are not square, the whole building may suffer. This is especially true in building with concrete slab floors. Where floor joists rest on sill plates, some leveling can be done above the plate.

Treated lumber is brand stamped by the American Wood-Preservers Association and the American Wood Preservers Institute. A typical stamp will indicate the supplier's brand, the plant designation, the year of treatment, the species of the lumber, the type of treatment material, the amount retained or class of treatment, and the length of the piece.

Treated lumber may also be stamped to indicate the toxic chemicals used, the vehicle in which they are suspended or dissolved, and the amount of material deposited. The depth of penetration depends on the end-use. Industry standards have been developed for various usages and exposures. The quality stamp may therefore indicate only the type of preservative, the vehicle, and usage (for above ground use, for ground contact use, etc.). After lumber has been treated, in order to obtain complete protection, any new surface that has been exposed by sawing,

planing, or boring should be re-treated with the same type of preservative or fire-retardant material.

Some metals may not be compatible with some chemicals used in preservative treatments. Aluminum, for example, should not come in contact with preservatives other than pentachlorophenol, fluor chrome arsenate phenol (FCAP) creosote, or chromated zinc chloride (CZC).

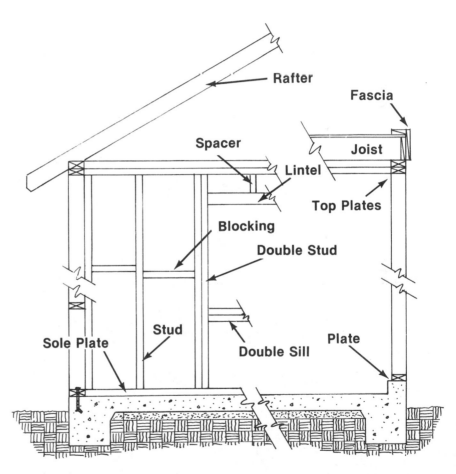

Board Feet Per Piece of Lumber

Size of Lumber in Inches	Length of Lumber							
	10'	12'	14'	16'	18'	20'	22'	24'
1 x 2	1-2/3	2	2-1/3	2-2/3	3	3-1/3	3-2/3	4
1 x 3	2-1/2	3	3-1/2	4	4-1/2	5	5-1/2	6
1 x 4	3-1/3	4	4-2/3	5-1/3	6	6-2/3	7-1/3	8
1 x 5	4-1/6	5	5-5/6	6-2/3	7-1/2	8-1/3	9-1/6	10
1 x 6	5	6	7	8	9	10	11	12
1 x 8	6-2/3	8	9-1/3	10-2/3	12	13-1/3	14-2/3	16
1 x 10	8-1/3	10	11-2/3	13-1/3	15	16-2/3	18-1/3	20
1 x 12	10	12	14	16	18	20	22	24
1 x 14	11-2/3	14	16-1/3	18-2/3	21	23-1/3	25-2/3	28
1 x 16	13-1/3	16	18-2/3	21-1/3	24	26-2/3	29-1/3	32
1 x 20	16-2/3	20	23-1/3	26-2/3	30	33-1/3	36-2/3	40
1¼ x 4	4-1/6	5	5-5/6	6-2/3	7-1/2	8-1/3	9-1/6	10
1¼ x 6	6-1/4	7-1/2	8-3/4	10	11-1/4	12-1/2	13-3/4	15
1¼ x 8	8-1/3	10	11-2/3	13-1/3	15	16-2/3	18-1/3	20
1¼ x 10	10-1/3	12-1/2	14-1/2	16-2/3	18-2/3	20-5/6	22-5/6	25
1¼ x 12	12-1/2	15	17-1/2	20	22-1/2	25	27-1/2	30
1½ x 4	5	6	7	8	9	10	11	12
1½ x 6	7-1/2	9	10-1/2	12	13-1/2	15	16-1/2	18
1½ x 8	10	12	14	16	18	20	22	24
1½ x 10	12-1/2	15	17-1/2	20	22-1/2	25	27-1/2	30
1½ x 12	15	18	21	24	27	30	33	36
2 x 4	6-2/3	8	9-1/3	10-2/3	12	13-1/3	14-2/3	16
2 x 6	10	12	14	16	18	20	22	24
2 x 8	13-1/3	16	18-2/3	21-1/3	24	26-2/3	29-1/3	32
2 x 10	16-2/3	20	23-1/3	26-2/3	30	33-1/3	36-2/3	40
2 x 12	20	24	28	32	36	40	44	48
2 x 14	23-1/3	28	32-2/3	37-1/3	42	46-2/3	51-1/3	56
2 x 16	26-2/3	32	37-1/2	42-2/3	48	53-1/3	58-2/3	64
2½ x 12	25	30	35	40	45	50	55	60
2½ x 14	29-1/6	35	40-5/6	46-2/3	52-1/2	58-1/3	64-1/6	70
2½ x 16	33-1/3	40	46-2/3	53-1/3	60	66-2/3	73-1/3	80
3 x 6	15	18	21	24	27	30	33	36
3 x 8	20	24	28	32	36	40	44	48
3 x 10	25	30	35	40	45	50	55	60
3 x 12	30	36	42	48	54	60	66	72
3 x 14	35	42	49	56	63	70	77	84
3 x 16	40	48	56	64	72	80	88	96
4 x 4	13-1/3	16	18-2/3	21-1/3	24	26-2/3	29-1/3	32
4 x 6	20	24	28	32	36	40	44	48
4 x 8	26-2/3	32	17-1/3	42-2/3	48	53-1/3	58-2/3	64
4 x 10	33-1/3	40	46-2/3	53-1/3	60	66-2/3	73-1/3	80
4 x 12	40	48	56	64	72	80	88	96
4 x 14	46-1/3	56	65-1/3	74-2/3	84	93-1/3	102-1/2	112

Board Feet Per Foot of Length

Width In Inches	Thickness in Inches							
	1	2	3	4	5	6	7	8
1	0.0833	0.1667	0.2500	0.3333	0.4167	0.5000	0.5833	0.6667
1¼	0.1042	0.2083	0.3125	0.4167	0.5208	0.6250	0.7292	0.8333
1½	0.1250	0.2500	0.3750	0.5000	0.6250	0.7500	0.8750	1.000
1¾	0.1458	0.2917	0.4375	0.5833	0.7292	0.8750	1.021	1.167
2	0.1667	0.3333	0.5000	0.6667	0.8333	1.000	1.167	1.333
2¼	0.1875	0.3750	0.5625	0.7500	0.9375	1.125	1.313	1.500
2½	0.2083	0.4167	0.6250	0.8333	1.042	1.250	1.459	1.667
2¾	0.2292	0.4583	0.6875	0.9167	1.146	1.375	1.604	1.833
3	0.2500	0.5000	0.7500	1.000	1.250	1.500	1.750	2.000
3¼	0.2708	0.5417	0.8125	1.085	1.354	1.625	1.896	2.167
3½	0.2917	0.5833	0.8750	1.167	1.458	1.750	2.042	2.333
3¾	0.3125	0.6250	0.9375	1.250	1.563	1.875	2.188	2.500
4	0.3333	0.6667	1.000	1.333	1.667	2.000	2.333	2.667
4¼	0.3541	0.7083	1.062	1.062	1.416	2.125	2.479	2.833
4½	0.3750	0.7500	1.125	1.500	1.875	2.250	2.625	3.000
4¾	0.3953	0.7917	1.188	1.584	1.979	2.375	2.776	3.167
5	0.4167	0.8333	1.250	1.667	2.085	2.500	2.917	3.333
5½	0.4583	0.9167	1.375	1.833	2.292	2.750	3.208	3.667
6	0.5000	1.000	1.500	2.000	2.500	3.000	3.500	4.000
6½	0.5417	1.083	1.625	2.167	2.708	3.250	3.792	4.333
7	0.5833	1.167	1.750	2.333	2.917	3.500	4.083	4.667
7½	0.6250	1.250	1.875	2.500	3.125	3.750	4.375	5.000
8	0.6667	1.333	2.000	2.667	3.333	4.000	4.667	5.333
8½	0.7083	1.417	2.125	2.833	3.542	4.250	4.958	5.667
9	0.7500	1.500	2.250	3.000	3.750	4.500	5.250	6.000
9½	0.7917	1.583	2.375	3.167	3.908	4.750	5.542	6.333
10	0.8333	1.667	2.500	3.333	4.167	5.000	5.833	6.667
10½	0.8750	1.750	2.625	3.500	4.375	5.250	6.125	7.000
11	0.9167	1.833	2.750	3.667	4.583	5.500	6.417	7.333
11½	0.9583	1.917	2.875	3.833	4.792	5.750	6.708	7.667
12	1.000	2.000	3.000	4.000	5.000	6.000	7.000	8.000
13	1.083	2.167	3.250	4.333	5.417	6.500	7.583	8.666
14	1.167	2.333	2.500	4.667	5.833	7.000	8.167	9.333
15	1.250	2.250	3.750	5.000	6.250	7.500	8.750	10.00
16	1.333	2.667	4.000	5.333	6.667	8.000	9.333	10.67
17	1.417	2.833	4.250	5.667	7.083	8.500	9.917	11.33
18	1.500	3.000	4.500	6.000	7.500	9.000	10.50	12.00
19	1.583	3.167	4.750	6.333	7.917	9.500	11.08	12.67
20	1.667	3.333	5.000	6.667	8.333	10.00	11.67	13.33

Softwood Lumber Sizes and Grades

Independent agencies throughout the country publish detailed grades and specifications in their grading rule manuals, and all association grading rules comply with the American Lumber Standards and National Grading Rules Committees. Regional grading associations include the following:

Western Wood Products Association
West Coast Lumber Inspection Bureau
Canadian Lumber Standards Grading Rules
Southern Forest Products Association
California Redwood Inspection Service

National Grading Rules for Dimension Lumber classify dimension into two width categories. Dimension up to 4'' wide is classified as "Structural Light Framing," "Light Framing," and "Studs." Dimension 6'' and wider is classified as "Structural Joists and Planks." "Appearance Framing" is designed for high bending stress ratio requirements.

The following grades are generally common for framing lumber:

Construction [Light Framing]: Highest grade of light framing.
Standard: Most commonly used in 2x4's as standard and better. Strength and load-bearing capacity is slightly less than construction grade lumber.
#2 [Structural Joists and Planks]: Commonly applied to 2x6 and wider in grade terminology as #2 and better.
#3 [2x4 Utility] and [2x6 and wider #3]: Used where economical construction is desired. Studding, blocking, plates, bracing and rafters.
Economy: Low grade of lumber, not intended for use where strength and appearance is a consideration. Material often cut up for use in pallets and crating and for general industrial use.
Stud Grade: Referring to basic 2x4's, 10' and shorter. This grade is suitable for all stud uses, including use in load-bearing walls. Western softwoods other than dimension are generally graded into two classes: **Select and Finish Grades** and **Board Grades.**
Select and Finish Grades are widely used for interior walls, mouldings, cabinets, siding, woodwork and trim, and are usually sold as B.BTR, C.BTR and D. The boards are graded as #2. BTR, #3, #4 and #5.
#2.BTR is used as paneling and shelving as well as for general industrial uses.
#3 is used frequently for paneling and shelving as well as for industrial use.
#4 is used for subfloor, roof and wall sheathing and for fencing.
#5 is for construction and industrial use where appearance and strength are not basic requirements.
Board Grades are general purpose items usually of 1'' nominal thickness.

Softwood Lumber Grade Stamps

An official grading agency mark on a piece of lumber insures that it has been inspected and graded for the use intended.

A number of lumber groups have grading rules and marks that comply with the National Grading Rule for Softwood Dimension Lumber: Southern Pine Inspection Bureau (SPIB), Western Wood Products Assn. (WWPA), West Coast Lumber Inspection Bureau, Canadian Lumber Standards Grading Rule and California Redwood Inspection Service.

The element of the grade stamps issued by two grading agencies, WWPA and SPIB, are as follows:

A is the mark that identifies the grading agency (SPIB or WWPA) under whose rules the piece was inspected,
B is a permanent number assigned to each mill for grade purposes,
C is an example of an official grade name abbreviation,
D indicates the moisture content of the lumber when manufactured,
E identifies the wood species in the case of lumber graded under WWPA rules. No species mark is indicated on SPIB grade stamps since all Southern pine is graded under common rules.

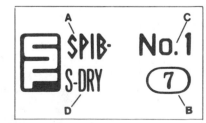

Redwood Lumber

Redwood lumber is available in a wide range of standard and specialized grades, sizes and patterns for a variety of end uses. Each piece is graded at the mill according to appearance and durability.

Grades
The seven most popular grades of redwood are: clear all heart, clear, select heart, select, construction heart, construction common, and merchantable. Of these seven, the first three are all-heartwood grades. Heartwood comes from the inner portion of the tree and contains colored extractives which render durability against termite and decay attack.

The remaining four grades contain some sapwood, including occasional all-sapwood pieces. Cream-colored sapwood is from the outer growth layers of the redwood and does not have inherent resistance to decay and insects.

The two top grades (clear all heart and clear) are normally kiln-dried. This lumber is used for the finest exterior and interior architectural applications such as siding, paneling, trim and cabinetry.

The remaining grades are not normally kiln dried and are generally specified for decks, shelters, garden structures and fences where knots and other characteristics have no effect on their application.

Patterns for profile lumber have been established by the California Redwood Association. These include beveled, channel rustic, shiplap and tongue-and-groove.

Sizes
Redwood, like softwood lumber, is available in the full range of lengths and dimensions from 1'' lath to posts and timbers.

Grade Stamps
Standard grade marks include the grade designation and the symbol of an authorized grading agency, such as the California Redwood Association (CRA). Grade marks may appear on either seasoned or unseasoned lumber on the face, edge or end of a piece.

Lumber that has been kiln dried according to accepted standards includes the words "Certified Kiln Dried" in the grade mark.

Uses
Where wood will be in contact with the soil (such as fenceposts, patio paving, or garden stairs) it is essential to use an all heart grade redwood.

When working with boards containing knots, place large knots over joists in load-bearing applications.

Use only hot-dipped galvanized, aluminum alloy, or stainless steel nails and fastenings. Any other type of metal can corrode and stain wood when exposed to moisture. When nailing at the ends of boards, predrill holes to avoid splitting wood.

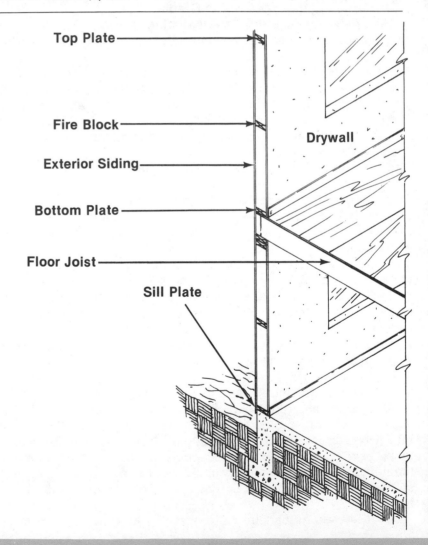

Nails

Nail sizes start at 2d which is 1'' long and range up to 60d which is 6'' long. The 2d through 16d nails increase in length in ¼'' increments: 2d is 1'' long, 3d is 1¼'', etc. Nails above 16d increase by ½'' increments.

Nails are distinguished by their heads, shanks, points, and surface finishes.

Bright After manufacture, nails are tumbled to remove dirt and ''chips'' and packed uncoated.

Galvanized A zinc coating is applied, via either hot-tumbler method or electrolysis to protect against atmospheric corrosion.

Cement-coated An adhesive is applied via tumbler to provide increased holding power.

Blued Nails are sterilized by heating them until an oxidation layer is formed.

Phosphate-coated A zinc-phoshate coating is applied to give limited corrosion protection and some holding capacity.

Standard Nail Sizes and Weights

Size	Length Inches	Common Diameter Inch	Common Number of Lbs.	Box Diameter Inch	Box Number of Lbs.
4d	1½	.102	316	.083	473
5d	1¾	.102	271	.083	406
6d	2	.115	181	.102	236
7d	2¼	.115	161	.102	210
8d	2½	.131	106	.115	145
10d	3	.148	69	.127	94
12d	3¼	.148	63	.127	88
16d	3½	.165	49	.134	71
20d	4	.203	31	.148	52
30d	4½	.220	24	.148	46
40d	5	.238	18	.165	35

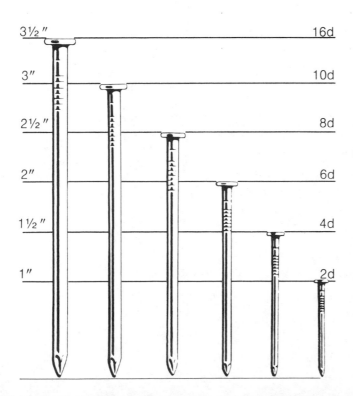

Types of Fasteners

Common steel wire nails are most commonly used in rough carpentry. Similar nails with smaller diameter shanks are called ''box nails''. Nails 6½'' in length or more are called spikes. Other special nails include finishing nails, casing nails, brads, roofing nails, flooring nails, drywall nails, and a host of others with special size or shaped heads, lead heads, washered heads, or specially shaped or sized shanks and points.

Nails used in pneumatic guns generally have hooded or ''T'' shaped heads so that they can be packed closer together in the magazine. Staples are also used in pneumatic or mechanically driven hammers. Special hardened nails and drive pins are used to penetrate steel or concrete. They are driven manually with a heavy hammer and nail holding device or with a pistol-like power driven tool.

Threaded fasteners include screws, lagscrews, and bolts. Metal washers are used under nuts and bolt heads or lagscrew heads bearing on wood. To increase joint strength in heavy timber connections, split rings or shear plates are placed in prepared grooves around the bolt hole in mating members before assembly. Spiked grids, toothed rings and clamping plates are also used in similar fashion except that they are forced into the wood fibers by the bolt as it is being tightened.

Standard Nail Requirements

Material	Unit	Size and Kind of Nail	Number of Nails Per Unit	Lbs. of Nails Per Unit
Wood shingles	1000 S.F.	3d common	2,560	4 lbs.
Individual asphalt shingles	100 S.F.	7/8" roofing	848	4 lbs.
Strip asphalt shingles	100 S.F	7/8" roofing	320	1 lb.
Bevel or lap siding, ½" x 4"	1000 S.F.	6d coated	2,250	*15 lbs
Bevel or lap siding, ½" x 6"	1000 S.F.	6d coated	1,500	*10 lbs.
Drop siding, 1" x 6"	1000 S.F.	8d common	3,000	25 lbs.
3/8" hardwood flooring	1000 S.F.	4d finish	9,300	16 lbs.
13/16" hardwood flooring	1000 S.F.	8d casing	9,300	64 lbs.
Softwood flooring, 1" x 3"	1000 S.F.	8d casing	3,350	23 lbs.
Softwood flooring, 1" x 4"	1000 S.F.	8d casing	2,500	17 lbs.
Softwood flooring, 1" x 6"	1000 S.F.	8d casing	2,600	18 lbs.
Ceiling, 5/8" x 4"	1000 S.F.	6d casing	2,250	10 lbs.
Sheathing boards, 1" x 4"	1000 S.F.	8d common	4,500	40 lbs.
Sheathing boards, 1" x 6"	1000 S.F.	8d common	3,000	25 lbs.
Sheathing boards, 1" x 8"	1000 S.F.	8d common	2,250	20 lbs.
Sheathing boards, 1" x 10"	1000 S.F.	8d common	1,800	15 lbs.
Sheathing boards, 1" x 12"	1000 S.F.	8d common	1,500	12½ lbs.
Studding, 2" x 4"	1000 S.F.	16d common	500	10 lbs.
Joist, 2" x 6"	1000 S.F.	16d common	332	7 lbs.
Joist, 2" x 8"	1000 S.F.	16d common	252	5 lbs.
Joist, 2" x 10"	1000 S.F.	16d common	200	4 lbs.
Joist, 2" x 12"	1000 S.F.	16d common	168	3½ lbs.
Interior trim, 5/8" thick	1000 L.F.	6d finish	2,250	7 lbs.
Interior trim, 3/4" thick	1000 L.F.	8d finish	3,000	14 lbs.
5/8" trim where nailed to jamb	1000 L.F.	4d finish	2,250	3 lbs.
1" x 2" furring or bridging	1000 L.F.	6d common	2,400	15 lbs.
1" x 1" grounds	1000 L.F.	6d common	4,800	30 lbs.

*Cement coated nails sold as two-thirds of pound equals 1 pound of common nails.

Aluminum Nail Requirements

No. Nails Per Lb.	Size and Type of Nail	No. Nails Per Box	Coverage
604	6d wood siding — sinker hd.	575	500 B.F. ½" x 8" bevel sdg.
468	7d wood siding — sinker hd.	575	500 B.F. ¾" x 8" bevel sdg.
319	8d wood siding — sinker hd.	575	500 B.F. ¾" x 8" bevel sdg.
185	10d wood siding — sinker hd.	290	250 B.F. ¾" x 8" bevel sdg.
604	6d wood siding — casing hd.	575	500 B.F. ½" x 8" bevel sdg.
468	7d wood siding — casing hd.	575	500 B.F. ¾" x 8" bevel sdg.
319	8d wood siding — casing hd.	575	500 B.F. ¾" x 8" bevel sdg.
185	10d wood siding — casing hd.	290	250 B.F. ¾" x 8" bevel sdg.
1230	1¼" asbestos siding	885	500 S.F. asb. sdg. face nailing
720	1¾" asbestos siding	885	500 S.F. asb. sdg. face nailing
785	1¼" asbestos shingle	885	500 S.F. asb. sdg. conc. nailing
659	1½" asbestos shingle	885	500 S.F. asb. sdg. conc. nailing
544	1¾" asbestos shingle	885	500 S.F. asb. sdg. conc. nailing
1300	1¼" cedar shake	1680	300 S.F. single course
724	1¾" cedar shake	1680	300 S.F. double course
1480	3d cedar shingle	3150	300 S.F. with 5" exposure
1313	7/8" standard shingle	2600	General purpose
1009	3d standard shingle	2000	Barn battens, joist lining, etc.
1048	1-3/8" dry wall	1530	1000 S.F. 3/8" gypsum board
900	1-5/8" dry wall	1530	1000 S.F. 1/2" gypsum board
988	1-1/8" rock lath	2666	35 S.Y.
939	1-1/4" rock lath	2666	35 S.Y.
725	1-1/2" rock lath	1900	25 S.Y.
495	2" insulated siding	1680	500 S.F.
295	2½" insulating siding	600	60 buttress corners
663	7/8" roofing	860	500 S.F. roll roofing
605	1" roofing	980	300 S.F. sq. tab. shingles
491	1¼" roofing	980	300 S.F. sq. tab. shingles
017	1½" roofing	980	300 S.F. sq. tab. shingles
368	1¾" roofing	980	300 S.F. sq. tab. shingles
336	2" roofing	980	300 S.F. sq. tab. shingles
274	2½" roofing	650	200 S.F. sq. tab. shingles
318	1¾" roofing w/w/attached	1050	1000 S.F. aluminum roofing
285	2" roofing w/w/attached	1050	1000 S.F. aluminum roofing
242	2½" roofing w/w/attached	1050	1000 S.F. aluminum roofing

6 Carpentry

Labor for Rough Carpentry

Work Element	Unit	Man-Hours Per Unit
Mudsill, 2" x 6"		
Bolted	1000 B.F.	21
Shot	1000 B.F.	18
Basement beams (girders)		
2" x 8", built-up	1000 B.F.	33
2" x 10", built-up	1000 B.F.	25
Basement posts	1000 B.F.	18
Box sills	1000 B.F.	29
Floor joists		
2" x 6" to 2" x 8"	1000 B.F.	16
2" x 10" to 2" x 12"	1000 B.F.	14
Headers, tail joists and trimmers		
2" x 6" to 2" x 8"	1000 B.F.	16
2" x 10" to 2" x 12"	1000 B.F.	14
Bridging 2" x 3"	50 sets of 2	4
Subflooring, boards		
Straight	1000 B.F.	13
Diagonal	1000 B.F.	15
Subflooring, plywood, 4' x 8'	1000 S.F.	10
Stud walls, including plates, blocks and bracing		
2" x 4"	1000 B.F.	22
3" x 4"	1000 B.F.	21
2" x 6"	1000 B.F.	20
Ceiling joists		
2" x 6" to 2" x 8"	1000 B.F.	16
2" x 10" to 2" x 12"	1000 B.F.	14
Ceiling backing		
2" x 6" to 2" x 8"	1000 B.F.	15
2" x 10" to 2" x 12"	1000 B.F.	14
Attic floor	1000 B.F.	13
Headers for wall openings		
2" x 4"	1000 B.F.	25
2" x 6"	1000 B.F.	20
Gable-end studs	1000 B.F.	20
Fire stop wall blocks	1000 B.F.	20
Corner braces	1000 B.F.	20

Labor For Rough Carpentry (continued)

Work Element	Unit	Man-Hours Per Unit
Partition plates and shoe	1000 B.F.	20
Partition studs	1000 B.F.	20
Wall backing	1000 B.F.	20
Grounds	1000 L.F.	12
Knee wall plates and studs		
2" x 4"	1000 B.F.	25
2" x 6"	1000 B.F.	25
Wall sheathing, boards		
1" x 6" diag., includes paper	1000 B.F.	15
1" x 8" diag., includes paper	1000 B.F.	14
1" x 10" diag., includes paper	1000 B.F.	13
Wall sheathing, plywood		
4' x 8' sheets, includes paper	1000 S.F.	11
Wall sheathing, composition		
½"	1000 S.F.	9
¾"	1000 S.F.	10
1"	1000 S.F.	11
Siding, plywood, 4' x 8' sheets	1000 S.F.	13
Corner boards	1000 B.F.	40
Common rafters	1000 B.F.	17
Hip, valley, jack rafters	1000 B.F.	29
Roof sheathing, boards		
1" x 6", S4S	1000 B.F.	15
1" x 6", center match	1000 B.F.	18
1" x 8", shiplap	1000 B.F.	17
1" x 10", shiplap	1000 B.F.	13
Roof sheathing, plywood, 4' x 8' sheets	1000 S.F.	12
Window and door headers	Each	.6
Make and install rough door buck	Each	1.2
Furring concrete or masonry walls	1000 L.F.	46
Wood plaster grounds on masonry	1000 L.F.	38
Attic stairways	Each	10
Basement stairways	Each	7

Time includes layout, all precutting, stacking, repairing and clean-up as required.
Suggested Crew: 2 carpenters and 1 laborer

Labor For Commercial Grade Rough Carpentry

Work Element	Unit	Man-Hours Per Unit
Beams		
6" x 8" to 6" x 10"	1000 B.F.	25.0
8" x 12" to 8" x 16"	1000 B.F.	27.0
10" x 12" to 10" x 16"	1000 B.F.	28.0
Columns		
6" x 6"	1000 B.F.	24.0
8" x 8"	1000 B.F.	25.0
10" x 10"	1000 B.F.	26.0
12" x 12"	1000 B.F.	30.0
14" x 14"	1000 B.F.	32.0
Light canopy framing		
2" x 4"	1000 B.F.	21.0
2" x 8"	1000 B.F.	20.0
3" x 8"	1000 B.F.	19.0
Steel joist bridging		
Cross strapping	100	0.7
Compressible type bridging	100	0.9

Time includes layout, all precutting where necessary, repairing and clean-up.

Suggested Crew: 1 carpenter and 1 laborer

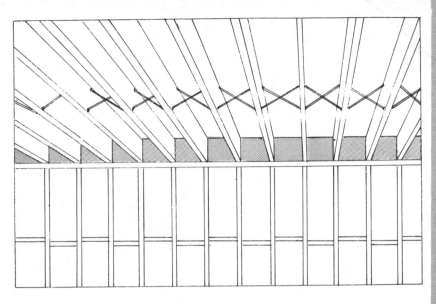

Bracing Steel Joist Bridging

Materials For Floor Joists

Joist Size	Material B.F. Required for 100 S.F. of Floor				Lbs. of Nails Per 1000 B.F.
	12" O.C.	16" O.C.	20" O.C.	24" O.C.	
2 x 6	128	102	88	78	10
2 x 8	171	136	117	103	8
2 x 10	214	171	148	130	6
2 x 12	256	205	177	156	5

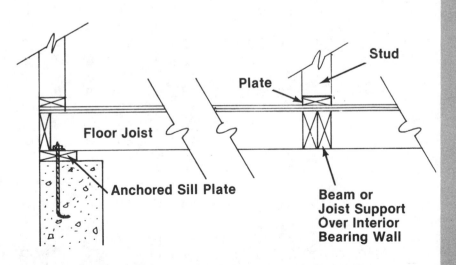

Stud

Plate

Floor Joist

Anchored Sill Plate

Beam or Joist Support Over Interior Bearing Wall

Number of Joists Required

Length of Span	Spacing of Joists									
	12"	16"	20"	24"	30"	36"	42"	48"	54"	60"
6	7	6	5	4	3	3	3	3	2	2
7	8	6	5	5	4	4	3	3	3	2
8	9	7	6	5	4	4	3	3	3	3
9	10	8	6	6	5	4	4	3	3	3
10	11	9	7	6	5	4	4	4	3	3
11	12	9	8	7	5	5	4	4	3	3
12	13	10	8	7	6	5	4	4	4	3
13	14	11	9	8	6	5	5	4	4	4
14	15	12	9	8	7	6	5	5	4	4
15	16	12	10	9	7	6	5	5	4	4
16	17	13	11	9	7	6	6	5	5	4
17	18	14	11	10	8	7	6	5	5	4
18	19	15	12	10	8	7	6	6	5	4
19	20	15	12	11	9	7	6	6	5	5
20	21	16	13	11	9	8	7	6	5	5
21	22	17	14	12	9	8	7	6	6	5
22	23	18	14	12	10	8	7	7	6	5
23	24	18	15	13	10	9	8	7	6	6
24	25	19	15	13	11	9	8	7	6	6
25	26	20	16	14	11	9	8	7	7	6
26	27	21	17	14	11	10	8	8	7	6
27	28	21	17	15	12	10	9	8	7	6
28	29	22	18	15	12	10	9	8	7	7
29	30	23	18	16	13	11	9	8	7	7
30	31	24	19	16	13	11	10	9	8	7
31	32	24	20	17	13	11	10	9	8	7
32	33	25	20	17	14	12	10	9	8	7
33	34	26	21	18	14	12	10	9	8	8
34	35	27	21	18	15	12	11	10	9	8
35	36	27	22	19	15	13	11	10	9	8
36	37	28	23	19	15	13	11	10	9	8
37	38	29	23	20	16	13	12	10	9	8
38	39	30	24	20	16	14	12	11	9	9
39	40	30	24	21	17	14	12	11	10	9
40	41	31	25	21	17	14	12	11	10	9

One joist has been added to each of the above to provide an extra joist at the end of span. Add for doubling joists

Materials For Built-Up Girders

Size of Girders (Inches)	B.F. Required Per L.F.	Nails Required Per 1000 B.F.
4 x 6	2.15	53
4 x 8	2.85	40
4 x 10	3.58	32
4 x 12	4.28	26
6 x 6	3.21	43
6 x 8	4.28	32
6 x 10	5.35	26
6 x 12	6.42	22
8 x 8	5.71	30
8 x 10	7.13	24
8 x 12	8.56	20

Materials For Partition Walls

Size of Studs	Spacing On Center	B.F. Per S.F. of Wall	Lbs. of Nails Per 1000 B.F.
2" x 3"	12"	.91	25
2" x 3"	16"	.83	25
2" x 3"	24"	.76	25
2" x 4"	12"	1.22	19
2" x 4"	16"	1.12	19
2" x 4"	24"	1.02	19
2" x 6"	16"	1.48	16
2" x 6"	24"	1.22	16

Includes top and bottom plate, end studs, blocks backing, framing around openings and normal waste.

Materials For Exterior Walls

Size of Studs	Spacing On Center	B.F. Per S.F. of Wall	Lbs. of Nails Per 1000 B.F.
2" x 3"	12"	.83	30
2" x 3"	16"	.78	30
2" x 3"	20"	.74	30
2" x 3"	24"	.71	30
2" x 4"	12"	1.09	22
2" x 4"	16"	1.05	22
2" x 4"	20"	.98	22
2" x 4"	24"	.94	22
2" x 6"	12"	1.66	15
2" x 6"	16"	1.51	15
2" x 6"	20"	1.44	15
2" x 6"	24"	1.38	15

Includes corner bracing, 3 plates, end studs, blocks, framing around windows and doors and normal waste.

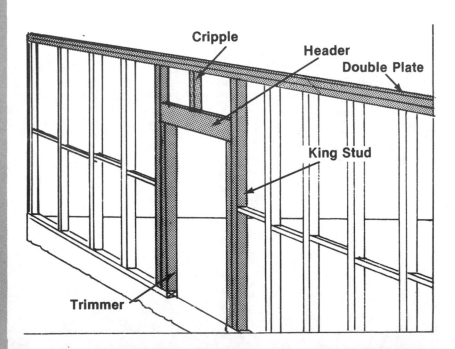

Cripple

Header

Double Plate

King Stud

Trimmer

Materials For Partition Walls

Partition Length in Feet	No. Studs Required	Ceiling Heights in Feet			
		8'-0"	9'-0"	10'-0"	12'-0"
2	3	1.25	1.167	1.13	1.13
3	3	0.833	.812	.80	.80
4	4	0.833	.812	.80	.80
5	5	0.833	.812	.80	.80
6	6	0.833	.812	.80	.80
7	6	0.833	.75	.75	.80
8	7	0.75	.75	.75	.70
9	8	0.75	.75	.75	.70
10	9	0.75	.75	.75	.70
11	9	0.75	.70	.70	.67
12	10	0.75	.70	.70	.67
13	11	0.75	.70	.70	.67
14	12	0.75	.70	.70	.67
15	12	0.70	.70	.70	.67
16	13	0.70	.70	.70	.67
17	14	0.70	.70	.70	.67
18	15	0.70	.70	.67	.67
19	15	0.70	.70	.67	.67
20	16	0.70	.70	.67	.67
For a double plate, add per S.F.		0.13	.11	.10	.083

This table shows the number of feet of lumber required per S.F. of wood stud partition using 2" x 4" studs, and assumes studs are spaced 16" on centers, with single top and bottom plates. For 2" x 8" studs, double the above quantities. For 2" x 6" studs, increase quantities 50%. Example: Find the number of board feet of lumber required for a stud partition 18'-0" long and 9'-0" high. This partition would contain 18 x 9 = 162 S.F. The table gives 0.70 B.F. per S.F. of partition. Multiply 162 by 0.70 equals 113.4 B.F. No waste, blocks, or extra framing members included.

Materials For Ceiling Joists

Joist Size	Material B.F. Required for 100 S.F. of Ceiling				Lbs. of Nails Per 1000 B.F.	B.F. Per Man-Hour
	12" O.C.	16" O.C.	20" O.C.	24" O.C.		
2 x 4	78	59	48	42	19	60
2 x 6	115	88	72	63	13	65
2 x 8	153	117	96	84	9	65
2 x 10	194	147	121	104	7	70
2 x 12	230	176	144	126	6	70

Labor and Materials for Framing and Sheathing

Work Element	Number of B.F. of Framing Per S.F. of House	Man-Hours Per S.F. of House	
		Skilled	Unskilled
Plain gable roof	9.8	.16	.05
Hip roof, no dormers or gables	10.4	.17	.05
Hip roof, with dormers and gables	11.0	.18	.06

Example: To determine the number of man-hours and lumber required to frame a one story rectangular house (including subfloor, wall, and roof sheathing) that has a plain gable roof and an over-all area of 1200 square feet:

For lumber, multiply 1200 S.F. by 9.8 = 11,760 B.F.

For skilled labor, multiply 1200 S.F. by .16 = 192 man-hours.

For unskilled labor, multiply 1200 square feet by .05 = 60 man-hours.

Rafter Lengths by Building Width

Building Width	3" in 12" Rise	4" in 12" Rise	5" in 12" Rise	6" in 12" Rise
10'	5'- 2"	5'- 3"	5'- 5"	5'- 7"
12'	6'- 2"	6'- 4"	6'- 6"	6'- 8"
14'	7'- 3"	7'- 5"	7'- 7"	7'-10"
16'	8'- 3"	8'- 5"	8'- 8"	9'-10"
18'	9'- 3"	9'- 6"	9'- 9"	10'- 1"
20'	10'- 4"	10'- 7"	10'-10"	11'- 2"
22'	11'- 4"	11'- 7"	11'-11"	12'- 4"
24'	12'- 4"	12'- 8"	13'- 0"	13'- 5"
26'	13'- 5"	13'- 8"	14'- 1"	14'- 6"
28'	14'- 5"	14'- 9"	15'- 2"	15'- 8"
30'	15'- 6"	15'-10"	16'- 3"	16'- 9"
32'	16'- 6"	16'-10"	17'- 4"	17'-11"

Table is accurate only to the nearest inch.

Ratio of Common Rafter Length to Run

Rise	Run	Ratio	Rise	Run	Ratio
3	12	1.0308	9	12	1.2500
4	12	1.0541	10	12	1.3017
4.5	12	1.0680	11	12	1.3566
5	12	1.0833	12	12	1.4142
6	12	1.1180	13	12	1.4743
7	12	1.1577	14	12	1.5366
8	12	1.2019	15	12	1.6008

Ratio of Hip or Valley Rafter Length to Run of Common Rafter

Roof Slope			Roof Slope		
Rise	Run	Ratio	Rise	Run	Ratio
3	12	1.4361	9	12	1.6008
4	12	1.4530	10	12	1.6415
4.5	12	1.4631	11	12	1.6853
5	12	1.4743	12	12	1.7321
6	12	1.5000	13	12	1.7815
7	12	1.5298	14	12	1.8333
8	12	1.5635	15	12	1.8875

Materials for Roof Framing

Rafter Size	B.F. per 100 S.F.			Lbs. of Nails Per 1000 B.F.
	12" O.C.	16" O.C.	24" O.C.	
2 x 4	89	71	53	17
2 x 6	129	102	75	12
2 x 8	171	134	112	9
2 x 10	212	167	121	7
2 x 12	252	197	143	6

Includes common rafters, hip and valley rafters, ridge boards, and collar beams.

Labor For Roof Trusses

Span in Feet	Unit	Man-Hours Assembly	Man-Hours Placement
20, placed by hand	Each	2	3
30, placed by hand	Each	4	4
40, placed by crane	Each	10	3
50, placed by crane	Each	16	3
60, placed by crane	Each	19	3
80, placed by crane	Each	25	4

Suggested Crew: Hand placement - 2 carpenters and 2 laborers.
Crane placement - 1 operator, 2 to 3 men on guylines.

Softwood Plywood

Made from thin sheets of veneer peeled from the log, dried, and bonded under heat and pressure with adhesive. The primary species used are Douglas fir and Southern pine. The grain of each ply runs at right angles to adjacent plies, creating an extremely strong, durable panel.

Appearance Grades [Interior]
A-A: For interior applications where both sides will be on view. Built-ins, cabinets and partitions. Face is smooth and suitable for painting.
A-B: For uses similar to Interior A-A but where the appearance of one side is less important and two smooth, solid surfaces are necessary.
A-D: For interior uses where the appearance of only one side is important. Paneling, built-ins, shelving, partitions and racks.
B-B: Interior utility panel used where two smooth sides are desired. Permits circular plugs.
B-D: Interior utility panel for use where one smooth side is required. Good for backing, sides of built-ins.

Appearance Grades [Exterior]
A-A: Use where the appearance of both sides is important. Fences, built-ins, cabinets, commercial refrigerators.
A-B: For use similar to A-A EXT panels but where the appearance of one side is less important.
A-C: Exterior use where the appearance of one side only is important. Sidings, soffits, fences, structural uses.
B-C: An outdoor utility panel for farm service and work buildings.
B-B: An outdoor utility panel with solid paintable faces.

Engineering Grades [Interior]
C-D: Unsanded sheathing grade for wall and roof sheathing, subflooring. Also available with exterior glue (CDX).
Underlayment: For underlayment or combination subfloor-underlayment under resilient floor coverings and carpeting. Used in homes, apartments, mobile homes, commercial buildings. Ply beneath face is C or better veneer. Sanded or touch-sanded is available.
C-D Plugged: For utility built-ins, backing for wall and ceiling tile. Not a substitute for underlayment. Ply beneath face permits D grade veneer.

2.4.1, 1-1/8" Thickness: Combination subfloor-underlayment. Quality base for resilient floor coverings, carpeting, wood strip flooring. Use 2.4.1 with exterior glue in areas subject to excessive moisture. Unsanded or touch-sanded is available.

Engineered Grades [Exterior]
C-C: Unsanded grade with waterproof bond for subflooring and roof decking, siding on service and farm buildings, crating, pallets, pallet bins, cable reels.
Underlayment C-C Plugged: For underlayment or combination subfloor underlayment under resilient floor coverings where excessive moisture conditions may be present such as bathrooms or utility rooms. Sanded or touch-sanded as specified.
C-C Plugged: Also used for tile backing where unusual moisture conditions exist. For refrigerated or controlled atmosphere rooms. Also for pallets, bins, reusable cargo containers, tanks and boxcar and truck floors and linings. Sanded or touch-sanded as specified.
Structural I C-C: For engineered applications in construction and industry where full Exterior-Type panels made with all Group I woods are required. Unsanded.
Plyform Class I and II B-B: Concrete form grades with high reuse factor. Sanded both sides. Edge-sealed. Mill-oiled unless otherwise specified. Special restrictions on species. Also available with a high density paper overlay.
Sturd-I-Floor Ext-APA: For combination subfloor-underlayment under resilient floor coverings where severe moisture conditions may be present, as in balcony decks. Possesses high concentrated- and impact-load resistance during construction and occupancy. Touch-sanded. Available square edge or tongue-and-groove.

Plywood is back-stamped or edge-branded to identify grade, type and other applicable information. On page 153 is an example of a typical grade/trademark back stamp.

Materials for Plywood Roof Sheathing

Recommended Thickness	Maximum O.C. Spacing of Supports at Designed Roof Loads			Nail Size and Type	Nail Spacing	
	20 PSF[2]	30 PSF	40 PSF		Panel Edge	Intermediate
5/16"	20"	20"	20"	6d Common	6"	12"
3/8"	24"	24"	24"	6d Common	6"	12"
1/2"[1]	32"	32"	30"	6d Common	6"	12"
5/8"[1]	42"	42"	39"	8d Common	6"	12"
3/4"[1]	48"	47"	42"	8d Common	6"	12"

[1] Provide blocking or other means of suitable edge support when span exceeds 28" for ½"; 32" for 5/8" and 36" for ¾" plywood.

[2] For the special case of two span continuous beams, plywood spans can be increased 6½% except for roofs in this column.

Materials For Plywood Floor Sheathing

Application	Recommended Thickness	Maximum O.C. Spacing of Supports	Nail Size and Type	Nail Spacing	
				Panel Edges	Intermediate
Subflooring	1/2"[1]	16"[2]	6d common[3]	6"	10"
Subflooring	5/8"[1]	20"	8d common[3]	6"	10"
Subflooring	3/4"[1]	24"	8d common[3]	6"	10"
Subflooring	2.4.1	48"	8d ring shank[3]	6"	6"
Underlayment	3/8"[3]		6d ring shank or cement coated	6"	8" ea. way
Underlayment	5/8"		8d flathead	6"	8" ea. way

[1] Provide blocking at panel edges for carpet, tile, linoleum or other non-structural flooring. No blocking required for 25/32" strip flooring.

[2] If strip flooring is perpendicular to supports, ½" can be used on 24" span.

[3] If resilient flooring is to be applied without underlayment, set nails 1/16".

[3] FHA accepts ¼" plywood.

If supports are not well seasoned, use ring shank nails.

Materials For Plywood Wall Sheathing

Panel Identification Index	Minimum Thickness (Inches)	Maximum Stud Spacing (Inches)	
		Exterior covering nailed to:	
		Stud	Sheathing
12/0, 16/0, 20/0,	5/16	16	16*
16/0, 20/0, 24/0	3/8	24	{16 24*
24/0, 32/16	1/2	24	24

When plywood sheathing is used, building paper and diagonal wall bracing can be omitted.

*When sidings such as shingles are nailed only to the plywood sheathing, apply plywood with face grain across studs.

Installing Softwood Plywood

Sheathing Commonly, lumber framing is sheathed with plywood for roofs, walls and floor.

Joint Treatment For subfloor, roof and wall sheathing, a space of ⅛" should be left between panel edges and ¹⁄₁₆" between ends to allow for expansion. Exterior siding joints may be sealed with building paper or caulking. Edges may be shiplapped, or horizontal butt joints may be sealed with Z-flashing. Battens may be used with vertically applied sidings to double as a decorative finish. In all cases, exterior joints should be backed with solid lumber framing.

Roof Sheathing Plywood provides a light roof deck with no wind, dust or snow infiltration and high resistance to cracking. Roof systems include plywood roof sheathing under conventional shingles, preframed and folded plate systems. Plyclips, which are special aluminum H-clips, may be used as a substitute for lumber blocking in roof construction.

Soffit Construction Plywood is often used as a finishing material for soffits.

Floor Construction With conventional floor construction, plywood underlayment covers subflooring applied to joists.

Coverage of Square Edge Boards

Measured Size Inches	Finished Width, Inches	Add For Shrinkage, Percent	Quantity Required, Multiply Area by	B.F. of Lumber Required Per 100 S.F.
1 x 3	2½	25	1.25	125
1 x 4	3½	20	1.20	120
1 x 6	5½	14	1.14	114
1 x 8	7½	12	1.12	112
1 x 10	9½	10	1.10	110
1 x 12	11½	9½	1.095	109½
2 x 4	3½	20	2.40	240
2 x 6	5½	14	2.28	228
2 x 8	7½	12	2.25	225
2 x 10	9½	10	2.20	220
2 x 12	11½	9½	2.19	219
3 x 6	5½	14	3.43	343
3 x 8	7½	12	3.375	337½
3 x 10	9½	10	3.30	330
3 x 12	11½	9½	3.29	329

This data is based on lumber surfaced one or two sides and one edge. The waste allowance shown includes width lost in dressing plus 5% waste in end-cutting. If laid diagonally, add 5% additional waste.

Typical Back-Stamp

Grade of veneer on panel face

Grade of veneer on panel back

A-C

Grading Agency

Species Group Number Designates the type of plywood

GROUP 2 EXTERIOR PS 1-74 000

APA®

Product Standard governing manufacture

Mill Number

Framing Spacing For Plywood, Fiberboard and Paneling

	Thickness		
Framing Spaced	Plywood	Fiberboard	Paneling
16"	1/4"	1/2"	3/8"
20"	3/8"	3/4"	1/2"
24"	3/8"	3/4"	5/6"

Particleboard

Particleboard is an engineered product composed of wood particles combined with resin binders and hot pressed into panels.

Types:
Type I — Urea bonded for interior application
Type II — Phenolic binder for both interior and exterior applications.

Installing Particleboard Floor Underlayment
Store panels flat in a dry place away from moisture. Do not bring underlayment panels to the job site before they are needed. Do not install before plaster, concrete, and lumber are dried to the approximate conditions that will exist in the structure during occupancy. Underlayment should be installed just before the floor covering and after other interior finishing work is completed.
Plywood subfloors must be at least 5/8'' thick with a minimum of 32/16 panel identification index when particleboard is to be nailed or stapled. If glue nailing is used, plywood subfloors should be at least ½'' thick with a minimum of 32/16 panel identification index.
The plywood face grain should be perpendicular to the joist system.
Board subfloors must be at least 1'' nominal thickness and not over 8'' wide. A vapor barrier with a maximum rating of 1.0 perm should be used over board subfloors (except when underlayment is glue nailed) and as a ground cover in all basementless spaces.
Areas over furnaces should be insulated and ventilated and hot air ducts should be insulated to prevent localized drying and shrinking of floor components.
With plywood subfloors, offset underlayment panel joints and plywood panel joints that are at right angles to the joists at least two inches. Offset underlayment panel joints and plywood panel joints that are parallel to the joists at least one joist. When ¼'' or 5/16'' particleboard underlayment is used, the floor thickness (subfloor plus underlayment) must not be less than one inch.
When nailing, use ring-grooved underlayment nails. Start nailing in the center of each panel and work toward edges. Drive nails perpendicular to the surface and set flush. Drive nails no closer than ½'' or further than ¾'' from the panel edges. Nail each panel completely before starting the next.

Glue nailing results in a superior floor system. Make sure the subfloor is free of all dust, dirt and debris.
Fill gouges, gaps and any chipped edges with a hardsetting patch compound made for this purpose. Allow patches to dry thoroughly, then sand flush with a wide belt sander. Sand any uneven joints between panels.
Cover particleboard underlayment with carpeting, resilient floorings, or seamless floor coverings.

Fiberboard Sheathing

Fiberboard sheathing is a low density pressed board made of wood fibers. The fibers are impregnated with asphalt for strength and moisture resistance.
Regular density (½'' thick) Used as insulating sheathing on wood-framed wall construction. Plywood or "let-in" corner bracing is needed.
Regular density ($^{25}\!/_{32}$'' thick) Use is the same as regular density except that codes approve use without corner bracing.
Intermediate density (½'' thick) Use is the same as regular density, but the extra density is high enough to pass code requirements for racking strength. No corner bracing is needed.
Nail base is a higher density product for use as a nail base for direct application of shingles as well as other types of siding.
Sound deadening board is not impregnated. Use only in wall constructions designed to resist sound transmission.
Thickness: ½''. (Regular Density also available in $^{25}\!/_{32}$'' thickness.)
Width: 2' and 4'. Length: 8' and 9'.

"R" Factors For Fiberboard Sheathing

Fiberboard	"R" Factor
½'' regular density	1.32
$^{25}\!/_{32}$'' regular density	2.06
½'' intermediate density	1.22
½'' nail base	1.14
½'' sound deadening board	1.39
½'' plywood (comparison)	.68

Finish Carpentry

Wood exterior finish may be of plywood sheet, wood shingles, wood siding boards, bevel siding or many other materials. Even if other materials are used for wall and roof covering, wood is ordinarily used for the trim of the building. Trim is available in standard moulding patterns throughout the United States from retail lumber dealers, millwork jobbers or building material distributors.

The choice of wood for exterior finish must consider paint-holding quality, resistance to decay, and the grade of lumber. The cost and availability of finish lumber will also affect selection.

Common woods for outside finish are Douglas fir, redwood, western cedar, and white pine. Medium-density overlaid plywood siding is very common. This material provides a tough, durable painting surface.

Wood shingle roofing is available in pre-assembled 8-foot panels combining ½" exterior plywood sheathing, 30-pound saturated felt paper and hand split cedar shakes in a single pre-assembled, ready-to-install unit. The best woods for shingles are western red cedar, redwood and cypress. Shingles are graded by thickness and length. The length is given in inches and the thickness in the number of butts which, when placed together, measure 2 inches. Best grade shingles are vertical-grain with annual rings appearing on the surface instead of straight lines.

Most manufactured wood materials used for exterior finish carry the manufacturer's recommendations for correct installation. The manufacturer probably also recommends the type of finish, whether exterior or interior. The specifications may cover such items as grades, moisture content, size and pattern, surface texture, and grain.

The back of paneling should be sealed with a resin sealer or a coat of the same material to be used for face finish to prevent moisture damage. Prepare the wall for paneling by installing blocking or furring strips.

Apply sheet vapor barrier or insulation to the inner face of the stud wall or furred masonry wall on perimeter walls. Take care to avoid damaging the vapor barrier once it has been installed. Paneling may be applied with nails or adhesive to walls. Nails are normally used for ceiling installation.

Prefinished wood moldings are usually used to finish paneling installations. They trim door and window openings to complement paneled walls, cover seams and joints at ceilings, floors, corners and other areas and protect paneling from kicks and bumps. The casings and stops of the doors and windows, and stools and aprons are sometimes delivered to a job cut to rough lengths. This saves time to assort, select, and place the various members at each opening. Group the pieces so the correct lengths can be located easily when trimming an opening. If the materials for these members come in random lengths, cut them to the rough lengths and then sort them by size. This practice will help tradesmen find the proper lengths for the various openings. All base and moldings come in random lengths.

Labor For Finish Carpentry

Work Element	Unit	Man-Hours Per Unit
Exterior trim		
Corner boards, verge, fascia	100 L.F.	4
Cornice, 3 member	100 L.F.	12
Porch post, plain	Each	1.0
Porch post, built-up	Each	2.0
Clothes closets		
One shelf, hookstrip, hook and pole	Each	2.0
Open shelving and cleats	Each	0.5
Linen closet, shelving and cleats	Each	3.0
Baseboard		
Two member, ordinary work	100 L.F.	4 to 6
Two member, hardwood, first class difficult work	100 L.F.	6 to 8
Three member, ordinary work	100 L.F.	5 to 7
Three member, hardwood, first class or difficult work	100 L.F.	7 to 9
Picture molding		
Ordinary work	100 L.F.	2.7 to 3
Hardwood, first class or difficult work	100 L.F.	4 to 5
Chair rail		
Ordinary work	100 L.F.	2.5 to 3
Hardwood, first class or difficult work	100 L.F.	3 to 4
Plate rail		
Two member, ordinary work	100 L.F.	10 to 12
Two member, first class work	100 L.F.	12 to 15
Interior cornice		
Ordinary work	100 L.F.	10 to 16
First class or difficult work	100 L.F.	18 to 22

Time includes layout, all pre-cutting, stacking, repair and clean-up.
Suggested Crew: 1 carpenter and 1 laborer

4' x 8' Panel Requirements For a Room

Perimeter	Number of 4' x 8' Panels Needed
36'	9
40'	10
44'	11
48'	12
52'	13
56'	14
60'	15
64'	16
68'	17
72'	18
92'	23

First, find the perimeter, the length of all walls in the room. For example, if room walls measure 14' + 14' + 16' + 16', this would equal 60' and require 16 panels. To allow for areas such as windows, doors, fireplaces, etc., use the following deductions:

Door	1/2 panel (A)
Window	1/4 panel (B)
Fireplace	1/2 panel (C)

Thus, the actual number of panels for this room would be 13 pieces (15 pieces minus two total deductions). If the perimeter of the room falls between the figures in the table, use the next highest number to determine panels required. These figures are for rooms with 8' ceiling heights or less.

Labor Installing Wall Paneling

Work Element	Unit	Man-Hours Per Unit
Plastic faced hardboard, including moulding and trim		
1/8''	100 S.F.	2.8
1/4''	100 S.F.	2.9
Plywood, 4' x 8' panels, including trim		
¼''	100 S.F.	3.5
½''	100 S.F.	4.4
Plank paneling		
¼''	100 S.F.	3.9
¾''	100 S.F.	5.0
¾'', random width	100 S.F.	5.4
Cedar closet lining		
1'' x 4'' plank	100 S.F.	5.9
¼'' plywood	100 S.F.	4.5

Allow about 25% more time for ceiling installation. Deduct 5 to 15% when 9' or 12' high plywood panels can be used. If installtion is on metal studs, add 10% to the times listed.

Material Required For Furring

Size of Strips	O.C. Spacing of Furring	B.F. Per S.F. of Wall	Lbs. Nails Per 1000 B.F.
1'' x 2''	12''	.18	55
	16''	.14	
	20''	.11	
	24''	.10	
1'' x 3''	12''	.28	37
	16''	.21	
	20''	.17	
	24''	.14	
1'' x 4''	12''	.36	30
	16''	.28	
	20''	.22	
	24''	.20	

Hardboard Siding

Hardboard siding is made by bonding wood fibers under heat and pressure to form a uniformly dense material that resists abuse. Both unfinished and prefinished styles are available in smooth or textured surfaces.

Storage Siding must be protected from the elements prior to installation and priming. If kept outside, it must be stored off the ground on a flat surface and protected on top and sides with a waterproof cover. Adequate ventilation is necessary. Siding must be protected from dirt, grease and other foreign substances. Handle siding carefully. During storage it is important to align stickers directly when loads are stacked to prevent warpage.

Nailing Use only corrosion-resistant nails. Nails should penetrate framing members a minimum of 1½''. Nails with a head diameter of at least 3/16'' must be used when siding is applied directly to studs. Lap siding requires 2½'' (8d) nails when applied directly to studs and 3'' (10d) nails when applied over sheathing. Panel siding requires 6d nails when applied over studs and 8d nails when applied over sheathing.

Vapor barriers must be properly installed on the warm side of the wall in all insulated or heated buildings. A continuous barrier (1 perm or less) such as polyethylene film or foil-backed gypsum is recommended. A ground cover of polyethylene film must be used in all crawl space areas

Ground clearance Allow a minimum of 8'' between the siding and the ground or any area where water may collect.

Factory primed hardboard siding must be painted within 30 days of installation. If exposed for a longer period of time, reprime the surface with a good-quality exterior grade primer compatible with the finishing system.

Unprimed hardboard siding must be finished within 30 days of installation. If a paint system is to be used, the siding must be field primed with a good quality exterior primer.

Painting hardboard siding Use good quality exterior oil or water base paints. The number of coats of finish paint is dictated by the type of paint, method of application, and performance desired. Follow the paint manufacturer's recommendations concerning film thickness, requirements for use of special primers or undercoats, rate of spread, and application procedures. **Do not use flat alkyd type house paints.** All cut edges must be primed and painted.

Caution As with all quality wood products, hardboard siding should never be applied to a structure having excess moisture conditions such as drying concrete or plaster or wet sheathing materials. If such conditions exist, the building should be well ventilated to allow drying before application of siding.

Labor Installing Siding

Work Element	Unit	Man-Hours Per Unit
Bevel siding		
½'' x 6'', 3' to 7' long	1000 S.F.	19.0
½'' x 6'', 6' to 18' long	1000 S.F.	17.0
½'' x 8'', 3' to 7' long	1000 S.F.	17.0
¾'' x 8'', 3' to 7' long	1000 S.F.	18.0
¾'' x 10'', 6' to 18' long	1000 S.F.	18.5
Tongue and groove siding		
1'' x 4''	1000 S.F.	21.5
1'' x 6''	1000 S.F.	20.5
Board and batten siding, 1'' x 12''	1000 S.F.	29.0
Coved channel siding, 1'' x 8''	1000 S.F.	24.5
Plywood siding, 4' x 8' panels		
3/8''	1000 S.F.	14.0
5/8''	1000 S.F.	16.0
Trim pieces		
Edging, 1'' x 3''	10 L.F.	0.4
Corner, 1'' x 3'' x 3''	10 L.F.	0.5
Base trim, 1'' x 4''	10 L.F.	0.35

No waste or allowance for coverage is included.
Suggested Crew: 1 carpenter and 1 laborer

Coverage of T&G and Shiplap Boards

Measured Size, Inches	Finished Width, Inches	Add For Shrinkage, Percent	Quantity Required, Multiply Area by	S. F. of Lumber Required, Per 100 S.F.
1 x 2	1-3/8	50	1.50	150
1 x 2¾	2	42½	1.425	142½
1 x 3	2½	38-1/3	1.383	138
1 x 4	3¼	28	1.28	128
1 x 6	5¼	20	1.20	120
1 x 8	7¼	16	1.15	115
1¼ x 3	2¼	38-1/3	1.73	173
1¼ x 4	3¼	28	1.60	160
1¼ x 6	5¼	20	1.50	150
1½ x 3	2¼	38-1/3	2.08	208
1½ x 4	3¼	28	1.92	192
1½ x 6	5¼	20	1.80	180
2 x 4	3¼	28	2.60	260
2 x 6	5¼	20	2.40	240
2 x 8	7¼	16	2.32	232
2 x 10	9¼	13	2.25	225
2 x 12	11¼	12	2.24	224
3 x 6	5¼	20	3.60	360
3 x 8	7¼	16	3.48	348
3 x 10	9¼	13	3.39	339
3 x 12	11¼	12	3.36	336

This data applies to most dressed and matched lumber. Waste allowance shown includes width lost in dressing and lapping. Add 5% for end-cutting and matching.

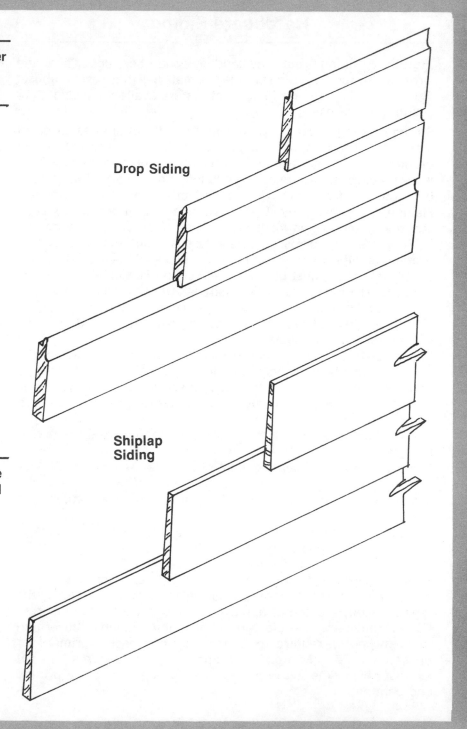

Drop Siding

Shiplap Siding

Labor and Materials For Bevel Siding

Size	Material			Lbs. Nails Per 100 S.F.	Man-Hours Per 100 S.F.
	Exposed to Weatner	Add for Lap	S.F. of Siding per 100 S.F. of Wall		
1/2″ x 4″	2¾	46%	151	1½	3.2
1/2″ x 5″	3¾	33%	138	1½	2.5
1/2″ x 6″	4¾	26%	131	1	1.9
1/2″ x 8″	6¾	18%	123	¾	1.7
5/8″ x 8″	6¾	18%	123	¾	1.8
3/4″ x 8″	6¾	18%	123	¾	1.8
5/8″ x 10″	8¾	14%	119	½	2.1
3/4″ x 10″	8¾	14%	119	½	2.2
3/4″ x 12″	10¾	12%	117	½	2.1

Quantities include 5% for end cutting and waste. Deduct for all openings over ten square feet.

Labor and Materials For Drop Siding

Size	Material			Lbs. Nails Per 100 S.F.	Man-Hours Per 100 S.F.
	Exposed to Weather	Add for Lap	S.F. of Siding Per 100 S.F. of Wall		
1″ x 6″	5¼″	14%	119	2½	2.0
1″ x 8″	7¼″	10%	115	2	1.9

Quantities include 5% for end cutting and waste. Deduct for all openings over ten square feet.

Board Siding Application

Some wood siding patterns are used only horizontally and others only vertically. Some may be used in either manner if adequate nailing areas are provided.

Plain bevel siding can be obtained in sizes from ½ by 4 inches to ½ by 8 inches, and also in sizes of ¾ by 8 inches and ¾ by 10 inches. "Anzac" siding is ¾ by 12 inches. Usually the finished width of bevel siding is about ½ inch less than the size listed. One side of bevel siding has a smooth planed surface, while the other has a rough resawn surface. For a stained finish, the rough or sawn side is exposed because wood stain is most successful and longer lasting on rough wood surfaces.

Dolly Varden siding is similar to true bevel siding except that shiplap edges are used, resulting in a constant exposure distance. Because it lies flat against the studs, it is sometimes used for garages and similar buildings without sheathing. Diagonal bracing is then needed to provide racking resistance to the wall.

Regular **drop sidings** can be obtained in several patterns. This siding, with matched or shiplap edges, can be obtained in 1 by 6 and 1 by 8 inch sizes. This type is commonly used for lower cost dwellings and for garages, usually without benefit of sheathing. Tests have shown that the tongued-and-grooved (matched) patterns have greater resistance to the penetration of wind-driven rain than the shiplap patterns, when both are treated with a water-repellent preservative.

A number of siding or paneling patterns can be used both horizontally and vertically. These are manufactured in nominal 1 inch thicknesses and in widths from 4 to 12 inches. Both dressed and matched and shiplapped edges are available. The narrow and medium-width patterns will likely be more satisfactory when there are moderate moisture content changes. Wide patterns are more successful if they are vertical grain to keep shrinkage to a minimum. The correct moisture content is also important when tongued-and-grooved siding is wide, to prevent shrinkage to a point where the tongue is exposed.

Treating the edges of both drop and the matched and shiplapped sidings with water-repellent preservative usually prevents wind-driven rain from penetrating the joints if exposed to weather. In areas under wide overhangs, or in porches or other protected sections, this treatment is not as important. Some manufacturers provide siding with this treatment applied at the factory.

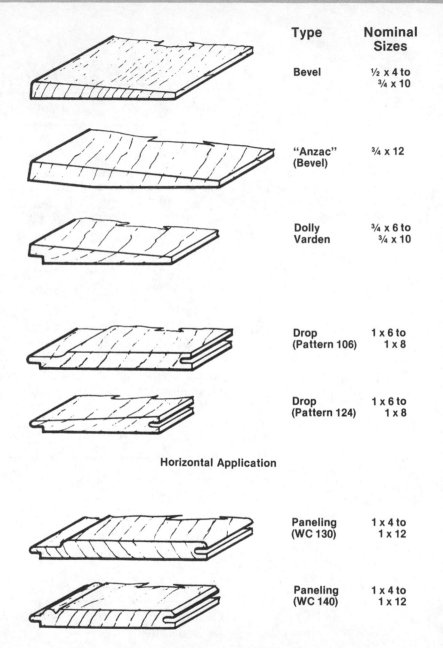

Type	Nominal Sizes
Bevel	½ x 4 to ¾ x 10
"Anzac" (Bevel)	¾ x 12
Dolly Varden	¾ x 6 to ¾ x 10
Drop (Pattern 106)	1 x 6 to 1 x 8
Drop (Pattern 124)	1 x 6 to 1 x 8

Horizontal Application

Type	Nominal Sizes
Paneling (WC 130)	1 x 4 to 1 x 12
Paneling (WC 140)	1 x 4 to 1 x 12

Horizontal or Vertical Application

Board Siding Types

Labor Installing Fireplace Mantels

Work Element	Unit	Man-Hours Per Unit
Prefabricated milled decorative unit, 42" high, 6' wide	Each	3.2
Bracket mounted 10" wide, 3" thick hardwood beam		
6' long	Each	1.7
8' long	Each	2.0
Rough sawn oak or pine beam		
4" x 8"	L.F.	0.25
4" x 10"	L.F.	0.27
4" x 12"	L.F.	0.31

Time includes layout, cutting, drilling and placing of shields where required, repairs and clean-up.
Suggested Crew: 1 carpenter

Labor Installing Counter Tops

Work Element	Unit	Man-Hours Per Unit
Setting flat tops	10 S.F.	1.0
Setting factory formed, self-edge tops with backsplash	10 S.F.	2.4
Setting backsplash only	10 L.F.	1.0

Use these figures for setting factory made, laminated plastic counter tops on base cabinets. Larger "L" and "U" shape tops will require more time if access to the point of installation is a problem. **Suggested Crew:** 1 carpenter and 1 helper for tops under 25 S.F.

Stair Construction

Stair Construction
Stairwork is considered a special field of carpentry. More intricate and decorative stairs are built in a stair shop which may be a part of a mill. Stairs usually built by carpenters on the job include porch, basement, and any stairs on the outside of buildings. The straight flight stair is simplest to build but not necessarily the easiest to use or best of all applications. A landing is needed somewhere in most flights near the halfway point to reduce climbing fatigue and improve safety. The length of all treads and all risers should be equal in any one flight. The sum of tread and one riser, exclusive of the nosing, should not be more than 18 inches nor less than 17 inches. The nosing should not exceed 1¾ inches. The terms **stringers, strings, horses,** and **carriages** are applied to the supporting members of the stairs and mean the same thing.

Stair treads are usually oak or birch 1⅛ to 1¼ inches thick. It is important to provide headroom of 7 feet from every tread on the stairway.

Labor Erecting Stairs

Work Element	Unit	Man-Hours Per Unit
Erecting stairwork, hours per 9' rise		
Building ordinary plain box stairs on the job	Each	8 to 16
Rails, balusters and newel post for above	Each	4 to 8
Erecting plain flight of stairs built-up in shop	Each	6 to 8
Erecting two short flights	Each	10 to 12
Erecting open stairs	Each	10 to 12
Erecting open stairs with two flights	Each	12 to 16
Newels, balusters and hand rail for the above	Each	6 to 8
Erecting prefabricated wood stairs, hours per 9' rise		
Circular, 6' diameter, oak	Each	23.0
Circular, 9' diameter, oak	Each	31.0
Straight, 3' wide, assembled	Each	3.0
Straight, 4' wide, assembled	Each	3.2

Dimensions For Straight Stairs

Height Floor to Floor H	Number of Risers	Height of Risers R	Width of Treads T	Total Run L	Minimum Head Room Y	Well Opening U
8'-0''	12	8''	9''	8'- 3''	6'-6''	8'- 1''
8'-0''	13	7-3/8''+	9½''	9'- 6''	6'-6''	9'- 2½''
8'-0''	13	7-3/8''+	10''	10'- 0''	6'-6''	9'- 8½''
8'-6''	13	7-7/8''−	9''	9'- 0''	6'-6''	8'- 3''
8'-6''	14	7-5/16''−	9½''	10'- 3½''	6'-6''	9'- 4''
8'-6''	14	7-5/16''−	10''	10'-10''	6'-6''	9'-10''
9'-0''	14	7-11/16''+	9''	9'- 9''	6'-6''	8'- 5''
9'-0''	15	7-3/16''+	9½''	11'- 1''	6'-6''	9'- 6½''
9'-0''	15	7-3/16''+	10''	11'- 8''	6'-6''	9'-11½''
9'-6''	15	7-5/8''−	9''	10'- 6''	6'-6''	8'- 6½''
9'-6''	16	7-1/8''	9½''	11'-10½''	6'-6''	9'- 7''
9'-6''	16	7-1/8''	10''	12'- 6''	6'-6''	10'- 1''

Dimensions shown under well opening "U" are based on 6'-6" minimum headroom. If headroom is increased well opening also increases.

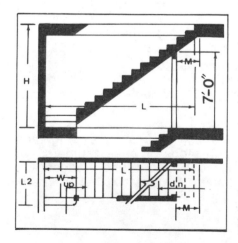

Dimensions For Stairs With Landings

Height Floor to Floor H	Number of Risers	Height of Risers R	Width of Tread T	Run		Run	
				Number of Risers	L	Number of Risers	L2
8'-0''	13	7-3/8''+	10''	11	8'-4''+W	2	0'-10''+W
8'-6''	14	7-5/16''−	10''	12	9'-2''+W	2	0'-10''+W
9'-0''	15	7-3/16''+	10''	13	10'-0''+W	2	0'-10''+W
9'-6''	16	7-1/8''	10''	14	10'-10''+W	2	0'-10''+W

Stairs with landings are safer and reduce the required stair space. The landing provides a resting point and a logical place for a right angle turn.

7

thermal and moisture protection section

section contents

Membrane Waterproofing

Membrane waterproofing is used to protect interior surfaces from water seepage from the soil. Membranes are applied to below-grade concrete slabs and below-grade concrete and masonry walls. Membranes under concrete slabs or slightly below grade on sites with good drainage may be referred to as "moisture barriers" rather than waterproof membranes. The materials and construction are the same.

Membrane waterproofing is also used to protect building interiors from overhead moisture from natural sources or plumbing fixtures. These membranes are generally concealed under concrete, terrazzo, ceramic tile, clay pavers, or soil where underground structures are involved.

Several materials are used as waterproof membranes. These include liquid applied and sheet elastomerics plastic films, and built-up membranes of alternating coatings of hot or cold applied bitumens (asphalt) and saturated felts. Other materials include sprayed-on glass fiber reinforced bituminous material, sheet metal and composites. Composites are metal foils or plastic films laminated to fabrics, paper, or felts of various kinds. Waterproof membranes are always protected from sunlight and oxidation by toppings or backfill.

Built-up membranes are made up of hot moppings of asphalt or coal tar pitch and felts or cold applied cutback of emulsified asphalt and felts. Felts are organic and inorganic fibers saturated or coated with asphalt or coal tar pitch. Organic felts are generally called "rag" felts although they may contain little or no rag fibers. Inorganic felts are made from asbestos and glass fibers. Thicknesses of felts are specified in terms of weight per 100 square feet (one square). The saturant or coating on the felt should always be of the same material as material in which it is embedded. Cutback or emulsified bitumens are applied over a thinned prime coat.

Fibered asphalt membranes are made of cutback or emulsified asphalts in liquid, semi-liquid, or paste form. Mineral fibers are either mixed with the asphalt at the factory, or are run through a separate hose and mixed with liquified asphalt after it leaves the spray nozzle during application. Factory mixed fibered asphalt is generally trowel applied. Both trowel and spray applied fibered asphalts are generally applied over a brushed on thinned asphaltic prime coat. They may also be reinforced with glass fibric embedded in an asphaltic undercoat prior to the application of the fibered asphalt membrane coat.

Elastomeric membrane systems may be divided into two groups: Liquid applied elastomerics and cured preformed elastomerics.

Liquid applied elastomerics include the synthetic rubbers: chloroprene (neoprene), polyurethane, and polysulfide (Thiokol). Other synthetic rubbers are suitable for membrane waterproofing but are seldom used. Their cost does not generally justify their use in the exposure conditions to which they will be subjected.

Liquid applied elastomerics may be modified by blending with other materials. Some are one-part and some are two or more part mixes.

Some require primers for proper bonds to some substrates.

Cured, preformed sheet elastomerics include Butyl (isbutylene -isoprene), chloroprene (neoprene), and ethylene propylene (EPDM).

Composites are made of two or more materials laminated together to take advantage of lower cost or improved properties or both. Composites generally include metals and plastic films as the waterproofing agent and paper, fiber, or felts as the strengthening or protecting agent. One composite employs a rubberized asphalt layer laminated to a plastic face. The rubberized asphalt acts as an adhesive as well as a sealer. The soft rubberized asphalt is self healing against small punctures. Although not considered a composite, some metals are plated, coated, or clad with other metals to achieve similar results.

Application

The surface on which membranes are applied should be clean, dry, and reasonably smooth. Projections and depressions may rupture the membrane by puncturing or stretching it beyond its elastic limits when subjected to the pressure of backfill or toppings.

Heads of nails or other fasteners used to secure the membrane should be sealed by patching with a piece of membrane held in place by adhesive-sealant or other means. Some plastic sheets are made with fairly closely spaced projections that form knobs or hooks. When used as a form liner for concrete, these projections engage in the concrete as it is poured, holding it in place. Tie rod holes, of course, must be grouted and sealed. Membranes on walls are generally not fastened or otherwise secured. They are held in place by whatever topping or fill goes over them. They are usually held in place at the edges by foundation walls, room walls, partitions, or curbs.

Membranes on which reinforced steel and concrete slabs are to be placed are easily punctured by foot traffic, tools, or reinforcing steel supports. Slab membrane should be patched just before placing concrete.

Wrinkles, puckers, entrapped air, and similar defects are the common causes of failure. Sheet material, especially the elastromerics, should not be stretched as it is being cemented in place. Watch for foaming when hot asphalt is brought into contact with concrete or masonry. This is an indication that moisture content in the concrete or masonry too high. The first course of membrane should be held in place with spots of adhesive or embedded in a solid coating of adhesive, and should be nailed along the top edge. The manufacturer probably recommends certain types of nails and adhesives. Spacing, location, and sizes are all important to the durability of the membrane.

Liquid, semi-liquid, and paste consistency membranes may require primers on some surfaces. Check manufacturer's recommendations. These membranes are applied in one or more coats by brush, spray, roller, or trowel. Membranes that have little or no elasticity when cured are generally reinforced with woven, felted, or chopped fibers. It is difficult to obtain a uniform coating thickness with this type of membrane.

Working cracks and expansion joints require special attention. Even elastomeric membranes may be stretched beyond their elastic limit if the membrane is adhered to the surface directly adjacent to both sides of the crack or joint. The manufacturer's printed recommendations should be followed.

Both horizontal and vertical membranes are vulnerable to damage if there is movement in the surface due to thermal changes, settlement, or other causes. Preformed elastomeric sheet membranes that are spot adhered rather than solidly cemented in place are least susceptible to this type of damage. Backfilling and compacting operations, even against membranes with protective coverings, should be done with care to prevent displacement of the protective covering or damage to the membrane.

Sealing around anything that projects through a membrane is critical. It is especially difficult where piping penetrates a membrane below a floor slab because the seal may be broken by construction operations up until the time the slab is poured. Even if a break is detected and repaired before the concrete is poured, it may be re-broken during concrete placement. Once the slab is poured, the break may not be detected for several months. By then repair has become quite expensive.

Labor For Membrane Waterproofing

Work Element	Unit	Man-Hours Per Unit
Asphalt felt (15 lb. mopped on)[1]		
One ply	100 S.F.	2.6
Two ply	100 S.F.	3.5
Three ply	100 S.F.	4.0
Four ply	100 S.F.	4.2
Five ply	100 S.F.	4.5
Fibrous asphalt felt (mopped on)		
One ply	100 S.F.	2.7
Two ply	100 S.F.	3.4
Three ply	100 S.F.	3.9
Four ply	100 S.F.	4.4
Five ply	100 S.F.	4.6
Rubberized membrane sheet		
40 gauge	100 S.F.	3.9
30 gauge	100 S.F.	3.5
22 gauge	100 S.F.	3.2
Plastic vapor barrier		
.004 thick, with plastic mastic	100 S.F.	.314
.006 thick, with plastic mastic	100 S.F.	.315
.010 thick, with plastic mastic	100 S.F.	.393

Time includes move-on and off-site, set-up and job preparation.
Suggested Crew: 2 roofers. **Note:** All time is for vertical surfaces only. No joint taping included.
[1] Add 10% for 30 pound felt.

Labor Applying Liquid Plastic Waterproofing

Work Element	Unit	Man-Hours Per Unit
Metallic oxide waterproofing Iron compound, troweled flat		
⅝" thick	100 S.F.	2.5
¾" thick	100 S.F.	3.0
1" thick	100 S.F.	3.6

Time includes move-on and off-site, set-up and job preparation.
Suggested Crew: 2 roofers.
Note: For vertical surface application, add 100 percent to time. If surface preparation and cleaning are required, add .4 man-hours per 100 square feet.

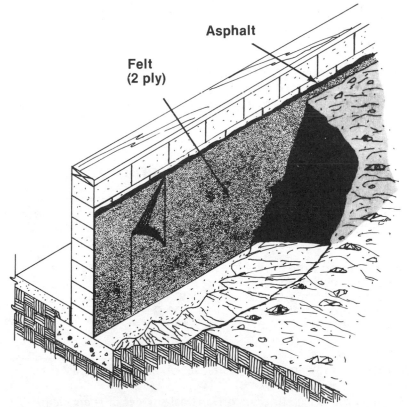

Backfill Against A Building Wall

Labor Applying Waterproofing Panels

Work Element	Unit	Man-Hours Per Unit
Asphalt coated protective board		
¼" thick	100 S.F.	2.5
½" thick	100 S.F.	2.6
¾" thick	100 S.F.	2.7
1" thick	100 S.F.	2.9
Cement asbestos protective board		
⅜" thick	100 S.F.	3.9
½" thick	100 S.F.	4.3
Laminated plastic panels		
Standard types	100 S.F.	3.4
Bentonite		
Panels ³⁄₁₆" thick	100 S.F.	2.0
Panels ¼" thick	100 S.F.	2.2
Granular admixtures — trowel on ⅜" thick	100 S.F.	2.8

Time includes move on and off, set-up and job preparation.
Suggested Crew: 2 roofers
Note: All time is for vertical surfaces only.

Labor For Silicone Waterproofing

Work Element	Unit	Man-Hours Per Unit
1 coat on concrete	100 S.F.	2.0
2 coats on concrete	100 S.F.	2.4
1 coat on concrete block	100 S.F.	2.0
2 coats on concrete block	100 S.F.	2.6
1 coat on brick	100 S.F.	2.0
2 coats on brick	100 S.F.	2.4

Time includes move-on and off-site, set-up and job preparation.
Suggested Crew: 2 roofers
Note: All time is for spraying on vertical surfaces. If application is by brush, add 0.7 man-hours per 100 S.F.

Bituminous Dampproofing

Natural tar and pitch seepage and resinous materials have long been used as a waterproofing material. Materials commonly used for dampproofing are asphaltic materials made from petroleum and coal tar pitch or creosote from bituminous coal. Both asphalt and coal tar pitch are applied over a penetrating liquid primer. Creosote is the primer for coal tar pitch and cut-back asphalt asphalts. Asphalts are available in liquid, semi-liquid or paste form at room temperature. Asphalt and coal tar pitch are also available in a solid state at room temperature and must be liquified by heating before application.

Dampproofing should not be confused with waterproofing. The purpose of dampproofing is to significantly reduce entry of water by capillary action resulting from occasional exposure to moisture. Waterproofing provides protection like the bottom of a boat does. Dampproofing can be applied to the outside of below grade exterior walls where good drainage and low soil moisture exists, and to the inside of exterior walls that are furred. Dampproofing is not meant to protect against intrusion of water under pressure or where unusually wet soil conditions exist. These conditions require waterproof membranes.

Application
Concrete or masonry surfaces should be clean, reasonably smooth, and dry before coatings are applied. Form fins should be rubbed off and voids, rock pockets and form tie depressions should be filled with mortar and troweled smooth and flush. Rough or porous masonry below grade should be given a cement mortar parge coat. Temperature of the surface as well as the air temperature should be within a range acceptable to the manufacturer of the dampproofing material.

The prime coat is applied by spray, roller, or brush as uniformly as possible and at the recommended coverage rate. Porous lightweight concrete block masonry will take more material for a given area than smooth-formed concrete. Porous or rough surfaces below grade should be parged before primer is applied.

After the prime coat has dried, two or more coats of cold asphalt or hot asphalt or coal tar pitch are applied. Allow each coat to cure or dry before subsequent coats are applied. Apply each coat uniformly at the recommended coverage rate until the film thickness is reached. Hot bitumen should not be heated above a certain temperature or applied below a certain temperature. These temperatures are different for asphalt and coal tar pitch.

Below grade dampproofing should be protected from damage during backfilling. This is done by sticking saturated felt or other suitable covering material to the last coat while it is in the semi-liquid state. If saturated felts are used, the type of asphalt or pitch should be the same as the material used in dampproofing.

The prime coat and each dampproofing coat should be dry before subsequent coats are applied. All coatings should be applied to clean, moist-free surfaces. Each dampproofing coat should dry to a uniform glossy appearance. Touch up dull or porous spots before re-coating and on completion of the final coat. If a minimum amount of material was delivered to the jobsite to coat all surfaces required to be dampproofed at the minimum rate of coverage, there should be no material left over upon completion. All coats should be applied when the previous coat has cured. Backfilling should be done as soon as the final coat has cured. In hot application work, use a thermometer to monitor the kettle temperature and check the temperature at the time of application. The kettle nearly has to be close to the work. Since the fumes are toxic, the kettle should be located in a well ventilated place. Hot bitumen is generally applied by hand mop.

Cold asphalt is usually used in confined spaces where hot material would be hazardous or where the area to be coated is not large enough to bring in a kettle. Liquified asphalt is generally applied by brush over the primer, and by brush or spray for coats.

If dampproofing is applied below grade and no protective covering is required the backfill should be sand or loam with no stones or gravel. It should be placed and compacted to avoid damage to the dampproofing.

Although the plans or specifications will indicate the areas to be dampproofed, they may not always indicate the height on walls where the material is to be applied. Dampproofing, to be effective, should extend from about two inches above finish grade to several inches below the lowest floor line or to the spread footing. Felt or other protective material should extend from the bottom of the dampproofing to just below the finish grade elevation.

Labor for Cementitious Dampproofing

Work Element	Unit	Man-Hours Per Unit
Cement parging		
½" for interior faces	100 S.F.	1.00
¾" for interior faces	100 S.F.	1.00
1" for interior faces	100 S.F.	1.40
Cement parging to exposed wall		
½" for exposed faces	100 S.F.	1.40
¾" for exposed faces	100 S.F.	1.40
1" for exposed faces	100 S.F.	1.70

Time includes move on and off site, set-up, preparation, application, repair and clean-up.

Suggested Crew: 1 plasterer, 1 laborer, 1 foreman. On small jobs, add 5 to 10 percent to time shown

Labor for Bituminous Dampproofing

Work Element	Unit	Man-Hours Per Unit
Asphalt dampproofing		
Primer plus one coat	100 S.F.	1.5
Primer plus two coats	100 S.F.	2.3
Primer plus three coats	100 S.F.	3.1
Troweled on cold asphalt		
One coat	100 S.F.	1.1
Troweled on hot fibrous asphalt		
Primer plus one coat	100 S.F.	1.7
Primer plus two coats	100 S.F.	2.8
Primer plus three coats	100 S.F.	4.0
Asphaltic paint dampproofing		
Per coat — brush on	100 S.F.	1.1
Per coat — spray on	100 S.F.	1.0

Time includes move-on and off-site, set-up and job preparation.
Suggested Crew: 2 roofers.
Note: All time is for vertical surfaces only.

Coverage of Materials

Work Element	Unit	Coverage In S.F.
Asphalt primer	Gallon	75
Cold emulsion asphalt	Gallon	67
Asbestos fibre coating	Gallon	50
Roofing asphalt (hot)	100 lbs.	160
15 lb. asphalt felt	324 S.F. Roll	300
30 lb. asphalt felt	216 S.F. Roll	200

7 Thermal and Moisture Protection

Building Insulation

Heat is transmitted in three ways: (1) By conduction in which heat travels through a solid body and is transmitted from one solid body to another at contact points. More dense bodies such as metal are more efficient conductors than less dense materials, such as wood. (2) By radiation in which heat flows at a high speed in a direct line from a warmer body to a cooler one. Radiated heat has no appreciable effect on the air it passes through. (3) By convection in which heat is transmitted through a flow of air.

A good insulating material will impede heat flow by conduction because of its light weight in relation to its bulk; by radiation either by its reflective surface or by providing barriers between surfaces of different temperature; and by convection by eliminating air currents. A good insulating material should resist moisture, not support combustion, not harbor vermin, not sustain fungus, and not change volume. It should not deteriorate or lose other properties due to aging or exposure to temperature extremes.

Thermal insulation is manufactured in several forms from a number of natural and synthetic materials. Some are designed to serve more than one purpose. For example, structural insulating boards serve as sheathing, roof decks, and permanent concrete form boards. Other thermal insulating materials serve as sound insulators and decorative panels or tiles.

Common thermal insulating materials include mineral fibers made from glass, rock, slag, and asbestos, vegetable fibers, such as wood, cane, cotton and redwood bark; expanded mineral granules, such as perlite and vermiculite; vegetable granules such as ground cork; foamed materials such as glass and synthetic resins (styrenes and urethanes); and aluminum foil.

Insulating batts, blankets, and boards of various degrees of rigidity, stiffness, and density are made from fibers or granules combined with binders and formed or felted into desired widths, lengths, and thicknesses. The binder may also make the insulation material more water-resistant, mildew resistant, and lend structural or other properties to the finished product. Vegetable fibers are often chemically treated for fire and mildew resistance. Glass and synthetic resins are formed into blocks, sheets, or boards of various sizes by carefully controlling the foaming process to obtain the desired cell size and density. Batts or blankets are generally placed between joists or studs, and are manufactured in widths to suit standard stud and joists spacings. Batts or blankets may be unfaced or faced on one or both sides with paper, aluminum foil or plastic. One side will have a lower vapor permeability rate to act as a vapor barrier. Paper faced batts or blankets are manufactured with continuous paper flanges along their long edges for nailing or stapling to studs and joists. Some insulation boards intended for roof deck use kraft paper which serves as moisture and bitumen barriers.

Reflective insulators are made with mirror-bright aluminum foil, either plain or reinforced with paper backing. They are generally made to provide two or more reflective surfaces with air spaces between each layer. Like batts or blankets, they are used between studs and joists, and are provided with nailing or stapling flanges. Aluminum foil insulation-vapor barrier combination may also be incorporated into batts, blankets and gypsum board.

Loose fill is made with the mineral or vegetable fibers or mineral or vegetable granules, and is placed by blowing or hand packing. It is used primarily in masonry cavity walls and over ceilings in attic spaces. It is also blown in cavities between studs in existing structures. Masonry walls should be water-repellent treated before insulation is placed in the cavity.

Sprayed-on insulation is made with mineral fibers or mineral granules. The insulation material is usually combined with a liquid binder in the spraying apparatus. The binders and insulating material may also be factory premixed. Spray-on insulation is applied much the same way concrete is pneumatically placed. Sprayed-on insulation is designed primarily to protect steel members from failure due to high temperatures that occur in a fire.

Foamed-in-place insulation is two-part synthetic resin material in liquid form which can be mixed together and deposited in a cavity. Within a certain time after mixing, a foaming action takes place, filling the cavity.

Thermal conductivity is measured in terms of the amount of heat (British thermal units, or Btu's) that will pass through one square foot of a material in one hour for a difference in temperature of 1°F between the two surfaces.

The "K" factor measures the amount of heat that will pass through a one-inch thick homogeneous material. For example, "K" for mineral wool blanket is 0.27. This means that for 1-inch thickness there is a heat transfer of 0.27 BTU per hour per square foot for each degree difference in temperature between its two surfaces. Thermal conductance or "C" factor measures the amount of heat that will pass through a homogeneous or non-homogeneous material or a combination of materials. For example if the average "C" value of an 8-inch concrete block is 0.90, this means that 0.90 BTU will pass through the face shells, webs, and hollow air spaces of a one square foot section of an 8-inch wall.

Thermal resistance or "R" factor is the reciprocal of conductivity "K" or conductance "C".

The "U" value represents the overall transmission of BTU's in one hour per square foot of area to create a difference in temperature of 1 degree between the air on one side and the air on the other side of a structure. The "U" of a wall consisting of 2x4 studs, wood sheathing, wood siding, air space, insulation, gypsum lath, and plaster is calculated on the "K" and "C" factors. The lower the value of "K", "C", and "U", the more resistance to heat transfer is indicated. The greater the value of "R", the more resistance to heat transfer.

Vapor barriers are used to prevent condensation of moisture from the air in insulated spaces. Warm air is capable of holding more moisture than cold air. When warm moisture-laden air comes in contact with a cold surface, little droplets of moisture are deposited on that surface. If moisture were allowed to collect within the insulation material, the insulation value would soon be lost. To prevent this, a vapor barrier is placed on the warm side of the insulation material. The warm side of an ordinary wall is the room side. But it is the exterior side in the case of a cold storage room.

Acceptable vapor barrier materials with vapor permeable ratings of 1 perm or less include aluminum foil, plastic sheet material and coated or laminated paper assemblies. Ordinary 15 or 30 pound asphalt-saturated felts are not acceptable. The greater the temperature difference from one side of the wall to the other and the greater the relative humidity of the air on the warm side, the more effective the vapor barrier must be. Batts and blankets will often have vapor barriers attached to one side of the material. Other insulating materials may require a separate application of membrane barrier.

Application
Batts and blankets are laid between runner channels on suspended ceilings. They are installed between wood studs, ceiling and floor joists, rafters and furring by nailing or stapling edge flanges or tabs to the edges or sides of the wood members. It is also possible to friction-fit uncovered material. Two air spaces are more effective as a heat barrier than one. Therefore the best practice is to install the batt or blanket so that it does not come in contact with either wall surface. All spaces are completely covered.

Flanges should be smooth and in continuous contact with the framing member and stapled at close intervals - not over 6 inches. At insulation ends and joints, about 2 inches of insulation material should be cut off or folded back to allow a vapor barrier overlap on to the sill or plate or the end of the adjacent batt or blanket. At joints, insulation material should be pressed tightly together. The vapor barrier should be overlapped and taped together, or double-folded, drawn tight, and stapled to the wood support on both ends of the fold. Insulation should be contour-fitted around pipes, conduit, wiring, and other obstacles in the wall cavity. The vapor barrier must be continuous and well-sealed obstacles in the wall cavity. The vapor barrier must be continuous and well-sealed at joints and patched where torn. Reflective insulation is usually nailed or stapled to joints and studs similar to the way batts or blankets are installed. To be effective, each layer of reflective foil should be separated by an air space, including the two outer layers. Reflective insulation should be installed between wood members so that foil faces are at least ¾ inch from the nearest parallel surface. Aluminum-faced batt or blanket insulation is installed with the aluminum on the room (or warm) side of the wall.

Insulation boards or blocks are primarily used for roof deck insulation. They are laid over concrete, gypsum, wood, and metal decks. On ribbed steel decks, the boards serve both as thermal insulators and span across the ribs to provide a smooth flat surface that carries the roofing. The boards are butted loosely together with joints staggered in adjacent rows, and secured by embedding with hot asphalt. Insulation board on sloping decks should be secured by nailing or with some other mechanical fastening devices. On ribbed steel decks, joints paralleling ribs are centered over solid bearing on the rib. The two-layer construction, joints in the upper layer are offset from those in the lower layer. On concrete and gypsum decks, insulation is not installed until decks are cured and dry. Cant strips, wedges, and other special shapes are frequently fabricated from the same material as the insulation boards and then installed in much the same way.

Insulation board or blocks for insulating cold storage room floors, walls, and ceilings are laid up in two or more layers, with joints offset from the layer below. Insulation is secured in place with asphalt or other adhesive material, wood skewers or mechanical fasteners. Inside faces of concrete or masonry walls should be insulated by fastening boards directly to the wall. When an air space is desirable, the board can be attached to furring which is secured to the wall. Insulation board can also be installed in the cavity space in masonry walls.

Loose fill fiber or granular insulation is placed by blowing, pouring, or hand-packing. Granular insulation is poured into a cavity or between joists and screeded off to the required thickness. Fibers are generally blown or hand-packed to the required thickness and density.

Sprayed-on insulation requires special equipment specifically designed to place insulating materials by air spray. Urethane foam and mineral insulating materials mixed with binders can be placed in this manner. Some mineral formulations must be sprayed directly over the base material. Spray application is the easiest method to insulate irregular configurations and otherwise inaccessable areas. Sprayed-on insulation is usually used to "fireproof" the underside of metal roof decks and structural steel.

Vapor barriers can be installed separately if not attached to the insulation material. The greater the difference in the temperature between one side of the wall and the other, the more effective the vapor barrier must be. Very low temperature cold storage rooms require special care to assure that all seams and punctures are sealed, including taping over fasteners if mechanical fasteners are used. Anything penetrating cold storage room vapor barriers, such as wiring and piping, presents special sealing problems.

Labor Applying Spray-On Fireproofing

Work Element	Unit	Man-Hours Per Unit
Sprayed on metal deck ceiling		
1 hour fire rating	100 S.F.	2.0
2 hour fire rating	100 S.F.	2.1
3 hour fire rating	100 S.F.	2.4
4 hour fire rating	100 S.F.	3.6

Time includes move-on and off-site, set-up, masking, clean-up and repairs as needed.
Suggested Crew: small jobs - 1 plasterer, 1 plasterer helper; for larger jobs - 2 plasterers, 1 helper. Add 10% for fireproofing steel beams.

Insulation Recommendations

Walls, using 2" x 6" framing
R-19: R-19 (6") mineral fiber batts

Walls, using 2" x 4" framing
R-19: R-11 (3½") mineral fiber batts and 1" polystyrene
R-13: R-13 (3-5/8") mineral fiber batts
R-11: R-11 (3½") mineral fiber batts

Floors
R-22: R-22 (6½") mineral fiber
R-19: R-19 (6") mineral fiber
R-13: R-13 (3-5/8") mineral fiber
R-11: R-11 (3½") mineral fiber

Ceilings, double layers of batts
R-38: Two layers of R-19 (6") mineral fiber
R-33: One layer of R-22 (6½") and one layer of R-11 (3½") mineral fiber
R-30: One layer of R-19 (6") and one layer of R-11 (3½") mineral fiber
R-26: Two layers of R-13 (3-5/8") mineral fiber

Ceilings, loose fill mineral wool and batts
R-38: R-19 (6") mineral fiber and 20 bags of wool per 1,000 S.F. (8¾")
R-33: R-22 (6½") mineral fiber and 11 bags of wool per 1,000 S.F. (5")
R-30: R-19 (6") mineral fiber and 11 bags of wool per 1,000 S.F. (5")
R-26: R-19 (6") mineral fiber and 8 bags of wool per 1,000 S.F. (3¼")

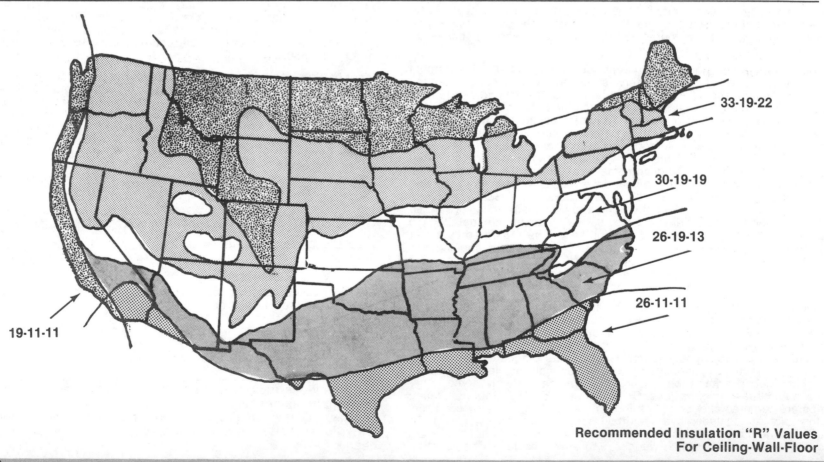

33-19-22

30-19-19

26-19-13

26-11-11

19-11-11

**Recommended Insulation "R" Values
For Ceiling-Wall-Floor**

Insulation Value of Common Materials

Material	Thickness	"R" Value
Air film and spaces		
Air space		
Bounded by paper or wood	¾" or more	0.91
Bounded by aluminum foil	¾" or more	2.17
Exterior surface resistance	---	0.17
Interior surface resistance	---	0.68
Masonry		
Sand and gravel concrete block	8"	1.11
Sand and gravel concrete block	12"	1.28
Lightweight concrete block	8"	2.00
Lightweight concrete block	12"	2.13
Face brick	4"	0.44
Concrete cast-in-place	8"	0.64
Building materials		
Wood sheathing or subfloor	¾"	1.00
Fiberboard insulating sheathing	¾"	2.10
Plywood	5/8"	0.79
Plywood	½"	0.63
Plywood	3/8"	0.47
Bevel lapped siding	½" x 8"	0.81
Bevel lapped siding	¾" x 10"	1.05
Vertical tongue & groove board	¾"	1.00
Drop siding	¾"	0.94
Asbestos board	¼"	0.13
3/8" gypsum lath and 3/8" plaster	¾"	0.42
Gypsum board	3/8"	0.32
Interior plywood panel	¼"	0.31
Building paper	---	0.06
Vapor barrier	---	0.00
Wood shingles	---	0.87
Asphalt shingles	---	0.44
Linoleum	---	0.08
Carpet with fiber pad	---	2.08
Hardwood floor	---	0.71
Windows and doors		
Single window	---	Approximately 1.00
Double window	---	Approximately 2.00
Exterior door	---	Approximately 2.00

Thermal Conductivity Values of Insulating Materials

Insulation Group		"k" Range (Conductivity)
General	Specific Type	
Flexible	Various	0.25 — 0.27
Fill	Standard materials	0.28 — 0.30
	Vermiculite	0.45 — 0.48
Reflective	Foil 2 sides	(*)
Rigid	Insulating fiberboard	0.35 — 0.36
	Sheathing fiberboard	0.42 — 0.55
Foam	Polystyrene	0.25 — 0.29
	Urethane	0.15 — 0.17
Wood	Low density	0.60 — 0.65

*Insulating value is equal to slightly more than one inch of flexible insulation. (Resistance "R" = 4.3)

These values measure heat conductivity. The "k" value is defined as the amount of heat, in British thermal units, that will pass in one hour through one square foot of material one inch thick per 1° F temperature difference between faces of the material. Simply expressed, "k" represents heat loss; the lower this numerical value, the better the insulating qualities.

Insulation is also rated on its resistance, or "R" value, which is merely another expression of insulating value. The "R" value is usually expressed as the total resistance of the wall or of a thick insulating blanket or batt, whereas "k" is the rating per inch of thickness. For example, a "k" value of one inch of insulation is $\frac{1}{0.25}$ Then the resistance, "R" is 0.25 of 4.0. If there is three inches of this insulation, the total "R" is three times 4.0, or 12.0.

Labor Installing Batt or Blanket Insulation

Work Element	Unit	Man-Hours Per Unit
Fiberglass blankets		
2", paper backed	100 S.F.	.88
2½", paper backed	100 S.F.	.88
4", paper backed	100 S.F.	1.05
6", paper backed	100 S.F.	1.40
2", foil backed 1 side	100 S.F.	.88
3", foil backed 1 side	100 S.F.	.96
4", foil backed 1 side	100 S.F.	1.05
6", foil backed 1 side	100 S.F.	1.40
2", foil backed 2 sides	100 S.F.	.88
3", foil backed 2 sides	100 S.F.	.96
4", foil backed 2 sides	100 S.F.	1.05
6", foil backed 2 sides	100 S.F.	1.40
Fiberglass batts		
2", plain, unbacked batts	100 S.F.	.79
3", plain, unbacked batts	100 S.F.	.96
4", plain, unbacked batts	100 S.F.	1.14
6", plain, unbacked batts	100 S.F.	1.49
Mineral wool batts		
2", paper backed	100 S.F.	1.00
4", paper backed	100 S.F.	1.30
6", paper backed	100 S.F.	1.80

Time includes move-on and off-site, unloading and stacking, installing by staple only.

Suggested Crew: 1 installer, 1 helper

Materials for Batt Insulation

Size	S.F.	Number of Batts Per 100 S.F.	Number of S.F. for 100 S.F. of Wall	Staples Per 100 S.F.
15 x 24	2.5	40	95	160
15 x 48	5.	20	95	160
19 x 24	3.7	32	96	160
19 x 48	6.33	16	95	160
23 x 24	3.84	26	95	160
23 x 48	7.67	13	100	160

Rating and Facings for Insulation

"R" Rating	Thickness	Kraft Paper Faced	Foil Faced	Unfaced
		Fiberglass		
3.5	1			x
4	1-1/8			x
5	1-1/2			x
7	2¼-2¾	x	x	
11	3½-4	x	x	x
13	3-5/8			x
14	5			x
19	6-6½	x	x	x
21	7	x	x	
22	6-1/2	x		x
		Rockwool		
7	2	x	x	
11	3	x	x	
13	3-5/8	x		
19	5-1/4	x	x	
22	6	x	x	

Fiberglass Batts Between Studs

Foamboard Sheathing

Energy-efficient foamboard sheathing panels are designed as an alternative to conventional plywood sheathing systems.

Types

Block molded polystyrene board is made from tiny polystyrene beads. The beads are first expanded in a hot air and steam chamber, then fused under pressure and steam into a large block, which is cut to the desired size by hot wires.

Extruded polystyrene is made by Dow Chemical Company. The brand name is Styrofoam®. The extrusion process leaves a board with a tough, dense skin on both sides. The four edges of the product are machined with a tongue and groove for airtight installation. (The tongue and groove product is known as "Styrofoam TG.")

Urethane boards are typically a layer of foam between two sheets of paper or aluminum foil.

Uses

Energy-efficient walls Use as sheathing on the outside of studs, under siding, in conjunction with 3½" mineral wood batts (or equivalent) between studs. Install vapor barrier on the inside and ½" gypsum over the vapor barrier to finish the inside of the wall. This will result in an R-15 to R-19 wall system, depending on the kind of siding and foamboard sheathing used.

Ceilings Use above decking on exposed beam cathedral ceilings to achieve an R-19 rating. On conventional ceiling, foamboard can be applied to ceiling joists before applying gypsum board to ceiling.

Basement and foundation walls Use on the exterior of basement and foundation walls down to at least the frost level in new construction. Styrofoam® is the preferred product for this application since it absorbs less moisture than other materials. In existing construction, use foam sheathing inside concrete basement walls for insulation. ½" gypsum is recommended as a fire barrier on the room side of the basement application.

Foamboard Physical Properties

Property	Extruded Styrofoam	Molded Polystyrene	Urethane
Density, lbs./C.F.	2.2	1.0	1.5
Compressive strength (p.s.i. perpendicular to face @ 5% deformation)	40	1.2	23
Water absorption (% by volume)	0.7	3.9	1.2
Water vapor transmission (perm/inch)	0.6	2.0	2.0
Thermal conductivity BTU/hr/S.F.	0.20	0.28	0.16
"R" factor per 1"	5.00	3.57	6.25

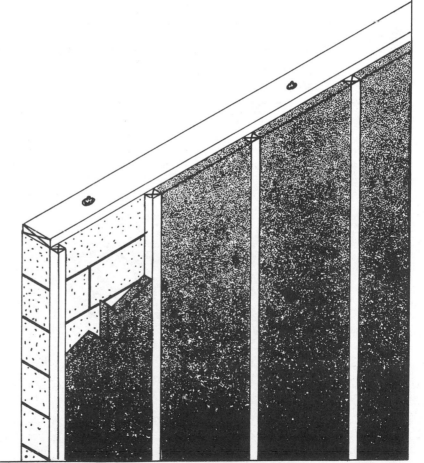

Styrofoam Insulation Between Furring Strips on a Concrete Block Wall

Labor and Materials for Poured Ceiling Insulation

Fill Thickness	Material Number of S.F. Covered By C.F. @ Density Ratings					Man-Hours Per 100 S.F. of Ceiling
	6 Lbs.	7 Lbs.	8 Lbs.	9 Lbs.	10 Lbs.	
1"	21.1	18.0	15.9	14.1	13.0	.6
2"	10.6	9.1	8.0	7.1	6.4	.6
3"	7.1	6.1	5.3	4.7	4.2	.7
4"	5.3	4.6	4.0	3.5	3.2	.8
5"	4.2	3.6	3.2	2.8	2.6	.9
6"	3.6	3.0	2.7	2.4	2.2	1.0
7"	3.1	2.6	2.3	2.0	1.9	1.1
8"	2.6	2.3	2.0	1.8	1.6	1.2

Labor Pouring Cavity Wall Insulation

Work Element	Unit	Man-Hours Per Unit
Loose insulation (poured in cavity)		
Expanded glass beads	100 C.F.	4.8
Expanded shale	100 C.F.	4.7
Vermiculite or perlite	100 C.F.	4.7
Polystyrene	100 C.F.	4.7

These figures assume insulation is poured by hand from bags into 6" or more concrete block cavities. Figure 5 S.F. of wall per C.F. of insulation. No scaffolding included.

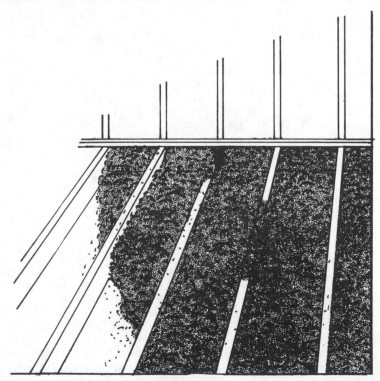

Poured Rockwool in an Attic

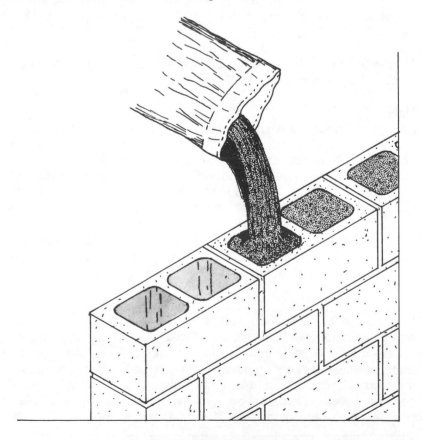

Cavity Insulation Poured

Roofing Insulation

Insulation normally used in built-up roofing systems includes rigid insulation boards applied directly to the deck surfaces, structural decks with insulating planks, poured-in-place insulating lightweight concrete fills, and sprayed-in-place plastic foams. All of these may or may not be provided with vapor barriers and venting systems.

The primary advantage of insulation is that it saves heat in the winter and lowers the air-conditioning load in the summer. Roof insulation also includes:

1. Prevents condensation on interior surfaces.

2. Provides a better surface or underlayment (than over a steel deck for example) for the application of built-up roofing.

3. Stabilizes deck components by reducing the deck temperature variations thus reducing expansion and contraction.

4. Retards heat flow from the building interior through the built-up membranes. This slows the drying out of the felts in hot climates for certain buildings, particularly at night.

There are disadvantages to roof insulation:

1. If no vapor barrier or ventilation system is present in buildings with high humidity, the insulation may shift the dew point from under the roofing system to within the roofing system. The dew point is the surface temperature at which water vapor will condense on the surface.

2. A built-up roofing membrane without roof insulation lasts longer than a roof with insulation. On a hot day the roofing membrane next to the insulation gets much hotter because the heat cannot escape through the insulation. This causes the membrane next to the insulation to become over 60 degrees hotter than the exposed surface of the roof and probably 40 degrees hotter than the air temperature. When the sun goes down the built-up membrane on an insulated roof quickly returns to the air temperature since the building heat does not come through to the membrane. Rapid temperature change produces greater expansion and contraction which accelerates the aging process.

Roof insulation thickness is usually specified in a "C" factor. "C" is the thermal conductance measured in BTUs per hour per square foot per degree Fahrenheit. The lower the "C" number, the greater the insulation value, the thicker the insulation. The thermal conductance factor for roof insulation has been standardized by the industry. Thus certain "C" factors apply to each inch of material used.

Fire rated insulation is used on steel or wood decks to provide a fire rated roof assembly.

All roof insulation boards must be strong enough to withstand normal traffic loads, expected snow loads and hailstone impact.

Roof insulation boards should have good sheathing strength so that uplift forces do not split membrane away from the insulation.

Poured-In-Place Insulation
Insulating concrete fills and lightweight insulating concrete are mixtures of lightweight aggregates that conform to ASTM C332 and Portland cement that conforms to ASTM C150. The ratio of cement to aggregate is based on the desired insulation value, the required compressive strength and the ability of the mixture to hold fasteners or special nails. Asphaltic concrete fills are also available, although they can not provide as much insulation value per unit thickness and they can burn.

These lightweight aggregates generally produce concrete weighing from 15 to 50 pounds per cubic foot and have a thermal conductivity or "K" factor of 0.70 to 1.15 for the 20 to 40 pounds per cubic foot range. The lightweight aggregates are expanded products such as perlite or vermiculite with a thermal conductivity of 0.45 to 1.50. both non-burning minerals that have been expanded 4 to 20 times by heating to between 1400 degrees to 2000 degrees. There are three different grading requirements: coarse vermiculite, fine vermiculite and perlite.

Perlite is a volcanic glass, like obsidian, that has taken on a concentric shell structure due to curved cracks produced by contraction in cooling. Perlite tends to absorb water on the surface and requires less mixing water than vermiculite.

Vermiculite can be any number of micaceous minerals that are hydrous silicates occurring in tiny leafy scales. This expanded product is highly water-absorbent material made up of many layers.

A lightweight concrete mix must be no lighter than 1:4 to be considered nailable. A typical 1:4 mix will weigh 53 to 63 pounds per cubic foot when wet and 31 to 40 pounds per cubic foot when dry. It will have a compressive strength of 300 to 500 pounds per square inch. A 1:8 mix, one bag of Portland cement to eight cubic feet of aggregate, is not suitable as a base for roofing.

Perlite and vermiculite produce fills with different characteristics. They are not interchangeable. The concrete made with one aggregate requires a different type of venting system or galvanized steel mesh than the other. The other requires crushable insulation or a one inch air space around the perimeter of the roof to allow horizontal expansion.

These heavier materials stabilize roof surface temperatures. One of the chief advantages of lightweight concretes is that they form monolithic masses of insulating material that have a greater heat storage capacity than lighter more efficient materials in small units. The mass of lightweight insulating concrete stores and releases heat slower than thin, light separate units.

Lightweight insulating concretes or insulating concrete fills are non-structural concretes that can be placed on steel decks, structural concrete decks, precast, preformed decks, plank decks of various descriptions, poured gypsum and other gypsum and other relatively flat surfaces. All insulating concretes have excellent compressive strengths and are incombustible.

Sprayed-in-place plastic foams are excellent lightweight insulation materials. They require less labor to apply than insulating concrete. When properly applied, they provide the necessary slope for a roofing system. Sprayed-in-place foams can be shaped or formed around surfaces to any contour. They have no joints to be sealed.

Sprayed-in-place plastic foams release dense smoke if they catch fire and have high rates of expansion and contraction. Some jobs require sealing or caulking at vertical edges and perimeters to prevent water entering small shrinkage cracks. Only an experienced applicator working in ideal weather conditions can produce a smooth surface for the subsequent application of roofing felts. All sprayed-in-place urethane foams turn brown, shrink and become brittle if exposed to the sun's rays unless they are properly covered or coated. Some of the coatings exposed to the weather will not hold up two years.

Sprayed-in-place plastic foams can be cut with a pocket knife. However the material can be patched nearly as easily. Small repair kits of pressurized foam are available. Cracks may develop in cold climates and these can also be easily repaired. Exposed urethane foams begin to collapse at 250 degrees.

Other Roof Insulation

Urethane boards are a different product altogether. Urethane is a highly efficient insulating material and coated sheets can take 425 degree F. asphalt for very short periods. Styrofoam cannot take hot mopping. The boards beginning to curl at 175 degrees. Styrofoam must be secured with cold-process adhesives or mechanical fasteners or skewers.

Vapor barriers are normally placed on the structural deck below the insulation. The effectiveness of vapor barrier materials is measured in perms. A 0.2 perm rating for a vapor barrier is usually required for good quality roofing and can be obtained with two plies of felt and three courses of plastic cement or three moppings of bitumen.

Vapor barriers should be used only in roofing systems where the insulation is sandwiched between the structural deck and the built-up membrane.

The function of a vapor barrier is to prevent condensation from forming within the built-up roofing system. The vapor barrier does this by blocking the upward flow of water vapor from the heated, high humidity of the building interior. Insulation swells when it absorbs water vapor. When the water vapor decreases, the insulation shrinks. Some cellular insulation disintegrates when water vapor destroys the surface. Movement of insulation caused by dimensional instability results in splitting or cracking of the roofing felts above the insulation.

If insulation contains moisture for any reason, the vapor barrier will prevent the water vapor from escaping. This is the reason for venting systems or for deleting the vapor barrier altogether. If the water vapor does not escape, it expands to form blisters in the built-up roofing membrane.

Polyethylene film is a suitable vapor barrier. Unfortunately, it is very difficult to get the material to adhere to anything. The plastic sheets are also subject to tearing. Hot bitumen will soften plastic sheets and some plastics are destroyed by hot bitumen. Since polyethylene and similar materials do not stick to most surfaces, they cannot resist wind uplift unless they are punctured by mechanical fasteners. The fasteners destroy the barrier properties.

Insulation Anchoring Methods

Insulations or vapor barriers may be anchored with asphalts or cold-applied adhesives. These adhesives include clay-filled bituminous emulsions, chlorinated rubber based compounds and other chemical setting cements.

Factory Mutual Research Corporation (FM) approves only two mechanical fasteners for Class I steel-deck assemblies. One has a serrated shank that grips the penetrated steel and the other has a locking tongue that springs after the fastener point penetrates the steel deck.

Gypsum, preformed wood fiber and lightweight insulating concretes may be fastened with tube nails, or various proprietary devices that have difficult locking mechanisms.

On wood decks use regular large headed nails long enough to go through the insulation and penetrate the deck ½-inch. Plywood decks require serrated nails.

Cant Strips consists of preservative treated wood, rigid fiberboard, perlite board, or concrete. They are used at intersections of the roof with walls, parapets and curbs extending above the roof. The faces of the cants provide a short incline of 45 degrees between the roof at the vertical surface. Cants bear on the insulation and fit flush against the vertical surfaces. Where possible, cants should be nailed to adjoining surfaces. When installed against non-nailable materials, the cants are set in a heavy mopping of steep asphalt or plastic cement. No projections, such as vent pipes, braces, etc., should be constructed through cants or within 10 inches from cants.

Edge strips are preservative treated wood, rigid fiberboard, perlite board or concrete. They are installed in the right-angle formed by the junction of the roof and wood nailing strips that extend above the level of the roof. Edge strips are tapered a maximum of 1½ inches one foot from the top of the wood nailing strips. Edge strips fit flush against the vertical surfaces of the wood nailing strips. Where possible, edge strips should be nailed to adjoining surfaces. Where installed against non-nailable materials, the edge strips are set in a heavy mopping of steep asphalt or plastic cement.

Treated wood nailers are required at eave edgings, sides of roofs, around curbs, and at other material intersection points to provide nailing for gravel stops, flashing and roof felts. Nailers are usually specified and estimated with carpentry items because they are installed by carpenters. Roof decks with slopes greater than 1 inch per foot that have non-nailable insulation require roof nailers of the same thickness as the roof insulation.

Application

Insulation is applied either in direct contact with the roof deck or over a vapor barrier. The insulation must be securely fastened to the deck to resist wind uplift. Insulation is applied to the deck so that the long dimension of the board or sheet is across the short dimension of the roof. End joints should be staggered. To allow for expansion, the insulation must be kept ½ inch back from vertical surfaces.

Solid bitumen mopping is the best method for anchoring insulation to the deck. Solid mopping provides the best protection against moisture infiltration and gives the highest horizontal shearing strength for preventing wind uplift. All decks are not suitable for solid bitumen mopping. On some decks bitumen can drip thru roof cracks. A layer of rosin sized paper can be used to keep bitumen out of the joints.

When using multiple layers of insulation, joints of each succeeding layer are parallel and overlap in both directions the layer below. Each layer should be firmly embedded in a solid steep asphalt mopping. Mop only far enough ahead to provide the complete embedment of one board. Except where mechanical fasteners are required, use at least 25 pounds of asphalt per 100 square feet of roof deck for the second and all succeeding layers of insulation.

All roof insulating materials must be kept dry before, during and after application. If the work stops for any reason, a glaze coat and cut-off detail will be required. This is explained in the built-up roofing section.

Mechanical fasteners are required to secure insulation or nailers to roof decks if the slope exceeds 1 inch per foot for asphalt built-up roofing and ½ inch per foot for coal-tar built-up roofing. Nailable insulation over a nailable roof deck must be nailed with a minimum of 6 nails per 8 square feet of insulation. Nails should not project through the roof deck. Insulation over a non-nailable roof deck must be installed between wood nailers running parallel to the slope. The nailers should be the same thickness as the insulation and be spaced the width of the insulation board plus ¼ inch. Mechanical fastening systems provides horizontal passages for the escape of water vapor that is trapped between the deck and the insulation above it. This is an important advantage. Some insulation is manufactured with beveled edges to reduce the problems of vapor pressure build-up.

Labor Placing Cementitious Insulation

Work Element	Unit	Man-Hours Per Unit
Lightweight insulating cement		
Cellular-portland cement	C.Y.	1.20
Perlite or vermiculite	C.Y.	1.22
With sand aggregate, 1-3-2 mix	C.Y.	1.20
Gypsum concrete, no wire mesh included		
Poured in place — 2" thick	100 S.F.	2.4
Sprayed on — 1" thick	100 S.F.	2.5
2" thick with wood fibre binder	100 S.F.	2.4
1" thick with wood fibre binder	100 S.F.	1.6
2" thick with glass bead binder	100 S.F.	2.3

Time includes move on and off site, set-up, place, finish, clean-up and repairs as necessary.
Suggested Crew: 2 cement finishers, 1 laborer. Add 50% if smooth trowel finish is required.

Labor for Sprayed Insulation

Work Element	Unit	Man-Hours Per Unit
Spray on insulation		
Polystyrene	100 S.F.	2.8
Urethane	100 S.F.	2.8
Fibre spray-on insulation		
Asbestos 1" thick	100 S.F.	2.6
Gypsum 1" thick	100 S.F.	2.3
Perlite 1" thick	100 S.F.	2.4

Note: Asbestos insulation can only be used under special conditions and requires unique safety measures for application.

Time includes move on and off site, equipment set up and application of insulation.
Suggested Crew: 1 installer, 1 helper

Labor Installing Rigid Board Insulation

Work Element	Unit	Man-Hours Per Unit
1" mineral fibre	100 S.F.	1.20
2" mineral fibre	100 S.F.	1.50
1" glass fibre	100 S.F.	1.20
1" polystyrene	100 S.F.	1.50
¾" urethane	100 S.F.	1.10
1" urethane	100 S.F.	1.20
1½" urethane	100 S.F.	1.30
2" urethane	100 S.F.	1.50
1" foamed glass	100 S.F.	1.10
1½" foamed glass	100 S.F.	1.20
2" foamed glass	100 S.F.	1.40
1" wood fibre boards	100 S.F.	1.10
2" wood fibre boards	100 S.F.	1.30
¾" particle board, compressed	100 S.F.	1.10
1" particle board, compressed	100 S.F.	1.10
2" particle board, compressed	100 S.F.	1.40
1" cork insulation board	100 S.F.	1.40
2" cork insulation	100 S.F.	2.90

Time includes move on and off site, unloading, stacking, installing by nail, staple, mastic or glue.

Suggested Crew: 1 installer, 1 helper. For overhead application, add 10% to time; for deck application, deduct 10%.

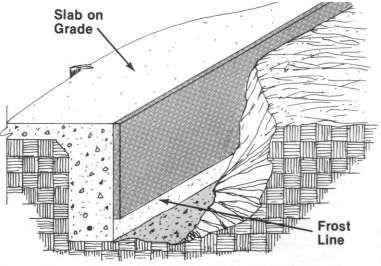

Slab on Grade

Frost Line

Cellular Plastic Molded Polystyrene

Converting Floor Area to Roof Area

Rise in 12" Run	Factor	Rise in 12" Run	Factor
3"	1.031	8"	1.202
3½"	1.042	8½"	1.225
4"	1.054	9"	1.250
4½"	1.068	9½"	1.275
5"	1.083	10"	1.302
5½"	1.100	10½"	1.329
6"	1.118	11"	1.357
6½"	1.137	11½"	1.385
7"	1.158	12"	1.414
7½"	1.179		

Multiply the horizontal or plan area (including overhangs) by the factor shown in the table opposite the rise. The result is the roof area.

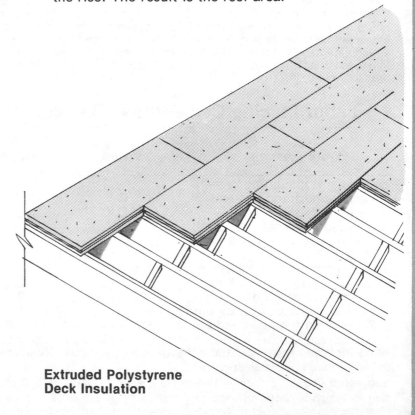

Extruded Polystyrene Deck Insulation

Preformed Roofing and Siding

The materials under consideration here are metal, fiberglass and cement-asbestos preformed panels designed to span up to several feet of wall or roof. The corrugations serve to stiffen the panel, provide concealed or semi-concealed side joints, and improve the looks of the finished installation. Corrugation can take many forms. The continuous serpentine type is most common. Ribs or flute-spaced configurations on a flat background are also common. The usual shape has valleys from ½ inch to 3 inches deep spaced from about 2 inches to 12 inches or more on center.

Some designs will have major ridges or valleys spaced about 12 inches on center with one or two minor ridges or valleys centered between them. Some designs incorporate a trough at or near the crown of the inside sidelap flute. The crown acts as a siphon breaker to drain off any water that might be blown up under the overlapping configuration. Width and shape of the ribs and valleys run all the way from slender ribs to broad flutes several inches across at the base.

Manufacturers of preformed roofing and siding panels publish tables showing the load carrying capacities for their products. These are generally expressed in terms of pounds per square foot of uniform load for various spans (center-to-center purlin or girt spacing). Load carrying capacity depends on the stiffness of the material, the thickness of the material, and the depth and spacing of the corrugations.

Cement asbestos panels are made with Portland cement and asbestos fibers. They are formed under high pressure and steam cured. The standard color is natural gray. The material does not need to be coated for weather protection, but is sometimes painted for aesthetic reasons. It is also available with factory applied paint coating. Corrugated panels are available in $\frac{3}{16}$ and $\frac{3}{8}$ inch thicknesses, 42 inches wide, and in lengths up to 12 feet. Cement asbestos panels must be drilled for fasteners.

Fiberglass panels are made from reinforced polyester resin. This is a translucent material. For this reason fiberglass panels are often used in skylights with metal panels of the same design. Fiberglass panels are available in a wide variety of designs and colors. They are made in both clear and opaque colors, and in both smooth and textured finishes. For added weather protection, a thin clear film of polyvinylfluoride (Tedlar) may be laminated to the exposed face.

Aluminum roofing and siding is made in several alloys. The most popular is alclad 3004, an aluminum alloyed with manganese and magnesium and thin coated with pure aluminum. This produces a sheet that can be easily formed, has adequate stiffness, and is relatively low in cost. The pure aluminum coating provides high corrosive resistance and reflectivity for thermal insulation. The smooth surface is usually embossed to break up the light reflecting pattern.

Although aluminum needs no protective coating for corrosion resistance, it is often painted to create a more pleasing appearance. Most manufacturers produce factory painted panels in several stock colors. Some of these are very high quality finishes applied under ideal conditions with factory quality control. Factory painted panels are shipped nested with paper sheets between all panels to prevent direct contact which might mar the finish. **Steel roofing** and siding requires a protective coating for corrosion resistance and to improve its appearance. Galvanized steel roofing is generally specified with a "commercial class" zinc coating on cold rolled carbon or copper bearing steel. This grade has a coating of 1.25 ounces of zinc per square foot of flat surface on both sides. This amounts to .625 ounce of zinc per square foot on one side only. Galvanized steel panels are also furnished in embossed finishes. They are available with a factory treatment for field painting in lengths of up to 40 feet.

Many suppliers furnish factory painted galvanized roof and wall panels in several stock colors, applied to one or both faces. Although other paint systems are used, alkyd, vinyl, and acrylic resin based enamels are the most popular. Silicones and fluoropolymers are most durable, but also more expensive.

The **coatings** are generally applied to the flat sheet by roller or spray before the sheet is roll formed to its final shape. These baked on finishes, applied under careful quality control conditions, are generally superior to the best field applied paints, and at a lower cost. Sheets should be protected from rubbing against each other in shipment.

Aluminized steel has an aluminum coating applied to the steel sheet by the hot dip method. Aluminum coatings perform as well as or better than zinc coatings in some conditions.

Built-up composite coatings are sometimes used on steel. Asbestos fiber and bituminous compounds are applied after the steel is dipped in molten zinc. While the zinc is still liquid, asbestos felt is rolled on at high pressure, mechanically locking the felt to the steel as the zinc cools. The felt is then saturated with hot asphalt and coated with a colored synthetic resin finish.

Asbestos fiber and alkyd resin compounds use asbestos felt bonded with acrylic adhesive to the steel core under heat and pressure. The felt is then saturated with a colored alkyd resin weather coat.

Colored mineral surfacing can be fused to the steel core. Ceramic frit is fused to enameling steel in one or two coats, each fired at about 1,500 degrees. This is the most expensive and long lasting fade-free coating for exterior paneling. It is not normally used as a roofing material except where the roof is exposed and aesthetics is a prime consideration. It is used primarily on curtain walls, wide decorative fascias, and mansard roofs. Manufacturers do not recommend use in walls near the ground where it is subject to impact chipping.

Joint sealant is required on nearly all sheet roofing and siding. Butyl sealant is generally used in concealed joints. It is relatively inexpensive, is permanently tacky (non-curing, non-drying type), and has adequate adhesion and cohesion for the limited movement apt to occur in such joints. It is available in cartridges and in preformed tape.

Where semi-transparent fiberglass sidelights or skylights are used, clear sealants are generally recommended to keep the dirty gray butyl sealant from showing through. Two types of transparent sealant are available. Synthetic rubber, such as silicone, will serve the purpose. But it is expensive and its high adhesion and cohesion properties make replacing a panel difficult. Extruded vinyl tape is less expensive and is generally furnished by fiberglass panel manufacturers for this purpose.

Concealed closures are generally used to close the open spaces on the underside of corrugations where the panel is secured to a flat surface such as at the floor, foundation wall or curb. Since they are concealed, the closure strip color or appearance is unimportant. They may be fabricated from solid or sponge rubber or plastic. The most popular strips are black solid rubber about ½ to 1 inch wide. They are die cut to fit the contours of the underside of a panel on one side and are straight on the other side.

Exposed closures may also be referred to as trim pieces or flashings. They are used to seal off non-mating surfaces where concealed closures cannot be used. Internal and external corners, openings and ridges would use exposed closure strips. Sometimes a combination of exposed and concealed closures are required to make a joint weatherproof. Exposed closures are generally fabricated from the same material as the sheet and have the same color and finish.

Fastening Preformed Siding and Roofing

Some panels fit together and are secured to the supporting structure with clips or other fasteners that are completely concealed from the exterior. These fastening systems are most frequently used on metal curtain walls, particularly the insulated panel types. They are also common in some uninsulated roof panel systems.

Most fasteners penetrate the panels and are exposed on the exterior. They are driven from the exterior. This saves scaffolding the interior of the building.

Exposed fasteners should be spaced evenly and aligned carefully. They should have a low silhouette and be the same color as the panels to which they are attached. Some manufacturers supply color-matched snap-on caps that go over the fastener head. Others supply color-matched enameled fasteners.

All exposed fasteners penetrating roofing and siding should have neoprene gaskets under the fastener head.

To avoid galvanic corrosion or corrosion stains, fasteners should be stainless steel, galvanized steel, cadmium plated steel, or aluminum alloy. Where there is direct metal-to-metal contact on weathering surfaces, the metal or metal coatings in contact should be of the same alloy.

Nails are used for securing metal panels to wood supports. Shanks should be the barbed, ringed, or spiral screw thread type. Long nails with smooth shanks (straw nails) are sometimes used to fasten panels to steel supports. The shank is bent around the supporting members.

Thread cutting fasteners cut through the metal as they are driven. They have one or more slots part way up the shank through which bits of metal may escape as the fastener is driven. Thread cutting fasteners are used primarily for securing roofing and siding to hot rolled structural steel or heavy gage cold rolled members. Their threads are fairly fine and closely spaced.

Thread forming fasteners do not remove metal; they displace it. Their threads are much coarser and more widely spaced than thread cutting devices. They are used primarily for fastening roofing or siding to light gage metal or for fastening light gage panels together at lapped joints. Thread forming fasteners are available in three points: pointed, blunt, and self-drilling. Combination or stand-off fasteners have a self drilling tip for drilling through the panel and the structural steel support, a thread cutting lower shank for securing the fastener to the steel, and thread forming upper shank for securing the panel to the fastener head. They are used where thermal insulation is installed between the panel and the structural support.

Threaded fasteners should be driven with electric or air power tools with adjustable clutches which permit control of the compression on the neoprene seal.

Studs are tapped or welded to the steel supporting members or secured to a steel strap which is bent around the steel support. Studs support panels at the crown of the corrugation, and therefore are designed with a shoulder for underside support. A nut or a rivet on the shank protruding up from the shoulder fastens the stud. Metal panels are impaled on the shank by a blow with a rubber mallet.

Rivets used in securing rigid roof and wall panels to supporting members are driven from only one side of the panel and are called "blind rivets". Several types of blind rivets are used for fastening panels together at lapped joints and for fastening panels to steel supports.

Where thermal insulation is placed between purlins and metal roof panels, the roof panels should be supported on the underside by the fasteners. The reason for this is that even rigid insulation boards compress under moderate concentrated loads. If the fastener is put under enough tension to provide an efficient seal, the metal will be deformed into a dimple around the fastener head. As the roof panel goes through cycles of thermal expansion and constraction, the fastener hole is likely to enlarge and the seal may be broken. This leaves a funnel through which moisture can pass. Several fasteners have been designed to provide this support. One is the combination or stand-off fastener. Another is the stud fastener. A third is a self tapping fastener with a spacer assembled on its shank. The lower end of the spacer is threaded. The upper end is split into 4 equal sections. The upper end of the shank is thickened into a conical shape. When the fastener is driven, the spacer comes to rest on the steel supports, and is driven up over the thickened upper portion of the shank, which spreads the split portion of the spacer out to provide underside support to the roof panel.

If thermal insulation is required, it may be included in the project specifications dealing with preformed roofing and siding. The roofing and siding may be in the form of insulated sandwich panels, or exposed insulation inserted between the exterior panels and structural support. If factory assembled sandwich panels are called for, these will generally use a system of concealed fastenings. If the insulation is placed between panels and supports, a system of exposed fastenings will more than likely be used.

Application
Before beginning application of the siding or roofing, be sure that all structural members are in place, properly fitted, positioned, aligned, and fastened.

Panels should be layed out so that the prevailing wind blows over, not into the sidelaps. All endlaps should be over continuous supports.

Panels should be carefully positioned and secured to structural supports. Fasteners of the recommended type and size are driven in the locations and spacings indicated. Use tools that will exert the right amount of compression on the seal under the fastener head. If fasteners are driven through the crown of metal panels, be careful that the fasteners are not drawn down so tight that the corrugation is spread open, causing "panel creep". Panel creep is particularly troublesome when it is not uniform. The panels become misaligned with supporting purlins and girts as application proceeds from one end of the building to the other.

Some specifications may not be specific on location of panel-to-structure fasteners. Panels with no flat surfaces, such as continuous corrugated cement asbestos, fiberglass, or metal, should be fastened through the crown of the corrugation. Sidelap fasteners should also be placed in the crown of the corrugation. Nails and other fastenings into wood should be placed in the crown of the corrugation. Even dry wood is apt to undergo some additional shrinkage as it is exposed to high heat adjacent to a wall or the underside of a roof. This shrinkage could cause the wood to pull away from the fastener head and break the seal.

Panels with two inches or more of flat surface bearing on steel supports should be fastened through the valley or low part of the corrugation. In this position, the fastener provides maximum clamping action and permits a good long-lasting seal. This tight clamp also guarantees that thermal stresses will be absorbed within the panel. Expansion and contraction won't break the seal and enlarge the hole as is likely when the panel moves in relation to its support.

If specifications call for fasteners in the low part of continuous corrugated panels, the resilient washer under the fastener head should be cone shaped to provide a good seal on the curved surface. One of the problems in placing fasteners in the low point of narrow corrugations is that the protruding fastener head catches debris and forms a dam. If the seal is broken, water from the dam will seep through.

There are four likely mistakes when driving sheet metal screws. Underdriving is the most frequent. (2) Failure of the fastener to clamp the panel to its support is caused either by the two pieces not being in contact when the fastener is driven, or the hole in the panel not being large enough to let the fastener spin free and engage only the support. If this is a problem, have someone stand with one foot on each side of the fastener on the panel being fastened as the fastener is driven. (3) Overdriving is probably the third most common failure. Overdriving will result in stripped threads or a ruptured seal, or both. Ruptured seals are most common directly under the spinning head. Fasteners designed with a non-threaded shoulder near the head will automatically trip the clutch when the threads run out. This puts just the right amount of compression on the seal. Fasteners with a cupped free spinning metal washer between the seal and the head will generally prevent a ruptured seal. (4) Fasteners that are driven at an angle to the surface of the panel result in an imperfect seal, or damage to the panel surface. Fasteners with thick conical seals or cupped free spinning metal washers between the seal and the fastener head tolerate more deviation from perpendicular.

Factory finished panels require more care in handling and installing than do unfinished panels. Organic coatings are more easily stained or physically damaged than mineral surfaces. Workmen on the roof should wear shoes with soft light colored rubber soles. If roof panels are field drilled in place, drill shavings or filings should be swept up and removed. They are very abrasive underfoot and can scratch the surface very easily.

Most factory finishes are applied to the flat material before panels are rolled and cut to length. The factory finish is not always applied in the same mill that rolls the coil into corrugated panels. The mill that produces corrugated panels may not produce matching factory finished accessories other than the most common closures and trim shapes. Gutters, downspouts, and custom made accessories not be available with the same factory finish. Some of the panel fabricators can supply paint of the same type, color and gloss as the factory finished panels. If available, this is probably the best way to get a good color match.

If thermal insulation is included as a part of the work, bats or blankets are pulled taut and temporarily clamped to purlins and grits until panels are installed. Rigid boards are laid over supports before panels are installed. All joints that do not occur behind structural supports should be supported.

Joint sealant is placed on the overlapping portion of the lower panel. Then each upper panel is positioned over the sealant.

Closures, flashings, and accessories are then installed as required by the contract documents to create neat weatherproof coverage.

Flashing and sealing around openings on the slope of a roof below the ridge presents a difficult problem, especially on the upslope. Crickets are difficult if not impossible to make, fit, and seal watertight in some corrugated materials. Making a neat, watertight transition from a corrugated sloping surface to a flat vertical surface is a problem that taxes the imagination. Some companies make specially formed metal or molded synthetic rubber flashings for this purpose.

Labor for Installing Preformed Steel Roofing

Work Element	Unit	Man-Hours Per Unit
Galvanized steel 2 oz. per S.F.		
22 gauge pressed panels	SQ.	3.85
24 gauge pressed panels	SQ.	3.80
26 gauge pressed panels	SQ.	3.60
28 gauge pressed panels	SQ.	3.55
26 gauge standing seams	SQ.	3.50
28 gauge standing seams	SQ.	3.50
24 gauge batten seams	SQ.	4.40
26 gauge batten seams	SQ.	4.30
28 gauge batten seams	SQ.	4.18
Stainless steel roofing (Type 304)		
24 gauge (any seam)	SQ.	7.70
26 gauge (any seam)	SQ.	7.90
28 gauge (any seam)	SQ.	7.80
30 gauge (any seam)	SQ.	7.20
32 gauge (any seam)	SQ.	7.03
Weathering steel roofing (preformed) ASTM-588		
22 gauge (standing seam)	SQ.	4.15
24 gauge (standing seam)	SQ.	4.03
26 gauge (standing seam)	SQ.	3.90

Note: If flat seams are specified, add 10% to time shown. Time includes move on and off site, unloading, stacking, place, trim and seam seal as required.
Suggested Crew: 2 roofers, 2 laborers

Labor Installing Corrugated Asbestos Cement Roofing

Work Element	Unit	Man-Hours Per Unit
On wood purling	SQ.	5.0
On metal framing	SQ.	5.4

Suggested Crew: 2 roofers, 1 laborer.

Labor for Installing Corrugated Aluminum Roofing

Work Element	Unit	Man-Hours Per Unit
Corrugated aluminum roofing — preformed		
.032" thick	SQ.	4.57
.040" thick	SQ.	4.67
Colored aluminum — bonderized		
.032" thick	SQ.	4.52
.040" thick	SQ.	4.72
Anodized aluminum		
.032" thick	SQ.	4.55
.040" thick	SQ.	4.63

Time includes move on and off site, unloading and stacking, trim, install and repair as required.
Suggested Crew: 2 roofers, 1 laborer

S.F. per Sheet of Corrugated Plastic Panels

Length of Sheet	Width of Sheet							
	26"	27½"	33"	33¾"	35"	36"	40"	42"
* 6"	1.08	1.15	1.37	1.41	1.46	1.50	1.67	1.75
3'	6.50	6.87	8.25	8.44	8.75	9.00	10.00	10.50
4'	8.67	9.17	11.00	11.25	11.67	12.00	13.33	14.00
5'	10.83	11.46	13.75	14.06	14.58	15.00	16.67	17.50
6'	13.00	13.75	16.50	16.87	17.50	18.00	20.00	21.00
7'	15.17	16.04	19.25	19.69	20.42	21.00	23.33	24.50
8'	17.33	18.33	22.00	22.50	23.33	24.00	26.67	28.00
* 9'	19.50	20.62	24.75	25.31	26.25	27.00	30.00	31.50
10'	21.67	22.91	27.50	28.12	29.17	30.00	33.33	35.00
*11'	23.83	25.21	30.25	30.93	32.08	33.00	36.67	38.50
12'	26.00	27.50	33.00	33.75	35.00	36.00	40.00	42.00
*13'	28.17	29.79	35.75	36.56	37.92	39.00	43.33	45.50
14'	30.34	32.08	38.50	39.37	40.84	42.00	46.66	49.00

*Non-stock lengths.

Labor for Installing Copper Roofing

Work Element	Unit	Man-Hours Per Unit
Flat seam copper roofing		
16 oz. copper	SQ.	7.78
18 oz. copper	SQ.	8.02
20 oz. copper	SQ.	8.09
Standing seam copper roof		
16 oz. copper	SQ.	7.18
18 oz. copper	SQ.	7.33
20 oz. copper	SQ.	7.34
Batten seam copper		
16 oz. copper	SQ.	8.56
18 oz. copper	SQ.	8.72
20 oz. copper	SQ.	8.87

Time includes move-on and off-site, unloading, stacking, place, trim and seam, seal as required, and 1 layer felt underlayment.
Suggested Crew: 2 roofers, 2 laborers

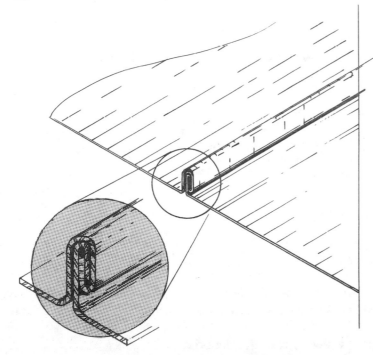

Standing Seam Metal Roof

Elastomeric Roofing

An elastomer is any material which, at room temperature, can be stretched repeatedly to at least twice its original length, and upon release will return to its original size. Elastomers obviously have applications as roofing materials where expansion and contraction can create serious problems.

Although synthetic rubbers have been around since the early part of this century, it was not until World War II that high quality substitutes for natural rubber were developed. Most synthetics do not have the elasticity of natural rubbers, but they have other properties that make them more desirable as roofing.

There are at least 18 distinct elastomers on the market today. The ones most commonly used for roofing are silicone (polysiloxane) and a combination of chloroprene (neoprene) and chlorosulfonated polyethylene (hypalon). The neoprene is generally applied in liquid form direct to the roof deck. It can also be applied in sheet form and secured by adhesion. The hypalon coating is applied in liquid form over neoprene. Hypalon is used over the neoprene as a color coat (neoprene is only available in dark colors) because it has slightly better sunlight aging characteristics. Silicone is used alone. The liquid is applied directly to the deck. Both silicone and hypalon are available in white, black and a number of colors.

Elastomeric roofing may be selected over other materials for its ability to conform to any contour or shape, because it offers a wide choice of color without changing the basic shape or texture, because it provides a reflective surface which deflects solar heat, or because of its light weight. Elastomeric roofing can be patched or completely reconditioned or color changed quickly and without a lot of preparatory work. It has enough elasticity to expand and contract without loss of bond to the deck. Another advantage is that it does not require aggregate shielding from the destructive rays of the sun. The principal disadvantage of elastomeric roofing is the cost.

Application
Elastomeric roofing can be applied over smooth, clean, dry concrete or exterior grade plywood decks. Other surfaces may be acceptable, but special techniques or procedures may be necessary. Most manufacturers recommended consultation with their experts before proceeding.

Joints, recessed nailheads in plywood, voids or rock pockets in concrete should be spackled or filled with a material compatible with the elastomer. Follow the recommendation of the manufacturer. Fins and other abrupt protrusions must be ground off smooth and flush with adjacent surfaces.

Where movement in deck joints is expected or likely, the joint can either be stabilized or allowed to work. If allowed to work, the area adjacent to the joint is not bonded to the elastomeric roofing. This way the membrane will have room to stretch as working takes place. Where considerable movement is anticipated, the joint or crack should be covered with sheet neoprene strips or tapes especially made for this purpose.

Where only nominal movement is likely, the joint can be reinforced with fiberglass embedded in the liquid elastomeric roofing material.

Some manufacturers do not recommend application on surfaces below 40 degrees because the surface may contain ice crystals or other moisture. Also some manufacturers do not recommend application on lightweight or foam concrete or gypsum decks because of the residual moisture inherent in these decks. Vapor cannot escape through the membrane and blistering results.

The first or base coat is generally a specially factory prepared primer or a thinned basecoat elastomer to provide maximum bond and penetration. Coatings are applied by brush, pressure fed roller, or spray. Each coat is a slightly different color. This provides visual evidence of complete coverage. Each coat is allowed to dry until it is tack free and dry enough for foot traffic without damage. Then subsequent coats are applied. Coats are applied until the minimum dry film thickness is reached. The dry film thickness will generally be somewhat less than the wet film due to shrinkage in drying.

Preformed sheet elastomers are cemented to the roof deck and at joints with adhesives recommended by the manufacturer. It is not necessary or desirable that the sheet be totally bonded over the entire area. The sheet should, in fact, not be bonded to the deck directly over joints or cracks where movement might occur. The material should not be stretched as it is being cemented down. At lapped joints the mating surfaces should be cleaned with thinner-cleaner. Then both surfaces are coated with adhesive, allowed to dry to the touch, and then brought together and rolled to apply pressure. Material used as roofing or flashing is protected and colored by one or two coats of liquid hypalon. When used for below grade waterproofing, it should be protected from damage in backfilling. Backfill should not contain rocks or gravel with sharp edges.

The hypalon top coat directly over neoprene base will sometimes soften and discolor the neoprene. A second coat of hypalon will generally run color true. Best practice is to delay the top hypalon coats as long as possible until all other work on the roof is complete. Colors may vary slightly from batch to batch. Therefore all containers from a single batch should be used on a single plane surface. Where two batches are needed to cover a single plane, intermixing might be required if slight variation of hue is detected.

Elastomeric sheet material is often used as flashing material, either with fluid-applied elastomeric roofing or conventional built-up roofs.

Balconies and walkways on roof decks are made with liquid or sheet membranes protected with a troweled-on wearing surface. For heavy duty decks supporting vehicular traffic, a Portland cement topping is often used. For foot traffic, the topping may be a fiberglass mat reinforced liquid neoprene or hypalon. Liquid applied elastomeric toppings are often filled with sand, cork particles or ground walnut shell aggregates.

Both taped and overlapped joints in sheet material and liquid applied systems may be visible at close range, especially with lighter colors. In liquid applied systems, this can be overcome or made less conspicuous by adding texture granules or covering the entire area with fiberglass instead of tape at joints or cracks.

Labor for Installing Sheet Roofing

Work Element	Unit	Man-Hours Per Unit
Butyl-rubber sheet roofing, self flashing		
$1/32$"	SQ.	4.38
$1/16$"	SQ.	4.52
$3/32$"	SQ.	4.68
$1/8$"	SQ.	4.79
Neoprene hypalon sheet roofing		
One ply, $3/16$"	SQ.	8.12
Extra heavy $3/8$" traffic deck	SQ.	9.75

Time includes move on and off site, unloading, stacking, cleanup and final sheet material trimwork as required. For splicing and taped joints, add 5 to 10% to time shown.
Suggested Crew: Asphalt, felt and rolled roofing, 2 roofers, 1 laborer. For elastic, neoprene, 2 roofers, 2 laborers.

Labor Installing Liquid Roofing

Work Element	Unit	Man-Hours Per Unit
Neoprene hypalon liquid roofing		
5 coat spray-on application	SQ.	8.90
Vinyl liquid		
2 mills	SQ.	9.20
4 mills	SQ.	10.40
Silicon liquid roofing		
Brushed on	SQ.	4.00
Rolled on	SQ.	3.10
Sprayed on	SQ.	1.80

Time includes move-on and off-site, unloading, stacking, clean-up and repairs as necessary. Time does not include surface preparation.
Suggested Crew: 1 roofer, 2 laborers

Built-Up Roofing

Built-up roofing is made of a base sheet and various felts which are strengthened and waterproofed by alternate moppings or coatings of coal tar or asphalt, which are then topped with a surfacing material. Each built-up roof is custom-made on the site.

Roofing materials are measured in "squares" or fractions of "squares". A square is 100 square feet of roofing. A 15-pound felt means that 100 square feet of that material weigh 15 pounds, or .15 pound per square foot. Weights of asphalt are also given in squares. A 30-pound coating means 30 pounds per square or .30 pound per square foot.

Built-up roofing is made from either roofing asphalt or coal tar pitch: both are heavy hydrocarbons produced by fractional distillation. These bitumens are the materials that make the roof waterproof. Felts do not keep water out. The sun's rays dry out the bitumens and change their chemical structure so that the bitumen becomes brittle with age and cracks. The felts and the top coating of aggregate surfacing or mineral surfacing protect the bitumen from the sun and delay the aging process. Water standing on asphalts accelerates the degrading of asphalts. Bitumens that are to be applied hot should never be diluted with solvents to soften the material.

Bitumens flow as fluids when they are heated. On warm days bitumens are elastic. In winter they become hard. All bitumens have good adhesion and uniform deformation under shearing stress. The chief disadvantage of bitumens is that they are flammable when sufficient heat is applied.

Roofing bitumens are shipped as solids. They must be broken up and placed in kettles and melted. Material weights must often be converted into gallons per square. The average asphalt weighs 8.72 pounds per gallon when liquid. The average coal tar pitch weighs 10.62 pounds per gallon when liquid. Divide the weight of the bitumen used into the total specified weights per square to find the number of gallons per square. For example, 30 pounds of asphalt at 8.72 pounds per gallon will yield 3.44 gallons of asphalt.

Manufacturers and roofers use two terms to describe diluted bituminous products. **Cutback** bituminous roofing has been thinned with an organic solvent. It has been made to flow more freely by the addition of lighter oils. Cutback products can usually be applied cold. **Emulsion** roofing is made from asphalt and fine droplets of water in dispersed in the asphalt with the aid of an emulsifier. Asphalt emulsions are used in "cold process" roofs. These thinned bitumens can be applied by spraying.

Roofing asphalts are cut, mineral stabilized, blended and processed into various types that are appropriate for various roof scopes and functions. All roofing asphalts have a flash point of about 437 degrees. The flash point is the lowest temperature at which it gives off vapor sufficient to form an ignitable mixture. Steep asphalts can safely be heated to 475 degrees.

Asphalt for use in built-up roofing is classified into four types or categories by ASTM D312. These four types have four different melting or softening points. In general, the softening point determines where the asphalt can be applied. Or to put it another way, the slope of the roof dictates the type of asphalt to be used. Manufacturers of asphalt group their products by softening or melting point and brand name. Low slope asphalts or low melt asphalts are used for "dead level" roofs, and roofs having slopes up to ½ inch per foot. These asphalts are in the lower softening-point range, 135 degrees to 148 degrees F. In hot climates these lower softening-point asphalts must not be overheated since overheating can lower the normal softening point. In a built-up roof, the lower layers of felt adjacent to the base sheet can easily reach 180 degrees F on a hot day. At this temperature low slope asphalt becomes liquid. It is possible for the roof to slowly slide away. A low softening point does have "self healing" properties when it is warm. ASTM D312 classifies this as Type I asphalt.

ASTM D312, Type II asphalt is used for inclines greater than ½ inch up to and including 3 inches per foot. It has a softening point of 160 degrees to 175 degrees. ASTM D312, Type III asphalt is used for inclines greater than ½ inch up to and including 6 inches per foot and has a softening point of 180 degrees to 200 degrees.

Many companies do not market asphalts conforming to ASTM standards. The contractor may have to special order material that will meet job specifications. Steep asphalt has a softening point from 205 degrees to 225 degrees and is used in areas with a relatively high year-round temperature. This harder grade of asphalt can also be used for smooth roof surfaces. However, it lacks the self-healing properties of low melt asphalts.

Asphalts for dampproofing and waterproofing are thicker and may be applied to vertical surfaces.

Asphalt primers are very thin liquids for application on concrete and masonry surfaces above or below ground level. They are used to improve the adhesion of heavier applications of bitumen.

Asphalt cements are thick pure asphalts with no fillers and only enough solvent to permit application at the recommended spreading rate. Cements are used for sealing laps of flashing and to seal around nails.

Plastic cements, flashing cements, all purpose roofing cements, and cut back cements are trowelable mixtures of asphalt or coal tar with fillers like asbestos, glass fibers, powdered aluminum, or various rubbers and a solvent. They can be applied cold and are generally used to secure flashings.

Roofing materials manufacturers sell various emulsions and roof coatings. These products are basically roofing asphalt with sufficient water added to make them liquid and some clay (bentonite) or chemical emulsifying agent added to make the bitumen globules disperse uniformly throughout the mass. They may contain asbestos or glass fibers or other compounds. Emulsions must not be subjected to freezing temperatures because of their water content. Freezing will make emulsions useless.

Coal-tar pitch is often specified as an option for roofs with a slope up to one inch in twelve inches. Coal-tar pitch is seldom used because it costs more. Coal-tar pitch is virtually unaffected by water even under prolonged immersion. It has excellent self-healing properties. After being punctured, the pitch flows together to close cracks even in moderate weather conditions. Pitch is from blast furnaces after the destructive distillation of bituminous coal or the manufacturer of coke or gas. It has a narrow range of softening points from 120 degrees to 155 degrees. Coal-tar pitch for roofing is specified in ASTM Specification D450 as Type A. It can be used in waterproofing applications when not exposed to temperatures exceeding 125 degrees. Type A coal-tar pitch has a softening point from 140 degrees to 155 degrees. After cooling, coal-tar pitch is much softer than asphalt. When solid pitch is broken, it has a shiny glass-like appearance.

Roofing Felts

The flexible sheet materials provide strength for the membrane system and help protect the bitumen in the lower layers from the drying and weathering effects of the sun. The felts also protect the bitumen from dirt, debris, water, and some of the oxidizing effects of the air. Roofing felts are usually perforated. They are also absorbent so that they hold the bitumen in place and keep from flowing away in warm weather. The perforations let trapped air out from under the sheet during application. Glass fiber felts are porous and thus do not require perforation. If the felts are made of fire resistant materials or are coated with fire resistant materials, they help keep the bitumen from burning or at least slow down the spread of fire.

Flexible sheet materials for roofing come in 36 inch wide rolls in a wide variety of weights and materials and perform several functions in the built-up membrane and flashing system. Felts are often coated with a fine mineral powder on one or both sides to prevent sticking in the roll.

Flexible sheet materials for built-up roofs include roofing ply felts, coated sheets, rosin sized-papers or unsaturated felts, mineral surfaced sheets and special purpose sheets such as those made especially for ventilation purposes. The center of each roll of roofing materials often contains the cement, nails or tin caps required for that roll. Each roll has lateral lines or markings to help locate the lap point for subsequent layers of felt.

Flashing felts include roofing ply felts, finishing felts, mineral surfaced cap sheets, coated sheets or other specially reinforced sheets made for flashing purposes.

Base sheets may be coated sheets, roofing ply felts, or other special purpose sheets. Base sheets are usually heavier than roofing ply sheets. Base sheets may be laid in a complete layer of bitumen, laid over a spot mopping of bitumen, or nailed. The base sheet forms a base for the complete membrane system. When it is secured to the roof it receives a complete coating of bitumen over its top surface.

A four-ply roof usually means a base sheet and three plies of roofing felt. Five-ply roofing systems are usually made from two lower plies nailed to the deck and three plies mopped on. However, nailing two plies tends to reduce the amount of expansion and contraction the roof can accept without failure. Roofing manufacturers have developed three-ply and even two-ply roofing systems that give reasonably good service.

Mineral-surfaced systems, smooth-surfaced systems and three-ply systems are not used in permanent construction. More expensive one, two and three-ply systems use materials other than asphalt or coal-tar pitch. They may provide as good a roof as the conventional four and five-ply built-up membrane systems, but they are used only under special conditions where the additional cost can be justified.

Roofing felts or ply felts are non-woven fabrics produced in two basic types: organic and inorganic. The most commonly used roofing felts weigh 15 pounds per square. Equivalent glass fiber roofing felt weighs only 7½ pounds per square. Heavier roofing felts weigh 30 pounds per square. Manufacturers frequently designate their felts by weight. "Type 8" means a 7½ pound glass fiber felt. "Type 30" means a 30 pound felt.

Organic or rag felts are made of felted papers and shredded wood fibers which are saturated with asphalt, tar and other waterproofing compounds. Rag felts have little or no rag content, although they may contain vegetable or animal fibers. Organic felts are subject to deterioration by oxidation and wicking. Asphalt saturated organic felts are used more than any other roofing felt because they are the cheapest felts available. Inorganic felts are made in two different types. Asbestos felts are manufactured from asphalt saturated thread-like fibers of magnesium and calcium silicate. Glass-fiber felts are made from asphalt and glass-fiber mats. A light coating of asphalt adds strength to the glass fibers. Inorganic felts are non-rotting, non-wicking, and have a much longer life than organic felts. In addition the glass fiber are fire resistant.

Rosin-sized sheathing papers or unsaturated felts are applied over wooden decks and other constructions where bitumen could drip down thru the roof deck. These sheets also keep the roof from drying out through open joints. The nailed in place rosin-sized papers act as a slip sheet to prevent the membrane from adhering to the roof deck. This helps relieve the expansion - contraction problem. Rosin-sized papers or unsaturated felts are not at all waterproof. They weigh about 5 pounds per square, have a hard, smooth surface and are clean to handle.

Coated roofing sheets are felts which are coated on both sides with hot bitumen and finished with a heavy coat of non-sticking powdered mineral matter. Fiberglass material weighs 10 or 18 pounds per square while organic materials weigh 30, 35, 37, 40, 43, 45, 53, 55, or 58 pounds per square. Coated base sheets act as vapor barriers, preventing the air from entering the underside of the roof. Coated sheets are also used as the top ply in smooth surface built-up roofs.

Mineral surfaced sheets, sometimes called cap sheets or mineral surfaced split sheets, are heavy sheets that are coated on both sides and receive an embedded coating of mineral granules. The mineral surfaced side is exposed to the weather and provides a finished roof surface that does not require aggregate. Cap sheets usually have a 4" selvage edge with 32" of mineral surfacing. The mineral surfaced split sheet is 36" wide with a 177" wide band of mineral surfacing exposed to the weather overlapping 19" across the unsurfaced portion of the roll. Mineral surfaced sheets are not very durable, but they are a quickly installed and cheap roofing material. Mineral surfaced sheets vary in weight from 58 to 110 pounds per square. Since they are the visible portion of the finished roof, mineral surfaced sheets are available in a variety of colors including white and aluminum.

One type of heavy sheet material is used on roofs where walkways around equipment would be too expensive. This material consists of a mineral surfaced sheet on top, a center core of asphalt and fillers, and a bottom sheet of organic felt. These sheets are usually ½" or ¾" thick and are bonded to the top roofing ply with bitumen.

Securing Built-Up Roofing

Roof fastening machines are rayon cord tape ⅜" wide which holds large staples at specified intervals. A machine simply rolls tape over the edges and centers of the felts, securing the felts to the deck with the staples as it travels. This leaves the tape exposed to be covered with bitumen. These machines hold the roof down better than individually spaced fasteners because the tape provides additional holding power. The Asphalt Roofing Manufacturers' Association Committee on Built-up Roofing recommends a minimum 40 pound pullout strength per fastener to prevent wind uplift.

Nails are the most commonly used fasteners. They are usually galvanized steel. Wood decks are nailed with large head nails. The head may be integral with the nail, or the nail may be driven through a loose nail cap. The shape of the head or cap can be square or round as long as it has an area equivalent to a 1" diameter head. This area is required to prevent tearing the felt. Plywood decks require annular ring shank nails in addition to large heads. The common "roofing nail" is not usually satisfactory for good roof construction. Nails must be long enough to penetrate wood at least 5/8 inch and plywood ½ inch. Nails must not penetrate the underside of the decking because the metal point provides a place for drops of water to form in the attic.

Built-up roofing can be used over gypsum decks, insulating concretes, lightweight concrete, insulating boards, preformed wood fiber boards and similar "nailable decks". Various proprietary nails or fasteners are made especially for use over particular materials. These fasteners are not interchangeable. The chief concern is to get enough holding power to prevent wind uplift and slippage. In many cases it is necessary to use both fasteners and adhesive. Lightweight (insulating) concrete is considered nailable if the mix is between 1:4 and 1:6 and has been cured.

On steel roof decks the fastener must be specially made to penetrate both the insulation and the steel deck. "Friction locking" fasteners that rely on springing between the flutes in the steel deck may not be satisfactory. Mechanical fasteners must penetrate through or into the steel decking and remain fast in the steel decking. The rock surface is the last thing to go on the built-up roof. Aggregates perform four important functions. They shield the bitumen and the felts from the drying effects of the sun. They protect the felts from wind, hail, people and sharp objects. They act as ballast to resist wind uplift, they absorb heat and make the roof more fire resistant. Aggregate is set in bitumen that is flooded on top of a roof at a rate of about 60 pounds per square. The aggregate holds the bitumen in place while it congeals and until the aggregate itself is held in place by the bitumen. Aggregates for roofing must be clean, dry, opaque, free of dust and properly graded. Aggregates used include gravel, slag or crushed rock.

Other aggregates are used when appearance is important: marble chips (dolomite); colored rock; crushed tile, brick or limestone; scoria, pumice (volcanic rocks); asbestos rock; and lightweight aggregates peculiar to certain areas. All of these aggregates are usually more expensive than the standard aggregates. Some may be coated with dust or minute particles which don't bond to bitumen. Others are translucent so that the sun reaches the bitumen and dries it out. When the bitumen dries out it loses its adhesive value. It also cracks. The advantage of white chips is that they reflect more of the sun's rays and keep the roof cooler in summer. Some gravels and crushed rocks come close to the same degree of heat reflectance.

Smooth-surfaced roofing uses various weights of inorganic felts exposed to the weather. The felts may have a protective coating of special steep asphalt, cutback or emulsion, but not aggregate. Smooth surfaced roofing can be used on steeper slopes.

Mineral surfaced roofing sheets are lighter by 3 to 4 pounds per square foot. A properly built aggregate surfaced roof should last 20 years or more. Mineral or smooth surfaced roofs are considered temporary and do not last more than 6 years before requiring renewal or additional coatings. A smooth-surfaced roof can be worn away by the abrasive action of grit, cinders and other particles carried by high winds across the surface of the roof. Aggregate-surfaced roofs are subject to this same sandblast effect, but the slow wearing away of the aggregate doesn't make much difference to the life of the roof. It should also be noted that well over half of all failures are the result of faulty flashing and this can occur on both smooth-surface roofs and aggregate-surfaced roofs.

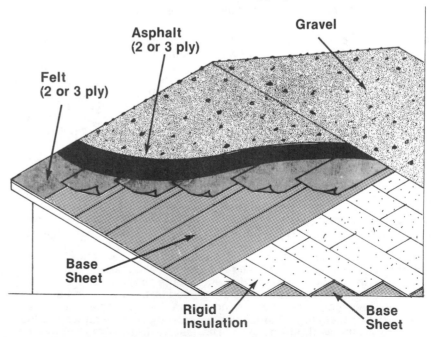

7 Thermal and Moisture Protection

Application

The built-up roofing with its own flashing drainage system is a unique item being built on top of the structure. A good built-up roof requires correct application of the proper materials, using the best methods or technique for that particular roof. The roofing system depends on everything coming together at the proper time to create a real work of art. The key to the successful roof is to keep the surfaces dry and clean before the roof is applied and keep the materials dry before use. Roofs do not wear out; they dry out.

Built-up roofing should not be used on a roof with a slope of over 3 inches in 12. The bitumens simply won't stay on the roof in hot weather.

For built-up roofing, the basic principle is that no felt should touch another felt. All felts, including the base sheet (if any), are separated from each other by a layer of bitumen.

Roofers sometimes refer to the "pitch" of the roof instead of the slope. The pitch is the angle or degree of the slope of a roof from the plate to the ridge and is expressed in degree of inclination or inches of vertical rise to a horizontal foot. The pitch is found by dividing the rise or height to the highest point by the span or total width of the roof. If the height is 8 feet and the span is 16 feet, the pitch is 8/16 or ½". The angle of pitch is 45 degrees.

Asphalt and coal-tar pitch are incompatible and should not generally be used together. There are certain exceptions to this rule. Asphalt flashing material and asphalt-coated base sheets with a top of stabilized asphalt and mineral surfacing can be used with coal-tar pitch. And a glaze coating of hot Type II asphalt can be used over coal-tar saturated roofing felts.

All felts are strongest in their longest dimension and weakest across their width. This built-up weakness is a result of the way felts are made. Built-up roofing should be applied with its long dimension parallel to the long dimension of the roof. The practice of going up and over the crown of the roof puts the greatest pull across the width of the felts, which pull the felts apart in time unless an extra roll of material is put down the long dimension over the crown and nailed or fastened securely with adhesive. Heavy mineral-surfaced split sheets and cap sheets can be applied over a crown if in accordance with the manufacturer's instructions.
If the roof is laid over insulation board, the insulation should have been applied with its long dimension parallel to the short dimension of the roof. Thus the roofing felts' longest dimension is perpendicular to the short dimension of the insulation. If the felts are laid parallel to the longest dimension of the insulation board, the felts will probably split and wrinkle, causing early roof failure. When the roofing ply joints coincide with the insulation joints, the felts above are almost sure to crack.

Considerable cooperation is required between the sheet metal workers and the roofers. To assemble the roof properly, the sheet metal workers should put in some of the metal flashings at the perimeter of the roof and elsewhere before the roofing plies are placed. After the roofing plies are in place, the sheet metal man should put on the remainder of the metal flashings and the metal counter flashings. The sheet metal worker may try to do all the work at one time. The result will be a roof with metal flashings improperly installed and a sure leak.

Adhesives for flashings and roofing plies tend to become too fluid in hot weather and too brittle in cold weather. There is always the big difference in the expansion-contraction rates of felts and metals. This is why some specifications insist on nailing or mechanically fastening metal flashings and flashing felts.

The deck or surface must be clean and free of projections. Holes in wood over 1" in diameter may require a small piece of sheet metal. Knot holes may need a little shellac to prevent them from rising or falling out Ridges between boards may have to be re-nailed or ground smooth. Nails should be set to prevent them from rising. Poured surfaces such as gypsum and all classes of concrete should be ground smooth if necessary. Any insulation on the roof must be in good condition before the roofing is applied.

The roof must slope to the drainage points or the entire roof structure will settle as the building ages, making low spots in the roof more pronounced. Low spots can be filled with Portland cement mortar, an epoxy filler or some other product that can be feathered.

Be alert to loads placed on the deck during construction. Storage loading on platforms should not normally exceed 40 pounds per square foot of uniformly distributed load. Mechanical equipment used on the roof should not deflect the roof and must not puncture the membrane.

If the roofing is to be placed on a wood or nailable deck, the first sheet will be a layer of rosin sized paper, lapped two inches, and nailed enough to hold it in place. Complete nailing is not particularly important for the dry sheet since the plies that follow will have many more nails. Then the base sheets gets nailed (not mopped) into place. A combination glass fiber sheet will not require rosin-sized paper. It can be nailed directly to the deck.

If the base sheet is placed over lightweight concrete, a ventilation sheet will be applied in lieu of the base sheet. This ventilation sheet is required to carry away moisture from the concrete to the perimeter of the roof. Ventilation sheets are like mineral-surfaced sheets except that the mineral-surfaced side has an incised waffle-like grid that forms escape passages for the vapor. The mineral-surfaced side is placed against the insulating concrete. The smooth side is up to receive bitumen. Ventilation sheet is nailed or clipped to the deck. Bitumen or adhesive would fill up the little channels and destroy the ability of the sheet to carry moisture away. This special ventilation sheet weighs about 72 pounds per square.

Moisture release vents, stack vents or "coolie hats" are not used for releasing vapor. These built-in holes in the roof are easy to damage or install wrong so the entire roof is undermined by the very water the roof is designed to keep out.

Once the surface has been properly prepared, with rosin-sheet in place if required, the coated base sheet, combination sheet or special ventilation sheet is applied. The usual base sheet laps 4 inches over the preceding sheet. The end laps should not be less than 4 inches. Base sheets are staggered so the end laps do not occur on the same line. The base sheet is nailed as specified, 8, 9 or 12 inch intervals on the laps and stagger nailed down the center as specified. If the deck is not nailable and the roof slopes more than 1" per foot, wooden roof nailers should have been provided. They are either set in concrete or attached to the surface with the insulation placed in between the wooden nailers. The base sheet is nailed to the wooden nailers and is the only roofing ply that is put down by itself.

It is better practice to allow for expansion and contraction between the roof deck and the base sheet. This isolates the built-up membrane from the structure. However, the membrane must be securely fastened to prevent wind uplift. Don't have the bitumen tightly bond two materials that are supposed to expand and contract at a different rate from one another. Another advantage of a nailed base sheet is that it allows horizontal passage of water vapor that is trapped between the deck and the membrane.

If the roof is to be placed on insulation, most specifications require 30 pounds of hot steep asphalt per square on top of the insulation. Then the base sheet or combination sheet is quickly installed in the hot asphalt. Asphalt may be used to place the base sheet in coal-tar-pitch systems.

If a porous insulation board is used under the roof surface, some jobs require that the base sheet be applied by an adhesive other than hot asphalt. Specifications may be vague on what adhesive to apply. If a "fire rated" adhesive must be used, that eliminates a lot of choices. For fibrous insulations that absorb adhesive, the adhesive should be applied in thick heavy ribbons so that the area for absorption by the insulation is less.

Regardless of what adhesive is used, the adhesive must not be permitted to dry out. This is most important. Where asphalt is used as an adhesive, even a ten minute wait will prevent formation of a good bond between the insulation and the base sheet.

The specifications may call for adhesive applied to the insulation in spots or rows rather than solid mopping. The spots or rows of bitumen will allow the built-up roofing some movement between the insulation and the built-up membrane. If the solid mopping technique is used, the base sheet will have greater bond and greater wind resistance. But solid mopping allows no movement of the base sheet whatever. The difference in the expansion and contraction rate of the upper membranes and the lower membrane may cause the membranes to split open.

After the base or combination sheet is applied, the 3 roofing plies are put down. All the plies are rolled out on top of the base sheet at the same rate, parallel to each other. The bitumen is applied to each ply so that the bitumen is sandwiched in between each ply of felt. The felts originate at an edge or a low point and work up so the edges shed water to the ply below like shingles on a roof. For three plies, each ply laps 24⅔ inches over the previous ply. This leaves 36 inches minus 2 divided by 3 or 11⅓ inches of each ply exposed. End laps should not be less than 6" and must be staggered.

If the roof is to be placed directly on a structural concrete deck, the deck should be coated uniformly with asphalt primer at a rate of a gallon a square or 7.5 pounds per 100 square feet. After the primer is dry (about 24 hours), the roof is coated with hot bitumen at a rate of 20 to 30 pounds per square or spot mopped if so specified. The first roofing ply is applied on top of the bitumen while it is hot. No base sheet is needed over structural concrete.

If no base sheet is to be applied, the four plies of roofing felt are applied by the shingling method in the same manner as the three plies and base sheet. But for four plies, each ply laps 27½ inches over the previous ply. This exposes 36 minus 2 divided by 4 or 8½ inches.

The overlap of plies in both cases (with or without the base sheet) assures that any vertical cross section will receive the minimum number of specified plies.

The last two items needed to complete the built-up membrane are the flood coat and the aggregate or other surfacing. The top ply is coated uniformly with hot bitumen (asphalt at 60 pounds per square, coal-tar-pitch at 70 pounds per square). While the bitumen is still hot, dry aggregate or slag is distributed on the roof.

Aggregate must not be stockpiled on the roof. Obviously it is likely to collapse the structure. Dusty aggregate is not acceptable because the bitumen will not bond properly. Normally the roof receives 400 pounds of gravel or crushed stone per square or 300 pounds of roofing slag per square. High wind areas (70 mph and up) should get a "double surfacing" of aggregate. This would be an additional 90 pounds of coal-tar-pitch or 80 pounds of asphalt and 200 pounds of slag or 300 pounds of gravel per square.

If the work is stopped at any point before all of the felts and aggregate can be applied, it is essential that the top surface of the exposed roofing be given a glaze-coating. This glaze-coat consists of 25 pounds of hot, steep, (180-200 degree softening point) asphalt per square. This protects the base sheet and the felts from water, moisture, dirt, and traffic until the work resumes. The roof should be immediately glaze-coated if the work stops for any reason. The roof in place can be ruined very easily if the glaze-coat is omitted. Organic felts exposed to air over night pick up moisture from the air, even on a dry night. Moisture causes them to expand and subsequent drying causes them to shrink. The result is curling and wrinkling. The glaze-coat seals the edges so water doesn't get under the felts.

If any moisture gets trapped in the plies of the roof, it will evaporate, expand and cause blisters. The lower plies of a roof can reach 180 degrees when the sun shines on the roof on a 100 degree day. In the winter the temperature of an insulated roof is near the outdoor temperature. Water vapor enters on a cool day, is raised 100 degrees and exerts a pressure of over 4 pounds per square inch, or almost 600 pounds per square foot on the membrane. This is sufficient to lift any part of the roof.

The glaze-coat requires extra labor and materials. In warm weather it leaves a sticky surface which is difficult to work on. It also gives subsequent plies a lumpy appearance since the bitumen required between the plys of felt go over the glaze coat when the work begins again. But the glaze-coat is necessary.

Base flashings below the aggregate usually consist of regular roofing felts. For the felt envelope at the perimeter it is best to use the same materials as are used in the built-up membrane because they have identical expansion characteristics. Asbestos felts and plastic cement are best. When base flashings are adjacent to metal, they should be nailed rather than fastened with bitumen to allow for movement. Roofing plies or flashing felts secured to metal with solid bitumen are sure to split during expansion and contraction. The base flashings are secured with plastic cement because it is likely to remain permanently soft and waterproof. These flashings should also allow for movement.

Labor Installing Built-up Bituminous Roofing

Work Element	Unit	Man-Hours Per Unit
Bitumen top built-up roofing with 15 lb. asphalt felts		
Three ply	SQ.	1.8
Four ply	SQ.	2.0
Five ply	SQ.	2.4
Gravel surfaced built-up roofing		
Four ply asphalt felt, 15 lb.	SQ.	2.8
Five ply asphalt felt, 15 lb.	SQ.	3.2
Gravel surfaced 90 lb. cap sheet on built-up roofing		
Three ply, 15 lb. felt	SQ.	2.3
Four ply, 15 lb. felt	SQ.	2.9
Cut and place cant strips	100 L.F.	2.0

Time includes move on and off site, unloading, stacking, cleanup and repairs as necessary. For sloped roofs over 2 in 12 pitch add 10 to 15% to time. Time does not include removal of old roofing.
Suggested crew: 2 roofers, 1 laborer.

Materials for Asbestos Shingle Roofing

Kind of Shingle	Size in Inches	Exposure in Inches		Number Per 100 S.F.	Weight Per 100 S.F.	Lbs. of Nails Per 100 S.F.
American	12 x 24	7	x 24	86	275	1.0
American	14 x 30	6	x 30	80	265	2.0
Dutch lap	12 x 24	9	x 20	80	260	1.0
Dutch lap	16 x 16	12	x 13	92	260	0.8
Dutch lap	16 x 16	10-2/3	x 13	104	295	1.0
Hexagonal	16 x 16	13	x 13	86	265	1.0

Labor Installing Rigid Roofing Tile and Shingles

Work Element	Unit	Man-Hours Per Unit
Glazed clay roofing tiles	SQ.	7.18
Concrete roofing tile		
Flat types	SQ.	4.60
Spanish "S" tile	SQ.	8.30
Mediterranean tile mortar set	SQ.	9.20
Porcelain enamel shingles	SQ.	6.12
Aluminum shingles, colored or plain	SQ.	4.90
Bonderized aluminum shingles	SQ.	5.30
Steel shingles	SQ.	4.50
Slate	SQ.	5.50

Time includes move on and off site, unloading and job stacking tile or shingles, installing complete with underlayment, ridge and valley units, flashing and mortar where required.
Suggested Crew: 2 roofers, 1 helper

Labor for Asbestos Cement Roofing Shingles

Work Element	Unit	Man-Hours Per Unit
14" x 30" x 5/32"	SQ.	3.45
9" x 16" x 1/4"	SQ.	3.20
8" x 16", 6" exposure	SQ.	4.10
16" hex, 8" exposure	SQ.	3.45
9" x 32", with underlay	SQ.	4.30
12" x 24" with underlay	SQ.	4.30

Time is man-hours per 100 square feet of roof and includes move on and off site, unloading and stacking. Shingles on sidewalls usually take 10 to 20% less time because exposures can be increased. If exposures are not increased, add 10% for shingling sidewalls.

Recommendations for Reroofing with Shingles

The New Shingles	Existing Shingles Prior to Reroofing			
	Wood	3-Tab and Strip Shingles With No Cut Outs	Lock Shingles	Dutch Lap and Hexagon Shingles
Wood	Yes, in all areas except wet climates. Then the old shingles should be removed.	Yes	Yes	Yes, but a tear off would make a smoother roof.
3-tab and strip asphalt shingles	Yes, if the butt-up method can be used. If not, the old shingles should be removed.	Yes, if the butt-up method can be used. If not, the old shingles should be removed.	No. Tear off the shingles or use another shingle.	No. Tear off the shingles.
T-Lock Asphalt Shingles	No. Use a 3-tab, strip shingle or a wood shingle.	Yes	Yes, but offset the headlaps.	Not desirable. A tear off would make a smoother job.
Shakes	Yes	Yes	Yes	Yes
Heavily textured asphalt strip shingles (shake appearance)	Yes, if the butt-up method can be used. Some brands have a different weather exposure, and with these the butt-up method can not be used.	Yes	Only the heavier shingles should be used.	Only the heavier shingles should be used.

In a "butt-up" reroofing job the head of each new shingle is placed flush against the butt of each existing shingle to offset shingle headlaps.

Labor and Materials for Asphalt Roofing

Type of Roofing	Shingles Per 100 S.F.	Nails Per Shingle	Length of Nail*	Nails Per 100 S.F.	Pounds Per 100 S.F. (Approximate)		Man-Hours Per 100 S.F.
					12 Ga. by 7/16" Head	11 Ga. by 7/16" Head	
Roll roofing on new deck	--	--	1"	252**	0.73	1.12	1.0
Roll roofing over old roofing	--	--	1¾"	252**	1.13	1.78	1.25
19" selvage over old shingles	--	--	1¾"	181	0.83	1.07	1.0
3 tab sq. butt on new deck	80	4	1¼"	336	1.22	1.44	1.5
3 tab sq. butt reroofing	80	4	1¾"	504	2.38	3.01	1.85
Hex strip on new deck	86	4	1¼"	361	1.28	1.68	1.5
Hex strip reroofing	86	4	1¾"	361	1.65	2.03	2.0
Giant American	226	2	1¼"	479	1.79	2.27	2.5
Giant Dutch lap	113	2	1¼"	236	1.07	1.39	1.5
Individual hex	82	2	1¾"	172	.79	1.03	1.5

*Length of nail should always be sufficient to penetrate at least ¾" into sound wood. Nails should show little, if any, below underside of deck.

**This is the number of nails required when spaced 2" apart.

Labor for Shingles and Shakes

Work Element	Unit	Man-Hours Per Unit
Wood shingles		
16" perfections, 5½" exposure	SQ.	2.3
18" perfections, 6" exposure	SQ.	2.0
Split or sawn cedar shakes, 10" exposure		
24" long ½" to ¾" thick	SQ.	2.9
White cedar, 5" exposure		
16" long	SQ.	2.5
Fire retardant cedar shingles		
16", 5" exposed	SQ.	2.6
18" perfections, 5½" exposed	SQ.	2.2

Time assumes pitch less than 6 in 12 roof. Sidewalls will take 10 to 20% longer unless exposures are increased.

Exposure for Wood Shingle Roofing

Shingle Length Inches	Shingle Thickness (Green)	Maximum Exposure Slope Less Than 4 in 12, Inches	Slope 5 in 12 and Over, Inches
16	5 butts in 2 in.	3¾	5
18	5 butts in 2¼ in.	4¼	5½
24	4 butts in 2 in.	5¾	7½

Minimum slope for main roofs is 4 in 12.
Minimum slope for porch roofs is 3 in 12.

S.F. Coverage of One Square (4 Bundles) of Wood Shingles

Weather Exposure	Length and Thickness 16" x 5/2"	18" x 5/2¼"	24" x 4/2"
3½"	70	--	--
4"	80	72½	--
4½"	90	81½	--
5"	100	90½	--
5½"	110	100	--
6"	120	109	80
6½"	130	118	86½
7"	140	127	93
7½"	150	136	100
8"	160	145½	106½
8½"	170	154½	113
9"	180	163½	120
9½"	190	172½	126½
10"	200	181½	133
10½"	210	191	140
11"	220	200	146½
11½"	230	209	153
12"	240	218	160
12½"	--	227	166½
13"	--	236	173
13½"	--	245½	180
14"	--	154½	186½
14½"	--	--	193
15"	--	--	200
15½"	--	--	206½
16"	--	--	213

Sheet Metal

Sheet metal included in division 7 of the Construction Specification Institute format is custom fabricated or stock architectural items used largely at the joint where materials meet or for control of moisture. Ductwork is usually listed under section 15 of the specifications with heating and cooling materials.

Sheet metal items are generally fabricated from cold rolled metals. Cold working, however, causes metals to harden. If softer, easier to form metal is desired, the material will be either hot rolled or cold rolled and annealed.

Some metals are coated with paints or other metals to improve corrosion resistance or appearance. Metal coatings on sheet material are generally applied by the hot dip method.

Steel and iron sheets are widely used because of their low cost, workability, solderability, and low thermal expansion. They do, however, require a protective coating against corrosion. For interior work, proper cleaning, treating, priming, and painting will generally provide enough protection. For other exposures, coating with less corrosive metals by the hot dip method provides the best protection. Ferrous metal sheet thickness is always given in U.S. Standard gage. Except for framing, stiffening members, or other incidental work, anything over 10 gage is considered to be sheet metal.

Galvanized steel is used for exterior sheet metal work. The zinc coating is available in several thicknesses and with differing characteristics. Zinc is deposited on steel sheet by either electroplating or the hot dip method.

Galvanizing combines high corrosion resistance with low cost. The zinc forms a tenacious bond with the base metal and can be easily soldered with lead solder. It does not stain adjacent surfaces, and can be used in direct contact with wood (except redwood and red cedar, both of which contain acids) concrete, mortar, lead, tin, and aluminum. It even protects steel that has been exposed on cut edges or where the coating is scuffed or scratched through. Zinc-coated steel used in architectural sheet metal work generally has a commercial prime coat of 1.25 ounces of zinc per square foot of sheet (.625 ounces per square foot of surface area on one side).

Terne metal or terneplate is the other common steel material used for exterior sheet metal. It is lightweight, strong, solderable with lead-tin solder, and can be severely formed. It is more expensive than galvanized steel, and is used primarily for painted metal roofing and flashing. Terne is a lead-tin alloy in which the base metal is coated by the hot dip process.

Modern terne is manufactured in two grades, short terne and long terne. Short terne is made in widths of 14, 20 24, and 28 inches in 50 foot rolls and in sheets of various lengths. The material is made from copper bearing steel alloy dipped in hot lead-tin alloy containing about 85% lead and 15% tin. This grade is used for roofing, flashing, gutters, and leaders.

Long terne is made in 90 foot rolls and is cut to larger sheets of up to 4 feet by 10 feet. It has a coating weight of 8 to 40 pounds per single base box unit and is available in 14 to 30 gage. The coating alloy ranges from 80 to 87½% lead to 12½ to 20% tin. Long terne is used for doors and frames and other fireproof construction as well as for roofing. **Copper** has a long history of successful use as architectural sheet metal. Uncoated copper sheet, on exposure to the elements, forms a thin tight coating of copper carbonate. This coating is pale green and may take several years to build up a uniform thickness and color. Once formed, it prevents the base metal from further corrosion. Copper sheet and strip used in sheet metal work is known commercially as tough pitch copper. It is 99.9% pure, and is available in hot rolled (known as soft, or dead soft) or cold rolled (known as hard or cornice temper) forms. Copper, unlike most other sheet metals, is specified in terms of ounces per square foot rather than thickness. Twenty-four ounce copper is about $\frac{1}{32}$ inch thick. Sheet copper is rolled in about 28 thicknesses ranging from 2 ounce (0.0027") to 96 ounce ($\frac{1}{8}$").

Copper and its alloys used in architectural sheet metal are easily worked and soldered, and resist corrosion in most atmospheric exposures. They are used primarily in severe exposure conditions or on structures built to last many years.

Aluminum, like copper, is made in many alloys in sheet form. Very little sheet metal used in construction is pure aluminum. To obtain maximum mechanical properties, lightness, and corrosion resistance, aluminum is always alloyed with copper, manganese, magnesium, chromium, nickel, or zinc.

The alloy most frequently used in sheet metal work is 3003 alclad or 3004 alclad. The word "alclad" after the alloy number means that thin slabs of aluminum alloy or commercially pure aluminum are fused to the core alloy by heat and pressure in a hot rolling operation. This cladding provides better corrosion resistance and finishing characteristics than the core alloy. The core alloy is designed for strength rather than wear resistance. Aluminum alloys cannot be coated with aluminum by the hot dip process because the melting point of the alloy is the same or lower than the cladding.

Anodizing is an artificial method of depositing a thin film of oxide on the surface of the metal. The anodic film is dense and hard and consists of billions of closely packed hexagonal cells, each containing a microscopic pore or opening.

Alclad aluminum is used most frequently in sheet metal work. Anodized aluminum is used mostly for ornamental work and trim, such as extruded gravel stop or fascia.

Aluminum is more expensive than galvanized steel but less expensive than copper. Unlike most other material used in architectural sheet metal work, aluminum is difficult to solder. Waterproof joints are generally made with a combination of sealants and mechanical fasteners. Aluminum thickness is expressed in decimals of an inch.

For exterior applications, the **chrome-nickel stainless steels** are most frequently used. The chrome-nickel AISI Types used in architecture are 302, containing 17 to 19 percent chromium and 8 to 10 percent nickel, known also as Grade 18-8. Type 430 contains no nickel, has less resistance to corrosion than the chrome-nickel (300 series) and the 200 series types. It is used primarily for interior exposures. Types 201, 202, and 301 have properties and characteristics quite similar to types 302 and 304. They are appropriate for certain architectural applications, but are not as widely used as 18-8, Type 302. Type 316, (18-12) is used in more corrosive seacoast or marine atmospheres. **Monel** is a white metal that looks like stainless steel. Its principal alloys are nickel (about 66%) and copper (about 30%) with small amounts of iron, manganese, silicon, and other elements. It has good coefficient of expansion, strength, ductility, and corrosion resistance characteristics. Monel is used in just about every application where stainless steel could be used.

Sheet lead is the softest and most dense material used in architectural sheet metal work. It resists corrosion and can be easily shaped. The properties that make lead the choice over other sheet metals are its density and softness. It is used extensively to line walls and doors to shield against x-rays and nuclear radiation, and to act as a sound barrier. Its softness allows hand forming on the job to meet almost any desired contour. It is popular as a flashing material on tile roofs. Pans in plumbing installations, other waterproof linings, laboratory sinks and table tops are often made of lead. It can be used as a roofing material. Lead is produced and sold in thicknesses calculated at pounds per square foot in sheet and strip form. Lead sheet weighs approximately 1 pound per square foot for each $\frac{1}{64}$ inch of thickness.

Solders used in sheet metal work are generally the common 50-50 lead-tin alloy. Stainless steel and nickel alloys are soldered with high (60-70%) tin material. Although there are special alloys for soldering aluminum, they are seldom used in architectural sheet metal work. Different fluxes are used with solder on different metals. Most are corrosive and all traces should be removed after the joint has been soldered. Soft solders are used to conceal joints or to make joints watertight. They are not meant to resist stresses.

Brazing and welding are seldom used as a method of joining architectural sheet metal. These methods do, however, produce strong joints which do not require reinforcement by mechanical fasteners or locked seams to resist stresses as soldered joints do. Where aesthetics or sanitation are prime considerations, make the joint by butting, welding with a similar color alloy rod, and grinding smooth.

Where soldered, brazed, or welded joints are impractical, waterproof joints are made by applying sealant between the two pieces of metal to be joined. Then the joint is finished with mechanical fasteners or by lock folding the metal. Several sealants are used for this purpose, but the non-curing butyl base type is probably the most popular and best choice.

In general, fasteners exposed to moist environments should be of an alloy similar to the metal with which it is in contact. Nails and screws are used to fasten sheet metal to wood. Nails should preferably have barbed, ringed, or screw shanks. Sheet metal screws and rivets are used to fasten sheet metal to itself. Where sheet metal screws are exposed to the exterior, and waterproof fastening is required, the screw should have a neoprene washer under its head. Metal backed neoprene washers are preferable inasmuch as this design allows the washer assembly to spin free under the screw head thereby avoiding torsional stresses in the neoprene as the screw is driven home.

Rigid, long lasting waterproof joints in sheet metal are made by a combination of soldering and mechanical fastening. Mechanical fastening is generally done by riveting or folding the material into a "lock". Good waterproof joints are also made by welding. But this technique is seldom used in sheet metal work because thin metals are difficult to weld. Welding tends to burn off the protective coating on ferrous metal. Brazing is seldom used in sheet metal work because it is expensive and because the dissimilar metals are likely to cause galvanic corrosion.

Soldering of sheet metal work is done with a lead-tin alloy, generally half lead and half tin. Since this alloy has a lower melting point than lead, even lead can be soldered. Almost all metals form a thin oxide film on exposure to the atmosphere. This film forms quite rapidly on some metals. Successful soldering requires that this film be removed and kept off the metal until it is wetted with molten solder. One of the functions of a flux is to remove this oxide. Fluxes, however, are not a substitute for cleaning the metal to a bright shine before soldering.

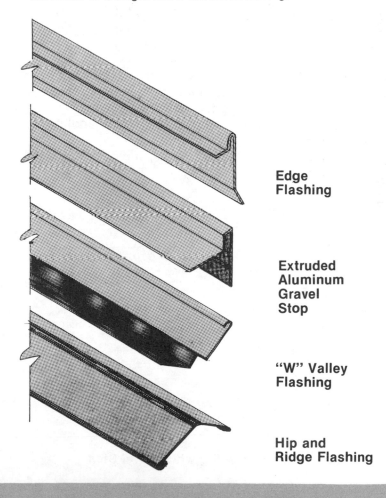

Edge Flashing

Extruded Aluminum Gravel Stop

"W" Valley Flashing

Hip and Ridge Flashing

Sheet Metal Work

Work Element	Unit	Man-Hours Per Unit
Fabrication (galvanized steel)		
Roof gutters	100 L.F.	2.5
Downspouts	100 L.F.	2.5
Roof ridges	100 L.F.	1.5
Roof valleys	100 L.F.	1.5
Flashing	100 L.F.	2.3
Installation (galvanized steel)		
Roof ridges	100 L.F.	2.5
Roof valleys	100 L.F.	2.5
Roof flashing	100 L.F.	8
Roof gutters	100 L.F.	4
Downspouts	100 L.F.	4

Fabrication is usually performed by a sheet metal shop and includes making patterns, cutting, forming, seaming, soldering, attaching stiffeners, and loading for delivery.

Installation includes unloading, storing on site, handling into place, hanging, fastening, and soldering.

Suggested Crew: two to six steel workers depending on the weight and length of the materials

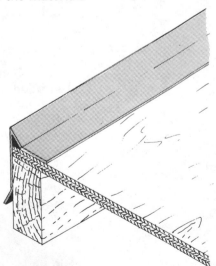

Installed Roof Flashing with Gravel Stop

Labor Installing Roof Flashing

Work Element	Unit	Man-Hours Per Unit
Copper flashing		
16 oz.	100 S.F.	9.0
18 oz.	100 S.F.	9.1
20 oz.	100 S.F.	9.3
22 oz.	100 S.F.	9.4
24 oz.	100 S.F.	9.8
3 oz. with paper back 2 sides	100 S.F.	4.7
3 oz. with mastic back 2 sides	100 S.F.	4.8
Aluminum flashing		
0.019 inch	100 S.F.	4.7
0.020 inch	100 S.F.	4.8
0.024 in.	100 S.F.	4.9
0.050 in.	100 S.F.	5.0
Stainless steel flashing (Type 304)		
No. 22 gauge	100 S.F.	8.8
No. 26 gauge	100 S.F.	8.6
No. 28 gauge	100 S.F.	8.3
No. 30 gauge	100 S.F.	8.1
0.005 in. thick with paper backing	100 S.F.	3.2
Lead flashing		
1.2 lbs. per S.F.	100 S.F.	7.9
2.5 lbs. per S.F.	100 S.F.	16.6
24 oz. lead coated copper	100 S.F.	9.9

Time includes unloading, stacking, cutting and clean-up.

Suggested Crew: 1 roofer

Labor Installing Commercial Metal Gutters and Downspouts

Work Element	Unit	Man-Hours Per Unit
Copper 18 oz.		
4"	100 L.F.	12.6
5"	100 L.F.	13.5
6"	100 L.F.	14.2
Stainless steel		
4"	100 L.F.	12.6
5"	100 L.F.	13.4
6"	100 L.F.	14.1

Time includes unloading, stacking, set-up, cutting and clean-up.
Suggested Crew: 1 roofer, 1 laborer

Labor Installing Roof Hatches

Work Element	Unit	Man-Hours Per Unit
Steel roof hatches. Not including mounting curbs		
Up to 6 S.F. each	S.F.	1.00
7 to 10 S.F. each	S.F.	.96
11 to 15 S.F. each	S.F.	.90
16 to 20 S.F. each	S.F.	.87

Time includes unloading, lifting and clean-up. Time required to install the mounting curb is not included.
Suggested crew: 1 glazer, 1 laborer.

Caulking and Sealants

Included under this heading is work that requires application of caulking compounds, glazing compounds, mastics, sealants, tapes, gaskets, and joint fillers. These materials fill joints between building materials. Consistencies vary from pouring grade for filling horizontal joints through gun grade, knife grade, and extruded or preformed material which can be used for either horizontal or vertical joints. Caulking and sealing materials have a wide range of prices and properties. Each product has its limitations and range of application.

In general, oil based compounds are known as mastics, caulking, or glazing compounds. Glazing materials may include gaskets, tapes, oil based putty, mastics and sealants. Window wall glazing on high-rise buildings may use a combination of materials in each installation to take full advantage of the properties of more than one material. The term **sealant** is generally associated with the elastomerics. Joint fillers are usually preformed impregnated fiber used in joints in concrete slabs on grade. Their function is to absorb compression stresses and to keep the joint free from impaction by gravel and other debris. Joint fillers are sometimes held slightly below the finished elevation, and then sealed flush with an elastomeric sealer after the concrete has cured.

Thermosetting compounds cure to their final consistency when heated, either by self generated heat or upon exposure to an outside source of heat, or both.

Chemical reaction thermosetting compounds shrink little or not at all as they cure. They have high bond strengths and are generally more elastic than other types. Examples include polysulfides, polyurethanes, and silicones. These are sometimes modified by the addition of other compounds, particularly coal tar and petroleum derivatives.

Solvent release thermosetting compounds cure by evaporation of the solvent. These materials shrink on curing in proportion to the volume of solvents released to the atmosphere. Examples would be neoprene, chlorosulfonated polyethylene (Hypalon), and butadiene styrene.

Thermoplastic compounds cure or harden from heat just as the thermosets do. But unlike the latter, thermoplastic materials may be changed from solid to liquid many times by repeated application of heat.

Hot melt thermoplastics include rubber-asphalt, rubber-coat tar and asphalt. Chemical cure thermoplastics are the epoxies. Thermoplastics that release solvent on exposure to the atmosphere include acrylics, vinyls, oleoresins, and butyls.

Field molded sealants are applied by pouring, caulking gun, putty knife, and similar tools. The uncured consistency ranges from syrup viscosity to stiff putty. They include all of the thermoplastic and thermosetting compounds.

Premolded sealants are available in a number of sizes, shapes and forms. They may also be called pre-cured because their properties do not materially change after they have been installed. They may be in the form of round ropes, hollow tubes, channel shaped beads, flat ribbons and other shapes. They are made from natural and synthetic rubbers (polysulfides, polyurethanes, silicones, neoprenes, hypalons, butyls, and vinyls), asphaltic and oleoresinous compounds. They are extruded, cast, or punched into final shape from flat goods. Soft or sticky materials such as butyl and oleoresinous compounds may be reinforced with cloth or felt. Asphaltic compounds are often used to impregnate fabrics, felts, and synthetic sponge material. Although some of these materials adhere fairly well to other materials and some are highly elastic, none of these materials when used as a preformed sealant have both properties. All preformed sealants are designed for use in compression or restrained joints.

The polybutenes and polyisobutylenes, in both the premolded and field molded forms, remain permanently tacky. They should be used only in concealed locations where they will not collect dirt. Joints such as cement asbestos or metal roofing and siding are good examples. In glazing and other applications, they are covered by other sealants or trim members.

Elastomeric sealing compounds may be classified by the way they cure. Two-part compounds are packaged so that the complete contents of each package are mixed together to start the curing cycle. Cure time depends on the air temperature and humidity. Two part sealants can also be factory mixed, tightly sealed, and refrigerated until used. In this form they are actually one-part sealants. Polysulphide base compounds are sometimes used this way. Epoxy compounds are also two-part mixes, but they are not elastomeric. They tend to restrain a joint and must be considered joint fillers or adhesives rather than materials that can seal a working joint.

One-part solvent-release compounds cure by evaporation of solvents. Materials in common use include butyls, neoprenes, hypalons and acrylics.

One-part chemical curing compounds cure by reaction with moisture or oxygen in the atmosphere. They include polysulfides, silicones, and polyurethanes.

Where sealants are used in working joints, such as expansion or control joints in unit masonry, the joint design as well as the sealant material deserve careful consideration. If this is a butt joint, the sealant will go through alternating cycles of tension and compression. In a lap joint there will be a shear force on the sealant. Some joints may even force the sealant material into three deformations.

Control the depth of butt joints in unit masonry or concrete with a sponge rope, rod, or tube of plastic or synthetic rubber. The sealant is inserted into the joint a uniform depth and held in place by friction.

Concrete sidewalks, roadways and other slabs on grade are generally designed with joints to absorb thermal movement. Such joints must be filled with some sort of resilient material to keep the void from being packed with dirt, sand, or gravel. In this situation, sealing out moisture or air is not important. Nor is adhesion important as long as the material remains in place. The important thing is how long the material will remain resilient yet rigid enough to resist impact by gravel or damage from spike heels. The material should resist damage by exposure to water, sunlight, oil and other spillage.

Adhesion is an important consideration in working joints where the sealant will be subject to tensile or shear stresses. The adhesion should be equal to or better than the pull developed at maximum elongation in the joint. Some sealant manufacturers have developed primers for use on construction materials before the sealant is placed. Porous surfaces, such as concrete and masonry, generally require primers that penetrate into the pores and create both a mechanical and chemical bond over a larger surface area.

Elastomeric sealants in working joints work better if they go through alternate cycles of stress in one direction only and then shear in the opposite direction. Constant less severe stress tends to shorten joint life. For this reason, joints should be sealed when they are at their average width (not fully extended as in the cold cycle or fully contracted as in the hot cycle). As the joint passes through expansion and contraction cycles, the sealant bead should be deformed in a symmetrical pattern. For this reason, butt joints should be adhered on two sides only. The back or bottom of the joint should be free to deform in the same concave or convex pattern as the face.

Preparation of the Joint. Joint sealant contact surfaces should be clean and dry. Concrete and masonry should be fully cured. All loose particles, dust, scum, laitence, efflorescence, and other unsound material should be removed. Bond breaking liquids should be removed with suitable solvents. Most materials can be removed with wire brushes followed by an air blast. Metal, plastic, and other non-porous surfaces should be cleaned with water, solvents, or commercial cleaners as recommended by the sealant manufacturer.

Protect adjacent exposed surfaces from smears or stains by masking with masking tape. This tape should be removed immediately after the sealant has been placed and tooled.

Working joints that are not simple through-wall butt joints (such as tongue-and-groove, splined, rabbeted, and other configurations with a "bottom") should have a bond breaker on the bottom or return surface. If the bond breaker is applied in liquid form, this should be done before the panels are assembled. This avoids contaminating the panel edges that receive the sealant. Pressure sensitive adhesive tape bond breakers are available for this purpose. They are not as likely to contaminate bonding surfaces as are liquids. But both types should be handled with care.

Working through-wall butt joints require a backing material to control joint depth and to support the sealant when it is tooled. Backing material, or backup filler as it is sometimes called, should be about ⅓ greater in diameter than the joint width to be sure that there is enough pressure against the joint walls to hold it firmly in place by friction. The material usually has a round cross section. It should not bond to the sealant and should be more compressible than the cured sealant. The filler should not be coated with a lubricant or other material that would contaminate the joint edges and act as a bond breaker. The backing material is forced into the joint to a uniform depth. Ends should be snugly butted.

Preparation of the Sealant

Prepackaged single component cartridges need no preparation except that the temperature should be in the range recommended by the manufacturer. Most cartridges are equipped with a conical plastic nozzle. This nozzle can be cut near the tip to extrude a fine bead, or up nearer the base to provide a coarse bead. The shape of the sealant bead can also be controlled by the way the nozzle is cut.

Bulk single-component sealants may or may not require mixing prior to use, depending on the type, viscosity, and tendency of the components to separate during storage. Thermoplastic hot melt sealants should be heated to the plasticity required for placing. Be careful to avoid overheating.

Preformed sealants generally require no preparation other than stripping off the waxed paper used in packaging.

Installation

After the joint and sealant have been properly prepared, the sealant is applied. Sealants in restrained joints are generally applied to one surface, then the mating surface is brought into contact and secured with mechanical fasteners. Metal, plastic, asbestos board and similar roof and wall panels are generally sealed by placing the sealant bead on the weather side.

Pouring grade (self-leveling) sealants are poured directly into level horizontal joints.

Gun grade sealants of stiff non-sag consistency can be used in vertical and horizontal floor and overhead joints. The nozzle of the cartridge or pressure fed hose is inserted into the joint. The sealant is extruded as the gun is drawn along the joint. If the gun is drawn too fast, the sealant may not completely fill the joint. If the gun is drawn two slowly, or if the sealant is fed too fast, the material will pile up in the joint and may smear or stain adjacent surfaces. Properly applied, the sealant should completely fill the joint down to the backing material and leave no voids. The joint is then tooled slightly concave like a masonry mortar joint. This tooling gives the joint a neat uniform appearance, tightly packs the sealant against the joint edges, and ensures better bond.

Bulk sealants of stiff consistency may be pumped by hose to the application nozzle. They also may be applied with a trowel, putty knife, squeegee or other hand tools.

Labor for Caulking and Sealing

Work Element	Unit	Man-Hours Per Unit
Oil-based caulking and sealants		
¼" x ½" joint	100 L.F.	1.7
½" x ½" joint	100 L.F.	1.8
¾" x ¾" joint	100 L.F.	2.1
¾" x 1" joint	100 L.F.	2.1
Polyurethane compounds		
½" x ½" joint	100 L.F.	2.0
½" x ¾" joint	100 L.F.	2.7
¾" x ¾" joint	100 L.F.	4.1
¼" x ¼" joint	100 L.F.	1.9
½" x ¼" joint	100 L.F.	2.0
¾" x ⅜" joint	100 L.F.	2.3
1" x ½" joint	100 L.F.	4.1

Time includes move on and off site, set-up, repair and clean-up as required.
Suggested crew: 1 painter.

8

doors
& windows
section

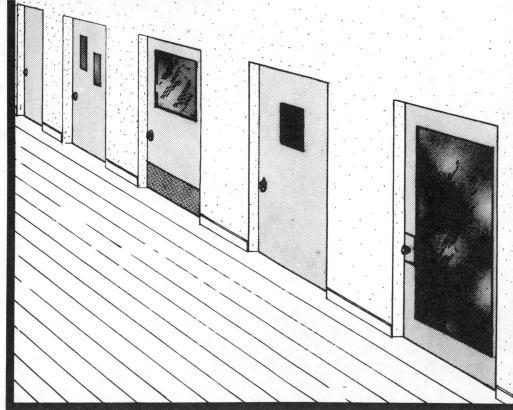

section contents

Doors

Hollow Metal Doors and Frames

These doors have a skin of face sheets (panels) attached to each side of steel strutting, honeycomb, or mineral core. When properly constructed, the assembly forms a rigid, light "sandwich" of metal around the core.

Hollow metal doors are made of cold-rolled low carbon steel on hot-rolled pickled and oiled steel. The material skin ranges from 22 gage to 16 gage, depending on the type and style.

Hollow metal doors are either 1⅜" or 1¾" thick. Standard widths are 2', 2'-4", 2'-6", 2'-8" and 3'. Standard heights are 6'-8", and 7'-0".

The Steel Door Institute has classified hollow metal doors into **types** which define construction, durability and use, and **styles,** which define appearance. The various classifications are listed below.

Types
Type I Light Commercial (Standard Duty)
Type II Commercial (Heavy Duty)
Type III Industrial (Extra Heavy Duty)
Type IV Custom (Special Construction)
Type V Bi-Fold (Closet Doors and Frames)

Styles
Flush
Semiflush
Recessed Panel
Full-Glazed Stile and Rail
Louvered
Combinations of these

Hollow metal doors (and their frames) are constructed for use as **fire doors.** Fire doors are designated as Class A, B, C, D, or E:

Class A - resists a specified amount of heat for 3 hours
Class B - 1½ hours
Class C - ¾ hours
Class D - 1½ hours (exterior)
Class E - ¾ hours (exterior)

The label on the door indicates the fire test rating for the door.

Steel door frames are also referred to as hollow metal frames. Actually steel frames are literally "bent" into shape. The terms **pressed metal, pressed steel,** or simply **steel** frames are better.

Steel frames are made with the same rolled processes used in hollow metal doors. Only 18 to 16 gages are used, however. The heavier 16 gage steel frames are used mostly with doors 1¾" thick.

Steel frames are available in many configurations which accommodate a variety of wall constructions. They are fabricated to fit standard door sizes and can be ordered with provision for sidelights, transoms, pairs of doors, or any combination of these.

Steel frames for doors, transoms, and sidelights are either the fully welded unit type (one piece assembly), or the knocked down field-assembled type. Generally, welded frames are recommended for custom grade work and buildings such as laboratories, hospitals, and other public facilities.

When used with fire labeled doors, steel frames must carry the same rating as the door.

Wood Doors

Wood is an ideal material for doors because of its durability and versatility. It has natural texture, grain, color, and is a natural insulator. Wood doors are produced in a variety of designs, sizes and types — panel, sash, flush, and louver. They can be adapted to all types of construction and architectural design. Various methods of installation, operation, and arrangement can produce special effects at a minimum cost.

Through scientific treatment and preservatives, wood doors are more durable and permanent than ever. They are highly resistant to decay and dimensional change such as swelling, shrinkage, and warping when treated with water preservatives. They do not rust or corrode and can withstand both acid fumes common in many industrial areas and salt air.

Most wood doors are manufactured under industry standards adopted by the National Woodwork Manufacturer's Association (NWMA) or the U.S. Commercial Standards (Dept. of Commerce). Wood doors meeting these standards are furnished with a stamp or certificate indentifying (1) the standard complied with, (2) the testing and inspection bureau, (3) the manufacturer and plant, and (4) a declaration of compliance by the plant. The standard guarantees performance of material and assembly for a prescribed period of time, usually one year.

The most common door is the flush type. Flush wood doors are either hollow core or solid core. They are available for exterior and interior use, with lights and louvers, with plastic faces, and other characteristics.

Solid core wood doors have a solid inner core of wood blocks, composition board, board, wood particleboard, or bonded mineral material.

Hollow core wood doors have a core of strips of wood or wood derivatives on insulation board. The space between these strips is hollow. The strips support the outer faces.

The most common species of flush door veneers are birch, ash, mahogany, gum, cherry, oak, elm, and walnut. Other species are available. The species, quality, and cut of face veneers can increase the price of otherwise identical doors by 75% or more.

Plastic Covered and Metal Clad Doors

Flush wood doors may also be furnished with plastic or metal facings. The plastic faces are 1/16-inch-thick plastic laminate. This material is available in an almost unlimited number of solid colors and patterns (including wood grains) and textures or finishes. Plastic faced doors are ideal where minimum maintenance is essential. Plastic laminate is always applied to wood on hardboard backing. Metal clad wood doors are used in fire hazard areas such as boiler rooms, or rooms where flammable materials are stored.

8 Doors and Windows

Special Purpose Doors

Acoustical doors should comply with all applicable requirements for solid-core flush doors. In addition, of course, they are constructed to comply with the specified Sound Transmission Class (STC) rating. Ratings from 35 to 55 STC are fairly common. X-Ray resistant doors comply with the requirements for solid core flush doors but have one or more continuous sheets of lead. A total minimum thickness of at least $\frac{1}{32}$" is built into the door. The lead sheet can be in the core, between the core and crossbanding, or between the crossbanding and face veneer. Each sheet must be continuous from edge to edge and from top to bottom, and must be an integral part of the door. Door faces can be of hardwood, hardboard, or plastic.

Prehung Doors

Both wood and hollow metal doors can be supplied by the manufacturer as a single prehung unit. The door arrives on the job already installed in the frame. All hardware is likewise factory installed. Factory supplied packaged units are ideal when field installation time is critical, or when skilled manpower is not available at reasonable cost.

Steel frames are the structural supports for steel doors. They must be set accurately, plumbed, aligned and securely braced until anchors are permanently set. When positioned, floor anchors can be placed and set (with either power fasteners or expansion shields and bolts). If the floor is to receive a topping or finishing material, the steel frame must be fitted with adjustable brackets which elevate the bottoms of frames to the top of the finished floor.

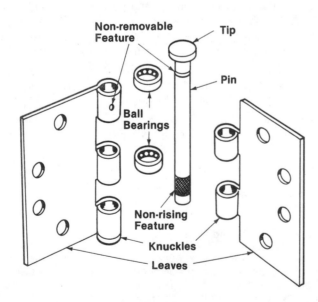

Parts of a Hinge
Figure 8-1

Door Hinges

Figure 8-1 shows the parts of a hinge and Figure 8-2 shows the four basic types of hinges. They are available from most manufacturers in wrought steel, stainless steel, brass, bronze and aluminum with all of the common finishes.

The standard design has five knuckles and button tips. Most manufacturers offer a premium priced modern design featuring two or three knuckles, flush tips, concealed bearings or slimmer barrels. Also ball, oval head, steeple and various other decorator tips are available at extra cost.

Hinges are manufactured in standard weight with plain bearings, standard weight with two anti-friction bearings and heavy weight which are approximately 50% thicker than leaves of standard hinges. The width, thickness, weight and frequency of use of the door determine the hinge weight and bearing type to be used. Doors with closers should have anti-friction bearings. Doors with overhead holders should be equipped with heavy weight hinges. Anti-friction bearings include ball bearings, oil impregnated (oilite) bearings and nylon bearings.

Non-rising pins have knurled surfaces or other means of preventing the pin from rising during normal operation of the door. Non-removable pins have a set screw which prevents the pin from being removed when the door is closed. Non-removable pins are required where the barrel of the hinge is exposed on the key side of locked doors. Fast pins are pins which are permanently anchored into the barrel at the factory and are provided with hospital tip hinges. Fast pins are used in high security applications.

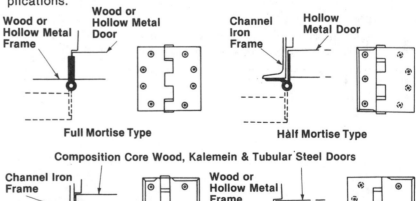

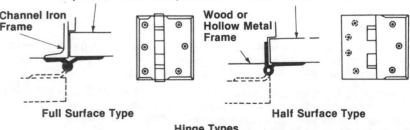

Hinge Types
Figure 8-2

The sizes of half mortise, half surface and full surface hinges are designated by height only. Width varies according to height. The sizes of full mortise hinges are designated by height and open width. Height is always given first. The required height of the hinge is determined by the width and thickness of the door. Full mortise hinges should be wide enough to provide clearance of trim or reveal when the door is opened to 180 degrees. Hinges wider than 5 inches are referred to as wide throw hinges.

Swing clear hinges (see Figure 8-3) have a barrel offset which allows the door edge to swing clear of the door opening when opened to 90 degrees. This is desirable in hospitals and other installations where gurneys, beds and carts must be moved through doorways. These hinges are manufactured of wrought steel in heavy weight only.

Anchor hinges and pivot reinforced hinges are mortise type top hinges with horizontal leaves or plates which are mortised into the top jamb of the door frame and into the top edge of the door. See Figure 8-4. The horizontal leaves eliminate bent leaves and loosened screws when doors are treated roughly or where overhead holders stop doors suddenly. The hinges are sold in sets with heavy weight mortise type intermediate and bottom hinges.

Thrust pivot units as in Figure 8-5 serve the same purpose as anchor hinges but are furnished in sets of three heavy weight mortise type hinges.

Slip-in hinges are full mortise type hinges with special swagging which places the leaves further apart when closed. They are used on aluminum doors and frames with special slots cut for the leaves.

Friction hinges have adjustable friction devices which hold the door open at any angle and prevent slamming. They are available in wrought steel, heavy weight, and full-mortise type only.

Double and triple weight hinges have leaves approximately two or three times the thickness of leaves of regular weight. They are manufactured from wrought steel in full surface type only and have two ball bearings. Lead lined doors and prison doors weighing up to 800 pounds require double weight. Doors up to 2,000 pounds require triple weight.

Concealed or invisible hinges as in Figure 8-6 are available from a limited number of manufacturers for use on doors in paneled walls where hinge barrels would interrupt the flush appearance of the wall.

Spring hinges as shown in Figure 8-7 can be used to hang, swing and close both double-acting and single-acting doors of almost any dimensions, weight and composition. The closing force is supplied by enclosed torsion springs. The closing speed is regulated by adjusting the spring tension. There is no checking of the closing action of the spring hinge. Therefore, their use is usually limited to areas where the noise of the door slamming is not objectionable.

Door Pivots

Pivots are sometimes used instead of hinges where extra heavy doors are involved because almost all of the weight of the door is transferred to the floor through the bottom pivot. All pivots are equipped with anti-friction bearings.

Pivots are available in two basic types: offset (Figure 8-8) and center hung (Figure 8-9) types. Only center hung pivots can be used on double-acting doors. Single-acting doors may have either center hung or offset pivots.

Where offset pivots are used, the bottom pivot for doors weighing 150 pounds or less can be jamb mounted with the door leaf mortised into the jamb of the frame. For heavier doors, use floor mounted bottom pivots installed in the floor with the door leaf mortised into the bottom edge of the door. Top and intermediate pivots are available for full mortise, half

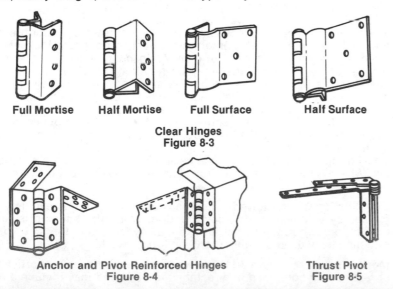

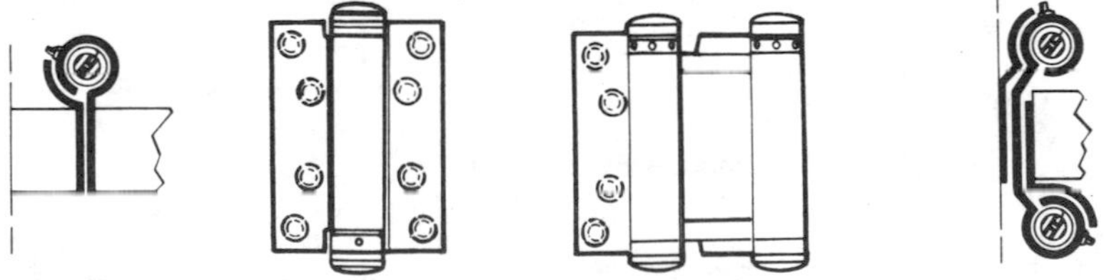

Full Mortise   **Half Mortise**   **Full Surface**   **Half Surface**

Clear Hinges
Figure 8-3

Anchor and Pivot Reinforced Hinges
Figure 8-4

Thrust Pivot
Figure 8-5

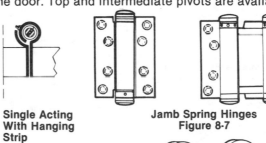

Single Acting With Hanging Strip

Jamb Spring Hinges
Figure 8-7

Double Acting No Hanging Strip

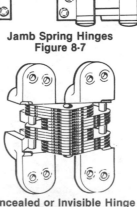

Concealed or Invisible Hinge
Figure 8-6

mortise, half surface and full surface applications. Mortises for top pivots are cut into the top edge of the door and the head of the frame. Generally an intermediate pivot will be required for doors between 60 and 90 inches high. An additional intermediate pivot will be required for each 30 inches of additional door height.

If a center hung pivot is used, the bottom pivot can be either jamb or floor mounted, as with offset pivots. For heavier doors, an adjustable lever arm located in a recess in the bottom of the door replaces the mortised door leaf. Top pivots are of the walking beam type and must be recessed into the head of the frame with the jamb leaf mortised into the top edge of the door. The walking beam mechanism allows the top pin to be lowered into place after the door has been set in the opening.

The spring pivots in Figure 8-10 are used with center hung top pivots when a double acting closing mechanism without checking action is needed. Spring pivots are available with horizontal or vertical springs with exposed parts of steel, brass or bronze. Spring pivots are mounted in a small cutout in the bottom heel of the door. As with pivots, spring pivots can be either floor or jamb mounted.

Gate gravity pivots and gate spring pivots may be used on gates, louvered bar doors and other similar light weight doors. Their most common use is on toilet compartment doors. They can be mortised into the bottom of the door or gate or set below gates where sufficient room exists.

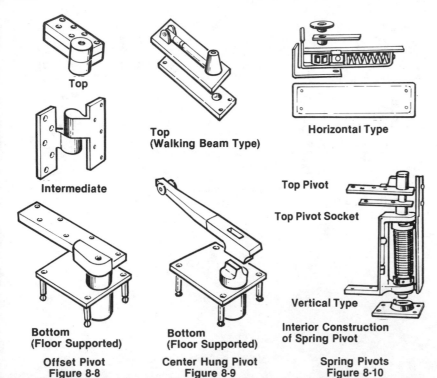

Top

Top (Walking Beam Type)

Horizontal Type

Intermediate

Top Pivot

Top Pivot Socket

Vertical Type

Bottom (Floor Supported)

Bottom (Floor Supported)

Interior Construction of Spring Pivot

Offset Pivot Figure 8-8

Center Hung Pivot Figure 8-9

Spring Pivots Figure 8-10

Door Latch Sets

Figures 8-11 thru 8-14 show the parts of common latch sets. The latch bolt is beveled and spring actuated to hold the door in a latched position when the door is closed. The size and throw of the latch bolt are good indicators of the grade of the latch set. Throw means the projection of the latch beyond the outside face of the lock front. There is some friction between the latch bolt and the strike. Easy latching of the door may require an anti-friction device the latch bolt. The device may be a split latch bolt, a pivoted latch bolt, a self-lubricating latch bolt or a plastic insert in either the bolt or the strike. These anti-friction devices are especially important when heavy duty lock and latch sets are used on doors with closing devices.

A dead bolt has no spring action or beveled face and must be operated by a key or thumbturn each time it is locked. In some types it can be retracted by turning the inside knob. Like the latch bolt, the size and throw determine the grade of the lock set.

The strike is a metal plate mortised into the door jamb to receive and hold the projected latch bolt and the dead bolt. It includes a lip which should be long enough to prevent the latch bolt from striking the door jamb. A box should be installed behind the strike to prevent foreign matter from filling the projection area for the bolt.

An electric strike has a pivoted portion, operated by a solenoid, which can be retracted to allow the locked latch bolt to pass out of the strike. The electric strike allows the door to be unlocked by electrical controls which may be located remote from the door.

Cylinders can be of the pin tumbler and wafer or disc tumbler types. Pin tumblers are most common because they offer greater security against picking and offer more possible key changes. Pin tumbler cylinders are also available in a wide variety of other types of locks including rim locks, padlocks, cabinet locks and panic exit hardware. Wafer tumblers are available from some maufacturers in residential grade locksets. Pin tumbler cylinders may have 5, 6 or 7 pin chambers to provide the number of key changes that may be required. See Figure 8-15. Each chamber has a change key pin and as many as 3 master key pins. The splits between these pins are lined up on the shear or parting line by notches in the key so that the plug will rotate to unlock the door. The more complex the keying system is, the greater the quantity of pins and splits necessary. The more splits there are, the easier it is to pick the lock since it offers more chances in each chamber to line up a split on the shear line. For this reason, keying systems should be kept as simple as possible. Whenever it is possible to key alike rather than use a master key, it should be done.

Construction master keyed cylinders provide an interim keying system for use by workmen during construction. When the building is turned over to the owner, the owner inserts his key to modify the lock so construction keys used previously will no longer work.

Mortise Type See Figure 8-11. The working mechanism of a mortise lock or latch set is contained in a rectangular case and may include a latch

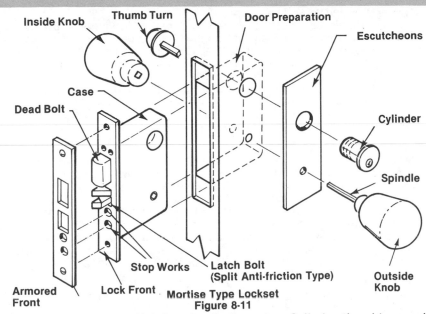

Mortise Type Lockset
Figure 8-11

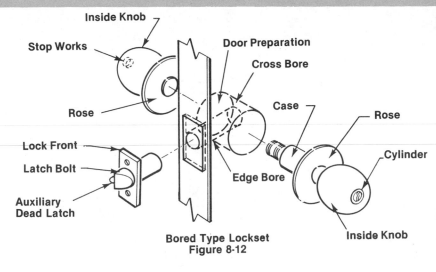

Bored Type Lockset
Figure 8-12

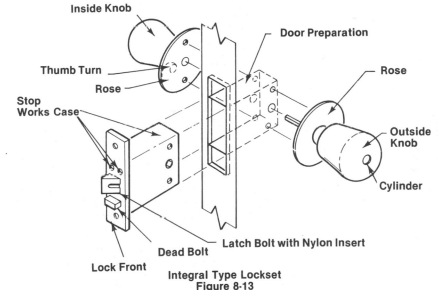

Integral Type Lockset
Figure 8-13

bolt, an auxiliary dead latch, a deadbolt, or stop. Cylinder, thumbturn and knobs, lever handles or pull handles are inserted into the appropriate holes in the case after the case has been installed in the mortise. The outside knob should be pinned to the spindle in the factory with pins that cannot be removed when the door is closed. Other knobs may be mounted on plain spindles with exposed screws through the shank, on threaded or simplex spindles using threaded shanks and set screws or with a screwless shank which eliminates use of screws and prevents unauthorized removal of knobs.

Heads of screws holding the lockset to the door edge as well as heads of screws retaining the cylinder in the lock are located in the lock front. Armored fronts are available to cover these screw heads to prevent tampering with them when the door is open.

Grades of mortise lock and latch sets are standard duty, heavy duty and extra heavy duty. Standard duty can be used in doors $1\frac{3}{8}$" thick. Heavy and extra heavy duty locksets are used on doors $1\frac{3}{4}$ inches thick.

Bored Type See Figure 8-12. The operating mechanism of a bored lock or latch set is located in a cylindrical or tubular housing which is connected to the outside knob and rose and is inserted in the cross bore from the lock side of the door. Cylinders and stop works, when required, are housed in the outside knobs. The inside knob and rose are assembled to the operating mechanism from the other side of the door. Latch bolt and auxiliary deadlatch are housed in a latch unit which is inserted through the edge bore and mates with the operating mechanism. There are no dead bolts in a bored type lockset.

Bored lock and latch sets are classified as light or residential duty, standard duty, and heavy duty. Light and standard duty grades can be used on doors $1\frac{3}{8}$ to 2 inches thick.

Integral latch sets as shown in Figure 8-13 are available from only one manufacturer. The working mechanism is contained in a rectangular case somewhat smaller than that of a mortise type. It may include a latch bolt, an auxiliary dead latch, a dead bolt and stop works. Knob and rose assemblies are inserted through the holes in the case after the case has been installed in the mortise. Cylinders are housed in the outside knob and thumbturns are located on the inside rose. Roses are through bolted to each other with screw heads exposed on the inside rose only. Knobs are affixed to split spindles in the factory and can be removed only by removal of the knob and rose assembly from the lockset.

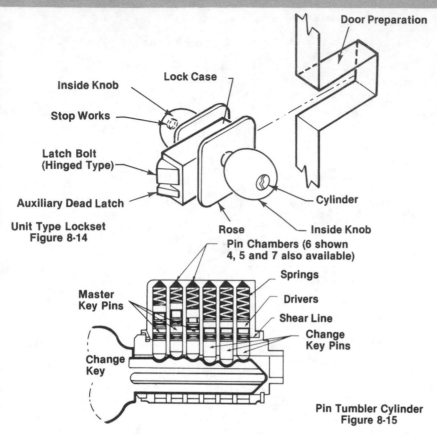

**Unit Type Lockset
Figure 8-14**

Inside Knob
Stop Works
Latch Bolt (Hinged Type)
Auxiliary Dead Latch
Lock Case
Door Preparation
Rose
Cylinder
Inside Knob

**Pin Tumbler Cylinder
Figure 8-15**

Pin Chambers (6 shown 4, 5 and 7 also available)
Springs
Drivers
Shear Line
Change Key Pins
Master Key Pins
Change Key

Mortise Deadlocks

This is a form of mortise lock set consisting of a dead bolt with outside cylinder or inside thumbturn. All components and materials are similar to those described for mortise-type lock and latch sets.

Bored Deadlocks

Bored deadlocks (Figure 8-16) consist of a dead bolt with outside cylinder and inside thumb-turn. The bored deadlock is installed in cross bore and edge bore similar to a bored-type lock or latch set. These locks may be used with a bored lock or latch set to provide deadlocking security or may be used separately where push-pull operation of the unlocked door is wanted.

Rim Locks

Rim locks as in Figure 8-17 consist of a dead bolt or a latch bolt with thumbturn enclosed in a cast bronze, iron or aluminum case for surface mounting on the inside face of the door. The lock includes a cast bronze, iron or aluminum surface mounted strike. The safety or jimmy-proof rim lock has a hardened steel bolt that interlocks with the strike to prevent forcing of the lock. Outside cylinders are installed through a drilled hole in the door and have a special tailpiece for operation from the exterior.

Narrow Stile Locks and Latches

Narrow stile glazed steel and aluminum doors (as in Figure 8-18) require locking and latching devices with a short backset and narrow cases so the mechanism can fit within the rail. Both mortised and bored deadlocks with narrow backsets are available for this situation. In addition, a limited number of manufacturers make a deadlock that pivots down into the stile in the unlocked position rather than retracting as with conventional dead bolts. This permits a bolt with a longer throw in the narrow stile. This type of dead bolt is also available in a lock set with a latch bolt.

Two and Three Point Locks

Generally swinging doors over 8 feet high used in locations requiring a labeled fire door must have a three point lock. Pairs of doors require a two point lock for the inactive leaf. These units have bolts at or near the top and bottom of the door. Another bolt is placed near knob height on three point types. The knob retracts all bolts simultaneously. All bolts are automatically projected when the door is closed. A cylinder on the outside of the door locks the outside knob. The rods and other mechanisms that operate the bolts must be installed within the door. Trim is similar to that provided with mortise type lock and latch sets.

Hospital latches (Figure 8-19) provide positive latching with the convenience of push-pull levers to retract the latch bolt. These latches are used where positive latching is required on a labeled fire door.

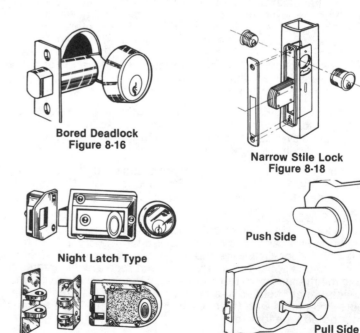

**Bored Deadlock
Figure 8-16**

**Narrow Stile Lock
Figure 8-18**

Night Latch Type

**Jimmy-Proof Type Rim Lock
Figure 8-17**

Push Side
Pull Side

**Hospital Latch
Figure 8-19**

Panic Exit Devices

Doors in buildings such as theaters and auditoriums are usually provided with exit devices. In case of panic, these doors open if the cross bar is pressed. All exit devices must meet the requirements of Underwriters' Laboratories for accident hazard.

Rim type devices (Figure 8-20) are entirely surface mounted on the inside face of the door. They are used on single doors and pairs of doors equipped with removable mullions.

Mortise type devices (Figure 8-21) have a mortise lock case and are generally used on the active leaf of pairs of doors.

Surface vertical rod type devices (Figure 8-22) have latch bolts located at the top and bottom of the door connected to an operating mechanism and cross bar. The entire unit is surface mounted on the inside face of the door. This type of device is used principally on the inactive leaves of pairs of doors or on both leaves of pairs of doors.

Concealed vertical rod devices (Figures 8-23) are similar to the surface vertical rod type except that the latch bolts, vertical rods and operating mechanism are concealed within the door. There are separate models of this type for use in hollow metal and wood doors.

Standard types of exit devices require door stile widths of between 4 and 4½ inches. Rim, surface vertical rod and concealed vertical rod types are made in special narrow stile types for application on glazed, narrow stile steel and aluminum doors. Types meeting the Underwriters' Laboratories requirements for fire exit devices are available in all four types.

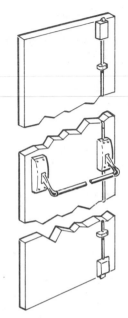

Surface Vertical Rod Exit Device
Figure 8-22

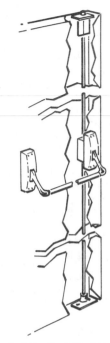

Concealed Vertical Rod Exit Device
Figure 8-23

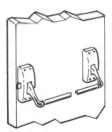

Rim Type Exit Device
Figure 8-20

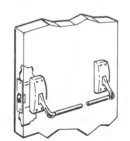

Mortise Type Exit Device
Figure 8-21

Labor Installing Panic Devices

Work Element	Unit	Man-Hours Per Unit
Panic bolts, complete set for single door		
Rim type with no outside trim	Each	1.4
Outside pull and cylinder	Each	1.8
Outside knob and cylinder	Each	1.9
Mullions for rim type	Set	1.2
Vertical rod panic bolts		
No outside trim	Each	1.6
Outside knob and cylinder	Each	1.7

Time includes move on and off site, unloading, cleanup and repair as needed.
Suggested Crew: 1 carpenter

Door Closers

All door closing devices have an adjustable valve, generally referred to as the sweep speed control, to control the closing speed. In addition many single acting models have a separate valve to control the speed within the last 4 to 6 inches of the closing cycle. This control, called latch speed control, permits adjustment to overcome the resistance of weatherstripping and ensure positive latching with the least noise possible. The following are optional features which are common to all types of door closing devices.

Adjustable closing force provides a means of pre-tensioning the spring, thereby increasing the closing force to overcome strong drafts or other abnormal conditions.

Backchecking is hydraulic or mechanical dampening of the opening speed through an arc of approximately 15 degrees of door opening. The amount of dampening is adjustable on models with hydraulic backchecking feature.

Floor closers (Figure 8-24) are installed in a recess in the floor. Because they are recessed, they are more difficult to install and repair than surface mounted closers. Like floor mounted pivots, they transfer most of the door weight to the floor and are preferred for extra heavy doors. The weight of the door is carried on ball bearings. Two sets of pistons and springs are provided in the larger sizes of double acting floor closers, one for each direction of swing.

A dead stop can be factory set in floor closers to stop the door at any specified point from 90 to 105 degrees.

A hold-open feature factory set to hold the door open at a specified point between 85 and 180 degrees is available in single acting off set hung types. Hold open can be provided from 85 to 105 degrees for center hung single and double acting types. Some models have a selector control which engages or disengages the hold open feature. The dead stop position must be 5 to 7 degrees beyond the hold position.

A positive centering adjustment is provided on many center hung double acting types to adjust the "at rest" position of the door within the jamb.

The degree of swing on center hung closers is usually limited to 95 to 110 degrees by the door trim. The standard degree of swing for offset hung closers is between 90 and 105 degrees. However, swings up to 180 degrees are possible on some models when required. The degree of swing on independently hung closers is generally limited to 105 degrees.

Overhead concealed closers (Figure 8-25) are similar in most respects to floor closers except they are designed to be installed in the frame or transom above the door. Like floor closers, overhead concealed closers can be single acting offset type, single acting center hung type, single acting independently hung type, or double acting center hung type. All except the independently hung type are furnished with bottom and intermediate pivots. A series of overhead concealed closers is manufactured with

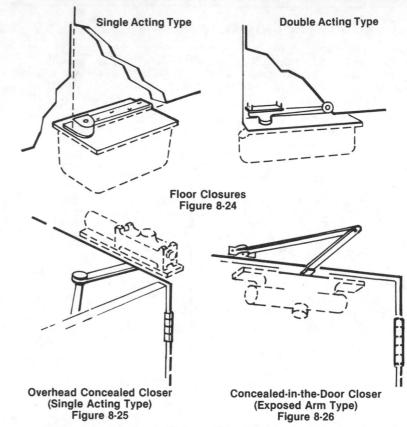

Single Acting Type **Double Acting Type**

Floor Closures
Figure 8-24

Overhead Concealed Closer
(Single Acting Type)
Figure 8-25

Concealed-in-the-Door Closer
(Exposed Arm Type)
Figure 8-26

cases suitable for mounting within a 1¾-inch wide transom bar for window wall and store front type construction.

Single acting offset hung and independently hung types can be furnished with a concealed arm operating in a slide mounted in the top of the door or with an exposed jointed arm. See Figure 8-26. Sizes are the same as for floor closers. Overhead closers have sweep speed control and separate latch speed control on single acting types. Some are equipped with adjustable back check and hold open features. Degree of swing is the same as with floor closers.

Surface mounted overhead closers (Figure 8-27) are available in traditional and modern styles. The modern style comes in two configurations. One has a 1¾-inch maximum case width suitable for mounting on narrow stile and rail doors. This closer can be furnished with or without a cover. The other closer has a 1⅝-inch case projection from the face of the door and always comes with a cover.

Mounting

The most common mounting for closers is on the face of the door on the same side as the hinge barrel with the arm attached to the door frame above the door. Where mounting on the opposite face of the door is desired, a closer with a parallel arm is used. As an alternative, the closer may be mounted on a corner bracket with the arm attached to the face of the door opposite the hinge barrel. Modern closers may be mounted on the frame or wall above the door with the arm attached to the door face opposite the hinge barrel. Brackets are available for mounting closers on a variety of special doors and frames.

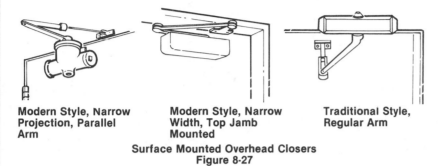

Modern Style, Narrow Projection, Parallel Arm **Modern Style, Narrow Width, Top Jamb Mounted** **Traditional Style, Regular Arm**

Surface Mounted Overhead Closers
Figure 8-27

Labor Installing Door Closers and Accessories

Work Element	Unit	Man-Hours Per Unit
Rack and pinion type		
2'8" hollow core	Each	1.2
3'0" hollow core	Each	1.2
3'4" hollow core	Each	1.2
3'4" solid core	Each	1.4
4'0" solid core	Each	1.4
Add 33% for concealed closers		
Door stops and checks		
Floor mounted bumper	Each	.2
Wall mounted bumper	Each	.2
Push bars		
24" long	Each	.7
30" long	Each	.8
36" long	Each	.9
Combined push-pull bars		
24" long	Each	.9
30" long	Each	1.0
36" long	Each	1.1
Door pulls		
Plain design	Each	.8
Modern design	Each	.8
Round or oval	Each	.8

Weatherstripping

Aluminum and bronze are the two most common metal weatherstrip materials. They are durable, extrude easily, and do not rust. Most pliable inserts are either neoprene or vinyl. Two basic types of weatherstripping are made from these materials: thin flat strips bent to create a spring action and thicker extrusions which act as a support for a pliable insert which actually performs the weatherproofing. Stainless steel is also popular because of its superior corrosion resistance and durability.

Weatherstripping is used to prevent entry of moisture, noise, dust or light through the cracks around windows and doors. The spring types are most effective around close tolerance openings such as double-hung windows. A smooth, flat surface is required for the spring types; they would be useless against rough concrete. Spring types can be easily dented or bent and thus should be used only where they are concealed or otherwise protected.

Where blowing dust and dirt is a problem, extruded thresholds with weep holes for water drainage should not be used. The small holes soon become clogged. Neoprene inserts are most effective in compression. If the opening receives rubbing action in use, vinyl should be used. Neoprene does not slide easily and will tear.

Garage doors often have large opening gaps and rough, uneven floors. A heavy neoprene weatherstrip is generally best. Specialty doors such as sliding and metal roll-up types are supplied with factory-installed weatherstripping.

Labor for Installing Weatherstripping and Trim

Work Element	Unit	Man-Hours Per Unit
Door weatherstripping, spring bronze or brass, 2 jambs and 1 head, wood door		
3' to 4' width	Each	1.0
Over 4' to 6' width	Each	1.2
Door weatherstripping, interlocking, metal, 2 jambs, head and sill seal, wood door		
3' to 4' width	Each	1.6
4' to 6' width	Each	2.0
Installing casing or trim, hours per side		
One member casing on door, ordinary work	Each	½ to ¾
Hardwood and first class work	Each	¾ to 1
Two member casing on doors, ordinary work	Each	¾ to 1
Hardwood and first class work	Each	1 to 1½

Door Installation

Locations for most door hardware have been standardized so that metal door frames can be prepared for hardware and furnished to the job site early in the construction period. The standard locations are shown in Figure 8-28. The location of door closers, floor hinges, overhead holders and specialized hardware should conform to the manufacturer's installation diagrams. When more than 2 hinges are required, the top and bottom hinge should be located as shown, with the remaining hinges spaced equally between them.

The hand of a door is a term used to indicate the direction of swing of the door. Strictly speaking, the door itself is only right or left hand. However, for locks and latches you must know whether the door swings inward (regular bevel) or outward (reverse bevel). This is especially important when different finishes are desired on opposite sides of the door. The outside of a door is the side from which security is necessary, normally the keyside. In a series of connecting rooms, the outside will be the side of each successive door as one comes to it proceeding from the entrance. Figure 8-29.

The use of the old fashioned hammer and chisel for installing locks and hinges in wood doors has largely disappeared. Special tools are available for installing many common products. Mortising tools, cutting and boring jigs for locksets, templates for mortising butt hinges, locating devices for floor closers and a variety of special tools for cutting in strikes and lock fronts are marketed by hardware manufacturers.

Full threaded wood screws or sheet metal screws are recommended for installation of hardware on wood doors because they provide greater pullout resistance than standard wood screws. Hardware surface applied to particle board or composition core doors should be anchored with through bolts and grommets.

Holes for installation of surface applied hardware on metal doors are usually drilled and tapped on the site. The installer may use a paper template or the device itself to locate hole spacing. A punch is a good tool to start the drilling for pilot holes. This prevents drill creep, off-center holes and improper screw alignment which would decrease holding strength considerably.

Through-bolting is normally used to attach hardware to doors which may not have been reinforced for surface applied hardware. Through bolts may also be used for attaching exit devices, overhead holders, surface mounted closers and pull plates and bars when they are subject to heavy abuse.

Single swing hollow metal doors should have not more than $\frac{1}{8}$" clearance at jambs and heads and $\frac{3}{4}$" clearance at the bottoms. Pairs of doors require $\frac{1}{4}$" clearance at their meeting edges.

A standard clearance for wood doors is $\frac{1}{8}$" at the heads and jambs and $\frac{1}{2}$" from the bottom of the door to the top of the finish floor. Pairs of doors require a minimum of $\frac{1}{8}$" clearance at meeting edges. Standard widths and heights may be changed where weatherstripping is required.

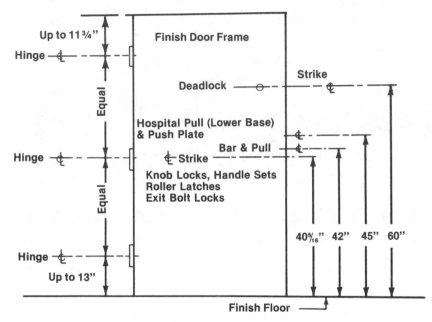

Standard Hardware Locations
Figure 8-28

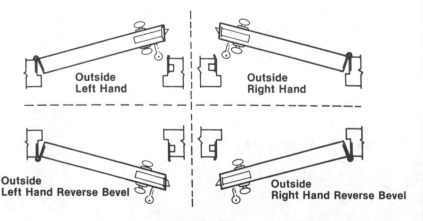

Hands of a Door
Figure 8-29

Labor for Installing Wood Door Frames

Work Element	Unit	Man-Hours Per Unit
Exterior wood door frame and sill		
3'0" x 7'0" opening	Each	1.9
6'0" x 7'0" opening	Each	2.4
Interior wood door frame, no sill		
3'0" x 7'0" opening	Each	1.0
6'0" x 7'0" opening	Each	1.5
Prefabricated interior wood door jambs		
2'6" to 3'6" x 6'8"	Each	.70
4'0" to 5'0" x 6'8"	Each	.83
6'0" x 6'8"	Each	1.0

Time includes move on and off site, unloading, stacking, repair, installation of assembled frame and cleanup as needed. Add ½ hour for installation of trim on 2 sides.
Suggested Crew: 1 carpenter

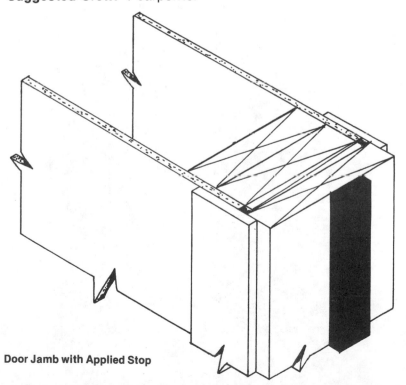

Door Jamb with Applied Stop

Labor Installing Steel Doors and Frames

Work Element	Unit	Man-Hours Per Unit
1 ¾" thick unrated doors		
2'8" x 6'8", 18 gauge	Each	2.3
2'8" x 7'0", 18 gauge	Each	2.5
3'0" x 6'8", 18 gauge	Each	3.0
3'0" x 7'0", 18 gauge	Each	3.1
3'4" x 7'0", 18 gauge	Each	3.7
Hollow metal door frames		
2'8" x 6'8", 18 gauge	Each	1.4
2'8" x 7'0", 18 gauge	Each	1.5
3'0" x 6'8", 18 gauge	Each	1.6
3'0" x 7'0", 18 gauge	Each	1.6
3'4" x 7'0", 16 gauge	Each	1.6
6'0" x 7'0", 16 gauge	Each	1.8

Time includes move on and off site, unloading, stacking, installing 1 pair template butts and latchset on each door, cleanup and repair as needed.
Suggested Crew: 1 carpenter

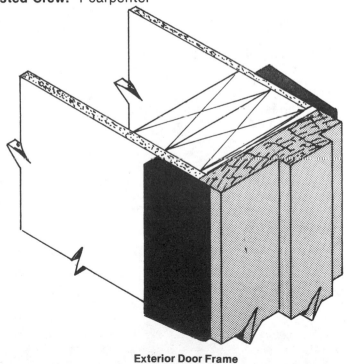

Exterior Door Frame

Labor for Installing Doors and Frames

Work Element	Unit	Man-Hours Per Unit
Solid core exterior doors		
3'0" x 7'0" x 1¾" thick	Each	3.4
3'0" x 7'0" x 2¼" thick	Each	3.5
Exterior fire doors		
3'0" x 7'0", 1 hour rating	Each	3.7
3'0" x 7'0", 1½ hour rating	Each	3.7
Hollow core 3'0" x 6'8" x 1¾" exterior doors with frames		
3'0" x 6'8"	Each	3.5
4'0" x 7'0"	Each	3.6
Interior doors		
3'0" x 7'0", hollow core	Each	3.5
3'0" x 7'0" solid core	Each	3.6
Architectural type doors with transom and panels		
3'0" x 7'0" x 1¾" thick	Each	5.2
Acoustical doors		
3'0" x 7'0" class 36 STC	Each	4.4
3'0" x 7'0" class 40 STC	Each	4.4
Exterior combination storm and screen doors, excluding frames		
3'0" x 6'8"	Each	2.2
French doors 1¾" fir, job hung		
3'0" x 6'8"	Each	4.0
Closet bi-passing, 1⅜" thick, complete with pulls and tracks		
4'0" x 6'8"	Each	2.2
6'0" x 6'8"	Each	2.6

Time includes move on and off site, unloading, stacking, setting frame and trim, hanging door, installing lockset, repair and cleanup as needed, but no finishing.
Suggested Crew: 1 carpenter

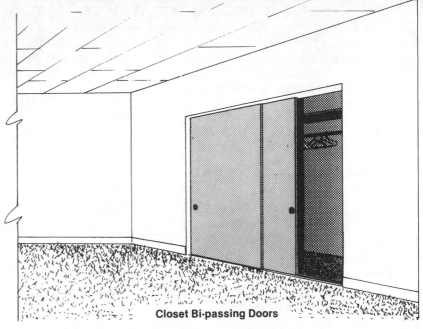

Closet Bi-passing Doors

French Doors

Folding Doors

Folding doors, sometimes better known as "bi-fold" doors, are made from a wide variety of materials. The following are the most common:

Wood folding doors are made like ordinary wood doors in either paneled construction, flush hollow core or flush solid core styles. Compressed hardboard is also used.

Metal folding doors come in two styles - all metal or veneered metal. The metal most often used is cold-rolled steel, although aluminum is available. All-metal doors are sometimes stamped to look like wood paneled doors or stamped into louvers or other decorative patterns. All-metal flush doors are also available. Veneered metal bi-fold doors are also available with a plastic surface. The veneer is usually vinyl.

Plastic is used as a veneer over both wood and metal. But solid plastic folding doors are extremely durable and easily cleaned. Plastic laminate and vinyl are the veneers most often used. One manufacturer uses a stamped, patterned plastic over a wood frame and honeycomb core.

Some bi-folding doors have interchangeable decorative inserts. The door consists of a wood or metal frame with a removable molding on the back side which holds the insert. Plate glass mirrors, louvers, colored vinyl panels, natural wood panels and peg board panels are a few of the inserts available.

Almost all folding door manufacturers offer factory prefinished doors. If the doors aren't "ready-for-use", they are least painted white which protects from rust and provides a good base for field-applied paint or veneers. Wood doors are generally the only types available unfinished.

The chief advantage of folding doors over ordinary doors is that they allow full-width opening of closets. Folding doors can also be used at regular door openings where swing clearance is limited. Other uses are over-counter pass-thru doors, storage room doors and washer-dryer alcove doors. They are especially practical at furnace closets where both an extra-wide opening and louvered ventilation are needed.

The more common sizes are 2', 4', and 5' widths and 6'-8", 8'-0" and 10'-0" heights. Combinations of these standard sizes can produce doors for openings of practically any width.

There are many types of operating mechanisms for folding doors, but all can be classified as either roller type or guide type. There are advantages and disadvantages to both. Rollers, especially the ball bearing type, operate more smoothly than guides, but guides do not jump the track as do rollers. An accurate and true installation will provide good results with either system.

Bi-folding doors are made in distinct quality levels. On paneled wood doors, for example, the frame will be mortised and tenoned on a good door and merely stapled together on cheaper brands. A quality wood door is made of hardwood rather than softwood and will have wide stiles. On hollow core doors, the core ribs of a high quality door will be of wood rather than corrugated cardboard. The thickness of the hollow core door's "skin" or the gage thickness on a metal door also indicates quality. Although two hinges may be adequate, better doors have three. Other items not found on lower quality doors are surface aligners which keep the doors closed tightly and flush, fiber washers on knobs to keep the surface from being marred, and provision for horizontal and vertical adjustment. Spring loaded pivots are used to compensate for frame irregularity and are standard on quality doors.

The opening for the door must be plumb and square. Folding door mechanisms have some provision for adjustment to out-of-square openings, but it is minimal. The maximum vertical adjustment is generally plus and minus ½". Some doors have no horizontal adjustment at all.

The tracks must be installed securely as well as level. This requires the right length of screws and substantial framing members. On doors over 5 feet wide, the ceiling track should be mounted on its own furring strip. Use a 1 x 2 rather than fastening through the ceiling to the joists or the plaster. If the floor is concrete, some manufacturers provide jamb mounting brackets with an adhesive strip under the floor track. Other floor tracks require concrete nails.

Labor for Installing Folding Doors

Work Element	Unit	Man-Hours Per Unit
Bi-folding wood doors		
2'4" x 6'8"	10 S.F.	1.2
2'8" x 6'8"	10 S.F.	1.1
3'0" x 6'8"	10 S.F.	1.3
4'0" x 6'8"	10 S.F.	.9
Bi-folding vinyl covered wood doors		
3'0" x 7'0"	10 S.F.	1.2
4'0" x 7'0"	10 S.F.	1.1
8'0" x 8'0"	10 S.F.	1.0
10'0" x 10'0"	10 S.F.	1.3
15'0" x 10'0"	10 S.F.	1.2
20'0" x 10'0"	10 S.F.	1.1
Add 10% for custom grade doors.		

Time includes move on and off site, unloading, stacking, hardware installation, cleanup and repair as needed, but no finishing.

Labor for Installing Bi-fold Hollow Metal Doors

Work Element	Unit	Man-Hours Per Unit
Two panel		
2'0" x 6'8" in 4' opening	Each	1.0
2'6" x 6'8" in 5' opening	Each	1.0
3'0" x 6'8" in 6' opening	Each	1.1
2'0" x 8'0" in 4' opening	Each	1.1
2'6" x 8'0" in 5' opening	Each	1.2
Four panel		
3'6" x 6'8" in 7' opening	Each	1.5
4'0" x 6'8" in 8' opening	Each	1.6
5'0" x 6'8" in 10' opening	Each	1.8
3'0" x 8'0" in 6' opening	Each	1.3
4'0" x 8'0" in 8' opening	Each	1.7
5'0" x 8'0" in 10' opening	Each	2.0

Time includes move on and off site, unloading stacking, cleanup and repair as needed.
Suggested Crew: 1 carpenter for doors up to 4' wide; over 4', 1 carpenter and 1 laborer

Labor Installing Metal Storm and Screen Doors

Work Element	Unit	Man-Hours Per Unit
Aluminum storm and screen frames, pre-hung in separate frame		
Residential grade, stock	10 S.F.	.5
Residential grade, custom	10 S.F.	.8
Commercial grade, stock	10 S.F.	.6
Commercial grade, custom	10 S.F.	.9

Time includes move on and off site, unloading, cleanup and repair as needed.
Suggested Crew: 1 carpenter

Labor Installing Sliding or Swinging Glass Patio Doors

Work Element	Unit	Man-Hours Per Unit
Standard weight doors and glass		
8' x 8'	Each	2.7
8' x 6'8"	Each	2.6
6' x 6'8"	Each	2.4
10' x 6'8"	Each	3.4
12' x 6'8"	Each	3.6
Heavier or better quality doors with insulating glass		
8' x 8'	Each	3.5
8' x 6'8"	Each	3.3
10' x 6'8"	Each	4.4
12' x 6'8"	Each	4.6
16' x 6'8"	Each	5.0

Time includes move on and off site, unloading, setup, installing in a frame opening, adjusting, cleanup and repair as needed.
Suggested Crew: 1 carpenter, 1 laborer

Sliding Glass Patio Door

Sliding Metal Fire Doors

Sliding metal fire doors are produced in two basic styles, the metal clad door (commonly called the tin clad), and the corrugated metal door. Although the metal clad door is more effective, both are suitable fire stops.

The metal clad fire door is fabricated of metal panels attached by interlocking seams, nailed to a core of well dried, nonresinous wood. The wood core is generally made of three layers of tongue and groove stock laid at right angles, and clinch nailed. The minimum covering is 30 gage steel sheet. All points between the sheets are locked with ½" seams and nailed under the seams to meet the requirements of Underwriters' Laboratories.

The corrugated metal variety is much lighter and easier to install. It has two layers of corrugated galvanized steel (2½" corrugations) attached to an inner layer of sheet asbestos. Corrugations on each side of the core run in opposite directions. The vertical sections are applied on the exposed side. Sheets are assembled in a substantial steel frame which allows for expansion and contraction along straight lines in a fire.

Both varieties are available with "A" labels from Underwriters' Laboratories, Factory Mutual, and other testing agencies. It is important to note however, that although all size sliding metal fire doors can be fabricated with construction techniques approved by testing agencies, doors that exceed 120 S.F. in area, or exceed 12' in any direction may be certified, but can not actually carry a label.

Sliding metal fire doors can be specified as single slide (one section), or double slide (two sections). The double slide is recommended only when there is not enough clearance on either side to permit the use of the single slide type. They can be operated on inclined tracks or flat (horizontal) tracks. In each case the door is moved by a system of weights attached to fusible links. In the first method, the door is held in the open position by the weight. When heat or fire releases the fusible link, the weight is released and the door slides shut by gravity. The horizontal slide operation employs two weights, one counterbalancing the other; when the fusible link is released, one weight remains to pull the door into the closed position.

Sliding metal fire doors can also be operated electrically. This method is used primarily for situations where periodic or constant operation of doors is contemplated.

Doors specified to carry "A" labels should have a minimum thickness of 2½" and require a minimum 3 ply construction. Doors carrying "B" labels or less should be of 1¾" minimum and may be of 2 ply construction. In both instances, testing agencies require construction that will limit heat transmission to 250 degrees after 30 minutes of exposure.

Doors with inclined tracks are more frequently specified because less apparatus is required for operation. Most manufacturers recommended no less than 14" of headroom plus 1½" of incline for each foot of opening width. This slope produces a minimum required speed of roughly 1 foot per second. If a greater closing speed is desirable, it can be achieved by increasing the track incline.

Labor Installing Metal Fire Doors

Work Element	Unit	Man-Hours Per Unit
3'0" x 6'8"	Each	5.0
3'8" x 6'8"	Each	6.4
4'0" x 7'0"	Each	7.1
5'0" x 8'0"	Each	9.1

Time includes track, hangers, hardware, move on and off site, unloading, stacking, cleanup and repair as needed.
Suggested Crew: 1 steelworker, 1 helper

Labor Installing Bronze or Stainless Steel Doors

Work Element	Unit	Man-Hours Per Unit
Single doors		
2'6" x 7'2"	Each	9.1
3'0" x 7'2"	Each	9.7
3'6" x 7'2"	Each	11.9
Double doors		
5'0" x 7'2"	Pair	17.6
6'0" x 7'2"	Pair	18.5
7'0" x 7'2"	Pair	19.4

Time includes move on and off site, unloading, stacking, cleanup and repair as needed. Add 10% for stainless steel or bronze finish.
Suggested Crew: 1 carpenter

Aluminum Doors and Frames

The basic unit in aluminum doors is the **extrusion.** Extrusions are produced by forcing hot metal through die openings. This is followed by finishing. They are available in a variety of shapes and sizes providing greater design, variety and flexibility. Sharp edges, hollow sections, and thin walls are easy to create in extruded aluminum.

Several factory applied coatings are available as aluminum finishes. The most widely used are the caustic etched finish, the satin finish, the anodized finish, and a protective coating of non-yellowing, clear lacquer. The latter is the least expensive and most widely used. The anodized finish is used for better classes of construction finishes and in main entrances. The caustic etched and satin finishes produce a richer "luster", but cost more.

Sliding glass doors were developed in California but are used throughout the country. The frame, sliding, and stationary panels are usually 6063 aluminum alloy. The extrusion sections may be tubular or H-shaped. Connection of panels and frames should be rigid, and designed to accept weatherstripping. Connectors, when used, should be noncorrosive and are usually of stainless steel or aluminum.

Door dimensions are based on unit panel sizes and panel combinations. Standard units are 6'8" high and 3, 4, 5, or 6' wide per section. Assemblies can be specified in groups of two, three or four sections, some operable and others stationary. Generally, a single track is commonly specified.

Many companies offer preglazed sash. In other cases glazing is done by the installer. Panels can be supplied with interchangeable glazing beads for use with either standard or insulating glass. Glazing should be installed with glazing strips of vinyl or channels rather than glazing compounds. When aluminum glazing beads are used, they may be the screwed-in-place type, or the snap-on type.

The most common sliding glass door has bottom rollers with adjustable sealed bearing sheaves on either aluminum or stainless steel tracks. When stainless steel tracks are specified or supplied, they can either be inserts in the aluminum sill, or "cap" for the sill track. The movable panel of the sliding glass unit is designed to slide inside the fixed panel, and should be removable when the door is unlocked.

Screens are available in either glass fiber or aluminum mesh, and rest in an aluminum frame. They should slide on rollers tracked in the sill and outside the stationary panel.

Labor Installing Aluminum Doors and Frames

Work Element	Unit	Man-Hours Per Unit
Narrow style, store front type		
2'6" x 7'0" x 1¾"	Each	8.6
3'0" x 7'0" x 1¾"	Each	8.8
3'6" x 7'0" x 1¾"	Each	9.2
5'0" x 7'0" x 1¾"	Pair	18.1
6'0" x 7'0" x 1¾"	Pair	18.4
7'0" x 7'0" x 1¾"	Pair	18.9
Wide style, office entry type		
2'6" x 7'0" x 1¾"	Each	8.8
3'0" x 7'0" x 1¾"	Each	9.2
3'6" x 7'0" x 1¾"	Each	9.6
5'0" x 7'0" x 1¾"	Pair	18.4
6'0" x 7'0" x 1¾"	Pair	19.1
7'0" x 7'0" x 1¾"	Pair	20.5

Time assumes a 1¾" I beam framing system with push-pull handles, butt or pivot hinges and flush bolts. Time includes move on and off site, unloading, stacking, cleanup and repair as needed.
Suggested Crew: 2 steelworkers

Rolling Doors and Grilles

Rolling doors are made of a curtain of interlocking metal slats attached to a drum or shaft. The shaft is contained by a metal housing, and rotates to coil the curtain into the housing space. The curtain jamb ends are held in place by guides of steel angles mounted on, and set flush with, the walls or jambs. A helical spring anchored to a tension rod relieves the weight of the metal curtain as it is extended.

Slats for rolling doors are produced in many shapes from flat to heavily shaped. The flatter sections provide more resistance to air infiltration. The heavily shaped sections offer more strength and are ideal for long spans. Slats are manufactured in gages ranging from 22 to 14.

Rolling doors are provided with three systems of operation: manual, mechanical, and power operated. Manually operated doors have hand pulls on the bottom rail. With assistance from the counter balancing mechanism, they are easily opened with minimum effort. This type of door is used in smaller openings.

Mechanically operated doors can be specified with either chain-gear or crank-gear operators. The chain-gear operator has an endless chain and sprocket and a series of gears on the roller shaft bracket. In the event of fire, an automatic release disengages to close the door.

Power operated doors use electric motors which replace the chain or crank gears. Power operation is recommended for very large doors and heavy gage metal curtains. They are controlled by push button switches and emergency stop buttons. They can be furnished with safety bottom bars which halt the downward movement of the door when contact is made with an obstruction.

Rolling doors are designed to withstand a wind pressure of 20 pounds per square foot regardless of the dimensions involved. To reach this standard, each increase in area requires additional "stiffness" from deeper configured sections, heavier gages, or both.

Rolling doors are not weather-tite. Although weather seals are reasonably effective, joints formed by the curtain slats can not be sealed. Manufacturers have developed flat slats which form a tighter joint than deeply configured sections. However, since the curtain slats must coil to operate, a true sealed joint is nearly impossible.

Labor Installing Rolling and Sliding Doors

Work Element	Unit	Man-Hours Per Unit
Rolling, manual operated	Each	8.0
Rolling, motor operated	Each	25.0
Sliding, manual operated	Each	8.0
Sliding, motor operated	Each	14.0

Labor for Installing Overhead Doors

Work Element	Unit	Man-Hours Per Unit
Wood panel overhead doors		
8'3" x 8'7"	Each	4.3
10'3" x 10'7"	Each	6.5
12'3" x 12'7"	Each	9.3
16'3" x 16'7"	Each	16.4
22'3" x 18'11"	Each	25.4
24'3" x 18'11"	Each	26.3
Steel overhead doors, chain operated		
10' x 10' high	100 S.F.	4.5
12' x 10' high	100 S.F.	5.5
12' x 20' high	100 S.F.	6.5
Heavy duty 2" thick stock doors with standard hardware and tracks		
18' x 18' x 2" thick	Each	30.3
20' x 20' x 2" thick	Each	45.5

Deduct 10% for aluminum doors. Time includes move on and off site, unloading, stacking, cleanup and repair as required.
Suggested Crew: 1 carpenter, 1 laborer on wood door up to 10' wide. Wood doors over 10' wide: 1 carpenter, 2 laborers. Use 2 steelworkers for steel doors.

Windows

Wood windows are usually supplied treated with toxic preservative and with a factory prime paint coat on the exterior surfaces. Some are available with all surfaces primed. The primed and unprimed surfaces require field-finishing with paint or stain.

There is a trend toward incorporating other materials into wood windows: aluminum, sometimes bronze or plastic for wearing surfaces and vinyl plastic for weatherstripping and for exterior weather protection.

Fabrication and Installation

Wood window parts are machined to shape. The joints are mortise and tenoned, slot and tenoned, dadoed or doweled, glued and sometimes nailed or screwed together. The glass is groove-glazed, face-glazed or glazed with wood stops.

Wood window installation in wood stud walls consists of inserting and shimming the window frame into the rough-framed wall opening, and nailing or screwing through the frame and shims into the rough wall framing. The job is completed with installation of trim, caulking, flashing, testing, adjusting and painting.

Metal window members are rolled, extruded or cold-formed to shape. Joints are mitered if the shapes are compatible, otherwise coped or tenoned and then fastened mechanically or welded and dressed. If jointing is mechanical, the joint must be sealed to exclude the weather.

Metal windows are installed in wood framing the same as wood windows - fastening through the frame and shims to the wall members. Designs that have a nailing flange are nailed through the flange to the face of the wall members. In masonry or concrete walls the metal window is anchored through the frame to anchor straps which are anchored to the sides of the opening. When the windows are to be installed as the masonry work progresses, the window is set in place and braced when the masonry work passes the sill level. The slot provided in the end of the masonry jamb units is placed around the outstanding leg of the window frame. The job is completed by closing all perimeter openings, caulking and sealing perimeter joints, testing, adjusting and applying specified field painting, if any.

Labor Installing Weatherstripping

Work Element	Unit	Man-Hours Per Unit
Spring bronze or brass, 4 sides, in double hung window, including sash removal		
To 3' x 2'	Each	1.0
To 3' x 4'	Each	1.8
Over 3' x 4'	Each	2.4

Labor Installing Windows

Work Element	Unit	Man-Hours Per Unit
Casement windows and screens		
1 leaf, 1'10" x 3'2"	Each	1.4
2 leaves, 3'10" x 4'2"	Each	1.9
3 leaves, 5'11" x 5'2"	Each	2.4
Picture windows		
4'6" x 4'6"	Each	3.0
5'8" x 4'6"	Each	3.2
9' x 5'	Each	3.7
10' x 5'	Each	4.0
11' x 5'	Each	4.4
Double or single hung windows and screens		
2'0" x 3'2"	Each	1.1
2'0" x 4'6"	Each	1.6
2'8" x 3'2"	Each	2.0
2'8" x 5'2"	Each	2.1
3'4" x 5'2"	Each	2.4
5'6" x 5'2"	Each	3.4
8'4" x 5'2"	Each	5.2
Bow bay windows and screens		
8' x 5'	Each	5.1
9'9" x 6'8"	Each	7.6
7' x 5'	Each	7.1
8'9" x 5'0"	Each	7.7
7'6" x 6'0"	Each	7.6
8'9" x 6'6"	Each	8.4
One member casing on windows, ordinary work	Each	1
Hardwood, first class work	Each	1½ to 2
Two member casing on windows, ordinary work	Each	1½ to 2
Hardwood and first class work	Each	2 to 4
Window trim on brick walls, ordinary work	Each	1½ to 2
First class or difficult work	Each	2 to 4

Time is for setting factory-made assembled windows in a prepared opening and includes move on and off site, unloading, stacking, installing, repairing and cleanup as needed, but no trim or framing.
Suggested Crew: 1 carpenter, 1 laborer

Glass & Glazing

The "Glass and Glazing" section of construction specifications may include a wide range of materials: sheet glass, plate glass, heat and glare reducing glass, insulating glass, tempered glass, laminated glass and various transparent or translucent plastics, ceramic coated, corrugated, figured, silvered and other decorative glass; and glaziers' points, setting pads, glazing compounds, and other installation materials.

Glare-reducing glass is available in two types (1) transparent glass with a neutral gray or other color tint, which lowers light transmission but preserves true color vision; and (2) translucent glass, usually white, which gives wide light diffusion and reduces glare. Both types absorb some of the sun's radiant energy and therefore have heat absorbing qualities. The physical characteristics of glare-reducing glass are quite similar to those of plate glass. Even though glare-reducing glass absorbs heat, it does not require the special precautions that heat-absorbing glass does.

Sheet or window glass is manufactured by the flat or vertically drawn process. Because of the manufacturing process, a wave or draw distortion runs in one direction through the sheet. The degree of distortion controls the usefulness of this type of glass. For best appearance, window glass should be drawn horizontal or parallel with the ground. To ensure this, the width dimension is given first when ordering.

Corrugated glass is $\frac{3}{8}$" thick glass with both sides formed into corrugations. The chemical composition is the same as that of plate or window glass. Corrugated glass is available in standard corrugations in either wired or unwired types.

Heat-absorbing glass contains controlled quantities of ferrous iron admixture that absorbs much of the energy of the sun. The glass holds a considerable amount of this heat. Heat-absorbing glass is available in plate, heavy plate, sheet, patterned, tempered, wired and laminated types.

Both heat-absorbing and glare-reducing glass require large clearances in installation. Because both types absorb heat, it is important to control the covering of edges. This minimizes cold edge effects and reduces temperature differences over the plate. Edge coverings that extend too far over the edge leave the covered portion much colder than the uncovered areas. Cracking will be the result.

All edges must be clean cut. Nipping the edges will cause stress concentration in the edges and result in breakage. This type of glass should be glazed with a permanently elastic glazing compound.

Heat-strengthened glass is plate glass or patterned glass with a ceramic glaze fused to one side of the glass. Preheating of the glass to apply the ceramic glaze strengthens the glass considerably, giving it characteristics similar to tempered glass. Heat-strengthened glass is about twice as strong as plate glass. Like tempered glass, it can not be drilled or cut.

Heat-strengthened glass is available in thicknesses of $\frac{1}{4}$" and $\frac{5}{16}$" and in limited standard sizes. It is opaque and is most often used for spandrel glazing in curtain wall systems. Framing members must be sturdy and rigid enough to support the perimeter of the tempered glass panels. Each panel should rest on resilient setting blocks. The glass can not be cut or drilled. When used in operating doors and windows, it must not be handled or opened until the glazing compound has set.

Insulating glass units consist of two or more sheets of glass separated by either $\frac{3}{16}$", $\frac{7}{32}$" or $\frac{1}{4}$" air space. These units are factory sealed. The captive air is dehydrated at atmospheric pressure. The edge seal can be made either by fusing the edges together or with metal spacing strips. A polyisobutylene mastic seal and metal edge support the glass.

Insulating glass requires special precautions: Openings into which insulating glass is installed must be plumb and square. Glazing must be free of paint and paper because they can cause a heat trap that may result in breakage. There must be no direct contact between insulating glass and the frame into which it is installed. The glazing compound must be a nonhardening type that does not contain any materials which will attack the metal-to-glass seal of the insulating glass. Never use putty. Resilient setting blocks and spacers should be provided for uniform clearances on all units set with face stops. Use metal glazing strips for $\frac{1}{2}$" thick sash without face stops. Use a full bed of glazing compound in the edge clearance on the bottom of the sash and enough at the sides and top to make a weathertight seal. It is essential that the metal channel at the perimeter of each unit be covered by at least $\frac{1}{8}$ inch of compound. This insures a lasting seal.

Laminated glass is composed of two or more layers of either sheet or polished plate glass with a layer or more of a transparent or pigmented plastic sandwiched between the layers. A vinyl plastic, such as plasticized polyvinyl resin butyl 0.015" to 0.025" thick is generally used. Only the highest quality sheet or polished plate glass is used in making laminated glass. When this type of glass breaks, the plastic holds the pieces of glass and prevents the sharp fragments from shattering. When four or more layers of glass are laminated with three or more layers of plastic, the product is known as bullet-resisting glass. Safety glass has only two layers of glass and one of plastic.

Safety glass is available with clear or pigmented plastic, and either clear glass or heat-absorbing and glare-reducing glass. Safety glass is used where strong impact may be encountered and the hazard of flying glass must be avoided. Exterior doors with a pane area greater than 6 square feet and shower tubs and enclosures are typical applications.

Glazing compounds must be compatible with the layers of laminated plastic. Some compounds cause deterioration of the plastic in safety glass.

Mirrors are made with polished plate, window, sheet and picture glass. The reflecting surface is a thin coat of metal, generally silver, gold, copper, bronze, or chromium, applied to one side of the glass. For special mirrors, lead, aluminum, platinum, rhodium or other metals may be used. The metal film can be semi-transparent or opaque and can be left unprotected or protected with a coat of shellac, varnish, paint or metal (usually copper). Mirrors used in building construction are usually either polished plate glass or tempered plate glass.

Proper installation requires that the weight of the mirror be supported at the bottom. Mastic installation is not recommended because it may cause silver spoilage.

Patterned glass has the same composition as window and plate glass. It is semi-transparent, with distinctive geometric or linear designs on one or both sides. The pattern can be impressed during the rolling process or sandblasted or etched. Some patterns are also available as wired glass.

Patterned glass allows entry of light while maintaining privacy. It is also used for decorative screens and windows. Patterned glass must be installed with the smooth side to the face of the putty.

Plate glass is similar to window and heavy sheet glass. The surface rather than the composition or thickness is the distinguishing feature. Plate is manufactured in a continuous ribbon and then cut in large sheets. Both sides of the sheet are ground and polished to a perfectly flat plane. Polished plate glass is furnished in thicknesses of from $\frac{1}{8}$" to $1\frac{1}{4}$". Thicknesses $\frac{5}{16}$" and over are termed heavy polished plate. Regular polished plate is available in three qualities: silvering, mirror glazing and glazing. The glazing quality is generally used where ordinary glazing is required. Heavy polished plate is generally available in commercial quality only.

Tempered glass is plate or patterned glass which has been reheated to just below its melting point and then cooled very quickly by subjecting both sides to jets of air. This leaves the outside surfaces, which cool faster, in a state of compression. The inner portions of the glass are in tension. As a result, fully tempered glass has 3 to 5 times the strength against impact forces and temperature changes. Tempered glass chipped or punctured on any edge or surface will shatter and disintegrate into small blunt pieces. Because of this, it can not be cut or drilled.

Glazing Materials
Wood sash putty is a cement composed of fine powdered chalk (whiting) or lead oxide (white lead) mixed with boiled or raw linseed oil. Putty may contain other drying oils such as soybean or perilla.

As the oil oxidizes, the putty hardens. Litharge or special driers may be added if rapid hardening is required. Putty is used in glazing to set sheets of glass into frames. Special putty mixtures are available for interior and exterior glazing of aluminum and steel window sash.

A good grade of wood sash putty won't stick too much to the putty knife or glazier's hands. Yet it should not be too dry to apply to the sash. In wood sash, apply a suitable primer, such as priming paints or boiled linseed oil.

Putty should not be painted until it is thoroughly set. Painting forms an air tight film which slows the drying. This may cause the surface of the paint to crack. All putty should be painted for proper protection.

Metal sash putty differs from wood putty in that it is formulated to adhere to a nonporous surface. It is used for glazing aluminum and steel sash either inside or outside. It should be applied as recommended by the manufacturer. Metal sash putty should be painted within two weeks after application, but should be thoroughly set and hard before painting begins.

There are two grades of metal sash putty, one for interior and one for exterior glazing. Both wood sash putty and metal sash putty are known as a oleoresinous caulking compound. Oleoresinous materials are the least expensive materials on the market. However, they have little or no adhesion, high shrinkage and an exposed life expectancy less than five years.

Elastic glazing compounds are specially formulated from selected processed oils and pigments which remain plastic and resilient over a longer period of time than the common hard putties. Butyl and acrylic compounds are the most common elastics. Butyl compounds tend to stain masonry and have a high shrinkage factor. Acrylic-based materials require heating to 110 degrees prior to application. Some shrinkage occurs during curing. At high temperatures, it sags considerably in vertical joints. At low temperatures, acrylic-based materials become hard and brittle. They are available in a wide range of colors and have good adhesion qualities.

Polybutene tape is a non-drying mastic which is available in extruded ribbon shapes. It has good adhesion qualities, but should not be used as a substitute or replacement for spacers. It can be used as a continuous bed material in conjunction with a polysulfide sealer compound. This tape must be pressure applied for proper adhesion.

Polysulfide-base products are two-part synthetic rubber compounds based on a polysulfide polymer. Its consistency after mixing is similar to a caulking compound. The activator must be thoroughly mixed with the base compound at the job. It may also be supplied pre-mixed, frozen in cartridges packed in dry air. The mixed compound is applied with either a caulking gun or spatula. The sealing surfaces must be extremely clean. Surrounding areas of glass should be protected with tape before glazing. Excess and spilled material must be removed and the surfaces cleaned promptly.

Once polysulfide elastomer glazing compound has cured, it is very difficult to remove. Any excess material left on the surfaces after glazing should be cleaned during the working time of the material (2 to 3 hours). Toluene and xylene are good solvents for this purpose.

Rubber compression materials are molded in various shapes. They are used as continuous gaskets and as intermittent spacer shims. A weather-tight joint requires that the gasket be compressed at least 15 percent.

Preformed materials reduces costs because careful cleaning of glass is not necessary and there is no waste of material.

Labor Installing Window Glass

Work Element	Unit	Man-Hours Per Unit
Common window glass		
3/16″ stock size up to 50 S.F.	S.F.	.08
7/32″ stock size up to 50 S.F.	S.F.	.09
1/4″ tempered plate glass		
Up to 15 S.F.	S.F.	.12
16 to 25 S.F.	S.F.	.13
26 to 50 S.F.	S.F.	.15
51 to 75 S.F.	S.F.	.14
76 to 100 S.F.	S.F.	.13
3/8″ tempered plate glass		
Up to 15 S.F.	S.F.	.14
16 to 25 S.F.	S.F.	.16
26 to 50 S.F.	S.F.	.18
51 to 75 S.F.	S.F.	.17
76 to 100 S.F.	S.F.	.16
1/2″ tempered plate glass		
Up to 15 S.F.	S.F.	.16
16 to 25 S.F.	S.F.	.18
26 to 50 S.F.	S.F.	.20
51 to 75 S.F.	S.F.	.19
76 to 100 S.F.	S.F.	.18

Ground floor work only. Time includes move on and off site, unloading, cutting where required, cleanup and repair.
Suggested Crew: 2 glazers. For large and heavy units crew size will be larger.

Labor Installing Insulating Glass

Work Element	Unit	Man-Hours Per Unit
1/4″ plate glass, 1/2″ airspace		
3′ x 4′	S.F.	.21
4′ x 7′	S.F.	.22
9′ x 4′-10″	S.F.	.25

Add 10% to man-hours for 3/4″ glass.
Add 25% to man-hours for 1″ glass.
Add 20% for panes smaller than two square feet.
Ground floor work only. Time includes move on and off site, unloading, cleanup and repair as needed.
Suggested Crew: 1 glazer and 1 helper

Conversion Factors for Glazing

Glass Size	Panes Per Box	Percent Waste
8 x 12	75	10
10 x 16	45	10
12 x 20	30	10
14 x 24	22	10
16 x 28	16	10
Glazing cups	—	10

Putty For Glass Size	Lbs. Per Pane	Percent Waste
8 x 12	0.6	20
10 x 16	0.8	20
12 x 20	0.9	20
14 x 24	1.1	20
16 x 28	1.4	20

Labor Installing Specialty Glass

Work Element	Unit	Man-Hours Per Unit
Wire glass ¼" thick, plain, hammered and stippled		
3' x 6' up to 15 S.F.	S.F.	.12
4' x 8' up to 25 S.F.	S.F.	.13
5' x 11' 26 to 50 S.F.	S.F.	.14
Solarbronze and solargray ¼" thick		
5' x 6'-8" up to 25 S.F.	S.F.	.12
7' x 8' 26 to 50 S.F.	S.F.	.13
8' x 12' 51 to 75 S.F.	S.F.	.14
10' x 14'-2" 76 to 100 S.F.	S.F.	.14
10' x 18' Over 100 S.F.	S.F.	.13
Solarbronze and solargray 3/8" thick		
5' x 6'-8" Up to 25 S.F.	S.F.	.14
7' x 8' 26 to 50 S.F.	S.F.	.16
8' x 12' 51 to 75 S.F.	S.F.	.18
10' x 14'-2" 76 to 100 S.F.	S.F.	.16
10' x 18' Over 100 S.F.	S.F.	.16
Solarbronze and solargray ½" thick		
5' x 6'-8" Up to 25 S.F.	S.F.	.16
7' x 8' 26 to 50 S.F.	S.F.	.18
8' x 12' 51 to 75 S.F.	S.F.	.20
10' x 14'-2" 76 to 100 S.F.	S.F.	.19
10' x 18' Over 100 S.F.	S.F.	.18

Ground floor work only. Time includes move on and off site, unloading, cutting where required, cleanup and repair as needed.
Suggested Crew: 2 glazers. For large thick windows crew size will vary.

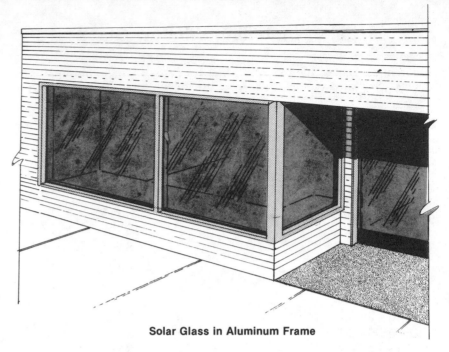

Solar Glass in Aluminum Frame

Labor Installing Laminated Safety Glass

Work Element	Unit	Man-Hours Per Unit
1/4" safety plate glass		
3' x 6' up to 15 S.F.	S.F.	1.0
5' x 6'-8" up to 25 S.F.	S.F.	1.2
6' x 11'-6" up to 50 S.F.	S.F.	1.4
3/8" safety plate glass		
3' x 6' up to 15 S.F.	S.F.	1.2
5' x 6'-8" up to 25 S.F.	S.F.	1.4
6' x 11'-6" up to 50 S.F.	S.F.	1.6

Ground floor work only. Time includes move on and off site, unloading, cutting where required, cleanup and repair.
Suggested Crew: 2 glazers. For large windows crew size will be larger.

Skylights

Skylights are mounted on the building roof in a more or less horizontal position or parallel to the roof surface. They serve a similar purpose to windows. Skylights come in 3 principal types: frame and glazing, frameless, and prefabricated plastic dome.

Frame and glazing skylights are usually mounted on a curb surrounding the roof opening and consist of glass or plastic lights and a system of extruded metal members and connectors. This frame holds the translucent material the same as the frame, mullions and muntins of a window. The lights are of wire glass or plastic. Glass may also be actinic. Actinic glass transmits lights of high visibility (such as green) but reduces the intensity of infrared and ultraviolet. This type of skylight includes a wide range of sizes and just about any shape of plan and profile.

Frameless skylights consist of corrugated plastic or wire glass installed in a corrugated metal roof. The corrugations of the roof and skylight match.

Prefabricated plastic dome skylights consist of plastic sheets formed into domes and mounted in metal frames. The varieties are almost infinite.

The plastic domes are available in single layers or double and occasionally triple layers, with air spaces between for insulating purposes. The plastic dome layer or layers may be clear, milky or opaque, or in variations from one layer to another to control the amount of light and heat transmitted. The shape of the dome itself may vary from a simple dome to a pyramid or various non-symmetrical shapes. The plan shape of the plastic dome skylights may vary from a simple square or rectangle, to a triangle, to various multi-sided shapes, to round. Others are designed in the shape of arches or barrel vaults so that a number of units can be installed end to end, with half-dome-shaped end units and metal or plastic connecting assemblies.

The frames that house the perimeters of plastic domes come in a number of shapes designed for various functions. One type fits on top of a previously built curb and has a vertical flange extending part way down the curb. Another type has an integral curb of several available heights, with or without insulation, with a horizontal nailing flange. This type can be installed directly on the roof deck in new construction or on top of roofing in a remodeling situation, and then stripped-in. Variations of the intregral curb type are available with hinges, fusible links and automatic lifting devices so that they can serve as smoke and heat vents and roof hatches. Another variation is an integral curb plastic dome skylight with an extra high curb and exhaust fan or blower built into the curb. The most readily available frame material is aluminum, though plastic frames are becoming available.

Another available skylight system consists of a number of plastic domes supported by a framework similar to those for the frame and glazing skylights.

Frameless skylights are built in with the corrugated metal roof as installation allows them to be anchored to supporting members.

Plastic dome skylights are set in mastic on the skylight curb or over the opening, secured in place, then flashed or stripped-in.

Labor Installing Plastic Skylights

Work Element	Unit	Man-Hours Per Unit
Plastic skylights, single or double thickness		
Up to 10 S.F. each	S.F.	.25
11 to 20 S.F. each	S.F.	.23
21 to 30 S.F. each	S.F.	.21
31 to 40 S.F. each	S.F.	.20
41 to 50 S.F. each	S.F.	.17

Time includes unloading, lifting and clean-up. Time required to install the mounting curb is not included.

Suggested Crew: 1 roofer, 1 laborer

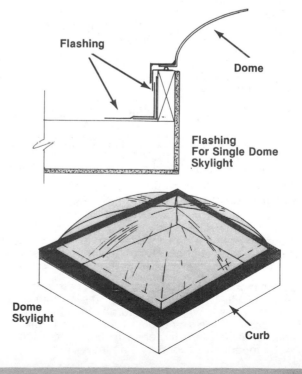

Flashing

Dome

Flashing For Single Dome Skylight

Dome Skylight

Curb

Curtain Wall Systems

Curtain walls have been used for centuries. The Greeks and Romans used post and lintel construction, infilled with brick, stone or concrete for enclosure. This is a type of curtain wall. Medieval cathedrals with arches supported on columns had curtain walls of stained glass. Early skyscrapers were built with skeleton construction and had curtain walls of glass, metal or masonry. Modern curtain walls replace masonry facing and infilling with a light, prefabricated exterior skin suspended in front of the structure like a curtain.

A curtain wall is any non-load bearing exterior wall with the following characteristics:
1. Suspended in front of the structural frame.
2. Dead weight and wind loads are transferred to the structural frame through point anchorages.
3. Wall elements and the fastening technique permit erection of continuous wall surfaces of any size.

Curtain walls may be distinguished from infilled walls by characteristics 1 and 2 above. But for purposes of this section, infilled walls using a prefabricated system will be considered curtain walls.

The major advantages of curtain wall include:

1. Thinner walls - 2 inches to 5 inches are most common.
2. Mass production - prefabrication and preassembly make use of modern factory production methods.
3. Efficient erection from inside the building eliminates scaffolding.
4. Easier transporting, handling and storage of large units.
5. Fewer joints thus fewer caulking or sealing problems.
6. Simple and positive attachment to the building of units which can be removed and replaced relatively easily.
7. The light weight of curtain walls reduces overall weight.

Curtain walls can be grouped in four categories: (1) mullion (2) spandrel (3) grid and (4) sheath.

Mullion design emphasizes the vertical member, the mullion and column cover. Spandrel design emphasizes the horizontal lines of the structure. Grid system design expresses both vertical and horizontal elements with equal emphasis.

Mullion, spandrel and grid systems may all be classed together in their construction method. Each consists of a grid of vertical or horizontal members framing openings filled with inserts or glass. The inserts or glass are the space-enclosing elements of the grid. The grid itself is the framework. The grid is fastened directly to the supporting structure and transfers the entire weight of the curtain wall and wind loads.

Sheath construction (some refer to this as panel construction) shows no structural elements at all. The components of the curtain wall are the panels themselves. Windows, where structurally feasible, are cut out of the middle of the panel. Panels are joined directly, without use of intermediate elements. Panels are fastened to the supporting structure either directly or by means of secondary framing.

Structural Problems

Curtain walls, though defined as non-loadbearing exterior walls, are loaded horizontally by the wind and vertically by their own dead weight. An average curtain wall will weigh between 12.5 and 15.5 pounds per square foot of wall surface, depending on the materials used. Wind loads are fixed by national and local building codes. Winds produce suction as well as pressure on the face of a building. These negative forces are important in the design of anchors, fasteners and the glazing beads used. All components subjected to wind load must be made of material capable of resisting bending action. In zones of high seismic risk, these walls should be attached to the structure to avoid shear forces generated by quakes.

Every curtain wall component is subject to changes in length due to fluctuations in temperature. Anchors must be adjustable and slip joints must be provided in vertical mullions. Window framing and insert fastening must permit movement.

Thermal Insulation

An important advantage of curtain wall construction is that walls can include lightweight thermal insulation. Materials used are low density, somewhat porous, sensitive to moisture and weather, and have little strength of their own. Therefore these materials must be weatherproofed and protected on their faces from damage.

The insulating portion of a curtain wall is composed of the following elements: an exterior weatherproof skin (outside facing), thermal insulation, and inside facing to protect the insulation from damage.

Thermal insulation can be installed outside the curtain wall. In this system the exterior weatherproof skin is the curtain wall itself. The thermal insulation is supported directly on the structure in the form of a back-up wall. The back-up wall material may be any fireproof or fire-resistant material such as gypsum, pumice, lightweight concrete block or poured-in-place or precast concrete.

Thermal insulation inside the curtain wall consists of material which is laminated, semi-laminated, mechanically fastened or independently mounted. These units eliminate the need for back-up walls.

Moisture Control

Assuming the materials used for the outside face are waterproof, rain can only penetrate the wall through joints. Joint sealing methods are explained later in this section. With curtain walls, condensation on the inside of the wall is a problem. Condensation must be collected and drained to the outside or it reduces insulating efficiency and causes corrosion. Water vapor condenses on surfaces that are cooler than the dew point of the air.

Four important rules in moisture control are to (1) provide a vapor barrier on or near the warm side of the wall (2) provide enough insulation to keep critical surfaces warmer than the dew point of indoor air (3) provide for vapor release on the cold side of the wall and (4) avoid using a tight vapor barrier on both sides of the wall.

In composite panel assemblies, laminated or hollow-pan type, ventilation is essential. The only exception is a laminated panel with a closed-cell core which in itself is a good vapor barrier. The edges of panels should be highly permeable to vapor and vented to prevent the entrance of rain or snow. Any edge seal applied to prevent damage during transit and handling should be removed or punctured before the panels are installed.

Materials Used in Curtain Walls

Metal is the primary structural material in curtain walls, though it may not be the most evident. Large portions of the wall may be sheathed in a non-metal material. But metal will be the supporting element.

Carbon steel is usually used in the form of structural shapes and thin sheets of ⅛ inch or less. Hot-rolled carbon steel is the least expensive curtain wall material. It must be protected from the weather by galvanizing, phosphatizing or painting.

Low-carbon alloy stainless steel with a chromium content of at least 12% resists weathering, water and most acids. This type of curtain wall material needs no additional surface protection except regular cleaning but is very expensive. Because of this, it is used primarily as cladding for regular steel, or for extrusions designed to enclose a reinforcing substructure. Highly polished finishes are not recommended as they reflect any unevenness in the surface. Brushed and matte-polished finishes are preferred.

Aluminum is the most common curtain wall material. Aluminum with alloys of manganese, magnesium and silicon are used. Alloys with copper are unsuitable. Aluminum in extruded shapes is used for mullion members, rail and sash sections, and in sheet form for column covers, fascias, soffits and panels.

When exposed to air, aluminum and its alloys become coated with a layer of oxide. If this layer is damaged, a new layer will form. This oxide gives aluminum a dull, light-gray appearance. Anodized, enamel and synthetic finishes are also used.

Glass

Many types of glass are used in curtain wall construction. Clear and tinted plate are used for transparent glazing and opaque colored "structural" sheet glass for panel material.

Flat glass used in windows, entrances and spandrels of curtain walls is non-structural or non-load bearing. Ordinary sheet or plate are not affected by the sun's heat and thermal breakage is uncommon. Heat-absorbing glass will develop thermal stress problems due to temperature differences of areas in the sun and areas in shade. These problems may be eliminated by careful attention to the edges of the glass. A clean cut edge is the strongest edge. Any chip or fissure of the edge is where a thermal break will start. Edges treated with an abrasive belt or stone are weak.

There must be adequate clearance all around the glass edge. The glass should in effect float in the frame. A good practice is to measure the openings on site before ordering the glass so that edge clearances are assured. A rule of thumb is that the lateral dimension of the glazing rabbet should be the glass thickness plus ¼ inch. This gives ⅛ inch clearance on each side of the glass. The depth should also be enough to assure ¼ inch clearance around the glass.

Sash or frame for the glass must be equipped with expansion joints both vertically and horizontally to permit movement from temperature changes and to prevent excessive pressure on the glass. The frame must also be strong enough to resist bending under wind loads.

Water leaks around glass are caused by poor glazing or defects in the sash or frame. The glazing material must form a tight bond with both sash and glass and must compensate for the movement in the sash due to temperature changes.

Glass spandrels should be heat strengthened or tempered to withstand thermal stresses. The construction behind a spandrel panel should be vented top and bottom to eliminate trapping of heat. Weep or vent holes will also carry off the condensation and water leakage. The color applied to a glass spandrel should be fused to the surface with heat. Annealed glass should not be used for spandrels.

Curtain Wall Inserts

A curtain wall is made up of these major components: the frame, the glazing, and the inserts. Inserts may be of a single layer of material known as a "skin panel", or it may be an assembly of components known as composite or "sandwich panel." All composite panels have one thing in common — they are fabricated from two or more materials into a single assembly. There are two main types: mechanical panels, held together by mechanical means, and laminated panels held together with adhesive.

Mechanical insert assemblies are sometimes referred to as hollow metal pans. They consist of an outer pan of metal, usually 16 gage. This outside surface may be flat, embossed or formed and will have some sort of surface coating applied. The inner pan is usually galvanized iron, often factory prime-painted. These two pans are attached together with screws, rivets or some other mechanical method. The edge detail varies but should always provide a means of venting the inside of the panel. The two pans form a box section which may be filled with lightweight inorganic insulation or with a cementitious fill When the panel is not reinforced by a filler, the face area is limited to 16 square feet. When a cement fill is used, the face area may be increased to about 24 square feet. Thickness of the panel may be from 1¼ inches to 2½ inches.

Laminated insert assemblies can be assembled in many different ways. Facings or skin materials may any of porcelainized metal, aluminum or steel sheet, ceramic tile on metal, polyester with glass fiber, cement asbestos, marble, or glass.

Core materials may be of paper, asbestos or metal honeycomb, foamed glass, foamed polystyrene, fiberboard, low density cementitious materials, cementasbestos board, plywood or particle board, or polyurethane foam.

The core's main function is insulation, though it may also give strength to the composite panel.

Non-insulating laminates such as metal faced plywood or cement-asbestos faced fiberboard are available in standard sheets and can be cut to desired dimensions.

8 Doors and Windows

Sealants

Many types of materials are used in sealing curtain wall joints. All of them should have the ability to maintain watertight contact regardless of dimensional changes in joint sizes. They must be durable under all weather conditions, and should be easy to apply.

In general there are two basic types of joint seals, rigid and non-rigid. Rigid joint seals have a very limited use in curtain wall construction. Non-rigid sealants are far more common. For non-rigid seals in closed joints the usual materials are bulk compounds and preformed shapes. For non-rigid seals in operable joints, the usual materials are preformed shapes, formed thin metallic strips, and pile weather seals. Bulk compounds are applied to the joint by gun or knife after the joining parts are brought together. They are among the oldest types of sealing materials and are usually referred to as "caulking". Oil-base compounds and synthetic rubber or plastic compounds are the most common. Conventional oil-base compounds have poor adhesion also, their elongation is less over a long period of time. But their cost is low. If there is no maintenance problem and the joint movement is small, they can be used.

Non-skinning compounds have a base of polybutenes or a blend of polybutenes and polyisobutylenes with fibrous inert materials added. They do not dry out but become stiffer with age. They are shrink-free, but have poor recovery properties. The surface remains tacky. Therefore their basic use is as a bedding compound in protected locations.

Two-part rubber-base compounds have superior adhesion and elasticity and cure to a tough rubber consistency. They are commonly referred to as "Thiokol." "Thiokol" is the name of the company which supplies the polysulfide liquid polymer used as the base ingredient. The initial cost is high and handling can be difficult. The parts are supplied in two packages. One contains the liquid polymer. The other contains the curing agent. They must be mixed properly and thoroughly. Time is essential. The work life of the mixed compound depends on the air temperature. The mixed compound may be kept longer if frozen. When thawed at room temperature, the curing process starts again within 3 to 6 hours. Applicators may mix the two parts on the site or pre-mix in the shop and bring it to the site in a cold box.

One-part compounds have a distinct advantage over the two-part compounds. No mixing is needed and storage of the material is easier. No freezing is required. Curing begins upon exposure to air or moisture. Therefore air tight containers are mandatory. Their properties are similar to two-part compounds though they may take longer to "set up" and cure.

Preformed shapes vary from a puttylike to a rubberlike consistency. They are used as gaskets applied to one or both of the joining parts or placed in the joint before assembly.

Nonresilient preformed shapes are made from mastic material similar to non-skinning bulk compounds and are referred to as "tape sealers." Two bases, polybutenes or butylenes are used. They are identified as "non-vulcanized" and "partially vulcanized" types. Another material is impregnated wool felt, supplied in rolls of different widths and thickness. One side will have a pressure-sensitive adhesive. These are used mostly for mechanically fastened joints in protected areas.

Resilient shapes are produced either in vinyl (polyvinyl chloride), neoprene or butyl rubber. Sealing depends on flexibility and elasticity. They are usually designed on a custom basis for specific applications. The plastic materials such as vinyl are affected more by temperature than the elastic materials, such as neoprene and butyl. Vinyl, being a thermoplastic material, will become stiffer at low temperatures and more fluid at high temperatures. The advantage is that it is heat-sealing. It is supplied in a continuous coil, cut to fit and welded at the corners with a hot blade. Elastic materials must be joined by vulcanizing.

Curtain Wall Installation

The grid system and the panel system have different erection procedures. There are two ways of erecting a grid, from elementary parts and from prefabricated frames. When constructed from elementary parts, the mullions are mounted at each floor by means of adjustable fasteners. Horizontal members are connected between them and spandrel panels inserted and secured from the outside. This usually requires scaffolding.

When elementary parts are erected in a horizontal grid, the horizontal members are mounted to the structural floor and columns. If the column spacing is more than 7 to 10 feet, special brackets must be provided to support the sill member from the floor. The principal horizontal members may be attached directly to a back-up wall. Secondary verticals are connected to the horizontal members.

The advantage of using elementary parts is that individual grid members and inserts are easy to handle. Also, the disassembled curtain wall can be transported and stored in a small space. The disadvantage is slow erection and that inserts have to be installed after the grid has been erected. Scaffolding may also be needed.

Erection of prefabricated frames is much easier. Only inserts are installed on the site to complete the assembly. Prefabricated frames can be installed from inside the building without scaffolding.

In a panel system the grid can act as a structural matrix. The story-high units of the panel system are jointed together without an exposed frame. Most panels are of sheet-metal facing, although lightweight concrete panels are becoming common.

When erected in layers, the inside face, thermal insulation and outside face are bolted or clamped together during erection.

Prefabricated panels are normally one story high but may be larger. Window openings are usually glazed on site even though the frames are incorporated in the panel in the shop. Windows are always of a type that can be glazed from the inside.

Joints and Connections The most critical element of a curtain wall system is the joint between elements. Wall assemblies include two basic types of joints, closed and operable. The closed type presents the most problems in design and construction. Joints must allow for movement due to thermal changes and still prevent air and moisture infiltration. Joints must be strong enough to transmit forces from one element to another.

Simple butt joints are seldom used except where horizontal and vertical members meet in grid construction of elementary parts.

Lap joints are used where the joining edges lap each other in full or partial thickness. They are limited almost entirely to systems using metal wall cladding.

A mating joint is where the edges of the joining elements are formed in "male and female" profiles. It is used primarily in panel systems to connect prefabricated mechanically fastened panels.

Rabbet joints are rarely used except where a mating joint may be hard to form.

A batten joint is where the joint opening is covered on one or both faces with a cover strip or batten. Suited for vertical joints more than horizontal, it is used in connecting frame and panel units. A spline joint is where an accessory piece bridges the joint gap by fitting into opposite grooves in the element edges or by providing back-to-back grooves in the spline itself to receive the element edges. This type of joint is suitable only for connecting frames of extruded sections.

Sealing the Joints

Joints must be sealed to guarantee a weathertight connection. Sealants may be either plastic or elastic materials or a combination of the two. Plastic sealants are packed or injected into the joint. Elastic sealants are inserted in the form of gaskets. Gaskets made of rubber or metal are suitable.

Sealing methods vary widely. Some systems use an insert of elastic sealant and a packing of plastic compound. Others require just filling the joint with plastic sealant. A few designers recommend building an effective drainage system into the wall as well as providing a plastic or elastic seal.

Labor Installing Curtain Wall

Work Element	Unit	Man-Hours Per Unit
Jamb, head or mullion sections		
2" x 4"	100 L.F.	68.0
2" x 5"	100 L.F.	69.5
2" x 6"	100 L.F.	71.0
3" x 5"	100 L.F.	85.0
3" x 6"	100 L.F.	87.0
3" x 7"	100 L.F.	90.4
4" x 6"	100 L.F.	118.0
4" x 8"	100 L.F.	125.0
Coping sections		
1/8" x 6" flat	100 L.F.	6.2
1/8" x 8" flat	100 L.F.	7.0
1/8" x 10" shaped	100 L.F.	7.6
1/8" x 12" shaped	100 L.F.	8.3
1/8" x 16" shaped	100 L.F.	9.7
Exterior sill sections		
1/8" x 6"	100 L.F.	6.6
1/8" x 8"	100 L.F.	7.5
1/8" x 12"	100 L.F.	9.7
1/8" x 14"	100 L.F.	12.5
1/8" x 16"	100 L.F.	16.0
Column and mullion covers		
1/8" x 6"	100 L.F.	2.8
1/8" x 8"	100 L.F.	3.3
1/8" x 12"	100 L.F.	3.5
1/8" x 16"	100 L.F.	4.9
1/8" x 24"	100 L.F.	7.0
Spandrel covers		
3/16" aluminum panel	100 S.F.	5.5
1/4" aluminum panel	100 S.F.	7.6
Porcelain enamel panel	100 S.F.	8.3
Insulated metal panel 2"	100 S.F.	6.9
Insulated metal panel 4"	100 S.F.	10.5
Spandolite Glass, 1/4"	100 S.F.	7.2

For stainless steel add 5% to labor man-hours. Time does not include setup for scaffolding if needed. Time does include move on, unloading, stacking, setup, repair and cleanup as needed. Crew size will vary with location and size of the job.

9

finishes section

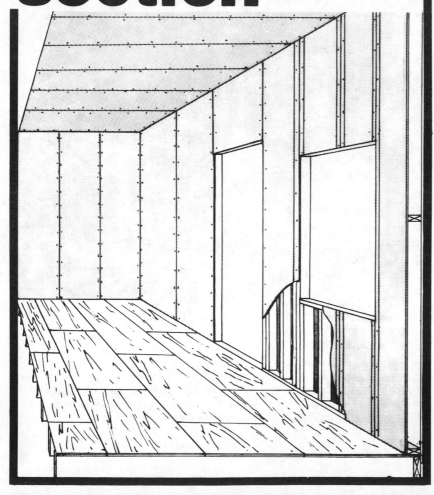

section contents

Lath and Plaster

Plaster is composed of cement binders, inert aggregate fillers, and water. Other materials may be added for color, workability, and to make the cured material harder, fire resistant, or to lend other properties. It is generally applied by trowel in one or more coats to form a durable wall or ceiling finish material, a backing material for tile, or as fireproofing for structural framing. The two types of plaster are made from either Portland cement or gypsum.

Portland cement plaster with integral color is called stucco in some parts of the country. Stucco finish interior coats are losing popularity in favor of painted plaster. Long lasting fade and alkali resistant paints have been developed that will provide a wide range of uniform colors. Stucco is available only in pastel shades and is often mottled in appearance.

Portland cement plaster is suitable for interior and exterior exposures. It should not be used over gypsum plaster, gypsum masonry, or gypsum lath. It is less stable than gypsum plaster and more susceptible to cracking. It cannot be finished as smooth as gypsum plaster, or with the same intricate detail. Portland cement plaster is not as fire resistant as gypsum plaster.

Gypsum plaster is not suitable for exterior exposures except in well protected areas. It should not be used in interior exposures subjected to wetting or extremely high humidity conditions. It can be applied over concrete, masonry, cement plaster and gypsum and metal laths.

Both cement and gypsum plasters may be made with sand, various lightweight aggregates, organic and inorganic fibers, and a wide variety of admixtures. Each coat in a multiple coat cement or gypsum plaster job is likely to have different ingredients or different proportions of the same ingredients.

A wide variety of finishes and textures can be made with both cement and gypsum plasters. Acoustical plaster uses a special spray applied formula. Various designs and textures are made on fresh plaster with trowels, brooms, brushes, sponges, and raking and scoring tools. Aggregates, glitter, and other colorful or decorative mixtures can be embedded in the finish coat while it is still plastic. Simulated brick and other masonry, travertine, or stone is also possible. Multi-colored, three dimensional designs can be made with two or more layers of pigmented plaster or stucco. While the mass is still green, plaster is scraped away to the required depth to reveal the desired color. Frescoes are made by painting a design on fresh plaster with water base paint.

Plaster can be applied directly to concrete or masonry. It can also be supported by lath attached to concrete, masonry, sheathing, wood or metal furring strips. It can also be applied to lath attached to a metal grid system suspended on wires from the structure.

Spacing of framing and furring members depends on the strength of the furring as well as the lath used, and whether the furring is applied vertically or horizontally. Spacing of all members should be indicated on the plans. Manufacturers, trade associations, and building codes publish tables showing maximum spacing for various components that meet fire and safety standards.

Lath
The earliest form of lath was unsurfaced wood strips, nailed to wood framing members. Wood lath is still manufactured but seldom used in modern construction. Modern lath is either gypsum or metal.

Gypsum lath is made of paper bonded to a gypsum core like gypsum wallboard except that long edges are generally rounded. It may be plain or perforated with ¾-inch diameter holes at about 4 inch centers in both directions. It is available in ⅜-, ½-, ⅝- and 1-inch thicknesses, 16-, 24-, and 48-inch widths, and in 48- and 96-inch lengths. Plain (unperforated) lath is available with aluminum foil backing for thermal insulation, and with lead sheet backing for x-ray shielding. Type X gypsum lath has special additives in the core to make it more fire resistant. Perforated lath provides mechanical as well as a natural chemical bond with gypsum plaster.

Expanded metal lath is available in several forms and in galvanized steel and factory painted copper bearing steel. Flat diamond mesh is a uniformly expanded flat sheet that is produced in two weights: 2.75, and 3.4 pounds per square yard. Self-furring diamond mesh is similar to diamond mesh (including weights) except that it is indented at regular intervals to hold the body of the lath away from sheathing or other flat surfaces. Flat rib lath has ⅛ inch deep ribs evenly spaced and parallel to the long dimension of the lath. Flat rib lath is available in 2.75 and 3.4 pound per square yard weights. Three-eighths inch rib lath has deeper ribs spaced at 4½-inch centers with reverse ⅛ inch ribs at midspan. This lath is available in 3.4 and 4.0 pound per square yard weights. Three-quarter inch rib lath has ribs spaced at 6 inch centers and is available in 5.40 and 6.75 pound per square yard weights.

Woven wire fabric has a hexagonal mesh pattern. It is also called stucco netting or poultry netting and is available in from 1-inch to 2¼-inch mesh with a maximum open area of 4 square inches per mesh. Minimum size wire gages are 18 for 1-inch mesh, 17 for 1½-inch mesh, and 16 for 2-inch mesh. Woven wire fabric is galvanized and generally used for exterior work. It is available dimpled at regular intervals to hold the body away from sheathing or other flat surfaces. Special nails with fibered spacer washers on the shank can also be used for this purpose. It is also available with paper backing for use over unsheathed framing.

Welded wire fabric lath is made from 16 gage wires, spaced not over 2 inches in either direction and stiffened in the longitudinal direction with 13-gage wires spaced at 6-inch centers. All wires are welded at all intersections. Most welded wire lath in use today is galvanized and provided with paper backing. The paper is slotted and laid between the wires before they are welded. Some lath has an additional backing sheet of paper glued to the sandwiched sheet. Some welded wire lath is also crimped at regular intervals to hold the body of the lath away from contact surfaces.

Wire cloth is sometimes used as a plaster base. It is generally 19 gage wires woven in a straight grid pattern and galvanized after weaving.

Metal Framing

Vertical metal framing consists of ceiling and floor runner tracks and other supports. Metal frames can be either load bearing and non-load bearing. Metal studs are available in a number of forms, sizes and load carrying capacities. The following are the most popular.

Hollow partition studs generally have perforated or truss-type webs to allow internal horizontal reinforcement and placement of utility lines. The principal difference between load bearing and non-load bearing studs is that the former are fabricated from heavier materials and may have wider flanges. Hollow partition studs are available in 1⅝, 2, 2½, 3¼, 3⅝, 4 and 6 inch widths, and in the following forms:

Perforated channels, designed for lath attachment by wire tying or clips.

Welded wire truss design with two straight wires forming each chord, welded to each side of diagonally bent wire flanges. This stud is designed for attachment of lath by clip or wire tying.

Combination studs are also designed for clip or wire tie attachment of lath. They have a continuous diagonally bent wire web, welded to cold rolled steel angle flanges.

Nailable studs are made by spot welding two channels back to back or spot welding two angles to a channel to form an I section. Nails are held either by bending to follow the deformation built into the web or by locking against a ring or helical deformation in the shank of the nail. Nailable studs designed to receive deformed shank nails will also accept screws. These studs may have solid or perforated webs.

Screw studs are light gage metal channels with either solid or peforated webs and with either serrated or dimpled flanges to receive sheet metal screws without drifting.

Studs for solid partitions and double studs for hollow partitions are cold rolled solid web channels. They are available in ¾, 1½ and 2 inch sizes, weighing 300, 475, and 590 pounds per 1000 lineal feet. They are available in factory applied painted and galvanized finishes.

Studless solid partitions are made with gypsum and expanded metal rib lath spanning from floor to ceiling, and secured with some form of metal angle track.

Horizontal metal framing consists of a suspended channel grid system on which lath and plaster may be applied to form a ceilng. Channels are installed with the webs in a vertical position. Splices are made by reversing the direction of the flanges, lapping, and wire tying with two loops of wire near each end of the lap. Runner channels (also called carrying channel) are generally 1½ to 2 inch solid web cold rolled steel. Hanger wires are secured to the construction above and saddle tied to runners.

Vertical and horizontal furring can serve many purposes. It is used to true an otherwise irregular surface, to provide an air space between an outside wall and plaster, to provide a utility space between a structural wall and the plaster, to span structural supports that would be too great to span by the lath alone, to provide a "floating" wall or ceiling that can reduce sound transmission.

All metal lath should be held away from the parallel contact surface by 1 inch or more to guarantee proper plaster embedment. Self-furring lath with dimples against supports, rib lath with ribs against supports, or flat lath with integral spacers are used for this purpose.

Furring members can be wood or metal strips attached directly to structural supports or attached to clips or braces which are attached to structural supports. Metal furring members come in several forms and in painted and galvanized finishes as follows.

1. Cold rolled solid web channels, generally ¾ inch thick, designed for wire tying or clip attachments to lath.

2. Nailing channels are designed to hold nails by friction or by bending the nail around a rod built into the channel.

3. Screw channels are actually thin gage hat shaped sections in which the face to receive the screw is perforated, dimpled, or scored to allow driving a sheet metal screw without drifting.

4. Resilient channels are Z-shaped sections in which some metal has been taken out of the web to give the member a spring action. They are used in assemblies designed to reduce sound transmission through the assembly.

Metal Accessories

Unprotected exposed edges and outside corners of plaster are vulnerable to chipping and spalling from impact. Driving nails into plaster will cause it to crack. Plaster, particularly cement plaster, expands and contracts with changes in temperature and moisture content. A number of accessories have been developed to combat these shortcomings. All accessories that are exposed should have a galvanized finish. Completely concealed items are available both in galvanized and painted finishes. Some accessories are available in solid zinc alloy, stainless steel, aluminum, and rigid plastics. The following are some of the common accessories used with lathing:

1. Corner beads are used to protect outside corners. They are available in small nose or radius bull nose type, and with expanded or perforated webs.

2. Casing beads are used to protect exposed edges of plaster, or to terminate plaster against other material. Casing beads are available in square and radius bull nose edges, and in solid or expanded flanges.

3. Picture molds resemble upside down reglets. When installed, the slot slopes downward from the lips. It is designed for hanging pictures and other objects on the wall without driving nails into the wall.

4. Control joints (also called expansion joints) are designed to break up large expanses of plaster areas by dividing them into smaller units. This allows each isolated unit to expand and contract without cracking. Various stock designs are available. Some are narrow and inconspicuous; others are wide and are incorporated into the design as an architectural feature. Others are perforated in the flanges between the grounds to provide ventilation into the attic space from soffits. Various designs include square nose casing beads, applied with a space between the nosings, bellows types, and channel shapes. Some designs in-

corporate a return flange at the lip edge to cover the edge of the plaster and hold it firmly against the base flange.

5. Base or parting screeds are used to provide a ground between cement plaster bases and gypsum plaster walls, between tile bases and plaster walls, or between walls and bases of different thicknesses where a step in the screed provides double grounding.

6. Partition terminals and caps are used in solid plaster partitions to provide a metal edging at the top of dwarf walls or wall ends that do not abut other construction. The exposed portion is generally convexed to a wide radius.

7. Drip screeds are designed to take the place of a corner bead on horizontal returns. They have a downward projecting lip which diverts water rather than allowing it to run back along the underside of the soffit or wall opening.

8. Partition bases, chair rails, and window stools are a group of accessories that come with a galvanized base and factory prime-coat finish. Various designs are available, generally all with concealed fasteners. Bases are available in grout-in, lock-in, and snap-in designs.

9. Corner and strip reinforcement (inside corners are called cornerite) are the only accessories in the group that are entirely concealed by the plaster. They are available in either painted or galvanized finish. They are made from expanded metal, woven wire, or welded wire and are used over both metal and gypsum lath where special reinforcement is needed.

Fasteners

A great variety of fasteners are used in the lathing and plastering industry. Some are common types found in a variety of assemblies. Others have been developed for specific purposes in the lathing and plastering industry. Following are some of the common fasteners:

1. Wire is perhaps the most common fastener. It is used to support horizontal grid frames, to tie vertical and horizontal framing and furring members together, and to secure lath to supporting members. Wire used is annealed galvanized steel.

2. Nails and staples of various wire gages, lengths, shank and head designs and finishes are used to secure metal and gypsum lath to wood and metal supports. Nails are available with galvanized, zinc plated, blued, cement coated, or bright finishes. Shanks come in smooth, barbed, ringed, or threaded design. Hardened stub nails are used to secure metal or wood to concrete or masonry.

3. Explosive powder-driven fasteners are used to secure metal runner tracks to concrete.

4. Self drilling and thread cutting or thread forming fasteners are used to attach both metal and gypsum lath to metal supports, and to connect metal framing members. Bolts are also used for some framing member connections and to secure framing members to concrete or masonry.

5. Clips of various designs, made of spring wire or sheet steel, are designed to secure cross furring channels to runner channels, gypsum lath to metal supports, and resilient furring to rigid supports. Clips are generally zinc plated.

Plaster Aggregate

Aggregates in plaster extend coverage, reduce shrinkage, increase strength, and lower cost. Aggregates for plaster include wood and asbestos fiber, sand, perlite and vermiculite.

Sand is the most common aggregate. It is denser, stronger and a better barrier against sound transmission. Vermiculite and perlite are used to improve fire resistance, insulation value, sound absorption, and reduce weight. Fiber aggregate, such as shredded wood, asbestos, or other fibers, are generally used with particle aggregates. Fibers increase plaster strength and fire resistance, and reduce weight. Fibers also aid placement in gun applied plaster.

Multiple Coat Applications

Gypsum basecoats are applied in one, or more often two, coats. In two coat work, the second coat generally has a higher aggregate content than the first. Basecoat material is available in factory mixed packages that require only mixing and water. Additional aggregate is sometimes added to the second (brown) coat. Sand is also sometimes added to the first (scratch) coat over gypsum lath or masonry bases.

Job mixed basecoats contain neat gypsum plaster mixed with graded sand or vermiculite or perlite.

Gypsum Finish Coats

Lime putty in a plaster mix provides whiteness, plasticity, and bulk. Lime, used alone, does not set hard, and shrinks as it dries. Quicklime requires a long slaking (soaking) period at the jobsite to develop the plasticity required for plastering. For these reasons, lime is blended with gypsum gauging plaster or Keene's cement to obtain the desired properties.

Hydrated lime is quicklime which has been partially slaked at the factory to reduce the field slaking period. The two basic types of hydrated lime are Normal (Type N, ASTM C-6) which requires 16 to 24 hours of soaking, and Special (Type S, ASTM C-206) which does not require soaking. It can be used immediately.

Gauging plasters are specially ground to provide controlled set, strength, and minimum shrinkage when mixed with lime. Gauging plaster is available in slow and quick set types. This allows control of the hardening process without the use of retarders or accelerators.

Keene's cement is used rather than gauging plaster when a dense, harder than average finish and more resistance to moisture are required. It is available in Type I regular slow setting (3 to 6 hours), and Type II quick setting (1 to 2 hours).

Molding plaster is a finely ground gypsum mix which produces a smooth, workable material suitable for intricate ornamental and decorative work.

Gypsum finish coats can be made from several possible mixes:
1. Gypsum gauging plaster and a lime putty trowel finish.
2. Keene's cement and a lime putty trowel finish.

3. Prepared (factory mixed) gypsum trowel finish.

4. Prepared high-strength gypsum white coat trowel finish.

5. Prepared and jobsite mixed lightweight aggregate spray-applied acoustical finish.

6. Keene's cement and lime sand float finish.

7. Gypsum sand float finish.

8. Special finishes that include integral coloring and receive special texturing.

9. Veneer or skin coat plaster is a thin finish coat applied directly over a suitable base. It can provide a smooth finish to an otherwise porous surface. It can also be used to add texture or to conceal joints or irregularities.

10. Concealed fireproofing is a special formulated spray-applied material containing asbestos fibers or expanded mineral aggregates. It may be applied in one or more coats to structural members in concealed locations.

Portland cement basecoats are generally applied in two courses: scratch, and brown. The scratch coat is made up of Portland cement, aggregate, and plasticizers. The cement may be Type I, II, or III. The plasticizers may be hydrated lime or lime putty. Rather than plasticizer additives, some or all of the cement may be masonry cement or plastic cement, which have plasticizers added at the factory. The aggregates may be graded sand, vermiculite, or perlite, or combination of sand and lightweight aggregates. Gun applied basecoats may have asbestos fibers added to the mix. The brown coat generally will have a higher aggregate content than the scratch coat. Single basecoats are applied to concrete or masonry by the double-back method. In both scratch and brown coats only enough water is added to produce a workable mixture.

The cement finish coat is similar to the scratch and brown coats except that the sand aggregates are finer and a higher ratio of cement is used. The plasticizers (also called buttering agents) may also differ. Smooth troweled cement finishes are difficult without craze cracking. For this reason most cement finishes are sand floated.

Bonding agents are used to ensure a good bond between plaster and hardened smooth concrete or similar surfaces. Factory prepared bonding agents replace sandblasting, bushhammering, or dashcoating as a preparatory measure under the first plaster coat.

Application

Vertical furring members are attached to channel or angle floor and ceiling runner tracks, which are held in place with friction or threaded fasteners. All members should be installed level and plumb unless shown otherwise on the drawings. Grouped multiple studs, bracing, or other means of reinforcing should be used in areas around openings, or where special loadings or stresses will be expected.

Suspension systems, main runners or carrying channels are secured to hanger wires, rods, or bars by tying, clamping, or special clips. Cross furring members are then attached to runner channels by wire tying, or with metal clips. The size, weight, and spacing of the hangers, runners and cross furring channels and the type and weight or thickness of the lath should be indicated in the plans and specs. Where control or expansion joints are indicated, the grid is broken at that point to form an independent suspension system.

Smooth concrete plaster bases are prepared for plaster by sandblasting, bushhammering, the use of a bonding agent, or the application of self-furring lath. Porous masonry is prepared for plaster by wetting with clean water. Allow the water to soak in. Surfaces must be damp when the scratch coat is applied.

Asphalt saturated felt should be applied over wood sheathing or unsheathed wood framing over which line wire has been stretched. The felt is applied with the long dimension horizontal. The horizontal joints should be lapped shingle fashion.

Lath is applied with the long dimension at right angles to supports. Ribbed and self-furring lath is applied with ribs or dimples against supports. Edge ribs are nested along longitudinal joints. Lath can be attached to supports with wire ties, nails, staples, clips or screws, depending on the type of lath and the material to which it is attached. Lath is generally secured to supports near each corner and at each framing member. Ribbed lath in studless solid partitions is installed with the ribs vertical, and with the ends secured to ceiling and floor angle runners. Lath can be interrupted by expansion or control joints. Ends should be secured to two closely spaced parallel supports.

Casings, corner beads, base screeds, picture molds, expansion or control joints, and other accessories are selected to fit the plaster thickness. In unusually corrosive conditions, zinc alloy accessories are used rather than galvanized steel.

Basecoats may be applied by trowel or by machine in either one or two (scratch and brown) coats. Single basecoat is enough for properly supported unperforated gypsum lath and on some concrete and masonry surfaces. Elsewhere two basecoats are generally used. The scratch coat is applied with enough pressure to produce a good mechanical bond. Where metal or perforated gypsum lath is used, the scratch coat plaster should be forced through the openings. If metal lath is used over a solid surface, the scratch coat should be embedded in the lath and fill all voids between the lath and the backing surface.

When applying the scratch coat to studless ribbed lath to build a solid plaster partition, the lath is temporarily back braced until the scratch coat has set. After the scratch coat has been applied, its surface is given a rough texture by raking (or scratching) in one or two directions to ensure a good mechanical bond with the brown coat.

In single basecoat work, the cross raking and brown coat is omitted. Plaster is applied to the base, then doubled back (generally within a few minutes) to bring the surface out to the required thickness and plane. The same plaster mix is used.

The brown coat is applied to a cured or partially cured scratch coat. Cured cement plaster scratch coat is dampened before applying the brown coat. The brown coat is rodded and floated to a true flat plane, and left with a rough sand texture to bond the finish coat.

The finish coat in both cement and gypsum plaster is about $\frac{1}{8}$ inch thick. It may be smooth troweled, sand floated, or finished in a wide variety of textures. Cement or stucco finish coats may be made from white cement, or a mixture of white and gray cement, or white cement and colored mineral oxide pigments.

Veneer plaster uses a skim coat similar to the finish coat except that it is applied directly to an unperforated flat base. It is generally a factory formulated and packaged product requiring only the addition of water and field mixing. Veneer plaster is applied to concrete, masonry, gypsum board (drywall) or plain gypsum lath. The thickness is generally about ⅛ inch.

Acoustical plaster is applied in one or two coats by trowel or by machine spray. It is a factory mixed and packaged product containing lightweight aggregates. Trowel applied plaster can be given an acoustical texture by stippling and piercing with a perforating tool. The texture can be varied in machine applied mixes by varying the mix and the application technique.

Sound retardant plastering depends on mass and density of the plaster or separation of through-wall framing. Resilient clips or furring strips isolate plastered areas. In addition, sound absorbing material such as mineral fiber blankets may be installed in hollow wall structures. All air leaks through which sound waves may travel should be sealed at wall perimeters, fixture outlets, and other through-wall openings with sealant or caulking material.

Labor Installing Gypsum Lath

Work Element	Unit	Man-Hours Per Unit
Gypsum lath on metal studs and furring		
⅜" plain	1000 S.F.	12.0
⅜" perforated	1000 S.F.	11.0
½" plain	1000 S.F.	13.0
½" perforated	1000 S.F.	12.5
Gypsum lath on wood studs and furring		
⅜" plain	1000 S.F.	9.5
⅜" perforated	1000 S.F.	9.0
½" plain	1000 S.F.	11.0
½" perforated	1000 S.F.	10.0

Add 5 man-hours per 100 S.F. for application on ceilings. Double these figures for application on beams and columns. Time includes unloading, stacking, cutting, handling material into place, nailing or clipping into place, installing beads and molding, and cleanup as needed.
Suggested Crew: 2 lathers

Labor Installing Metal Lath

Work Element	Unit	Man-Hours Per Unit
Metal lath on metal studs		
2.5 lb. diamond mesh	1000 S.F.	13.5
3.4 lb. diamond mesh	1000 S.F.	15.5
2.75 lb. rib lath	1000 S.F.	11.8
3.4 lb. rib lath	1000 S.F.	15.5
Metal lath wood studs		
2.5 lb. diamond mesh	1000 S.F.	10.9
3.4 lb. diamond mesh	1000 S.F.	12.7
2.75 lb. rib lath	1000 S.F.	10.9
3.4 lb. rib lath	1000 S.F.	12.7
Self-furring lath on solid surface		
2.5 lb. diamond mesh	1000 S.F.	15.5
3.4 lb. diamond mesh	1000 S.F.	17.0
Metal lath on columns and beams		
2.5 lb. diamond mesh	1000 S.F.	34.0
3.4 lb. diamond mesh	1000 S.F.	36.0
2.75 lb. rib lath	1000 S.F.	35.0
3.4 lb. rib lath	1000 S.F.	36.0
Metal lath on ceilings		
2.5 lb. diamond mesh	1000 S.F.	17.2
3.4 lb. diamond mesh	1000 S.F.	19.0
2.75 lb. rib lath	1000 S.F.	16.0
3.4 lb. rib lath	1000 S.F.	18.0

Lathing includes erecting scaffolding, handling material into place, cutting and installing wires and straps, cutting and fastening lathing, angles, beads, molding.
Suggested Crew: 2 lathers

Labor For Lath Accessories

Work Element	Unit	Man-Hours Per Unit
Miscellaneous lathing		
Corner beads	100 L.F.	2.3
Picture mold	100 L.F.	3.3
Metal base ground	100 L.F.	3.2
Metal plaster ground at door or window	Each	.8
Metal casing mold	100 L.F.	4.0
Wood plaster ground	100 L.F.	2.6
Stucco		
Vapor barrier, line wire, stucco netting	100 S.Y.	5.0

9 Finishes

Labor Installing Metal Framing For Plaster

Work Element	Unit	Man-Hours Per Unit
Wall systems, studs with tracks and runners		
¾" channels, 16" o.c.	100 S.Y.	30.0
1½" channels, 16" o.c.	100 S.Y.	31.0
2¼" channels, 16" o.c.	100 S.Y.	32.0
3½" channels, 16" o.c.	100 S.Y.	33.0
4" studs, 16" o.c.	100 S.Y.	34.0
6" studs, 16" o.c.	100 S.Y.	36.0
Ceiling systems		
¾" main channels 36" o.c. with ¾" furring channels 12" to 18" o.c.	100 S.Y.	50.0
1½" main channels 36" o.c. with ¾" furring channels 12" to 18" o.c.	100 S.Y.	48.0
1½" main channels 48" o.c. with 1½" furring channels 18" to 24" o.c.	100 S.Y.	47.0
2" main channels 60" o.c. with 1½" furring channels 24" to 36" o.c.	100 S.Y.	45.0
Wall channel furring @ 16" o.c.		
¾" channels	100 S.Y.	30.0
1½" channels	100 S.Y.	30.0
2¼" studs	100 S.Y.	31.0
3½" studs	100 S.Y.	34.0
Column furring		
¾" channels	100 L.F.	3.5
1½" channels	100 L.F.	4.0
2¼" studs	100 L.F.	4.5
Beam furring		
¾" channels	100 L.F.	7.2
1½" channels	100 L.F.	7.9
Ceiling channel furring		
¾" channels, 12" to 18" o.c.	100 S.Y.	49.0
¾" channels, 18" to 24" o.c.	100 S.Y.	43.0
1½" channels, 18" to 24" o.c.	100 S.Y.	49.0

Ceiling system assumes ceiling is suspended at 8' height by rod hangers from steel framing or metal decking. Time includes mobilization and de-mobilization, unloading, stacking, installation, repair and cleanup as needed.
Suggested Crew: 2 lathers

Materials for 100 S.Y. of Lath

Type of Construction	Lath	Other Materials
Metal lath on wood studs	105 S.Y.	8 lbs. nails or 15 lbs. staples
Metal lath on steel studs for 2" solid partition	1000 L.F. of ¾" channels 105 L.F. of lath	10 lbs. tie wire
Plasterboard on wood studs	950 S.F.	7 lbs. nails
Cornice furring	110 S.Y. of lathing	8 lbs. nails
Beam and cornice furring	1800 L.F. of ¾" channels 108 S.Y. of lath	27 lbs. tie wire
Suspended ceiling with cross channels 12" o.c.	285 L.F. of 1½" channel 1000 L.F. of ¾" channel 105 S.Y. of lath	85 ¼" hanger rods and 18 lbs. wire

Labor Plastering

Work Element	Unit	Man-Hours Per Unit
Solid plaster partitions, including lath and ¾" channel studs		
2" thick partitions	100 S.Y.	115.0
3" thick partitions	100 S.Y.	128.0
2 coat plaster, with finish indicated, on walls		
Gypsum or lime finish	100 S.Y.	42.0
Keene's cement finish	100 S.Y.	54.0
Wood fiber finish	100 S.Y.	47.0
Vermiculite finish	100 S.Y.	40.0
Portland cement finish	100 S.Y.	45.0
Acoustical finish	100 S.Y.	58.0
3 coat plaster, with finish indicated, on walls		
Gypsum or lime finish	100 S.Y.	55.0
Keene's cement finish	100 S.Y.	69.0
Wood fiber finish	100 S.Y.	68.0
Vermiculite finish	100 S.Y.	55.0
Portland cement finish	100 S.Y.	66.0
Acoustical finish	100 S.Y.	72.0
Scratch coat	100 S.Y.	12.0
Brown coat	100 S.Y.	13.0
3 coat stucco on metal lath		
¾", float finish	100 S.Y.	70.0
¾", trowel finish	100 S.Y.	74.0
Additional plastering requirements		
Add for ceiling work	100 S.Y.	4.5
Add for column work	100 S.Y.	23.0
Add for chases, fascia, recesses or soffits	100 S.Y.	33.5
Add for beams	100 S.Y.	26.0
Add for irregular surfaces	100 S.Y.	21.0
Deduct for plaster over gypsum lath	100 S.Y.	6.0

Plastering includes mixing plaster, installing and finishing plaster, scaffolding, curing and drying plaster.
Suggested Crew: 2 plasterers, 1 or 2 tenders

Square Feet Covered by One Bag of Cement in Various Mixes

Mix Parts by Volume		Thickness of Coat				
Cement	Sand	1/4"	3/8"	1/2"	3/4"	1"
1	1	266	177	133	89	66
1	1½	336	226	168	112	84
1	2	404	270	202	135	101
1	2½	472	314	236	157	118
*1	3	542	362	271	181	136
1	3½	612	408	306	204	153
1	4	682	455	341	227	171

* 1:3 is the mix most used for stucco work.

Material for 100 S.F. of Plaster or Stucco Walls

Wall Thickness Inches	C.F. Plaster	1:2½ Mortar		1:3 Mortar	
		Cement Sacks	Sand C.F.	Cement Sacks	Sand C.F.
1/4	2.08	0.79	1.95	0.68	2.06
3/8	3.13	1.19	2.95	1.03	3.10
1/2	4.17	1.58	3.94	1.37	4.12
5/8	5.21	1.98	4.92	1.71	5.15
3/4	6.25	2.37	5.90	2.06	6.18
1	8.33	3.16	7.86	2.74	8.24

Materials for 100 S.Y. Gypsum Wood Fiber Plaster

Construction	Proportion By Weight	100 Lb. Sacks Plaster	C.Y. Sand
Over metal lath	No sand	22 to 27	—
Over plasterboard or gypsum lath	No sand	13 to 16	—
Over brick and clay tile	1 to 1	18 to 20	.7 to .8
Over gypsum tile	1 to 1	14 to 16	.5 to .6

Proportions for gypsum plasters are by weight. Includes scratch and brown coat.

Materials for 100 S.Y. Finishing Coat

Kind of Finish	50 Lb. Sacks Hydrated Lime	C.Y. Sand	100 Lb. Sacks Plaster
Sand finish using lime and plaster	6 to 7	1/4	—
White finish using lime and plaster	6 to 9	—	1 to 1¼
Hard cement finish using Keene's cement	—	—	4 to 5
Smooth cement finish using Keene's cement and hydrated lime	2	—	3 to 4
Sand finish using special finishing plaster	—	.12 to .25	2 to 2.5
Lime putty	6 to 9	—	—

Materials for 100 S.Y. Gypsum Cement Plaster

Construction	Proportion By Weight	100 Lb. Sacks Plaster	C.Y. Sand
Over metal lath	1 to 2	17 to 20	1.3 to 1.5
Over plasterboard	1 to 2	8 to 9	.6 to .7
Over brick or clay tile	1 to 3	14 to 17	1.5 to 1.9
Over gypsum tile	1 to 3	10 to 12	1.1 to 1.3

Includes scratch and brown coat.

Materials for 100 S.Y. Gypsum Sanded Plaster

Construction	100 Lb. Sacks Plaster
Over metal lath	45 to 50
Over plasterboard	20 to 22
Over brick or clay tile	35 to 40
Over gypsum tile	25 to 28

Includes scratch and brown coat.

Yield for Hydrated Lime

Hydrated Lime	C.F. Putty
44 lbs.	1
50 lbs. (1 paper bag)	1.2
100 lbs. (1 cloth bag)	2.3
308 lbs. (6 sacks)	7.
456 lbs. (9 sacks)	10.5

Materials for 1 C.Y. of Mortar

Kind of Mortar	Proportion	Lime or Cement	C.F. Sand
Pure lime putty	--	44 lbs. hydrated lime	No sand
Lime mortar	1 to 2	19.6 lbs. hydrated lime	0.87
Lime mortar	1 to 3	15 lbs. hydrated lime	1.0
Portland cement mortar	1 to 2	41 lbs. cement	0.87
Portland cement mortar	1 to 3	32 lbs. cement	1.00
Gypsum cement plaster	1 to 2	33 lbs. plaster	0.66
Gypsum cement plaster	1 to 3	20 lbs. plaster	0.6
Gypsum wood fiber plaster	No sand	58 lbs. plaster	No sand
Gypsum sanded plaster	--	79 lbs. plaster	No sand

Proportions for gypsum plaster are by weight. The proportions for lime and Portland cement are by volume.

Materials for 100 S.Y. of Plaster

Thickness	Proportion by Volume	Sacks of Cement	C.Y. Sand
1/8''	1 to 1	6.75	0.25
1/8''	1 to 1½	5.36	0.30
1/8''	1 to 2	4.48	0.33
1/8''	1 to 2½	3.84	0.35
1/8''	1 to 3	3.32	0.37
1/4''	1 to 1	13.52	0.50
1/4''	1 to 2	8.96	0.66
1/4''	1 to 3	6.64	0.74

Plaster Required to Cover 100 S.Y.

Thickness	Quantity
1/8''	10 C.F.
1/4''	20 C.F.
3/8''	30 C.F.
1/2''	40 C.F.

Gypsum Drywall

Gypsum wallboard is a paper covered gypsum core material used as a finish wall or ceiling material, and as a rigid backing material for acoustical tile, ceramic tile, and wall coverings, or other finish materials. Similar products are also made as a base for gypsum plaster and sheathing for wood frame buildings.

Gypsum board is relatively inexpensive, fire resistant, has good sound resistant properties, is dimensionally stable, resists cracks caused by minor frame movement, and installs faster and with less mess than plaster. It also eliminates the need for drying out the building before other materials can be installed.

Gypsum board should not be used in areas of extreme or sustained moisture. Joints are difficult to completely conceal, especially if the wall is finished with a smooth high gloss paint.

Gypsum board may be divided into two board types. The first is wallboard, which is intended to be exposed after installation. It may be either job finished or factory finished. Backing board is intended for use as a backing material for gypsum board laminated assemblies, acoustical tile, ceramic tile, and other finish materials.

Wallboard and backing board are available in $\frac{1}{4}$, $\frac{3}{8}$, $\frac{1}{2}$ and $\frac{5}{8}$ inch thicknesses, 16, 24, 32, and 48 inch widths, and in lengths from 4 through 16 feet in 1 foot increments. Backing boards are also available in 1 inch thickness.

The core of gypsum wallboard can be regular gypsum, water repellent, or fire retardant which is specially formulated from gypsum and other materials to increase fire resistance.

Several finishes are available in gypsum board. Wallboard is available in plain manila paper face or with a factory applied finish consisting of sheet vinyl or special printed paper. Both plain gray liner paper and aluminum foil reflective backs are available.

Backing boards are available with a plain gray liner paper finish on both sides or gray paper on one side and aluminum foil on the other. Colored paper treated to repel water is an alternate finish.

Gypsum board is manufactured in a continuous process. Individual boards are cut to the desired length as it rolls off the production line. The ends are cut square and the core is exposed. The edges, however, are covered by paper or other facing material and may be tapered, rounded, eased, beveled, square, or V tongue-and-grooved. Backing boards generally have square or tongue-and-groove edges. Wallboard usually has tapered edges that can be taped and filled with cement. Tapered edges may be square, eased, rounded, or beveled. Factory decorated wallboard has untapered beveled, eased, or square edges.

Gypsum Board Fasteners

Threaded fasteners are used to fasten gypsum board to both wood and metal supports. They have a bugle shaped head which is recessed to take a Phillips type screwdriver. They are driven with a power tool which has a clutch to disengage the drive force at the right time. Screw heads should be set just below the surface without overdriving and tearing the paper or rupturing the core. Threaded fasteners come in different lengths and with different points and thread design. Fasteners for gypsum board to metal supports are self-drilling and thread-forming. The point design depends on thickness of the metal and the diameter of the fastener shank. For wood supports, screws should have a diamond point similar to a 6d common wire nail.

Nails for attaching gypsum board to wood members are either phosphate-treated or cement-coated smooth shank. Ring shank nails have the best holding power.

Staples are sometimes used to secure backing boards in place. They are mechanically driven.

Adhesives can be used for attaching gypsum board to wood or steel. Adhesives for laminating wallboard to backing board are specially formulated for this purpose.

Drywall Supports

Resilient metal furring strips installed over wood members are light-gage metal designed to be nailed to wood supports. Gypsum board is secured to the strips with sheet metal screws. Resilient strips are designed in several cross-section configurations. Their purpose is to break the rigid connection between the wallboard and the supporting member and allow the wallboard to "float" and vibrate independently of the support. This absorbs energy created by sound vibrations which would otherwise be transmitted through the wall.

Metal furring channels are clipped, wired, nailed, or secured with threaded fasteners to concrete or masonry walls, ceiling runners carrying channels, or open web joists. The gypsum board is then attached with screws.

For suspended ceilings, suspension wires are attached to the structure above and then saddle-tied to runner channels. Furring channels are attached to the runners to form a level, flat grid to which gypsum board may be attached.

Channel stud framing system consists of a channel floor and ceiling track to which channel studs are attached.

Solid and semi-solid gypsum partitions consist of metal channel or angle floor or ceiling tracks to which gypsum board is attached. Stiffening members may include a solid gypsum core, gypsum studs or ribs to which wallboard is laminated.

Movable gypsum board partitions are made by several manufacturers. Some use laminated gypsum board with tongue-and-grooved vertical edges in which a whole section of the wall (both faces) is installed at a time. Others have more or less conventional framing systems using kerf edged boards into which edges of steel framing members fit. Some use snap-on battens or other systems of securing wallboards to supports without through fastenings. Framing and trim for doors, windows, and other accessories are available in extruded aluminum or pressed steel.

Gypsum Wallboard Accessories

Exposed edges and exterior corners of gypsum wallboard should always be reinforced or protected from damage by metal trim. Where gypsum board abuts concrete walls or other dissimilar construction, the transition is often made with a metal casing bead. Inside corners and butted joints are taped and spackled, or finished with metal trim members to match factory finished gypsum board panels. Casing beads for field decorated wallboard are usually zinc plated steel. In the wrap around type, gypsum board edge is slipped into the installed bead. The surface applied type is installed after the gypsum wallboard is in place. The exposed flanges of both types are textured to hold joint compound for a smooth transitional finish.

Corner beads are either all metal or metal bead with paper flanges. Flanges of the all-metal type are textured to receive joint compound.

Trim for factory finish decorated panels includes casing, divider strips, inside corner beads, outside corner angles, and other trim shapes. They may be either extruded aluminum or rigid vinyl plastic with factory painted or factory laminated vinyl plastic sheet to match the wall panels.

Joint Treatment and Texturing
Reinforcing tape is made from strong fiber or high quality papers with feathered edges. It is either porous or perforated to allow air and water to escape from the joint below.

Several types of joint compounds are available. Most are in powder form and are packed in paper bags. They are ready to use after mixing with water. Others are ready mixed, generally vinyl based compounds, ready to use as they come from the metal container. "All-purpose" compound is suitable for bedding tape, for covering fastener heads and flanges of metal accessories, for second and finish coats over tape, and for texture finish over the entire wall. Other joint treatment compound systems use one formula for the bedding and first coat, a second for the finish coat and finish spackle treatment, and a third for overall texture.

Some compounds dry more rapidly than others. Some have a high shrinkage factor and can not be applied in thick layers. Some are not suitable for the finish coat because they can not easily be sanded smooth and flush with adjacent surfaces. Some materials have superior bonding strength to specific materials or are more compatible with certain reinforcing tapes or decorative finishes.

Texturing material is a dry powder which is mixed with water and used to create a variety of finishes. In addition to giving a texture to the finished wall, it helps conceal joints in the wallboard. Texture coats should not be used on walls or ceilings of kitchens or other rooms where scrubbing will be necessary.

Problem Areas
High joints or crowns that are apparent after finishing are the result of too much joint compound over the joint and not enough feathering of the compound beyond the joint. If non-tapered edges (wallboard ends) are butted together over a support, the joint must be built up thicker than adjacent wallboard. To conceal such a joint, the joint compound should be tapered out and feathered at least 8 inches on each side of the joint. It is possible to make a flush end joint without bulges. Wallboard ends are butted between supporting members, warped inward and held by temporary bracing while adhesive on gypsum board back-blocking sets.

After the bracing is removed, the shallow vee joint is taped, filled, and feathered flush.

Ridging or beading directly over the joint is more common when work is done in cold weather and high humidity. Movement of the framing system from settlement, expansion or contraction may cause ridging. Wallboard work should not be done at low temperatures or high humidity.

Nail popping is a protrusion in the smooth surface directly over the head of a nail. If nails do not hold the board firmly to supports, pressure on the gypsum board toward supports will cause the board to move and nailheads to pop. This can happen because of improper application or lumber shrinkage, or both.

Improper application can be the result of framing members out of alignment, bowed, twisted or warped. Wallboard improperly fit and forced into place or wallboard not nailed from the center of panels toward ends can result in nail pops. Nails that miss framing members or are not fully driven also cause popping.

Lumber shrinkage is caused by reduction of moisture content within the member. As the lumber shrinks, the nailhead remains in the same relative position. When lumber with relatively high moisture content is used, wallboard should be applied with short screws or with adhesive, using double headed nails for temporary support.

Depressed nails and puncturing of the face paper may be caused by overdriving, or driving against a framing member that is misaligned, twisted, bowed, or otherwise out of contact with the wallboard back. Swelling of framing members due to water absorption after the wallboard has been installed can also cause depressions.

Cracking and shrinkage at joints can be caused by mixing too much water with joint compound, applying compound too thick, applying a second coat over a previous coat that is not thoroughly dry, and improper drying conditions. Temperature, humidity, and ventilation should be adequate but not excessive.

Bond failure may be caused by using old joint cement, using cement too soon after mixing, using excessively cold water for mixing, using too much water in the mix, cement too dry when tape Is embedded, using insufficient cement under tape, failure to fully embed tape and work out all air bubbles, improper curing, and applying compound to wet or unclean surfaces.

Tearing of the paper backing from the core is the result of failure to cut through the entire sheet. It is acceptable to score both sides of the sheet before breaking, or score the face and carefully break through the core, then cut the back paper. Tearing is common around openings for electrical outlets where the outline of the hole is scored on the face, then the piece is knocked out with a hammer. This is poor practice. There are tools available that do a good job of cutting through both paper faces of these knockouts.

Holes for outlets, fixtures, pipes, conduits and the like are sometimes cut too large or are not accurately located and must be enlarged by drifting. The result is inadequate backing for cover plates or a large area to be sealed with caulking compound.

Labor Installing Gypsum Drywall

Work Element	Unit	Man-Hours Per Unit
Drywall on one face of metal or wood studs or furring		
1 layer, ⅜"	100 S.F.	1.8
1 layer, ½"	100 S.F.	1.9
1 layer, ⅝"	100 S.F.	2.1
2 layers, ⅜" (mastic)	100 S.F.	2.7
2 layers, ½" (mastic)	100 S.F.	3.0
2 layers, ⅝" (mastic	100 S.F.	3.4
Drywall for columns, pipe chases or fire partitions		
1 layer, ⅜", nailed	100 S.F.	4.4
1 layer, ½", nailed	100 S.F.	4.5
1 layer, ⅝", nailed	100 S.F.	4.6
2 layers, ½", mastic	100 S.F.	8.5
2 layers, ⅝", mastic	100 S.F.	8.9
3 layers, ½", mastic	100 S.F.	12.5
3 layers, ⅝", mastic	100 S.F.	13.0
1 layer, 1½", coreboard	100 S.F.	4.0
Drywall for beams and soffits		
1 layer, ½"	100 S.F.	4.0
1 layer, ⅝"	100 S.F.	3.9
2 layers, ½"	100 S.F.	7.3
2 layers, ⅝"	100 S.F.	8.0
Drywall, glued		
1 layer, ½"	100 S.F.	2.0
1 layer, ⅝"	100 S.F.	1.9
Screwed drywall		
1 layer, ½"	100 S.F.	1.9
1 layer, ⅝"	100 S.F.	2.2
Additional time requirements		
Add for ceiling work	100 S.F.	.6
Add for walls over 9' high	100 S.F.	.5
Add for resilient clip application	100 S.F.	.4
Add for vinyl covered drywall	100 S.F.	.4
Add for thincoat plaster finish	100 S.F.	1.4
Deduct for no taping, finish or sanding	100 S.F.	.9

Time includes move on and off site, unloading, stacking, installing drywall, repair and cleanup as needed. Taping, joint finishing and sanding are included.
Suggested Crew: 1 applicator and 1 laborer

Gypsum Wallboard Types

Regular is available in several thicknesses for both new and remodeling construction.

Fire rated is designed especially for fire resistance. Major additives are vermiculite and fiberglass.

Sound deadening board is usually applied in combination with other wallboard products to achieve higher sound and fire ratings.

Tile backer board is recommended as a base for adhesive application of ceramic, metal or plastic tile for interior areas where moisture and humidity are a problem. (Direct, continuous contact with moisture should be avoided.)

Sheathing is for exterior applications. Used as a substrate for siding, masonry, brick veneer and stucco.

Backerboard is recommended for backing paneling and other multi-layered applications. Adds strength and fire protection. Also can be used effectively with ceiling tile.

Vinyl-surfaced wallboard resists scuffs, cracks and chips. Ideal for commercial and institutional use.

Tapered edge inclines into the board from the long edge. With joint finishing results in a smooth, monolithic wall.

Square edge is used where an exposed joint is desired.

Tapered, round edge is for the same applications as tapered edge board. Designed to reduce beading and ridging problems often associated with poorly finished joints.

Beveled edge is used where a "panel" effect is desired. In this application the joints are left exposed.

Tongue and groove is available on 24" wide sheathing and backer boards.

Modified beveled edge needs no special joint finishing, though matching batten strips may be used if desired.

Thickness: ¼", 3/8", ½", 5/8". (Not all products are available in all thicknesses.) Width: 4'. Length: 6' through 16'.

Framing Spacing For Gypsum Wallboard

Board Thickness				Maximum Support Spacing		
					Two Layers	
Base Layer	Face Layer	Location	Application	One Layer Only	With Fastener Only	With Adhesive Between Layers
3/8''	None	Ceilings	Horizontal	16''	16''	16''
3/8''	3/8''	Ceilings	Horizontal	NA	16''	16''
3/8''	3/8''	Ceilings	Vertical	NA	NR	16''
1/2''	None	Ceilings	Horizontal	24''	24''	24''
1/2''	None	Ceilings	Vertical	16''	16''	16''
1/2''	3/8''	Ceilings	Horizontal	NA	16''	24''
1/2''	3/8''	Ceilings	Vertical	NA	NR	24''
1/2''	1/2''	Ceilings	Horizontal	NA	24''	24''
1/2''	1/2''	Ceilings	Vertical	NA	16''	24''
5/8''	None	Ceilings	Horizontal	24''	24''	24''
5/8''	None	Ceilings	Vertical	16''	16''	24''
5/8''	3/8''	Ceilings	Horizontal	NA	16''	24''
5/8''	3/8''	Ceilings	Vertical	NA	NR	24''
5/8''	1/2'', 5/8''	Ceilings	Horizontal	NA	24''	24''
5/8''	1/2'', 5/8''	Ceilings	Vertical	NA	16''	24''
1/4''	None	Walls	Vertical	NR	16''	16''
1/4''	3/8''	Walls	NR	NA	NR	NR
1/4''	1/2'', 5/8''	Walls	Horiz. or Vert.	NA	16''	16''
3/8''	None	Walls	Horiz. or Vert.	16''	16''	24''
3/8''	3/8'', 1/2'', 5/8''	Walls	Horiz. or Vert.	NA	16''	24''
1/2'' or 5/8''	None	Walls	Horiz. or Vert.	24''	24''	24''
1/2'' or 5/8''	3/8'', 1/2'', 5/8''	Walls	Horiz. or Vert.	NA	24''	24''

NA — Not applicable
NR — Not recommended
Framing spacing for two-ply, 3/8'' gypsum board wall may be 24'' if adhesive is used between plies.

Ceramic Tile

Ceramic tile is made from natural clays and finely ground ceramic materials such as silica, quartz, feldspar, marble, fluxes, cements, pigments, or acetylene black. The powders are either dust pressed to the desired shape or mixed with water and formed in a plastic state. Then they are heated to fuse the particles into a solid mass. Most tiles have grooves on the back side to provide more contact surface and a better bond.

There are only two grades of ceramic tile, standards and seconds. Standard grade tiles are as perfect as commercially possible. Tiles of this grade may vary in shade, but must be free from spots and defects. Second grade tile may have minor blemishes and defects not permissible in standard grade, but must be free from structural defects. Tiles are shipped in sealed cartons with grade and contents indicated by grade seals. Tile Council of America seals are grade color coded: blue for standard and yellow for seconds. Quarry tile is under the same rules as other ceramic tile except that the grade stamp is imprinted on the carton instead of on seals that seal the carton.

Ceramic tile can be divided into two general types, glazed and unglazed. Unglazed tiles are dense, hard, vitrified units that are available in a number of integral solid and mottled colors that go all the way through the tile body. They do not have slippery faces and are manufactured primarily for interior and exterior floor or paving. Unglazed tile is available in a number of sizes.

Unglazed porcelain tile is made from special clays, generally by the dust-pressed method, and is characterized by fine grain and sharp smooth faces. Unglazed natural clay tile is generally made by the dust-pressed method from selected natural clays. It is dense, coarse grained and has a distinctive flecked appearance. It is generally cushion edged, and vitreous.

Unglazed quarry tile is made by the extrusion process from selected natural clays or shale. It is dense, coarse grained, and available in several solid colors. Quarry tile is generally 6 square inches or more and from $\frac{1}{2}$ to $\frac{3}{4}$ inch thick.

Unglazed slip-resistant tile has an abrasive material added to the mix, abrasive particles embedded in the face, or cross-hatch grooves on the surface.

Unglazed mosaic tile has a face area of less than 6 square inches and a $\frac{1}{4}$ inch nominal thickness. It is usually factory mounted in sheets of about 2 square feet to speed setting. The mounting may be on a paper face sheet, a netting cemented to tile backs, or synthetic rubber on back edges. Mosaic tile is available in porcelain and natural clay tiles, and in slip-resistant and conductive grades.

Unglazed paver tile is made by the dust-pressed method and is available in porcelain and natural clay types. It is similar to mosaic tile except that it has an area of 6 square inches or more, is relatively thicker, and is not generally factory mounted.

Glazed tile is available in impervious porcelain, vitrified natural clay, quarry tile, nonvitreous white bodies, and many sizes. Ordinary glazed wall tile has an impervious glaze over a white nonvitreous body. It is intended for interior use, although it can be used to protect exterior locations not subjected to freezing. Exterior tile should be vitreous or the frostproof type. Glazed porcelain, natural clay, or quarry tile is generally used for walls or wainscots with similar unglazed floor tile. White bodied nonvitreous glazed wall tile is available in plain flat and sculptured three dimensional designed surfaces and scored to create false joints and the illusion of smaller units after grouting.

The glaze is a powdered glass-like material called **frit** which is fused to the tile body. The glaze may have a bright or matte finish, or it may be hard, lightly textured crystalline material suitable for drainboards and residential or light duty floors. Glazed tile is available in many special shapes and accessory units such as counter trim, curb caps, sills, toilet paper holders, soap trays, tumbler and toothbrush holders, and towel bars.

Glass tiles are called tessera. They are made by pouring molten glass over a waffle-like mold. After the glass has cooled, the individual pieces are broken away along the mold lines. The tile is about $\frac{1}{4}$ inch thick, $\frac{3}{4}$ inch square, and has an irregular, handcrafted look. Glass tessera is made to resemble mosaic murals and are available in hundreds of opaque color hues and shades. Individual pieces are cut, fit, and cemented to paper face sheets by the artist to create the design. Each sheet of mounted tiles is code marked to a design drawing to simplify field placement. Glass tessera is also available factory mounted in plain solid colors for wall application. It is not recommended for floors or other surfaces subject to abrasion.

Cement tile is made with white sand, white Portland cement, and mineral oxide pigments. The ingredients are mixed with water to a damp consistency, placed in highly polished metal molds and formed into geometric shapes. By using various moulding techniques, cement tile can be made in multi-colored designs. Cement tile has a smooth matte face and is available in a number of dark and pastel colors. It is suitable for exterior and interior walls and floors or pavers. Some fading is expected in most pigments in exterior exposures.

Setting Materials

Portland cement mortar is a mixture of Portland cement, sand, hydrated lime, and water. This is the traditional setting material for ceramic tile and is suitable for most interior and exterior horizontal and vertical surfaces.

Dry-set Portland cement mortar is a mixture of Portland cement with additives that aid water retention and sand. Dry-set mortar requires no presoaking of tile, and is applied about $\frac{1}{4}$ inch thick. It is suitable for most exterior and interior vertical and horizontal surfaces. It is available as a concentrate, as an unsanded mortar, and as a factory-sanded mortar, to which only water is added.

Latex-portland cement mortar is a mixture of Portland cement, sand, and special latex additive. It is similar to dry-set mortar except that it is somewhat more flexible.

Epoxy mortar is a two-part epoxy resin adhesive which may have sand or other fillers added. Epoxy mortar provides exceptionally high bond strength and chemical resistance, and is suitable for interior and exterior use on any sound horizontal or vertical surface. Some surfaces may require priming before application of an epoxy mortar setting bed. Epoxy is also used without sand as an adhesive to set tile on smooth, flat, sound, vertical and horizontal surfaces.

Furan mortar is a two-part furan resin mixed with sand or other fillers. It is used primarily as a bedding material for chemically resistant floors.

Organic adhesive is a water-resistant mastic which is applied with a notched trowel. It cures by solvent evaporation. Some porous surfaces may require priming before application of the adhesive.

Grouting Between Tiles

Tile grout is available in many forms to meet requirements of different kinds of tile and exposure conditions. Portland cement is the base for most grouts. It is modified to provide specific properties, such as whiteness, uniformity, hardness, flexibility, and water retention. Synthetic resin adhesive grouts are used with synthetic resin setting beds for their special properties.

Portland cement grout is available as a prepared commercial mix requiring only water. It also may be job mixed. It is composed of Portland cement and water to which ingredients such as color pigments, sand, and lime are added. Portland cement grout is the most widely used of all the grouts, and is suitable for most ordinary exposures and uses. It is available in white and any number of colors.

Dry-set grout is a factory-prepared mixture of Portland cement with water retention additives and sand.

Latex grout is similar to latex mortar and is generally used with latex mortar. It is suitable for ordinary exposures and use conditions.

Mastic grout is factory formulated and ready for use as it comes from the container. It is more flexible and stain resistant than Portland cement grout.

Synthetic resin grouts such as epoxy and furan resins are two-part systems that cure by chemical action after the two parts have been mixed. Resin grouts are used with resin setting beds for stain and chemical resistant installations.

Waterproof membranes are generally polyethylene film or other material with a low permeability rating. They are used over wood and concrete bases and under reinforced setting beds to prevent moisture from penetrating through the base. They may also serve as a cleavage or isolation membrane.

Cleavage or isolation membranes are generally 15-pound felts, but other materials are used as well. These membranes are used over structural slabs and under reinforced setting beds to prevent a bond between the setting bed and the base. This eliminates stresses in the tile due to horizontal movement in the base slab.

Application

Unbonded mortar setting beds on horizontal surfaces are the most common application. First, a membrane is laid over the surface to be tiled. Isolation or cleavage membranes are lapped at side and end joints or lapped and folded. Joints in waterproof membranes must be sealed and have turned up vertical surfaces at the edges.

The mortar setting bed is laid over the membrane and struck off with a temporary screed. If reinforcing wire is required, it is embedded near the center of the setting bed. The setting bed should be about $1\frac{1}{4}$ inch thick and should slope to a drain. While the setting bed is still plastic, tile is bedded with a thin neat Portland cement bond coat. This bond coat may be skimmed over the setting bed or applied to the back of larger tile.

Over concrete and wood subfloors reinforcing is mandatory. Expansion joints are recommended near the perimeter, especially with quarry tile. Wood subfloors should have waterproof membranes. Ceramic tile on floors require recessed subfloors to bring the finished tiled surface flush with other finish flooring materials.

Bonded mortar setting beds on horizontal surfaces are similar to unbonded mortar beds except that membranes and reinforcing are eliminated and the setting bed is bonded to a screeded but untroweled concrete subfloor. Bonding agents may be used, or the subfloor may be scored while it is in the plastic state for a better mechanical bond.

Portland cement mortar bed installation is often used on vertical surfaces. The mortar bed is about ¾ inch thick. If the base is irregular, the bed is placed over a cured and scratched leveling coat. The leveling coat and mortar bed can be placed over a dampened masonry, concrete, or Portland cement plaster scratch coat. Tile is embedded in a neat Portland cement skim coat while the setting bed is still plastic. Ordinary absorbent wall tile must be soaked in water and drained so that it is damp when applied. Tiles that are unmounted or without spacer lugs on their edges may require temporary spacers along horizontal joints to keep them from slipping until set. This method of installing tile is suitable over most sound solid bases. It should not be used directly over gypsum plaster or gypsum wallboard. Over smooth concrete surfaces that do not offer a good bond, metal lath should be firmly secured in place. Then scratch and bedding coats are applied.

Thin set application can be used on horizontal or vertical surfaces. This adhesive method requires a sound, true surface. It is similar to the Portland cement bond coat in the thick Portland cement mortar bed system, except that the thin bed is applied to cured dry surfaces. If the base is not flat, a Portland cement mortar leveling coat may be applied, moist cured and dried. Then the tile is applied using one of several thin-bed setting materials. Tile applied by the thin-bed method is not presoaked. It should be applied dry.

Thin-bed setting materials include dry set Portland cement, latex-Portland cement, organic adhesive, and epoxy and furan resins. Some thin-bed materials will require a prime coat over porous or absorbent surfaces. Thin beds are about $\frac{1}{16}$ to ¼ inch thick and should be applied uniformly with a notched trowel. Setting beds for chemically resistant floors should be at least ⅛ inch thick, and may contain fine sand fillers. Thin set mosaic floor tile is only about ¼ inch total thickness and may not require a recessed subfloor to finish within acceptable elevations of other finish flooring.

9 Finishes

The thin-bed adhesive method of setting tile is appropriate for most sound surfaces, including gypsum plaster and gypsum wallboard. It is not the method recommended by Tile Council of America for installing tile directly over wood subfloors, or on gypsum wallboard supported by wood studs.

Grout composition should be similar to the setting bed except that the filler, if any, is very fine sand. The grout may also have color pigments added to the mix. Factory mixed grouts are preferred. After the tile has set, grout is forced into the joints by trowel, squeegee or brush. Grout should completely fill all joints. Before grout sets, the joints are struck off with a tool to the depth of the cushion on cushion-edge tile, and flush with the surface of square-edged tile. All excess grout should be cleaned from tile faces with clean burlap, sponges, or other suitable material.

Application Problem Areas

Joints between tile and other materials, particularly wall mounted porcelain enameled plumbing fixtures, will often open up within a short time after the building is occupied. This is generally caused by different rates of expansion and contraction of different materials and by use of non-flexible materials for the transition grout joint. Bathtubs are a good example. When the tub is filled with hot water, the metal expands. If the tub has been mounted so that it rests on the floor adjacent to the tile, some of this movement will be upward against the wall tile. When the tub is emptied and cooled, it is apt to pull away from the tile. If the tub is mounted on a wood member nailed horizontally to the studs, shrinkage parallel to the grain will cause the tub to pull away from the tile by as much as 1/16 inch. Tubs should be supported on metal hangers located near the top of the tub so that expansion will be downward. If the plans call for flexible joints between the tile and other materials, joints should not be grouted with a Portland cement.

Tile can be applied to old sound plaster, concrete, masonry, and other surfaces by one of the adhesive or thin-bed setting systems. But particular care is necessary in surface preparation. Concrete is somewhat porous and may have absorbed oil or oily substances which prevent a good bond. Concrete surfaces on which tile is bonded should be thoroughly cleaned. Light sandblasting is an effective means of removing foreign material from the concrete surface. It provides "tooth" for good bond as well. Heavy sandblasting may cause an uneven surface which will show in most thin-bed applications.

Although it is possible to adhere tile to sound painted surfaces, it is not considered good practice. If the plans require such an installation, make sure the paint is well bonded and the surface gloss is removed by sanding. Test for bond strength by adhering several tile at various locations to the prepared painted surface. After the bedding material has thoroughly cured remove some test tiles by prying and some by side pressure or blows on the edge of the tile. If tiles are easily removed, note whether the failure occurred between the bedding material and the paint. If so, try using a different bedding material. If the failure was between the paint and the wall, it would be best to apply metal lath, a scratch coat and Portland cement setting bed. Portland cement setting beds (thick or thin-bed) should not be applied directly to gypsum plaster. The tile can be set with one of the other thin-bed adhesive materials. Otherwise, a membrane should be applied over the plaster, followed by metal lath, scratch coat, and Portland cement setting bed.

Wood expands, contracts, and warps as it absorbs and loses moisture. Almost all of this movement is parallel to the grain. For this reason, tile can not be laid directly on solid lumber. Plywood is more stable than lumber because the grains in the different plies run at right angles to each other. Tile can be installed directly over sound, rigid plywood subfloors with organic or epoxy adhesives.

If you have to apply tile over solid lumber, a membrane should be used between a reinforced setting bed and the wood. If the surface is a drainboard or floor in a room containing plumbing, the membrane should be waterproof. Moisture seeping into a wood base through tile joints or cracks cannot escape. Swelling of the wood will result in cracks or failure in bond with the tile. When applying metal lath to wood studs on a wall which is to be tiled, the studs should be covered by strips of felt to protect them from moisture from the scratch coat. When a wood member is wetted on one face or edge, that face or edge will swell, causing the member to bow. When it dries, perhaps after the tile has set, it will try to resume its original shape. This will cause cracks or tile popping in the finished work.

Labor Installing Ceramic Tile Base

Work Element	Unit	Man-Hours Per Unit
Portland cement bed with grout		
3" x 3" cove	10 L.F.	2.5
4¼" x 4¼" cove	10 L.F.	2.3
6" x 2" cove	10 L.F.	1.8
6" x 6" cove	10 L.F.	2.4
4¼" x 4¼" sanitary base	10 L.F.	2.3
6" x 6" sanitary base	10 L.F.	2.4
Adhesive bed with grout		
6" x 4¼" sanitary base	10 L.F.	1.7
6" x 6" sanitary base	10 L.F.	1.7

Time includes move-on and off-site, unloading, clean-up and repair as needed.

Suggested Crew: 1 tile setter, 1 helper

Labor Installing Ceramic Tile

Work Element	Unit	Man-Hours Per Unit
Floor tile in portland cement bed		
1" x 1", face mounted	100 S.F.	17.0
1" x 2", face mounted	100 S.F.	16.0
2" x 2", face mounted	100 S.F.	15.0
Floor tile in organic adhesive bed		
1" x 1", face mounted	100 S.F.	11.5
1" x 2", face mounted	100 S.F.	11.0
2" x 2", face mounted	100 S.F.	10.6
1" x 1", back mounted	100 S.F.	10.5
1" x 2", back mounted	100 S.F.	10.0
2" x 2", back mounted	100 S.F.	9.5
Wall tile in organic adhesive bed		
1" x 1", face mounted	100 S.F.	19.0
1" x 2", face mounted	100 S.F.	18.5
2" x 2", face mounted	100 S.F.	17.4
1" x 1", back mounted	100 S.F.	12.6
1" x 2", back mounted	100 S.F.	12.1
2" x 2", back mounted	100 S.F.	11.4
3" x 3", unmounted	100 S.F.	16.0
4¼" x 4¼", unmounted	100 S.F.	15.0
6" x 6", unmounted	100 S.F.	14.0
3" x 3", backmounted	100 S.F.	9.0
4¼" x 4¼", backmounted	100 S.F.	9.0
6" x 6", backmounted	100 S.F.	8.0
Wall tile and trim in portland cement bed		
3" x 3", unmounted	100 S.F.	20.0
4¼" x 4¼", unmounted	100 S.F.	19.0
6" x 6", unmounted	100 S.F.	18.0
3" x 3", mounted	100 S.F.	15.0
4¼" x 4¼", mounted	100 S.F.	15.0
6" x 6", mounted	100 S.F.	14.0
Ceramic tile cap	100 L.F.	19.0
Ceramic tile base	100 L.F.	23.0
Drainboards and counters, with backsplash, 4" x 4" tile in portland cement bed	100 S.F.	30.0
Shower walls and floor, 4" x 4" tile in portland cement bed	100 S.F.	19.0

These figures assume cushion edge tile are set with white grout. Time includes move on and off site, unloading, stacking, cleanup and repair as needed.
Suggested Crew: 1 tile setter

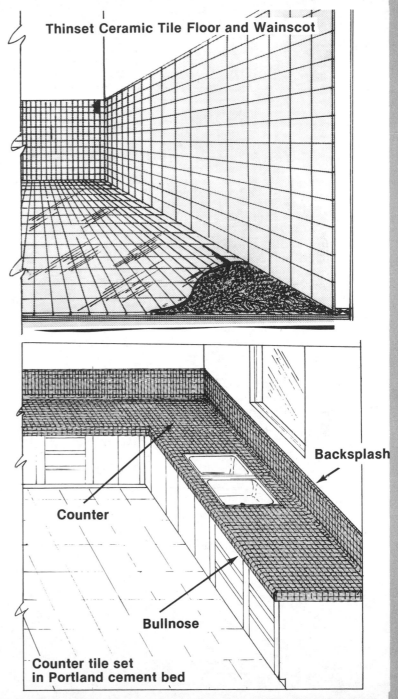

Thinset Ceramic Tile Floor and Wainscot

Backsplash
Counter
Bullnose
Counter tile set in Portland cement bed

Labor Installing Conductive Ceramic Tile

Work Element	Unit	Man-Hours Per Unit
Floor tile in conductive portland cement bed[1]		
1" x 1" x ¼"	100 S.F.	19.0
1⁹⁄₁₆" x 1⁹⁄₁₆" x ¼"	100 S.F.	18.0
Floor tile in conductive dry-set bed[1]		
1" x 1"	100 S.F.	17.8
1⁹⁄₁₆" x 1⁹⁄₁₆"	100 S.F.	16.8
Floor tile with epoxy bed and joints		
1" x 1" x ¼"	100 S.F.	20.0
1⁹⁄₁₆" x 1⁹⁄₁₆" x ¼"	100 S.F.	19.0
Trim pieces portland cement bed		
With grey grout	100 L.F.	13.0
With white grout	100 L.F.	11.6
Add for 2" x 2" wire mesh reinforcing	100 S.F.	.5

These figures assume square edge tile are set with flush joints.
[1] Joints are non-conductive portland cement. Time includes job startup, laying, cleanup and repair as needed.
Suggested Crew: 1 tile setter

Labor Installing Quarry Tile

Work Element	Unit	Man-Hours Per Unit
Floor quarry tile		
4" x 4" x ½", 1/8" straight joints	100 S.F.	24.0
6" x 6" x ½", 1/4" straight joints	100 S.F.	21.6
6" x 6" x ¾", 1/4" straight joints	100 S.F.	22.7
9" x 9" x ¾", 3/8" straight joints	100 S.F.	20.3
6" x 6" x ½", 1/4" hexagon joints	100 S.F.	24.0
Wall quarry tile		
4" x 4" x ½", 1/8" straight joints	100 S.F.	26.9
6" x 6" x ¾", 1/4" straight joints	100 S.F.	26.3
Quarry tile trim, cove base		
5" x 6" x ½", straight top	100 L.F.	18.0
6" x 6" x ¾", round top	100 L.F.	18.6
Stair treads, 6" x 6" x ¾", tile 12" wide tread	100 L.F.	26.4
Window sills, 6" x 6" x ¾", tile 6" wide	100 L.F.	16.7

All tile is assumed to be set in a ¾" portland cement bed. Deduct 33% from man-hours if thin set epoxy bed and grout are used. Time includes move on and off site, unloading, mixing and placing mortar or epoxy bed, setting and grouting tiles, cleanup and repairs as needed.
Suggested Crew: 1 tile setter and 1 helper

Labor For Suspended Ceiling Systems

Work Element	Unit	Man-Hours Per Unit
Concealed zee splines, 1½" carrying channels 4' o.c. with ³⁄₁₆" rod hangers		
Zee runners @12" o.c.	100 S.F.	3.3
Zee runners @24" o.c.	100 S.F.	3.0
Nailable channel system, 1½" carrying channels 4' o.c. with ³⁄₁₆" rod hangers		
Nailing channels 12" o.c.	100 S.F.	3.3
Nailing channels 16" o.c.	100 S.F.	2.7
Metal pan snap system, 1½" carrying channels 4' o.c. with ³⁄₁₆" rod hangers, snap type runners		
Runners 1' o.c.	100 S.F.	2.5
Runners 2' o.c.	100 S.F.	2.1
Runners 3' o.c.	100 S.F.	2.0
Runners 4' o.c.	100 S.F.	1.6
Runners 5' o.c., custom	100 S.F.	2.6
Exposed grid 1½" tee bar, with ³⁄₁₆" rod hangers and cross tee runners		
2' x 2' grid	100 S.F.	2.3
2' x 3' grid	100 S.F.	2.2
2' x 4' grid	100 S.F.	2.1
3' x 3' grid	100 S.F.	1.9
4' x 4' grid	100 S.F.	1.8
5' x 5' grid	100 S.F.	1.4
1' x 4' grid	100 S.F.	2.5
Add for rods hung from:		
Explosive set studs	100 S.F.	1.0
Drilled & set expansion shield	100 S.F.	4.7
Concrete embedded insert	100 S.F.	2.7

Suspension system includes layout, handling, fitting and placing ceiling framework under a steel deck, cleanup and repairs as needed. Heights over 9' will add about 10% to man-hours for each 2' of additional height. No tile or panel installation included in these figures.

Labor For Suspended Ceiling Panels

Work Element	Unit	Man-Hours Per Unit
Fiber tiles laid in suspension grid		
24" x 24"	100 S.F.	.8
24" x 48"	100 S.F.	.5
Asbestos board laid in grid, with 1" insulation blanket		
24" x 24"	100 S.F.	.9
24" x 48"	100 S.F.	.6
Adhesive mounted tiles		
12" x 12"	100 S.F.	1.3
12" x 24"	100 S.F.	.7
24" x 24"	100 S.F.	.6
Aluminum panels laid in grid		
12" x 12"	100 S.F.	1.6
12" x 24"	100 S.F.	1.4
Perforated steel panel in grid		
12" x 48" snap in	100 S.F.	1.4
24" x 24" snap in or lay in	100 S.F.	1.3
24" x 48" lay in	100 S.F.	1.1
48" x 24"	100 S.F.	1.3

Time includes handling, cutting panels as needed, installation, scaffolding and cleanup. Heights over 9' will add about 10% to man-hours for each 2' of additional height. No suspension system installation is included in these figures.

Terrazzo

Terrazzo is a hard mosaic material composed of marble or other aggregates held together in a suitable binder. Materials may be added to impart special properties. It is commonly used as a cast-in-place finish flooring material but can be used in precast form in tiles, panels, or special shapes. There are three principal methods of installing cast-in-place terrazzo:

Monolithic (also known as thinset or one-course) in which the terrazzo topping is bonded directly to the concrete slab. Thinset terrazzo ranges in thickness from 3/16 inch for synthetic resin type to ½ inch or more for Portland cement type.

Monolithic terrazzo is the least expensive installation, but it is vulnerable to cracking. Monolithic terrazzo does not require a recessed slab unless the small difference in finish floor elevations between terrazzo and other materials will be objectionable.

Bonded or two-course terrazzo uses a topping bonded to a Portland cement mortar underbed. The underbed in turn is bonded to a concrete slab, laid directly over steel deck or laid over waterproof membrane on wood. These installations vary in thickness from 1¾ inches when bonded to concrete to 3 inches or more for installation over wood or metal.

A bonded installation costs more than monolithic terrazzo, and requires a recessed slab to bring the finished floor flush with other flooring. It is, however, less susceptible to cracking because reinforcement is placed in underbeds over wood or metal bases. Cracks are less conspicuous because they tend to follow divider strips placed in the underbed. Reinforcement is not generally placed in underbeds bonded to concrete.

Sand cushion or three-course terrazzo layup includes a sand bed about ¼ inch thick over the concrete base slab and a membrane over the sand. The underbed and topping is similar to bonded or two-course terrazzo, except that reinforcing steel is always placed in the underbed. This is the most costly of the three installation methods, but is the least vulnerable to cracks caused by horizontal movement in the concrete base slab.

Although most terrazzo surfaces are ground and polished smooth, other finishes are also used. These include "rustic" or "textured," in which some of the matrix is washed away leaving the marble chips exposed similar to exposed aggregate concrete. Another textured finish is made by grinding only the tops of the exposed marble, leaving the matrix recessed.

Terrazzo will perform well under most exterior and interior exposure conditions. It is relatively expensive, but its low maintenance cost and long life expectancy justify use in many applications.

The best choice for exterior work is Portland cement terrazzo with non-fading pigments and aggregates. For walkways, rustic texture or abrasive aggregates in the topping are ideal because of their non-slip properties. In rooms or areas where explosive or highly flammable materials are kept, acetylene black should be added to the underbed and topping to provide an electrically conductive floor. Special aggregates, fillers and binders may be used in the design mix for chemically resistant terrazzo.

Divider strips are secured to the base slab or underbed before the terrazzo topping is poured. They are ground flush with the topping in the finishing operations. The strips are made from half hard brass, white alloy zinc, stainless steel, or plastic, and range in thickness from 18 gage through ½ inch. Divider strips serve several purposes: They act as screeds to which the toppings are struck and thus control the thickness and plane of the finished work. The dividers form weakened points in the terrazzo topping which tend to control cracks. The strips also form a neat termination from terrazzo to other materials. The finished terrazzo surface should normally show 70% or more marble chips, well distributed as to color and size gradation. Large marble chips require a thicker topping and more grinding, which increases the cost over toppings made with standard screenings of No. 1 (¼ inch) and No. 2 (⅜ inch).

Penetrating sealers are generally specified to be applied to the finished surface to reduce the possibility of staining. Marble chips and synthetic-resin binders have very low porosity. Portland cement binders, however, are porous and should always be sealed. Sealers are crystal clear, non-yellowing, penetrating type, and leave little surface residue. Sealers for electrically conductive floors are specially prepared for that use.

Wire mesh or welded wire fabric reinforcing is placed in the underbed when terrazzo is used on a steel deck. The reinforcement should be galvanized steel to prevent staining or spalling the terrazzo.

Terrazzo toppings are mixtures of aggregate and binders. Binders may be mixtures of Portland cement and water or synthetic resin and hardener. Color pigments, acid resistant fillers, aluminum oxide abrasive granules, and acetylene black may also be used in the mix. Marble chips or abrasive granules can be either mixed with the topping or broadcast over the freshly floated surface, or both. Toppings are struck off to screeds, troweled, and rolled with heavy rollers to compact the mass and to embed surface applied marble chips.

After curing, the topping is ground with abrasive stones of various grits, starting with coarse abrasives. After rough grinding, Portland cement toppings are cleaned and grouted with a mixture of Portland cement, filler, pigment, and water. This mix is applied in a creamy consistency and worked over the surface to fill all voids. After the grout has cured, finish grinding is completed. Synthetic topping may be given a squeegee coat of clear resin.

Careful curing minimizes shrinkage cracks. Portland cement underbeds and terrazzo toppings should be moist cured either by adding moisture to the surface as it cures or by preventing moisture from evaporating. Covering the surface with membrane right after it has been poured and rolled or troweled stops moisture loss. Synthetic resin terrazzo cures by the hardener in the mix. Curing time varies with the amount of hardener added, air temperature, and humidity.

Labor For Precast Terrazzo

Work Element	Unit	Man-Hours Per Unit
Wall tile		
½" x 6" x 6"	100 S.F.	11.5
½" x 9" x 9"	100 S.F.	10.3
½" x 12" x 12"	100 S.F.	9.2
1" x 9" x 9"	100 S.F.	14.3
1" x 12" x 12"	100 S.F.	12.6
1" x 18" x 18"	100 S.F.	11.5
1½" x 12" x 12"	100 S.F.	13.8
1½" x 18" x 18"	100 S.F.	12.0
1½" x 24" x 24"	100 S.F.	10.4
Base, coved or straight face		
4" high	100 L.F.	7.0
6" high	100 L.F.	7.5
8" high	100 L.F.	7.9
Curbing		
4" x 4"	100 L.F.	18.0
6" x 6"	100 L.F.	23.0
8" x 8"	100 L.F.	27.0
Wainscot		
1" x 12" x 12"	100 S.F.	14.4
1½" x 18" x 18"	100 S.F.	15.4
Stair treads, to 12" wide		
1½" thick, straight	100 L.F.	20.6
1½" thick, curved	100 L.F.	21.7
2" thick, straight	100 L.F.	21.7
2" thick, curved	100 L.F.	23.0
3" thick, straight	100 L.F.	22.9
3" thick, curved	100 L.F.	24.6
Stair risers, 6" high		
¾" thick, straight	100 L.F.	11.5
¾" thick, curved	100 L.F.	12.6
1" thick, straight	100 L.F.	12.6
1" thick, curved	100 L.F.	13.7
1½" thick, straight	100 L.F.	13.7
1½" thick, curved	100 L.F.	15.5
Tread & riser combinations, straight stairs		
1½" tread, ¾" riser	100 L.F.	29.0
2" tread, 1" riser	100 L.F.	31.0
3" tread, 1" riser	100 L.F.	33.0
Tread & riser combinations, curved stairs		
1½" tread, ¾" riser	100 L.F.	31.0

Labor For Precast Terrazzo (continued)

Work Element	Unit	Man-Hours Per Unit
2" tread, 1" riser	100 L.F.	34.0
3" tread, 1" riser	100 L.F.	36.0
Structural notched stair stringers, L.F. of carriage length		
1" thick	10 L.F.	2.6
1½" thick	10 L.F.	3.0
2" thick	10 L.F.	3.3
2½" thick	10 L.F.	4.0
3" thick	10 L.F.	4.6
Landings, structural		
1½' thick	10 S.F.	1.7
2" thick	10 S.F.	1.8
2½" thick	10 S.F.	2.0
3" thick	10 S.F.	2.1
Add for metal dividers, 4' x 4' grid, S.F. of floor	100 S.F.	1.6

Time includes moving materials into place, layout, setting, finishing, repair and cleanup as needed.
Suggested Crew: 1 mason and 1 helper

Labor Installing Epoxy Chip Flooring

Work Element	Unit	Man-Hours Per Unit
Clear epoxy up to ½" thick		
With sand, industrial type	100 S.F.	12.7
With plastic chips	100 S.F.	14.1
Colored quartz chips, mixed in	100 S.F.	15.3
Broadcast quartz chips, surface	100 S.F.	14.1
Marble chips, mixed in	100 S.F.	15.3
Marble chips, conductive type	100 S.F.	16.7

Time includes move on and off site, unloading, setup, area prep, apply epoxy and matrix, cleanup and repair as needed.
Suggested Crew: 2 applicators, 1 laborer on large jobs; small jobs, 1 applicator, 1 laborer

Labor For Cast-In-Place Terrazzo

Work Element	Unit	Man-Hours Per Unit
Flooring, 4' x 4' squares with metal divider strip		
Sand cushion, 2½" to 3" thick	100 S.F.	18.0
Bonded to concrete, 1¾" thick	100 S.F.	12.5
Epoxy thinset terrazzo, ¼" to ½" thick	100 S.F.	9.5
Monolithic ½" terrazzo cast with 3½" concrete base panels, approx. 25 S.F. each	100 S.F.	7.8
Bonded conductive, 1¾" thick	100 S.F.	13.7
Sand cushion conductive, 3", regular	100 S.F.	23.0
Sand cushion conductive, 3", mosaic textured	100 S.F.	32.0
Add for heavy-duty abrasive	100 S.F.	2.5
Add for venetian type topping	100 S.F.	3.4
Wainscot		
Bonded to concrete or masonry, 1½" thick	100 S.F.	19.4
Epoxy thinset terrazzo, ½" thick	100 S.F.	15.8
Base, 6" high with divider strips 16" o.c.		
Flush type	100 L.F.	20.0
Projecting, ¼" to ½"	100 L.F.	20.0
Splay edge type	100 L.F.	21.5
Stairs, 1½" to 2" on concrete or metal with dividers		
Treads, to 12" wide	100 L.F.	33.0
Treads and risers combined	100 L.F.	75.0
Stingers, curb and fascia	100 S.F.	31.0
Landings	100 S.F.	14.5
Add for heavy-duty homogenous abrasive surface	100 S.F.	1.0
Add for light-duty embedded abrasive surface	100 S.F.	2.0
Add for surface abrasive strips	100 L.F.	1.0
Add for abrasive metal nosing	100 L.F.	2.6

Time includes move on and off site, placing, grinding, cleanup and repair as needed.
Suggested Crew: 2 terrazzo workers

Wood Flooring

The most common hardwoods used in flooring are white oak, red oak, hard (sugar) maple, birch, beech, and pecan. Walnut, cherry, ash, hickory, and teak are used occasionally. The most common softwoods are Douglas fir and southern yellow pine (shortleaf).

Oak is by far the most common wood flooring material. It is hard, durable, attractive, and abundant. Maple is next in order of popularity. It lacks the beauty of oak but takes a better finish and is harder. Beech and birch are similar to maple but not as abundant. Pecan is scarce and expensive. The same is true of walnut, cherry, ash, hickory and teak.

Flooring lumber is produced in two ways — plain sawed and quarter sawed. Plain sawing produces pieces with annular rings of the wood at an angle to the face of from zero to 45 degrees. Quarter-sawed lumber has better wear and warp resistance. Plain-sawed lumber allows faster cutting and produces less waste at the mill. With rare exceptions, oak is the only wood generally available quarter sawed.

Six types of hardwood flooring are available: (1) strip flooring, (2) plank flooring, (3) parquet flooring, (4) laminated block flooring, (5) end-grain block flooring and (6) gymnasium flooring.

Strip flooring is the most common type. It consists of narrow strips 1½" to 3½" wide in various thicknesses. The most popular size is 2¼" x $\frac{25}{32}$". The thinner strips are used primarily in remodeling work where a new floor is being installed over an old one. Most strip flooring is tongue and groove although the smaller sizes are square-edged. End-matched strip flooring is also available. This term simply means the ends as well as the edges of each piece are tongued and grooved. Strip flooring is usually applied over a wood board or plywood subfloor. But if the finish flooring is at least 1" thick, it may be applied directly over the floor framing joists or over sleepers on concrete floors. Sleepers should be preservative-treated but should not be treated with creosote or heavy oils that can bleed through nail holes and stain the finish surface.

Plank flooring is a simulation of the Colonial original which consisted of hand-sawn planks secured to stringers with wooden pegs. Modern planks are machine planed, tongue and grooved. They are blind nailed and secured with countersunk screws covered by walnut plugs to simulate the pegs. Sometimes the edges are beveled to simulate the cracks of the old floors. Planks are generally made from oak although walnut and teak are available.

Parquet flooring is the most expensive style of hardwood flooring. It incorporates variations of species or shade of the same species in intricate patterns. Each tile is factory cut to exact size to match perfectly the adjoining pieces. Each piece is laid separately, either by nailing or setting in mastic. Exotic patterns are available, but the most popular are square, rectangular, and herringbone. Most parquet flooring is oak, although it also is produced in maple, beech, birch, walnut, mahogany and teak.

Laminated block flooring uses modern plywood technology. Each block is a 3-ply assembly. Many patterns and species are available in $\frac{5}{16}$",

⅜", ½" and ¹³⁄₁₆" thicknesses. Because of the dimensional stability inherent in the cross-ply lamination process, allowance for expansion of the flooring is not required as with solid wood flooring.

End-grain block flooring is widely used in industrial facilities. It consists of small blocks of preservative treated wood laid side by side in a bituminous mastic. The end grain of the pieces provide the wearing surface. Southern yellow pine, Douglas fir and oak are the most popular species. Sizes available vary from 1½" to 4" thick and 2½" to 4" wide. After the blocks are laid, they are finished with hot coal-tar pitch which is worked into the joints. Any pitch remaining on the surface soon wears off with traffic.

Gymnasium flooring is actually a form of strip flooring that uses one of several patented manufacturing and installation techniques. Patented procedures have also been developed for bowling alleys, handball courts, and roller skating rinks. The goal is to produce a floor which is warp-free and as level as is possible. The more common systems use metal channel screeds with interlocking metal clips that hold each strip in place. No nails or mastic are required.

Gymnasium floors are built to have resilience and shock absorbing qualities. Usually rubber or cork pads are placed under the sleepers or cork underlayment. Almost all gymnasium flooring is made of hard maple, although beech and birch are acceptable substitutes. On large floors, ventilation under the floor becomes very important. Mechanical ventilation is sometimes used, especially if the floor is below grade or subject to adverse moisture conditions.

Installation
Strip flooring should be laid at right angles to joists. When applied over old finish floors, it should cover the existing floor at right angles. Tongue and grooved strips should always be toe-nailed through the tongue. The flooring should also be blind-nailed; that is, each nail is set with a nail set. Square-edge strips are face-nailed. In plank flooring, the screws are generally placed 30 inches apart down the length of each piece and at the ends of each piece.

A good floor installer will not try to hammer each piece into Its final position as soon as it is nailed. After three or four pieces are laid, a short piece of straight-edged hardwood should be laid against the tongue of the outside piece. Tap the short piece lightly to drive the strips snugly together. Be careful not to break the tongue. Steel driving tools can be used to prevent damage to the tongues and faces.

In large areas such as gymnasiums, armories, roller rinks and factories, experienced floor layers consider it poor practice to drive flooring up too tight. Where high humidity is common, use metal spacers about 1/16" thick between the strips at intervals across the area. These spacers keep the flooring just loose enough. The cracks prevent buckling when the flooring swells. The cracks will close and remain closed after one season's swelling. Flooring should never be laid tight against walls, columns or floor inserts. An expansion space of ½" - 1½" is generally specified. Some gymnasium floors will require as much as 2" for expansion. When the toe molding is installed, it must be secured to the wall or subfloor, not the finish floor. This ensures complete freedom of movement for the floor.

Finishing
Many strip and parquet flooring materials are prefinished and require no sanding on the job. Most manufacturers advise at least four sandings. Start with No. 2 sandpaper and work down to ½, No. 0 and No. 00. Electrically operated sanding machines are generally used. The sander should be equipped with a dust bag to protect the rest of the building from dust. If the humidity is above 65 percent, sanding and finishing should be delayed from two to three weeks after completion of laying.

After sanding, the floor should be swept and vacuumed clean. Rust can be removed with solvent. The area should be closed off to traffic and covered with a non-staining, breathing-type building paper until sealer is applied. Even better, delay the final sanding until immediately before finishing is to start.

Floor seal differs from other finishes in one important respect. Rather than forming a surface coating, it penetrates the wood fiber and seals the fibers together. It wears only as the wood wears. Generally it is applied across the grain, then smoothed out in the direction of the grain. After the excess is wiped off, the floor should be buffed with No. 2 steel wool. A second coat is frequently specified.

Varnish directions are usually given on the varnish can. Varnish made especially for floors should be used. Three coats are generally needed, although two coats are adequate if a wood filler has been used.

Wax should be applied to all hardwood floors after the finish has dried thoroughly. In some cases two or three coats are recommended. Wax forms a protective film that prevents dirt from penetrating the wood pores. Gymnasiums are finished with a resin type finish which improves the floor's appearance and offers good resistance to rubber sole shoes.

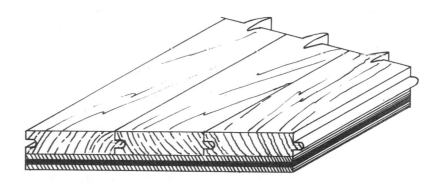

Wood Strip Flooring on Plywood

Grading and Descriptions for Strip Flooring

Species	Grain Orientation	Size Thickness Inches	Width Inches	First Grade	Second Grade	Third Grade
Softwood						
Douglas fir	Edge grain	25/32	2-3/8 - 5-3/16	B and Better	C	D
& hemlock	Flat grain	25/32	2-3/8 - 5-3/16	C and Better	D	---
Southern pine	Edge grain	5/16 - 1-5/16	1-3/4 - 5-7/16	B and Better	C and Better	D (and No. 1)
Southern pine	Flat grain	5/16 - 1-5/16	1-3/4 - 1-5/16	B and Better	C and Better	D (and No. 1)
Hardwood						
Oak	Edge grain	25/32	1-1/2 - 3-1/4	Clear	Select	---
Oak	Flat grain	3/8	1-1/2, 2	Clear	Select	No. 1 Com.
Oak	Flat grain	1/2	1-1/2, 2	Clear	Select	No. 1 Com.
Beech, birch,	Flat grain	25/32	1-1/2 - 3-1/4	First grade	Second grade	---
maple, and	Flat grain	3/8	1-1/2, 2	First grade	Second grade	---
pecan[1]	Flat grain	1/2	1-1/2, 2	First grade	Second grade	---

[1] Special grades are available in which uniformity of color is a requirement.

Materials For Strip Flooring

Size	B.F. Per 100 S.F.	S.F. Per 1000 B.F.	Nails, Lbs. Per 100 S.F.
25/32 x 1½	155.0	645.0	3.7
25/32 x 2	142.5	701.8	3.0
25/32 x 2¼	138.3	723.0	3.0
25/32 x 3¼	129.0	775.2	2.3
3/8 x 1½	138.3	723.0	3.7
3/8 x 2	130.0	769.2	3.0
1/2 x 1½	138.3	723.0	3.7
1/2 x 2	130.0	769.2	3.0

Labor for Installing Wood Strip Flooring

Work Element	Unit	Man-Hours Per Unit
Fir flooring (4' to 20' long 1" x 4")		
C and better vertical grain	100 S.F.	3.1
C and better flat grain	100 S.F.	3.0
Oak flooring (25/32" x 2¼")		
Clear quartered	100 S.F.	3.5
Clear plain	100 S.F.	3.6
Select plain	100 S.F.	3.9
Number one common	100 S.F.	3.9
Prefinished prime grade	100 S.F.	4.0
Maple flooring (25/32" x 2¼")		
First grade	100 S.F.	5.2
Second grade and better	100 S.F.	5.0
Maple flooring (25/32" x 3¼")		
First grade	100 S.F.	3.8
Second grade and better	100 S.F.	3.0
Pine flooring (25/32" x 2¼")		
First grade	100 S.F.	3.8
Second grade and better	100 S.F.	3.5

Time includes move on and off site, unloading, stacking, placing, repair and cleanup as needed, but no sanding or finishing. Figure sanding at 1 hour per 100 square feet and filling holes and 2 coats of lacquer at 1.5 hours per 100 square feet.
Suggested Crew: 1 laborer

Nails For Strip Flooring

Size Flooring	Type and Size of Nails	Spacing
(Tongued & grooved) 25/32 x 3¼	7d or 8d screw type, cut steel nails or 2" barbed fasteners*	10" to 12"
25/32 x 2¼	Same as above	10" to 12"
25/32 x 1½	Same as above	10" to 12"
½ x 2, ½ x 1½	5d screw, cut or wire nail. Or 1½" barbed fasteners*	8" to 10"
Following flooring must be laid on wood sub-floor		
(Tongued & grooved) 3/8 x 2, 3/8 x 1½	4d bright casing, wire, cut, screw nail or 1¼" barbed fasteners*	6" to 8"
(Square-edge) 5/16 x 2, 5/16 x 1½	1-in. 15 gauge fully barbed flooring brad, preferably cement coated.	2 nails every 7"

*If steel wire flooring nails are used they should be 8d, preferably cement coated. Machine-driven barbed fasteners used as recommended by the manufacturer are acceptable.
Tongued and grooved flooring must always be blind-nailed; square edge flooring must be face-nailed.

Labor for Installing Wood Block Flooring

Work Element	Unit	Man-Hours Per Unit
Parquet, prefinished		
5/16" thick	100 S.F.	6.0
1/2" thick	100 S.F.	6.1
Parquet, prefinished, top quality		
5/16" thick	100 S.F.	6.2
1/2" thick	100 S.F.	6.4
Simulated parquet and tile flooring, 5 ply, 5/8" thick plywood	100 S.F.	3.5
Wood block flooring (factory type)		
Creosoted 2" thick	100 S.F.	3.7
Creosoted 2½" thick	100 S.F.	4.0
Creosoted 3" thick	100 S.F.	4.3
Natural finish end grain 1½" thick	100 S.F.	3.8
Expansion strip 1" thick	100 L.F.	2.7
Gym floor (on shims, sleepers and screeds, with sub-floor included)		
25/32" thick maple	100 S.F.	13.0
25/32" thick maple plus extra sub-floor (2 ply)	100 S.F.	15.3
25/32" thick maple with steel springs and 2 ply sub-floor	100 S.F.	16.4

Time includes move on and off site, unloading, stacking, cleanup and repair as needed, but no sanding or finishing.
Suggested Crew: 1 carpenter, 1 laborer

Materials For Wood Block Flooring

Size	Block Per 100 S.F.	Adhesive Per 100 S.F.	Nails Lbs. Per 100 S.F.
8" x 8"	225 block	1 gallon	4.0
9" x 9"	178 block	1 gallon	3.5
12" x 12"	100 block	1 gallon	2.8

Resilient Flooring

Resilient floor covering includes asphalt tile, vinyl-asbestos tile, vinyl tile and vinyl sheet, rubber tile and rubber sheet, cork tile, and linoleum sheet goods. Accessories used with resilient flooring include rubber and vinyl base, stair treads, feature strips, edge nosings (reducer strips), thresholds (saddles), and stringer or stair skirting material.

The thickness of resilient floor covering ranges from less than 1/16 through 1/8 inch. Although other sizes and shapes are available, the most popular tiles are 9- and 12-inch squares. Sheet goods come in rolls, generally about 6 feet wide.

All types of resilient floor coverings, except cork, are available in a wide variety of plain colors, color patterns, and plain and textured finishes. Through combinations of color patterns and textures, remarkably accurate simulations of various natural and manufactured materials are possible. These include slate, marble, travertine, stone, terrazzo, brick, ceramic tile, wood, and cork. Some manufacturers make these tile with contact adhesive on the backside, protected by peel-off paper.

Asphalt tile is the least expensive of the resilient floor coverings. It is composed of asphalt or a combination of asphalt and resinous thermoplastic binders, asbestos fibers, mineral fillers, and pigments. It is brittle at temperatures below 60 degrees, softens when exposed to grease or oil, and is very pliable at temperatures of 120 degrees or more. It is somewhat porous, and thus does stain. Sustained loads over 25 pounds per square inch can leave a permanent indentation in asphalt tile, especially in tile over radiant heated floors.

Asphalt tile is suitable for on and below grade concrete floors. It is available in 3/32 and 1/8 inch thick and 9 by 9 and 12 by 12 inch squares. Marbleized color patterns penetrate the full thickness of the tile unlike simulated cork and some other designs.

The cost of asphalt tile varies with thickness, color, surface texture (embossing) and special properties. The industry has adopted three general price range groups: Group B dark marbleized colors have the lowest price; Group C medium marbleized colors and special color patterns or surface textures fall in the medium price range; Group D light colors and grease resistant or other special property tile have the highest price. Group D asphalt tile competes with vinyl asbestos tile which is grease resistant and available in light colors at no additional cost. Since vinyl asbestos tile are competitive with group D asphalt tile, some manufacturers have discontinued producing tile in this group. Vinyl asbestos tile has replaced asphalt tile in many applications because it has more desirable properties at only a nominal increase in cost.

Vinyl asbestos tile is composed of asbestos fibers, mineral fillers, pigments, vinyl resin binders, plasticizers and stabilizers. It is somewhat more flexible than asphalt tile at room temperature. It is not perfectly accurate to classify either asphalt or vinyl asbestos tile as "resilient" since neither product will recover from any indentation caused by concentrated loads. The recommended load limit for both products is 25 pounds per square inch. Vinyl asbestos tile is supplied in 1/16-, 3/32-, and 1/8-inch thicknesses (referred to as gage in the industry), and in 9 by 9 and 12 by

12 inch squares. Vinyl asbestos tile falls just above asphalt tile in price. Prices vary with tile thickness, and surface effects (embossing, special designs and colors). Like asphalt tile, some colors and surface designs do not penetrate through the entire tile body.

Linoleum is available in both tile and sheet form, although sheet is the most popular. Linoleum is composed of ground cork and wood floor fillers and pigments, held together by linseed oil or oleoresinous binders, and supported by a backing of burlap or felt. Sheet goods are manufactured in rolls 72 inches wide, and from 42 to 90 feet long. It is manufactured in 3 thicknesses as follows:

Grade	Overall Thickness	Typical Wear Surface Thickness
A (Light)	.125"	.110"
B (Standard)	.090"	.050"
C (Heavy)	.065"	.22"

Grades A and B (light and standard gages) are felt backed. Grade C is burlap backed. Linoleum generally becomes more expensive as the thickness increases. Installation costs for sheet goods are usually higher than for tile. This is because there is generally more waste in sheet goods and it is more difficult to handle, trim, and fit.

Linoleum can tolerate loads of up to 75 pounds per square inch without damage. It is resilient, quiet underfoot, resistant to staining and cigarette burns and durable.

Rubber flooring is composed of natural or synthetic rubber. The most common size is 1/8 inch in 9 by 9 or 12 by 12 inch squares. It is also available in 3/32, 3/16, 1/4 and 1/2-inch thicknesses and sheet form. The thicker material is used for high traffic areas, such as runners, landings, and in interlocking tile for industrial applications. Solid and linked rubber matting of various designs and thicknesses are also available. Ordinary 1/8 inch rubber tile is available in solid, marbleized and other mixed colors and embossed designs. Rubber flooring is resilient, quiet and comfortable underfoot, will withstand concentrated static loads over 150 pounds per square inch, is durable and easily maintained. Rubber flooring is made by very few companies and is not as widely used as vinyl asbestos or asphalt flooring.

Cork tile is losing popularity because of its high cost and the popularity of simulated cork in more durable flooring made from vinyl binders. Genuine cork tile is warm to the touch, quiet and comfortable underfoot. Cork tile is pure ground cork particles held together by a synthetic resin binder. Most popular sizes are 3/16 inch thick in 9 by 9 or 12 by 12 inch squares. Other sizes, including 1/8, 5/16, and 1/4 inch thickness and 6 x 6 inch squares are available. Cork will generally fall somewhere between rubber and vinyl in cost for similar thicknesses. Lowest cost cork has unfinished faces, followed by a factory applied wax, lacquer, or resin finish. Cork tile with a factory applied clear plastic film finish is the most expensive.

Vinyl is generally the most expensive of the ordinary resilient flooring materials. The terms vinyl, homogeneous vinyl, solid vinyl, and pure vinyl are used to describe this product. The essential difference between vinyl and vinyl asbestos is that vinyl contains a higher percentage of vinyl resins and less filler. Translucent vinyls have a higher resin content in relation to fillers and opaque pigments. This type of vinyl is most expensive. It has a rich luminous appearance. High purity vinyls however are not dimensionally stable. For this reason sheet goods are made with a relatively thin vinyl wearing surface bonded to a more stable backing material. Shrinkage of vinyl tile is evenly distributed throughout the floor in the joints.

Vinyl is more flexible and resilient than vinyl asbestos, but less resilient than linoleum and about the same as rubber. It can withstand 100 pounds per square inch without permanent indentation. It is very durable, stain resistant, and easily maintained. Tile is available in 1/16-, 0.080-, 3/32-, and 1/8-inch thicknesses, and 9 by 9 and 12 by 12 inch squares. Vinyl sheet goods backed by fibrous resin bonded felts are available in 72-inch widths and in rolls of from 42 to 100 feet long. Common overall thicknesses are 0.085", 0.065", and 0.055", with wearing surfaces of 0.055", 0.030", and 0.020" respectively. Some manufacturers produce a product with a rather thin vinyl wearing surface but with a cushion of foam vinyl between the wear surface and the backing material.

Static conductive resilient flooring is available in linoleum sheet, vinyl and rubber tile. Floor coverings of this type have electrical conductors in the mix. Face surfaces are smooth, and selections of colors or color patterns are limited. They are 1/8 inch thick.

Resilient bases are molded or extruded rubber or vinyl. There are 3 general types: (1) Butt-cove, in which the bottom of the base is concave and 1/8 inch thick at the floor end. This permits a butt joint and smooth transition between floor and base. (2) Top-set cove is similar to butt-cove except the bottom at the floor line is concave rounded and feather edged. This is the most popular type, and is installed after the resilient floor is in place. (3) Straight bases (without cove) are sometimes called carpet bases because they are generally used in carpeted rooms. They are set on top of the structural floor in carpeted rooms and on top of resilient floors elsewhere. All bases have rounded top edges and grooved or butted backs to allow bond with the adhesive.

Molded inside and outside corners and rounded end stops are also available. Bases are generally manufactured in 4-foot lengths and in heights ranging from 3/4 through 7 inches, with 2½, 3 and 4 inches being the most popular. Six inch base was more popular several years ago, and this size is often used in remodeling or addition work on older buildings. Seven inch high base is used almost exclusively to cover stair risers. Bases are made in a number of solid colors and marbleized or other multi-color patterns with both smooth and embossed face surfaces. Generally black is the least expensive, plain colors next, and marbleized or multi-color patterns the most expensive. Bases are manufactured to match some floor covering material.

Resilient stair treads are molded or extruded rubber or vinyl. Full treads are available from 9 to 13 inches deep and with square or rounded leading edges from 1¼ to 2¼ inches high. The wearing surface 4 to 5 inches back from the leading edge is generally thicker than the portion next to the risers. The wear surface may vary from 3/16 to 5/16 inch. The wearing surface next to the leading edge generally has abrasive mineral granule strips inlaid into tread to reduce chances of a fall. Nosings cover only 1½ to 4 inches of the stair leading edge. Treads and nosings are available in a limited number of solid and marbleized colors.

Reducer strips and threshold saddles are also molded or extruded vinyl or rubber and are available in all popular commercial floor covering thicknesses. They are usually about 1 inch wide and feather to a wedge or convex bullnose shape to change elevations at exposed edges.

Adhesives are specially formulated for resilient flooring. There is no single all-purpose flooring adhesive that is ideal for all jobs. Prices, ease of application, and desirable properties vary considerably. Adhesive should be selected for its resistance to moisture and alkali and its ability to bond the materials involved. Some adhesives dry by evaporation and have an "open time" that can be figured in hours. Others cure by chemical action and have an open time figured in minutes. Some adhesives harden over a period of time. Others remain plastic almost indefinitely. The most popular adhesive products include:

1. Asphalt primer is a hydrocarbon thinned asphalt used over dusting, porous, or damp concrete. It penetrates and seals pores to provide a sound surface over which adhesive may be applied.

2. Asphalt cut-back adhesive is a thinned asphalt heavier than primer. It is used over a primer. It is alkali and moisture resistant and suitable for use with asphalt and vinyl asbestos tile on concrete slabs on or below grade as well as on suspended slabs.

3. Asphalt emulsion is a water thinned asphalt adhesive. In general these adhesives can be used wherever cut-back asphalts can be used. Cutbacks are generally considered superior to emulsions for installing vinyl asbestos tile.

4. Linoleum paste is a water based water soluble emulsion. It is suitable for installing linoleum, rubber tile, some vinyls, cork tile and lining felt. It is not suitable for use on concrete slabs on or below grade, or on slabs that are not fully cured and dry.

Underlayments are used under some resilient flooring installations to flatten out surface irregularities that would otherwise show through the finished work. Plastic underlayments are made from very finely ground fillers and a bonding agent. Rigid board underlayment includes sanded and plugged plywood and tempered hardboard. Some grades of particleboard are used as underlayment.

Lining felt is generally 15-pound asphalt saturated rag felt. It is used primarily over tongue-and-groove wood floors. It serves as a cushioning agent to pad board edges that are not sanded absolutely flush and absorbs some of the movement in the floor system. Lining felt also makes it easy to remove the flooring after the adhesive has set.

Application
The subfloor should be smooth, firm, clean, and dry before starting application. Two layers of 1-inch boards (top layer should be tongue and groove), or a single layer of ¾ inch plywood make a good base for resilient flooring. It should be well nailed with all nails driven home. If the plywood or boards are not covered with an underlayment, hammer marks, knotholes, open joints, and the like should be filled with plastic wood and sanded smooth.

Where underlayment is required, it is applied so that joints do not occur over joints in the subfloor. Underlayment end joints should be staggered. End and side joints should be spaced about $\frac{1}{32}$ inch. Temporary sheet metal spacers are useful in maintaining proper clearance.

Lining felt should be laid at right angles to floor boards over the entire area, closely butting, but not lapping joints. Secure the felt with a continuous bed of adhesive rolled with a three-section 150-pound roller in several directions to assure good contact.

On old floors, the paint or varnish should be removed by sanding to the bare wood. Old resilient floor coverings are usually removed. If removed, the old floor should be sanded to remove any adhesive and felt liner, and to provide a smooth surface for the new installation. Top set resilient cove base is removed regardless of whether the old flooring is removed or left in place.

Adhesive is spread with a notched trowel or brush, depending on the consistency and type of adhesive and the amount required. Adhesives that dry by evaporation of a solvent vehicle are allowed to partially dry before being covered with flooring materials. Flooring materials should be installed before any type of adhesive becomes too dry.

Tile should be centered in each room so that not less than one-half a tile of equal size occurs at opposite sides of the room. Color and embossed pattern on many tiles are directional. These tile are usually laid at 90 degrees to each other to form a "basket weave" pattern. Even if the tile is non-directional, some installers turn each tile 90 degrees to avoid uneven joint lines which occur if tiles are not cut absolutely square. After placing, tiles are rolled to flatten them out and bring them into 100 percent contact with adhesive.

Sheet material is arranged in each room or area so that the lineal footage of seams is held to a minimum. Every other strip should be reversed so that the same sides abut at linear seams. Some sheet goods are installed with adhesive applied only at the perimeter of the floor area, at seam lines, and at columns and fixtures. After the material is in place, it is rolled to work out all air pockets and to bring the material into contact with the adhesive.

Edges of resilient floor covering that do not finish flush with other flooring or abut a return surface should be protected by metal or resilient reducer strip or covered by a threshold.

Cove base adhesive can be spread either on the wall or the back of the base. In either case, avoid getting adhesive on the wall surface above the top of the base line. Adhesives with a high tack can hold the base in place until the adhesive sets. Preformed corners and ends are preferred to wrap-around or cut ends.

Weights and Gauges for Resilient Flooring

Material	Approximate Thickness, Inches	Finished Gauge, Inches	Average Net Wt. Per S.F. In Lbs.	Roll Width In Feet
Asphalt tile	1/8	.125	1.16	—
	3/16	.187	1.75	—
Asphalt tile, grease proof	1/8	.125	1.17	—
	3/16	.188	1.74	—
Conductive asphalt tile, regular	1/8	.125	.97	—
and grease proof types	3/16	.187	1.45	—
Lining felt	1/25	.040	1.40	3
Industrial asphalt tile	1/8	.125	.90	—
	3/16	.187	1.35	—
Rubber tile	1/8	.125	1.24	—
Vinyl tile	3/16	.187	1.86	—
	3/32	.0925	.93	—
Sheet goods				
Battleship heavy gauge	1/8	.125	.83	6
Embossed inlaid standard gauge	3/32	.0925	.60	6
Jaspe				
Heavy gauge	1/8	.125	.83	6
Standard gauge	3/32	.0925	.65	6
Marbleized				
Heavy gauge	1/8	.125	.92	6
Standard gauge	3/32	.0925	.65	6
Light gauge	1/16	.070	.46	6
Plain				
Heavy gauge	1/8	.125	.83	6
Standard gauge	3/32	.0925	.60	6
Straight line inlaid				
Standard gauge	3/32	.0925	.62	6
Light gauge	1/16	.070	.46	6

The weights and gauges in this table are manufacturing standards; slight variations will occur.

Materials for Floor Tile

S.F. of Floor	Number of Tiles			
	9" x 9"	12" x 12"	6" x 6"	9" x 18"
1	2	1	4	1
2	4	2	8	2
3	6	3	12	3
4	8	4	16	4
5	9	5	20	5
6	11	6	24	6
7	13	7	28	7
8	15	8	32	8
9	16	9	36	8
10	18	10	40	9
20	36	20	80	18
30	54	30	120	27
40	72	40	160	36
50	89	50	200	45
60	107	60	240	54
70	125	70	280	63
80	143	80	320	72
90	160	90	360	80
100	178	100	400	90
200	356	200	800	178
300	534	300	1200	267
400	712	400	1600	356
500	890	500	2000	445

To find the number of tile required for an area not shown in this table, such as the number of 9 x 9 inch tile required for an area of 550 S.F., add the number of tile needed for 50 S.F. to the number of tile needed for 500 S.F. The result will be 979 tile, to which must be added 5% for waste. The total number of tile required will be 1.028.

Adhesive for Floor Tile

Type and Use	Approximate Coverage In S.F. Per Gallon
Primer — For treating on or below grade concrete subfloors before installing asphalt tile	250 to 350
Asphalt cement — For installing asphalt tile over primed concrete subfloors in direct contact with the ground	200
Emulsion adhesive — For installing asphalt tile over lining felt	130 to 150
Lining paste — For cementing lining felt to wood subfloor	160
Floor and wall size — For priming chalky or dusty suspended concrete sub-floors before installing resilient tile other than asphalt	200 to 300
Waterproof cement — Recommended for installing linoleum tile, rubber and cork tile over any type of suspended subfloor in areas where surface moisture is a problem	130 to 150

Waste for Floor Tile

1 to 50 S.F.	14%
50 to 100 S.F.	10%
100 to 200 S.F.	8%
200 to 300 S.F.	7%
300 to 1000 S.F.	5%
Over 1000 S.F.	3%

Labor for Installing Resilient Tile Flooring

Work Element	Unit	Man-Hours Per Unit
9" x 9" x 1/8", most colors, vinyl	100 S.F.	1.6
9" x 9" x 1/8", color group D, vinyl	100 S.F.	1.7
12" x 12" x 1/8", most colors and grades, vinyl	100 S.F.	1.5
9" x 9" x 3/16", cork tile	100 S.F.	2.2
9" x 9" x 5/16", cork tile	100 S.F.	2.3
12" x 12" x 3/16", cork tile	100 S.F.	1.9
12" x 12" x 5/16", cork tile	100 S.F.	2.0
9" x 9" x 1/8", rubber tile	100 S.F.	1.6
9" x 9" x 3/16", rubber tile	100 S.F.	1.6
12" x 12" x 1/8", rubber tile	100 S.F.	1.5
12" x 12" x 3/16", rubber tile	100 S.F.	1.5
Vinyl or rubber top set cove base	100 S.F.	2.8

Time includes move on and off site, typical area prep, unloading, stacking, installation, cleanup and repair as needed. Add 15% if less than 500 S.F., deduct 5% if over 5000 S.F.
Suggested Crew: 1 tile setter

Labor Installing Resilient Sheet Flooring

Work Element	Unit	Man-Hours Per Unit
Sheet vinyl or linoleum		
.070" thick	100 S.F.	1.8
.090" thick	100 S.F.	2.1
.125" thick	100 S.F.	2.3
.140" thick	100 S.F.	2.5
Natural cork sheets, vinyl faced		
1/8"	100 S.F.	3.4
1/4"	100 S.F.	3.5
Add for wall application	100 S.F.	1.3

Labor Installing Rubber Treads and Risers

Work Element	Unit	Man-Hours Per Unit
Rubber stair treads, 10" width		
3/16" thick	10 L.F.	.7
1/4" thick	10 L.F.	.7
Rubber stair treads, 12" width		
3/16" thick	10 L.F.	.8
1/4" thick	10 L.F.	.8
Cove type risers, 6" height	10 L.F.	.4

Labor includes handling materials into place, cutting and fitting, applying with mastic, cleanup and repairs as needed.
Suggested Crew: 1 soft tile layer

Labor Installing Conductive Resilient Floors

Work Element	Unit	Man-Hours Per Unit
Rubber tile, 12" x 12" x 1/8"	100 S.F.	3.6
Vinyl tile, 12" x 12" x 1/8"	100 S.F.	3.6
Sheet linoleum, 1/8"	100 S.F.	4.9
Sheet vinyl, .090 gauge	100 S.F.	4.9
Sheet vinyl, .075 gauge	100 S.F.	4.8

Time includes move on and off site, area prep, unloading, stacking, install, cleanup and repair as needed.
Suggested Crew: 1 tile setter, small jobs only

Paints

The liquid portion of a paint is known as the vehicle or binder. Its function is to bind the tiny solid particles together, and hold the film to the surface. Oil based paints are generally thinned with mineral spirits. Latex paints are emulsions and are commonly called water base paint because they are thinned with water. Portland cement is used as an additive to oil base paint and as a binder in cement base paint. Other thinners include volatile esters, ketones, styrene, alcohols, and special blended formulas.

The solids in a paint system are vehicle resins, pigments, extenders, and metallic oxides that give color, hiding power, gloss, mildew resistance, porosity, abrasion resistance, and other properties.

Varnishes and lacquers are blends of resins and solvents. They are comparable to the vehicle of pigmented opaque paints and lacquers. A small amount of powdered solids may be used to reduce the gloss or give other special characteristics.

Broadly speaking, lacquers cure by evaporation of quick drying solvents and are generally applied by spraying. Varnishes cure by oxidation, are slower drying, and can be applied by most of the conventional methods.

Modern paints, lacquers, and varnishes are manufactured to meet a wide variety of applications. In a three-coat system, it is not uncommon for each coat to be different. It is important not to mix products from different manufacturers, or even similar products of the same manufacturer. The formula will usually be different. It is also important to use only the thinner recommended by the paint manufacturer. When repainting, or when applying more than one coat, use the same brand of paint, or at least a paint system using the same type of vehicle, and follow the manufacturer's recommendations.

Paint failures are usually the result of improper surface preparation. Surfaces should be dry, clean, and free from foreign matter. Any depressions or irregularities should be filled, leveled, and sanded. Old paint should be sanded or washed with a solution of trisodium phosphate.

Knots and pitch pockets in wood should be sealed with a paint designed for this purpose. Surface resin should be scraped off and cleaned with alcohol before sealant is applied. Metals should have a coating before being painted.

Brush Application

Brushes are made with both synthetic and natural bristles. Some bristling material may absorb some of the vehicle in the paint system and change its stiffness and brushing qualities in use. Some synthetic bristles may even be dissolved by the vehicle. The size, shape, and design of a brush determine the job the brush is best suited to handle. Some of the common specific purpose brushes include wall brushes, sash and trim brushes, enameling and varnish brushes, and stucco and masonry brushes.

When painting is stopped for any length of time, the painted area should be brought to a corner, molding, or other natural break in the surface. Overbrushing will cause air bubbles in some paint systems. Overbrushing fast drying paints or attempting to overlap strokes onto previously painted tacky surfaces will cause pile-up of the paint film.

Roller Applications

Rollers are made with lamb's wool, angora hair, and several synthetic fibers. The nap ranges in lengths from $\frac{3}{16}$ to 1¼ inches. Roller sizes may be 1½ to 2¼ inches in diameter and from 1½ to 18 inches in length. Rollers, like brushes, are designed to do a particular job. Short fine naps are used for enamels and varnishes on smooth surfaces. Longer naps are used for porous, rough, or hard-to-reach surfaces, such as chain link fences.

Spray Application

A wide variety of spray equipment is available. One of the simplest rigs is a compressor which supplies air to a spray gun attached directly to a small paint container. Paint and air are delivered under pressure to the gun through separate hoses. Airless spray equipment uses a compressor to force the liquid under high hydraulic pressure through a specially shaped orifice on the gun. This causes atomization and driving without air. More sophisticated equipment for both air and airless spray use a heating element between the pot and the gun to heat the paint before it is atomized.

Spray painting allows more coverage per man-hour than other methods. It is also the most complicated and offers more opportunity for error. The operator must maintain proper viscosity of the paint, pot pressure, atomizing pressure, air-paint ratio, and spray pattern. Normally the gun is held at right angles to work. The ends are feathered out by pulling the gun trigger after the beginning of the stroke and releasing it before the stroke is completed. The return stroke overlaps the previous stroke at midpoint to achieve double coverage. Holding the gun too far from the work causes "dusting": the paint dries in mid-air and hits the surface in a nearly dry condition.

Priming and Sealing

Generally one coat does both priming and sealing. But where "bleeding" may occur, such as at knots, special spot sealing is applied prior to priming. Seal or isolation coats are also used where the base material is incompatible with the primer, or where the base material has staining substances that would dissolve under the primer or subsequent coatings. Primers are designed for deep penetration into the materials. This guarantees maximum mechanical bond. Friction from the brush results in maximum wetting and penetration of the base material. For this reason brush application of the prime coat is strongly recommended.

Wood siding, trim, and other finish woodwork should be primed on both sides, ends and edges soon after delivery to the jobsite. This minimizes absorption of moisture both until installation and from the back side after installation.

Metals that have been cleaned, and ferrous metals that have been sandblasted, should get a prime coat on the day they have been cleaned.

Of all the methods of applying paint, brushing is the most effective means of wetting the surface and bringing it into complete contact with the paint film. For this reason, brushing is the best way to apply primers, especially to porous surfaces.

Labor for Exterior Painting

Work Element	Unit	Man-Hours Per Unit
Brush paint, per coat		
Wood siding	1000 S.F.	7.5
Wood doors and windows, area of opening	1000 S.F.	9.5
Trim	1000 S.F.	8.5
Steel sash, area of opening	1000 S.F.	5.0
Flat metal	1000 S.F.	7.0
Metal roofing and siding	1000 S.F.	7.5
Masonry	1000 S.F.	7.5
Roller painting, per coat		
Masonry	1000 S.F.	5.5
Flat metal	1000 S.F.	4.5
Doors	1000 S.F.	7.0
Spray painting, per coat		
Wood siding	1000 S.F.	4.0
Doors	1000 S.F.	5.0
Masonry	1000 S.F.	6.0
Flat metal	1000 S.F.	5.0
Metal roofing and siding	1000 S.F.	6.0
Highway or airfield lines and symbols, including glass beads	1000 S.F.	8.5
Cementitious paint, including curing	1000 S.F.	10.0
Sandblasting steel	1000 S.F.	55.0
Wire brush cleaning of steel	1000 S.F.	17.5
Clean and spray waterproofing on masonry	1000 S.F.	10.0

Surface preparation for exterior painting includes removing mill scale from metal surfaces with wire brushes or by sandblasting, removing dust with brush or cloth, removing oil and grease, masking and taping adjacent surfaces, removing masking and taping. Sometimes it is necessary to lightly sand between coats or size and fill porous materials before painting, all of which is surface preparation.

Suggested Crew: one or two men spraying, one or two men tending (one man is used to mix and prepare paint for larger crews).

Labor For Interior Painting

Work Element	Unit	Man-Hours Per Unit
Brush painting, per coat		
Wood flat work	1000 S.F.	8.5
Doors and windows, area of opening	1000 S.F.	9.0
Trim	1000 S.F.	8.0
Plaster, sand finish	1000 S.F.	7.0
Plaster, smooth finish	1000 S.F.	6.0
Wallboard	1000 S.F.	5.5
Metal	1000 S.F.	8.5
Masonry	1000 S.F.	7.0
Varnish flat work	1000 S.F.	8.5
Enamel flat work	1000 S.F.	6.5
Enamel trim	1000 S.F.	8.0
Roller painting, per coat		
Wood flat work	1000 S.F.	6.0
Doors	1000 S.F.	8.5
Plaster, sand finish	1000 S.F.	2.5
Plaster, smooth finish	1000 S.F.	3.0
Wallboard	1000 S.F.	3.0
Metal	1000 S.F.	5.5
Masonry	1000 S.F.	3.0
Spray painting, per coat		
Wood flat work	1000 S.F.	2.0
Doors	1000 S.F.	3.0
Plaster, wallboard	1000 S.F.	2.5
Metal	1000 S.F.	3.5

The painting of interior surfaces includes minimum surface preparation, mixing paint materials, and application of paint to surface.

Labor for Specialized Painting

Work Element	Unit	Man-Hours Per Unit
Mechanical painting by pipe sizes		
Up to 7" outside diameter, 2 coats	100 L.F.	1.2
8" to 12" outside diameter, 2 coats	100 L.F.	1.4
13" to 15" outside diameter, 2 coats	100 L.F.	1.7
16" to 20" outside diameter, 2 coats	100 L.F.	2.0
21" to 25" outside diameter, 2 coats	100 L.F.	2.3
26" to 30" outside diameter, 2 coats	100 L.F.	2.5
Pipe lettering, to 1" high, per 100	100	1.3
Covered equipment, 3 coats	100 S.F.	1.8
Exposed equipment, 2 coats	100 S.F.	2.0
Radiators, per coat, 36" long	Each	0.4
Baseboard radiation, to 10" high	100 L.F.	0.9

Time includes move on and off site, setup, limited surface preparation, masking, a light sanding between coats where required, removing masking and tape, cleanup and touchup as needed. Man-hours will be 2 or 3 times more if access is a problem.
Suggested Crew: 1 painter

Labor Painting Chain Link Fencing

Work Element	Unit	Man-Hours Per Unit
Paint fence posts, 5' high	10 Each	1.0
Paint fence rails	100 L.F.	2.2
Paint fence gates	100 S.F.	2.6
Paint fence fabric		
With brush	100 S.F.	1.8
With roller	100 S.F.	1.0
Paint fence complete		
With brush		
4' high	100 L.F.	10.5
5' high	100 L.F.	12.5
6' high	100 L.F.	15.0
7' high	100 L.F.	16.5
With roller		
4' high	100 L.F.	7.0
5' high	100 L.F.	8.0
6' high	100 L.F.	9.5
7' high	100 L.F.	10.5

Fence painting includes removing rust, scale, oil, grease, and dirt, mixing and applying paint.
Suggested Crew: 2 painters and 1 helper

Labor Painting Millwork[1]

Work Element	Unit	Man-Hours Per Unit
Exterior wood trim, 3 coats	100 LF	1.7
Interior wood trim, 3 coats	100 LF	1.7
Kitchen cabinets, 3 coats	100 SF	2.7
Wood casework, 3 coats	100 SF	2.7
Metal casework, 2 coats	100 SF	1.7
Wardrobes, 3 coats	100 SF	2.7
Bookcases, 3 coats	100 SF	2.7

[1] Note to millwork painting table at the left:
Time includes move on and off site, set-up, surface preparation, masking and taping, light sanding between coats, remove masking and tape cleanup and touchup as required. These figures will apply on most jobs. But a wide variation does exist in painting of wood trim. The factors that govern the time required include the following: Location, height, surface area, area to be masked and taped, tie-back and covering, surface conditions. Surface preparation, quality, specifications, material used and supervision. All of these figures could be higher by as much as 3 times on different jobs.

S.F. of Wall and Ceiling for Room Sizes

Wall Length	Wall Length												
	6 Feet	8 Feet	10 Feet	12 Feet	14 Feet	16 Feet	18 Feet	20 Feet	22 Feet	24 Feet	26 Feet	28 Feet	30 Feet
6 Feet	C 36 W 192	C 48 W 224	C 60 W 256	C 72 W 288	C 84 W 320	C 96 W 352	C 108 W 384	C 120 W 416	C 132 W 448	C 144 W 480	C 156 W 512	C 168 W 544	C 180 W 576
8 Feet	C 48 W 224	C 64 W 256	C 80 W 288	C 96 W 320	C 112 W 352	C 128 W 384	C 144 W 416	C 160 W 448	C 176 W 480	C 182 W 512	C 198 W 544	C 224 W 576	C 240 W 608
10 Feet	C 60 W 256	C 80 W 288	C 100 W 320	C 120 W 352	C 140 W 384	C 160 W 416	C 180 W 448	C 200 W 480	C 220 W 512	C 240 W 544	C 260 W 576	C 280 W 608	C 300 W 640
12 Feet	C 72 W 288	C 96 W 320	C 120 W 352	C 144 W 384	C 168 W 416	C 192 W 488	C 216 W 480	C 240 W 512	C 264 W 544	C 288 W 576	C 312 W 608	C 336 W 640	C 360 W 672
14 Feet	C 84 W 320	C 112 W 352	C 140 W 384	C 168 W 416	C 196 W 448	C 224 W 480	C 252 W 512	C 280 W 544	C 308 W 576	C 336 W 608	C 364 W 640	C 392 W 672	C 420 W 704
16 Feet	C 96 W 352	C 128 W 384	C 160 W 416	C 192 W 448	C 224 W 480	C 256 W 512	C 288 W 544	C 320 W 576	C 352 W 608	C 384 W 640	C 416 W 672	C 448 W 704	C 480 W 736
18 Feet	C 108 W 384	C 144 W 416	C 180 W 448	C 216 W 480	C 252 W 512	C 288 W 544	C 324 W 576	C 360 W 608	C 396 W 640	C 432 W 672	C 468 W 704	C 504 W 736	C 540 W 768
20 Feet	C 120 W 416	C 160 W 448	C 200 W 480	C 240 W 512	C 280 W 544	C 320 W 576	C 360 W 608	C 400 W 640	C 440 W 672	C 480 W 704	C 520 W 736	C 560 W 768	C 600 W 800
22 Feet	C 132 W 448	C 176 W 480	C 220 W 512	C 264 W 544	C 308 W 576	C 352 W 608	C 396 W 640	C 440 W 672	C 484 W 704	C 528 W 736	C 572 W 768	C 616 W 800	C 660 W 832
24 Feet	C 144 W 480	C 182 W 512	C 240 W 544	C 288 W 576	C 336 W 608	C 384 W 640	C 432 W 672	C 480 W 704	C 528 W 736	C 576 W 768	C 624 W 800	C 672 W 832	C 720 W 864
26 Feet	C 156 W 512	C 198 W 544	C 260 W 576	C 312 W 608	C 364 W 640	C 416 W 672	C 468 W 704	C 520 W 736	C 572 W 768	C 624 W 800	C 676 W 832	C 728 W 864	C 780 W 896
28 Feet	C 168 W 544	C 224 W 576	C 280 W 608	C 336 W 640	C 392 W 672	C 448 W 704	C 504 W 736	C 560 W 768	C 616 W 800	C 672 W 832	C 728 W 864	C 784 W 896	C 840 W 928
30 Feet	C 180 W 576	C 240 W 608	C 300 W 640	C 360 W 672	C 420 W 704	C 480 W 736	C 540 W 768	C 600 W 800	C 660 W 832	C 720 W 864	C 780 W 896	C 840 W 928	C 900 W 960

Example: A 14' x 20' room has 280 square feet of ceiling and 544 square feet of wall area.

9 Finishes

Single Roll Wallpaper Requirements

Size of Room	Height of Ceiling 8'	9'	10'	Yards of Border	Rolls For Ceiling
4 x 8	6	7	8	9	2
4 x 10	7	8	9	11	2
4 x 12	8	9	10	12	2
6 x 10	8	9	10	12	2
6 x 12	9	10	11	13	3
8 x 12	10	11	13	15	4
8 x 14	11	12	14	16	4
10 x 14	12	14	15	18	5
10 x 16	13	15	16	19	6
12 x 16	14	16	17	20	7
12 x 18	15	17	19	22	8
14 x 18	16	18	20	23	8
14 x 22	18	20	22	26	10
15 x 16	15	17	19	23	8
15 x 18	16	18	20	24	9
15 x 20	17	20	22	25	10
15 x 23	19	21	23	28	11
16 x 18	17	19	21	25	10
16 x 20	18	20	22	26	10
16 x 22	19	21	23	28	11
16 x 24	20	22	25	29	12
16 x 26	21	23	26	31	13
17 x 22	19	22	24	23	12
17 x 25	21	23	26	31	13
17 x 28	22	25	28	32	15
17 x 32	24	27	30	35	17
17 x 35	26	29	32	37	18
18 x 22	20	22	25	29	12
18 x 25	21	24	27	31	14
18 x 28	23	26	28	33	16

This chart assumes use of the standard roll of wallpaper, eight yards long and 18'' wide. Deduct one roll of side wallpaper for every two doors or windows of ordinary dimensions, or for each 50 square feet of opening.

Labor Installing Wall Coverings

Work Element	Unit	Man-Hours Per Unit
Wallpaper		
Light to medium weight, butt joint	100 S.F.	1.4
Heavy weight, butt joint	100 S.F.	1.5
Vinyl wall covering		
Light to medium weight, butt joint	100 S.F.	1.9
Heavy weight, butt joint	100 S.F.	2.1
Special wall coatings	100 S.F.	3.4
Flexwood	100 S.F.	7.1
Flexi-wall	100 S.F.	8.1

Time includes move on and off site, unloading, limited surface and material preparation, cleanup and repair as needed.
Suggested Crew: 1 paper hanger, 1 laborer

10
specialties
section

section contents

Chalkboards and Directories

All man-made chalkboards have nearly replaced natural slate boards. Chalking and erasing characteristics, ease of maintenance, and life expectancy of many of the man-made materials are equal to or better than natural slate.

The monolithic chalkboard is made of asbestos fibers, fillers, pigments and binders, pressed and cured under controlled conditions to form a hard, dense board 3/16 or ¼ inch thick, and up to 8 feet long. The surface is composed of pigments and abrasives which are mixed with synthetic resins and cured by baking under moderate heat.

Porcelain writing surfaces are applied to steel sheet which is laminated to plywood or hardboard. Steel can be supplied in single-length pieces up to 20 feet long, although 12- to 16-foot lengths are more common. Steel boards can also be used as tackboards by using small magnets as "tacks." Steel, aluminum, or aluminum foil is usually laminated to the back of the board to prevent warping.

Resin writing surfaces are applied directly to smooth rigid boards such as particle boards, hardboards, cement asbestos boards, or steel sheets. The surface is then laminated to a rigid board. The maximum length of these boards is generally 16 feet.

Tackboards
Tackboard cork is made of finely ground natural cork particles and linseed oil or plastic binders, and cured under heat and pressure. If the cork is not to be covered with vinyl, pigment may be added to the mix for color, and to increase soil resistance. Tackboard cork is supplied in 1/8- and ¼-inch thicknesses, and is generally backed with burlap. When cork sheet goods 1/8 inch thick or less are used, they are backed with fiberboard to allow at least ¼-inch tack penetration.

Tackboard and chalkboard assemblies are sometimes incorporated into a single frame with H-section dividers. When this is done, both assemblies should have the same total thickness.

Bulletin Boards
Bulletin boards are tackboards enclosed behind glass doors, generally hung on piano hinges and provided with cylinder locks. In lieu of hinged doors, some bulletin boards are provided with by-passing sliding glass panels, either framed or unframed.

Directory Boards
Directory boards are enclosed behind hinged or sliding glass doors similar to bulletin boards except that the directory has a means of holding removable individual letters, numbers, and symbols. Although other methods can be used to hold the message to the background, the usual method uses a board finely grooved on ¼-inch centers horizontally, and covered with wool boardcloth. The metal or plastic letters are designed with flanges on the back which fit into the grooves and hold them in place. Directory boards are available with integral indirect illumination and permanent lettered header plates.

Frames and Accessories
Although wood, bronze, stainless steel, and other materials are used as framing material, aluminum is by far the most popular. It can be given a number of attractive finishes, it does not tarnish, stain, or soil easily, and is relatively inexpensive. Aluminum can be finished from dull to mirror bright by chemical or mechanical processes.

The least expensive aluminum frame is the screw-on type, in which all fastener heads are visible on the face of the members. The next most expensive is the slip-on type, in which fastener heads are concealed or semi-concealed. The most expensive is the snap-on type, with all fasteners completely concealed.

In addition to chalk trays, chalkboards are often fitted with a top rail on which flag holders, map hooks or map roller brackets and other accessories can be mounted. This top rail may also have a cork insert which is a tack holder for mounting maps, graphs, posters, or other papers.

Installation
Chalkboards and tackboards are available as factory assembled units or as components for field assembly. Larger units which are difficult to handle or move through passageways are assembled on the job. Smaller units, directory, and bulletin boards are factory assembled. All units are available for surface mounting, but directory and bulletin boards are also available in recess mounted units. Chalkboards and tackboards may be adhesive mounted to sound wall surfaces, or mechanically secured, or both. Moldings, trim, and framing members should be mechanically secured with concealed or inconspicuous fasteners, colored and finished to match adjacent surfaces. Chalkboard joints are either butted flush or separated by an aluminum H section colored to match the chalkboard.

Labor Installing Identifying Devices

Work Element	Unit	Man-Hours Per Unit
Open-face directory boards	S.F.	.13
Glass encased directory boards	S.F.	.65
Mounted plaques and nameplates	S.F.	1.10
Cast aluminum letters		
4" high, 1" deep	Each	.42
8" high, 1" deep	Each	.63
12" high, 2" deep	Each	.70
16" high, 2" deep	Each	.78

Use these figures for mounting on concrete or masonry surfaces. Add 50% for mounting on gypsum wallboard.

Labor Installing Chalkboards and Tackboards

Work Element	Unit	Man-Hours Per Unit
Chalkboards		
Most types and weights	100 S.F.	7.3
Tempered masonite	100 S.F.	8.3
Structoboard	100 S.F.	8.3
Porcelain enameled steel	100 S.F.	8.3
Nucite (tempered glass)	100 S.F.	11.3
Treated formica on plywood	100 S.F.	8.3
Slate, ⅜" thick	100 S.F.	9.3
Specially constructed types		
Vertical sliding, manual type	100 S.F.	12.5
Vertical sliding, electric	100 S.F.	15.6
Horizontal sliding, manual type	100 S.F.	12.5
Multi-leaf swinging boards	100 S.F.	9.5
Adjustable modular systems	100 S.F.	10.6
Special feature additives		
Tack strips, 2" wide	100 L.F.	8.1
Cork strips	100 L.F.	8.7
Tackboards, most types	100 S.F.	6.3

Time is for chalkboard and tackboards attached to masonry walls. Add 50% to labor if attached to gypsum wallboard. Time includes move on and off site, unloading, repairs and clean-up as needed.
Suggested crew: 1 carpenter, 1 laborer.

Folding Partitions

Folding, accordion, sliding, mobile, and coil partitions are available in a wide variety of sizes, operating principles, construction materials, and properties. Folding partitions may be classified by the type of panel, type of support, method of operation, sound rating or fire rating.

Folding partitions are used to divide room areas such as banquet rooms and classrooms or to convert recreational or other large areas into temporary smaller multiple use areas. They are generally supported on permanently installed overhead tracks, both with or without floor guides. Some are floor-supported with overhead guides.

Folding partitions can be classified into two types: rigid panel, and flexible accordion folding type.

Flexible accordion type partitions are supported by surface mounted or recessed semi-concealed overhead tracks. Also available is a track mounted on a portable integral truss or beam and post system. The curtain consists of a folding metal framework covered with fabric backed vinyl.

For better sound control, lead filled vinyl inner liners and sound absorbing fiberglass blankets are used. Most accordion folding partitions are light enough for manual operation. Where long runs are used, the partition will have two equal length curtains with latching lead posts and anchor posts secured at opposite sides of the room. In extremely long runs, power operation is needed. Flexible folding partitions may be subdivided into two types:

Pantograph-hinged double accordion-fold types have a hollow space between folds. This is the most popular type.

Single accordion-fold types with no hollow space between the two vinyl face sheets is lighter weight, less costly, and takes up less space in both the stacked and closed position.

There are many kinds of **rigid panels** used in folding partitions. They range from thin narrow extruded aluminum sections with integral hinges, weighing less than 1 pound per square foot, to laminated assemblies 4 feet or more wide, 3 inches thick and weighing 10 pounds or more per square foot. These larger panels are available with pedestrian pass doors and built-in chalkboards and tackboards. There are two basic panel types:

Monolithic panels are available in extruded aluminum, sheet steel, and solid wood. Various types of metal and extruded vinyl hinges are available. The bulbed edges of the full-length vinyl hinges are fitted into grooves in panel sides.

Composite assemblies are made up of solid wood, particleboard, or mineral cores, hollow wood latticework or honeycomb construction. Sound deadening material may be incorporated into the cores. Facing materials include wood, plastic laminate, hardboard, painted steel, porcelain enameled steel, vinyl, and various other wall coverings. Structural frames are made of wood, aluminum, or steel. Wood edges are often reinforced with extruded aluminum trim. Meeting edges can be interlocked on some panels and sealed with vinyl or synthetic rubber gasketing.

Sound rated partitions carry sound transmission class (STC) numbers to indicate the efficiency of the wall as a sound barrier. The higher the number, the better the sound barrier. These panels have resilient seals at the floor and ceiling as well as at panel edges. Edge seals are automatically compressed when the wall is in the closed position. Top and bottom seals are available in fixed and mechanically activated types.

Track mounted rigid panel partitions may be classified by the way they are hinged, folded, or stacked. Hinges are available in concealed, semi-concealed, and fully exposed types. Some track mounted panels are not hinged or are hinged together in pairs. This makes manual operation of long runs possible. Panels are arranged to fold accordion-like with single or double curtains, or fold in one direction only to form a coil. Other panels are folded parallel to adjacent walls, folded perpendicular to adjacent walls, or stored flat against the wall. Partitions are frequently biparting; about half of the run is stacked against each of the adjacent walls. Some systems allow stacking in several locations.

Track and Trolley
Overhead tracks are generally channels although I beams and other shapes are also used. They are made from extruded aluminum shapes and hot and cold rolled steel. In some installations electric or manual switch gear is built into the system for more flexibility in partition location or stacking. Floor tracks are designed with a narrow recessed slot in which a guide of flanged wheels travel. Floor bearing partitions are used where long runs of heavy panels make overhead structural supporting systems impractical.

Locking and Latching
Where panels are stacked close together and parallel, projecting hardware can not be used. Therefore handles, knobs, panic devices, and other hardware must be recessed into the panel. Some hardware is removable.

Structural supports for overhead supported folding partitions are generally designed into the building and are not the responsibility of the partition subcontractor. The partition subcontractor should, however, satisfy himself that the overhead construction is adequate to carry the weight of his partition.

Floor guides and tracks in concrete floors are usually installed as the floor is poured. Close coordination is required between the subcontractors and trades involved.

Overhead track is secured with brackets and fasteners. The track is adjusted for level and alignment by adjusting threaded suspending rods or bolts, or by shimming.

Labor Installing Folding Partitions

Work Element	Unit	Man-Hours Per Unit
Framed truss construction, manually operated		
Steel-lined vinyl, 12" profile	100 S.F.	8.6
Steel-lined vinyl, 8" profile	100 S.F.	10.6
Vinyl clad steel, 18" profile	100 S.F.	9.3
All vinyl profile	100 S.F.	9.3
All vinyl, residential type	100 S.F.	7.5
Single paneled folding type, manually operated		
Anodized aluminum 6" profile	100 S.F.	8.6
Bright aluminum	100 S.F.	8.7
Birch, 8", 12", 5" profile	100 S.F.	9.8
Sliding modular flat wall panels or manually operated vinyl covered steel folding partitions		
Single slider, top & bottom fixed	100 S.F.	20.0
Hinged pair, top & bottom fixed	100 S.F.	20.0
Add for electric operation per section		
To 20' long section	Each	19.0
Over 20' to 40' long	Each	23.0
Over 40' long section	Each	33.0

Time includes move on and off site, unloading, uncrating, installing, checking, clean-up and repair as needed.
Suggested Crew: 1 carpenter, 1 laborer

Demountable Partitions

Demountable partitions are flexible wall systems which divide interior space within a building. They are "demountable" because they can be moved to another location relatively easily. Components and properties are as follows:

The post or column carries the framing. It can be floor to ceiling or free standing.

Framing includes units enclosing the panel and attached to the post. Framing surrounds openings, doors and window areas.

The panel is the wall unit enclosed by framing and which encloses space. Various openings may be incorporated: framed openings, doors, grates, windows, etc. Panels can be gypsum board, steel, aluminum, hardboard or wood. The panel filler or core is usually particle board, honeycomb paper or insulating material.

Other components include framing, rail, cornice, cap and end plates. Attachment can be by brackets, clips and shoe. Raceway units are available to carry power and phone lines and leveling devices can be used to compensate for irregular surfaces.

Labor Installing Demountable Partitions

Work Element	Unit	Man-Hours Per Unit
Basic baked enamel, hardboard or gypsum wallboard unit, 9' high, average 2½" thick, with door frames at 20' O.C. and base	100 L.F.	70.0

Time does not include doors, hardware or locksets. Time does include move on and off site, unloading, stacking, placing, cleanup and repair as needed. Add 5.5 hours for each door and 1.0 hours for each automatic closer.
Suggested Crew: 1 carpenter and 1 laborer. On large jobs labor requirements will be less.

Labor Installing Office Cubicle Partitions

Work Element	Unit	Man-Hours Per Unit
Steel or aluminum framed panels 8' to 9' 6" high		
No glass	10 L.F.	4.0
25% glass	10 L.F.	5.5
50% glass	10 L.F.	6.5
100% glass	10 L.F.	8.1
Add for door	Each	6.0
Add for double acting closer	Each	2.0
Steel or aluminum framed dwarf partitions 5' high	10 L.F.	3.5
Add for gate	Each	5.8

Installation includes unloading, handling into place, installing fastening devices including drilling, fastening in place, and installing hardware and trim.
Suggested Crew: 2-7 men

Toilet Compartments

Prefabricated toilet compartments are made of steel, stainless steel, plastic sheet laminated to a structural core, or marble. Steel compartments are made with a baked enamel or porcelain enamel finish. Most manufacturers offer about 20 standard colors. Stainless steel is generally finished with a fine grain abrasive to produce a satin sheen, known as "No. 4 Finish" in the industry. Plastic laminate partitions are available in a wide variety of solid colors and patterns, including marbles and wood grains. They are also available in high gloss, satin, and wood grain texture. Marble partitions are available in several colors, and are polished on all exposed surfaces.

Toilet compartment partitions, and entrance and urinal screens are supported in four ways: (1) cantilevered from the building wall, (2) one end supported on the building wall and the other end supported on a stile or pilaster which is fastened to the floor only, (3) fastened to the floor on one end and to an overhead rail on the other end, (4) or fastened to the ceiling or overhead construction only.

Metal cantilevered partitions are wedge-shaped, generally about 6 inches thick at the wall, and taper to about 1½ inches at the unsupported pilaster. Except for small screens which can be cantilevered, marble is always supported on floor mounted pilasters. Plastic laminated partitions require pilaster support.

Metal toilet components are generally bonded to a resin impregnated honeycomb paper core, although some manufacturers use other core materials. Plastic laminated partitions are based on a plywood or particleboard as a core material for rigidity and screw holding power. Marbles for partition work are selected for strength as a first consideration. Marble compartments are fitted with metal, hardwood, or plastic laminated doors.

Toilet compartment hardware is generally white metal, either polished or finished in a satin sheen, and is furnished in anodized aluminum, stainless steel, chrome plated steel or brass. Unexposed hardware should be galvanized steel or non-ferrous metal.

Partitions, urinal screens and doors are normally 1 inch thick. Partitions and doors are normally set 12 inches above the finished floor. Headrails are normally set 12 inches above the top of doors. Junior size partitions and doors are available in 42- and 48-inch heights. The standard size is 58 inches. Some manufacturers set their partitions and doors over 12 inches above the finished floor and furnish 54 rather than 58 inch high panels and doors. Some manufacturers also make 60 inch high panels and doors. Pilasters are generally made in widths from 4 to 12 inches in 2 inch increments. Doors are made in widths from 22 to 30 inches and generally swing inward. Doors for paraplegics should swing outward and the compartment and door width should be ample to accommodate a wheel chair. These compartments should also be equipped with grab bars.

Of the various materials and finishes used in toilet compartment construction, painted steel is about the least expensive. It is also the most vulnerable to damage by vandals. But damage to a paint finish is the easiest and least costly to repair. In installations where vandalism is likely to be a problem, hardware can be installed with fasteners that require a special tool for removal. Overhead braced type partitions are also available with headrails which have a sharp edge projecting upward to discourage use as a chinning bar. Floor cleaning and sanitation is simplified when wall or ceiling hung units are used instead of floor mounted pilasters. However, wall or ceiling hung compartments are more costly, may be less rugged, and take up more space.

Installation
Pilasters should be secured to floors or overhead construction with anchors equipped with level adjusting devices. The anchor is concealed behind stainless steel or other white metal trim. Marble pilasters rest directly on concrete floor slabs.

Panels are secured to walls and pilasters with brackets designed for that purpose. Urinal screens that do not have floor or ceiling support are supported with wide wall-flange brackets.

Door hardware and accessories are installed with hex bolt expansion shields, toggle bolts, or other fasteners. Door hardware should be adjusted so that all inward opening doors remain ajar at the same angle when not latched.

Labor Installing Toilet Partitions

Work Element	Unit	Man-Hours Per Unit
Toilet partitions, marble with metal door	Each	3.8
Toilet partitions, corner type, floor mounted	Each	1.8
Toilet partitions, metal, ceiling hung	Each	2.0
Toilet partitions, free standing, metal	Each	1.9
Toilet partitions, floor mounted, metal	Each	1.8
Toilet partitions, stainless steel, school type	Each	2.8
Entrance screens, floor mounted, 36" wide	Each	0.5

Time is per enclosure installed and includes move-on and off-site, unload, place, clean-up and repair. Time assumes all anchors are placed on concrete.

Labor Installing Toilet Accessories

Work Element	Unit	Man-Hours Per Unit
Medicine cabinets	Each	1.3
Add for lighted unit	Each	.5
Ash trays, recessed wall urn	Each	1.2
Ash trays, surface-mounted wall urn		
Box type	Each	1.1
Compact	Each	.7
Bowl	Each	.8
Shower curtain rod	Each	.5
Mop rack and hook	Each	.3
Utility hook strip	Each	.7
Hook strip and shelf	Each	1.0
Pail & ladder hook	Each	.4
Hand dryer (electric)	Each	2.3
Bedpan holder & rack	Each	.5
Foot operated soap dispenser	Each	2.6

These figures assume installation on prepared mounting points. No electrical work is included.

Labor Installing Surface-mounted Bathroom Accessories

Work Element	Unit	Man-Hours Per Unit
Soap dish	Each	.24
Clothes hook, single	Each	.12
Clothes hook, double	Each	.12
Crystal shelf, 6" deep	L.F.	.26
Stainless steel shelf, 6" deep	L.F.	.26
Towel bar, 12" long	Each	.25
Towel bar, 18" long	Each	.26
Towel bar, 24" long	Each	.27
Soap & grab bar combined	Each	.20
Towel ring	Each	.13
Tumbler & toothbrush holder	Each	.24
Toilet paper holder, single roll	Each	.26
Toilet paper holder, double roll	Each	.26
Toilet paper dispenser, box type	Each	.39
Soap dispenser, globe type	Each	.20
Soap dispenser, box type	Each	.33
Towel pin	Each	.13
Towel ladder & bar	Each	.39
Mirror 18" x 24"	Each	.46
Mirror shelf, 18" x 24"	Each	.52
Government type mirror, 18" x 24"	Each	.53
Government type mirror with shelf 18" x 24"	Each	.60
Shelf, 18" wide x 6" deep	Each	.40
Pull down utility shelf	Each	.66
Straight grab bars, 24"	Each	.39
Angular grab bars, 24" x 36"	Each	.80
Wall to floor grab bars	Each	.65
Straddle grab bars	Each	1.00
Special shaped grab bars (custom)	Each	1.30
Towel & waste receptacle, 10.5 gal.	Each	.90
Sanitary napkin dispenser & receptor	Each	1.45
Sanitary napkin receptacle	Each	.92

These figures assume blocking has been installed where needed to carry each fixture if installation is on a frame wall.

Labor Installing Recessed Bathroom Accessories

Work Element	Unit	Man-Hours Per Unit
Toilet paper holder, single	Each	.30
Soap holder	Each	.18
Soap & grab bar combined	Each	.24
Tumbler holder	Each	.24
Tumbler & toothbrush holder	Each	.24
Paper towel dispenser	Each	.36
Concealed lavatory service unit	Each	.57
Facial tissue dispenser	Each	.26
Towel & soap dispenser with mirror	Each	1.20
Towel dispenser & receptacle	Each	1.20
Towel dispenser	Each	1.80
Towel & waste receptacle, 10.5 gal. capacity	Each	1.10
Sanitary napkin dispenser	Each	1.80

These figures assume fixtures are installed in prepared openings.

Labor Installing Shower Compartments

Work Element	Unit	Man-Hours Per Unit
Single entry metal compartment	Each	10.1
Single entry marble compartment	Each	13.0
Double entry metal compartment	Each	10.5
Double entry marble compartment	Each	13.6
Double entry & dressing steel compartment	Each	17.6
Double entry & dressing marble compartment	Each	33.0
Receptor only	Each	3.4
Shower door only	Each	1.0
Molded fiberglass three wall unit including receptor and door		
32" x 32"	Each	6.5
36" x 36"	Each	7.2
40" x 40"	Each	7.8

No plumbing hookup included in these figures. Compartments include receptors and floor-mounted walls. Ceiling-mounted walls will add about 40% to man-hours. Overhead bracing will add about 10% to man-hours. Time is based on 1 carpenter, 1 laborer. Time includes move on and off site, unloading, cleanup and repair as needed.
Suggested Crew: 1 carpenter, 1 laborer

Labor Installing Lockers

Work Element	Unit	Man-Hours Per Unit
Single tier 72" high steel athletic lockers		
9" width	Each	.65
12" width	Each	.70
15" width	Each	.75
18" width	Each	.80
Double tier 36" high steel athletic lockers		
9" width	Each	.40
12" width	Each	.45
15" width	Each	.50
18" width	Each	.55
Triple tier 24" high steel athletic lockers		
9" width	Each	.30
12" width	Each	.32
15" width	Each	.33
18" width	Each	.35
Add for closed base for lockers	10 L.F.	.75
Add for sloping steel locker top	10 L.F.	1.40
Add for interior partition shelf and rod	Each	.12
Box-type stackable steel lockers		
12" x 15" x 12" deep	Each	.42
15" x 15" x 12" deep	Each	.50
15" x 18" x 12" deep	Each	.66
15" x 18" x 15" deep	Each	.95
Enameled steel basket racks		
24 basket set	Set	4.20
30 basket set	Set	4.90
36 basket set	Set	5.75
42 basket set	Set	6.65
Wire mesh wardrobe lockers	Each	1.00
Locker benches	L.F.	.17

These figures assume units are mounted on a concrete floor and include hardware installation.

Labor Installing Storage Shelving

Work Element	Unit	Man-Hours Per Unit
72" high, 7 shelf baked enamel industrial shelf		
10" to 18" wide	L.F.	.76
24" to 36" wide	L.F.	.90
32" high, 3 shelf baked enamel industrial shelf		
10" to 15" wide	L.F.	.57
18" to 24" wide	L.F.	.62
Bookshelf, board or plywood, S.F. of shelf	S.F.	.09
Closet shelf, bracket mount	S.F.	.07

Industrial shelving assumes mounting on a concrete or masonry wall. Attachment to gypsum board walls may add 50% to the man-hours required.

Labor Installing Chutes

Work Element	Unit	Man-Hours Per Unit
Stainless steel linen chutes		
18" diameter	10 L.F.	4.0
24" diameter	10 L.F.	4.3
30" diameter	10 L.F.	4.5
Hopper	Each	3.5
Skylight	Each	5.6
Sprinkler unit at top	Each	6.2
Galvanized trash chutes		
18" diameter	10 L.F.	3.8
24" diameter	10 L.F.	4.2
30" diameter	10 L.F.	4.4
36" diameter	10 L.F.	4.9
Hopper	Each	2.8
Skylight	Each	5.6
Sprinkler unit at top	Each	6.2
Sprinkler unit at intake hopper	Each	2.1
Refuse bottom storage hopper	Each	7.8

Installation time includes handling into place, installing fastening devices, setting chute, doors and trim.

Labor Installing Flagpoles

Work Element	Unit	Man-Hours Per Unit
Fiberglass pole, free standing		
25'	Each	12.0
50'	Each	24.0
Steel pole, freestanding		
25'	Each	15.5
50'	Each	30.0
Aluminum pole, freestanding		
25'	Each	29.0
50'	Each	70.0
Aluminum pole, tapered, freestanding		
30'	Each	36.0
40'	Each	42.0
50'	Each	64.0
60'	Each	71.0
Aluminum wall mounted, with collar		
15'	Each	13.0
18'	Each	15.0
20'	Each	17.5
24'	Each	25.0

Freestanding poles are mounted in a 2 C.Y. concrete base. Times include excavation, concrete, erection and rigging.

Labor Installing Postal Equipment

Work Element	Unit	Man-Hours Per Unit
Residential letter slot	Each	.6
Post-mounted letter box	Each	.9
Apartment house keyed box group, per box, minimum of 12	Box	.1
Single mail chute, per floor	Each	4.8
Single chute receiving box	Each	4.4
Twin (double parallel) mail chute, per floor	Each	7.0
Twin chute receiving box	Each	11.0

These figures are based on installation of postal service approved units. Chutes are glass front units either recessed or surface-mounted. Floor to floor height is assumed to be 12'.

Labor Placing Fireplace Components

Work Element	Unit	Man-Hours Per Unit
Rotary control dampers		
30" long	Each	.88
50" long	Each	1.46
84" long	Each	2.34
96" long	Each	2.52
Plate type damper (cast iron)		
50" long	Each	1.61
50" long	Each	2.19
72" long	Each	2.64
84" long	Each	2.93
96" long	Each	3.36
Chain operated dampers		
32" x 20"	Each	1.72
48" x 24"	Each	2.64
Cast iron cleanout doors		
8" x 8"	Each	.41
12" x 12"	Each	.52
18" x 18"	Each	.64
18" x 24"	Each	1.05
24" x 30" with steel door	Each	1.61
Chimney screens for 8" x 8" flue		
Galvanized steel	Each	.35
Stainless steel	Each	.35
Chimney screens for 13" x 13" flue		
Galvanized steel	Each	.73
Stainless steel	Each	.73
Chimney screens for 24" x 24" flue		
Galvanized steel	Each	1.76
Stainless steel	Each	1.75

Time includes drilling, bolting or installing special brackets, bands, shields. Time does not include shop or field welding or modifications.
Suggested Crew: 1 mason and 1 helper

Pedestal Floors

Pedestal floors are also called total access floors, elevated or raised access floors, plenum floors, and computer floors. They provide space under the floor surface for routing service lines or conditioned air into a room where ducts, conduits, cellular floor systems, or conventional methods are impractical. Pedestal floors are most common in computer or data processing rooms. Other rooms where pedestal floors are used include communications rooms, laboratories, offices, and clean rooms.

The pedestal floor consists of a series of adjustable metal pedestals which support rigid floor panels. There are three types:

In the **gridless floor system,** the floor panels are supported at all four corners on the pedestal caps. The pedestal cap and panel corners are designed with a positive interlocking fit to prevent lateral movement of the panels. This type provides maximum underfloor access and is the least expensive of the three. This system has the least overall stability and is therefore not generally recommended for floors over 12 inches high or in large areas. It must be anchored around the perimeter by special bracing or solid masonry or concrete walls. This system provides the least airtight plenum, and requires special supports where cut panels are required as filler units or where openings are cut through the edge of a panel.

The **lay-in grid system** uses a series of beams running from pedestal to pedestal in both directions. The pedestal cap, grid, and panel all have an interlocking fit for lateral stability. This system is more expensive than the gridless type. But it is more stable and can be used in floors elevated up to 24 inches and in larger areas. It also provides a more airtight plenum because the floor panel is supported around its entire perimeter. Thus air leaks at panel joists are minimized. The grid beams can easily be lifted out for maximum underfloor access. This system also provides support for cut filler panels.

The **rigid grid system** is similar to the lay-in grid except that the grid beams are mechanically fastened to the pedestal cap for maximum stability. This is the most expensive system and can be used in larger areas, in floors elevated over 24 inches, and in areas that have earthquakes. It also provides the best electrical continuity for static bleed-off. Service lines must either be worked under the grid, or individual beams must be unfastened and removed in the path of the service run.

Components

Pedestals consist of a base, a column, and a cap. There are two basic types: cast aluminum alloy and welded steel. The welded steel is available in zinc coated and painted finishes.

The base may be square, round, hexagonal, or other shape, and may vary from 4 up to about 8 inches across, depending on the column height.

The column consists of two or more parts, basically a threaded stud and a hollow tube; one is integral with or welded to the base, and the other to the cap. The stud fits snugly into the tube. Fine height adjustments are made with a nut on the stud.

The cap is designed with bosses or recesses or both in which stringers and floor panels may be fitted.

Grid beams or stringers are made from hot or cold rolled steel with a zinc coated or painted finish, or extruded aluminum alloy. Some stringers are provided with continuous strips of resilient pads on top. These strips reduce noise, plenum air loss, and adjust for irregularities in mating surfaces. The ends of the stringers are designed to fit into a two-dimensional interlock or a three-dimensional rigid mechanical connection to the pedestal cap.

Floor panels are available in three basic types: (1) Die cast aluminum alloy which is made rigid by an integral waffle grid on the back. (2) Metal clad wood core in which rigidity results from the stressed skin held on a rigid core. The metal skins are galvanized steel. Generally the bottom and edge covering is of heavier gage than the top. (3) Cold rolled galvanized steel which is spot welded to a pressed grid beam or eggcrate member.

The most popular size for floor panels is 24 inches square. Some are made up to 3 feet square and some are available in modules of less than 2 feet square. All panels are equipped with continuous vinyl edge strips which are held in place in a slot formed or cast into the panel edge. This provides vinyl to vinyl contact at panel edges and makes a good plenum seal.

Several types of floor coverings may be used on floor panels. Where pedestal floors are used in computer rooms, floor coverings which dissipate static electricity are recommended. The four most frequently used floor coverings are vinyl asbestos, vinyl, high pressure laminate, and carpet.

Accessories Most pedestal floor manufacturers offer special panels, closures, and trim to divide underfloor plenums, ramps and steps from the subfloor elevation to the raised floor. Other accessories are supplied to bring plenum air and services through floor panels. Plenum air may be delivered through perforated floor panels or cutouts fitted with grilles, either of which may be equipped with dampers. Services are generally run through cutouts in panels, and fitted with metal edge trim and sponge rubber cushioning for a snug seal between the service line and the floor system.

Where a raised floor does not terminate against a wall, a safety railing should be installed along the exposed edge.

Tools

One tool needed will be a panel lifter. For smooth surfaced floor coverings, this will be two rubber suction cups connected by a handle. Each cup should be equipped with a vacuum release valve. For carpets, the lifting tool has two lifting points connected by a handle. A lever on the handle moves many rows of needlelike teeth on the lifting end to engage the teeth in the carpet tufts.

Installation

Subfloors on which pedestals will be installed must be clean, dry, and sound. Surfaces in contact with pedestal bases should be flat, level, and reasonably smooth. Pedestals may be installed over existing resilient floor covering if mechanical fasteners are used.

The floor area should be planned to avoid fractional size panels. Center-points of pedestal bases are established around the perimeter of the raised floor area. Snap a chalkline to connect corresponding points on opposite sides of the room. Except at the room or area perimeter, this chalkline grid will correspond to panel edges.

Final positioning and leveling of pedestals is generally done with the aid of a leveling target and surveying instrument. Grid beams and floors panel lock into pedestal caps with a fairly close tolerance. Pedestals must therefore be located precisely. Pedestal bases are set in a continuous bed of epoxy adhesive or secured with at least 3 mechanical fasteners. When pedestals are in place, pedestal caps are adjusted for level and secured with lock nuts, set screws or other positive means.

Grid beams, if used, are fitted to pedestal caps and secured if required. Floor panels are fitted into the supporting assembly. Check the joints for a snug, flush fit. Adjustments can be made as necessary to correct defects. Panels with cutouts for grilles, registers and underfloor electrical or plumbing services and steps, ramps, railings, and other accessories can then be located and installed.

Installations in older buildings may require that pedestals be installed on floors which are or have been covered with adhesive mounted floor covering or solid concrete. Adhesive mounting of the bases may be impossible in this case. Test a small area to see if adhesive can be used.

Installation of pedestal floors in earthquake areas may require special bracing, solid concrete or masonry walls around the perimeter, or a rigid grid system. A portion of the floor panels may require mechanical fastening to the support system.

Labor Installing Pedestal Floors

Work Element	Unit	Man-Hours Per Unit
Gridless system	100 S.F.	5.8
Lay-in grid system	100 S.F.	8.0
Rigid grid system	100 S.F.	11.5
Add for seismic bracing	100 S.F.	1.0
Floor grilles, custom fit	Each	1.0
Ramps	10 S.F.	1.2
Cut-outs, typical	Each	2.5

These figures assume a 400 S.F. minimum job with 24" x 24" panels and include typical trim. Time does not include leveling or preparation of the existing floor.

11 equipment section

section contents

Labor Installing Cabinets and Tops

Work Element	Unit	Man-Hours Per Unit
Base cabinets, 36" high		
24" wide	Each	1.0
36" wide	Each	1.2
Base corner cabinets, 36" wide	Each	3.2
Wall cabinets, 12" deep		
12" x 18"	Each	0.8
18" x 18"	Each	0.9
18" x 36"	Each	1.0
24" x 36"	Each	1.3
Cabinet stain finish	100 S.F.	1.0
Cabinet paint finish	100 S.F.	1.2
Cabinet vinyl finish	100 S.F.	1.4
Factory formed tops, 4" backsplash		
24" wide	L.F.	0.20
32" wide	L.F.	0.21
Backsplash only, 4" high	L.F.	0.08
Cutting blocks, custom sizes	S.F.	0.42
Broom closets, 7' high	Each	1.8

Time includes layout, unloading and all necessary trim work, cleanup and repairs as needed. Time does not include any cleanup prior to installing cabinets or demolition work.
Suggested Crew: 1 carpenter and 1 laborer, 1 painter for finishing.

Labor Installing Kitchen Equipment

Work Element	Unit	Man-Hours Per Unit
Office building kitchens		
Breakfast & lunch service only	100 S.F.	31.0
Full meal service	100 S.F.	40.0
Restaurant kitchens		
Low-cost or large kitchens	100 S.F.	32.0
Most installations	100 S.F.	43.0
Well-equipped or small kitchens	100 S.F.	57.0
Hospital kitchens		
Low-cost or large kitchens	100 S.F.	32.0
Most installations	100 S.F.	43.0
Well-equipped or small kitchens	100 S.F.	56.0
Frozen food service only	100 S.F.	66.0

Hours are per 100 square feet of kitchen excluding service and storage area. Time includes handling equipment into place, securing in place, finishing, cleanup and repairs as needed. Equipment includes work tables, racks and shelving, tables, griddles, fryers, sinks, disposers, pot washers and cutting boards appropriate for the type of kitchen. No mechanical or electrical work included.

Labor Installing Food Serving Line Equipment

Work Element	Unit	Man-Hours Per Unit
Breakfast and lunch service only	L.F.	2.8
Full course service	L.F.	2.8
Center island service	L.F.	2.8

Time includes handling equipment into place, securing, cleanup and repairs as needed. Hours are per linear foot of serving line and include installed stainless steel counters, serving shelving, racks, facing, steam tables, cup and plate dispensers, and cases. No mechanical or electrical work included.

Labor Installing Food Service Equipment

Work Element	Unit	Man-Hours Per Unit
Automated dishwashing equipment, wash unit only		
Minimum capacity, low speed	Each	22.0
Typical wash system	Each	28.0
High speed service unit	Each	35.0
Tray conveyor system, with single belt to 50' long, per L.F. of belt, 10' minimum		
12" wide	L.F.	1.5
15" wide	L.F.	1.7
18" wide	L.F.	1.8
24" wide	L.F.	1.9
Cold boxes, including insulation, framing and mechanical equipment, per S.F. of floor, 65 S.F. minimum, excluding door	S.F.	1.3
Cold box doors, including frame & hardware, 3' x 6'		
Hardwood door	Each	15.0
Stainless steel door	Each	17.0
Plastic constructed door	Each	15.0

Labor includes handling into place, mounting, electrical hook-up only, cleanup and repairs as needed.

Labor Installing Church Furnishings

Work Element	Unit	Man-Hours Per Unit
Lecturn	Each	6.5
Pulpit	Each	6.5
Single confessional, with curtain	Each	12.9
Double confessional, with curtain	Each	20.0
Single confessional, with door	Each	14.9
Double confessional, with doors	Each	24.0
Ark, 5' x 6', with curtain	Each	8.6
Ark, 5' x 6', with door	Each	14.0
Bench type pew	100 L.F.	53.0
Bench type pew with kneeler	100 L.F.	54.0
Seat pew	100 L.F.	56.0
Seat pew with kneeler	100 L.F.	57.0
Baptismal font	Each	13.0
Hardwood communion rail	10 L.F.	2.6
Bronze or stainless communion rail	10 L.F.	3.9
Aluminum or wrought iron rail	10 L.F.	3.3
Hardwood altar	Each	7.9
Marble altar	Each	26.0
Hardwood with marble base	Each	20.0
Stained glass window, lead frame	S.F.	.8

Labor for Installing Loading Dock Equipment

Work Element	Unit	Man-Hours Per Unit
Levelers, hinged, 10 ton capacity 6' x 8'	Each	4.0
Docking boards, 5' x 5' aluminum	Each	4.0
Docking bumpers, rubber, 4½" x 10" x 14"	Each	.25
Docking bumpers, 4½" x 10" x 36"	Each	.35

Time includes move-on and off-site, unloading, layout, clean-up and repair as needed.
Suggested Crew: 1 carpenter

Labor Installing Gymnasium Equipment

Work Element	Unit	Man-Hours Per Unit
Basketball backstops, stationary		
Wall mounted, wood	Each	5.2
Wall mounted, glass	Each	6.6
Ceiling mounted, wood	Each	18.0
Ceiling mounted, glass	Each	18.0
Pole & steel board, outdoor	Each	13.0
Wall mount steel board, outdoor	Each	6.0
Basketball backstops, motorized or manual operation		
Wall mount, wood	Each	11.0
Wall mount, glass	Each	12.5
Ceiling mount, wood	Each	23.0
Ceiling mount, glass	Each	25.0
Exercise equipment		
Wall mount parallel bars	Each	11.0
Wall mount wood ladders	10 S.F.	2.3
Eyelets & hooks	Each	.7
Floor sleeve inserts	Each	.7
Wall mounted mats		
2" thick	100 S.F.	7.0
3" thick	100 S.F.	7.8
Wall mount indoor scoreboard, with remote scorer cables and controls	Each	40.0
Steel framed benches, telescopic, 500 seat minimum, manual operation, per 100 seats	100 each	13.0

Time includes handling into place, attaching supports, electrical hook-up only, cleanup and repairs as needed.

Labor Installing Library Shelving

Work Element	Unit	Man-Hours Per Unit
Single faced book stack, 36" long sections 8" to 12" deep shelves or periodical rack		
3'6" high, 3 tiers	Each	.95
5'6" high, 5 tiers	Each	1.05
7'6" high, 7 tiers	Each	1.20
Add for double faced stack	Each	.60
Uncased metal shelving, single faced, 36" sections		
3'6" high tiered book stack, 10" to 12" deep	Each	1.20
3'6" high tiered book stack, 15" deep	Each	1.30
Add for sliding metal panel doors	Each	1.00
Add for sliding glass doors	Each	1.20
Add for hinged doors	Each	1.30
Add for 5'6" or 7'6" high units	Each	.40
Study carrels, multi-stacked modular carrels 3'6" wide with sides, front panel & shelf, per carrel	Each	2.00
Add for individual illumination, per carrel	Each	1.20
Charging counters		
Desk and cabinet without top	L.F.	1.0
Counter top, laminate	10 S.F.	2.00

Includes handling materials into place, installation in a prepared location, cleanup, electrical hook-up only, and repairs as needed.

Labor Installing Surgery Equipment

Work Element	Unit	Man-Hours Per Unit
Stationary surgery tables	Each	18.0
Service island for mobile table	Each	16.0
Scrub stations, floor mounted		
Manual, one bay	Each	20.0
Manual, two bay	Each	22.0
Manual, three bay	Each	40.0
Electric, one bay	Each	19.0
Electric, two bay	Each	30.0
Electric, three bay	Each	37.0
Minor operating lights		
Single light, fixed	Each	11.0
Single light, sliding type	Each	14.0
Major operating lights		
Single light, remote control	Each	15.0
Single light, electric operated	Each	22.0
Single light, sliding type	Each	23.0
Double light, remote control	Each	23.0
Double light, electric operated	Each	26.0
Double light, sliding type	Each	27.0
Multiple light, remote control	Each	34.0

Time includes handling equipment into place, installation in a prepared location, electrical and plumbing hook-up only, and cleanup.

Labor Installing Therapy Whirlpools

Work Element	Unit	Man-Hours Per Unit
25 gallon arm unit	Each	18.0
95 gallon hip & leg unit	Each	27.0
30 gallon foot & ankle unit	Each	19.0
120 gallon trainers-aid bath	Each	18.0
Full body immersion unit	Each	66.0
2000 gallon whirlpool tank	Each	80.0
Add for overhead carrier	Each	24.0

Time includes handling equipment into place, installation in a prepared location, electrical and plumbing hook-up only, cleanup, testing, and repairs as needed.

Labor Installing Patient Care Equipment

Work Element	Unit	Man-Hours Per Unit
Stainless steel medical preparation cabinets		
48" wide unit	Each	26.0
60" wide unit	Each	31.0
72" wide unit	Each	39.0
Nourishment and ice stations, 72" wide	Each	65
Ice stations, 36" wide unit		
Floor mounted	Each	12.0
Wall mounted	Each	20.0
Unitized janitor stations, 64" x 34" x 80"		
Without doors	Each	20.0
With doors	Each	21.0
Lavatory stations, recessed or wall mounted, with cabinet	Each	13.0

Time includes handling equipment into place, installation in a prepared location, electrical and plumbing hook-up only, cleanup and repairs as needed.

Labor Installing Dental Equipment

Work Element	Unit	Man-Hours Per Unit
Dental chair, electric	Each	32.0
Dental drill unit with accessories	Each	21.0
Floor mounted x-ray unit	Each	13.0
Wall mounted x-ray unit	Each	17.0
Ceiling mount dental light	Each	8.0
Wall mount dental light	Each	8.0
Floor mount dental light	Each	6.5
Sterilizer unit	Each	1.3
Dental tool & bit cabinet	Each	3.3
Plaster, boilout or casting bench	Each	2.6
Drawer bench	Each	1.0

Time includes handling material into place, installation in a prepared location, electrical and plumbing hook-up only, cleanup and repairs as needed.

Labor Installing Laboratory Equipment

Work Element	Unit	Man-Hours Per Unit
Fume hoods (excluding base cabinet)		
3' wide	Each	32.0
4' wide	Each	34.0
5' wide	Each	36.0
6' wide	Each	39.0
Fume hood with integrated base		
4' general purpose	Each	39.0
6' general purpose	Each	42.0
4' constant volume	Each	39.0
6' constant volume	Each	42.0
Walk in utility fume hood		
6' wide unit	Each	56.0
8' wide unit	Each	70.0
Sinks		
Standard, 24" x 18" x 8"	Each	6.0
Cuspidor type	Each	4.5
Safety shower unit	Each	6.0
Eye and face wash unit	Each	24.0
Refrigerator-freezer, undercounter	Each	2.2
Base cabinet & drawer units, L.F. of face		
36" high	10 L.F.	9.3
30" high	10 L.F.	8.5
Laboratory tables with open shelving below	10 L.F.	3.9
Wall storage cabinets, L.F. of face		
30" high	L.F.	.9
48" high	L.F.	1.0
84" high, swing door	L.F.	1.2
84" high, sliding door	L.F.	1.0
84" high, glass sliding door	L.F.	1.3
Top with backsplash, 30" deep	10 L.F.	4.0
Island tops, oversize splash, 10 S.F. of top and splash	10 S.F.	2.6
Splashbacks only, 6" to 12" high	10 L.F.	1.6
Reagent shelving, 4" to 8" deep	10 L.F.	1.9

Includes handling cases into place, mounting, and leveling in prepared positions, but no finishing or trim, plumbing or electrical runs.

Labor Installing Prison Equipment

Work Element	Unit	Man-Hours Per Unit
Steel plate wall lining	10 S.F.	3.0
Steel plate ceiling lining	10 S.F.	7.0
Sliding panel doors	10 S.F.	9.4
Sliding bar doors	10 S.F.	12.0
Hinged panel doors	10 S.F.	9.0
Hinged bar doors	10 S.F.	10.7
Bar walls, normal security	10 S.F.	3.1
Remote control door operator	Each	15.0
Wall hung single bunk	Each	6.8
Wall hung double bunk	Each	9.0
Lavatory, cell type	Each	8.7
Water closet, cell type	Each	13.0
Lavatory water closet	Each	17.0

Time includes handling into place, installation in a prepared location, electrical and plumbing hook-up only, cleanup and repairs as needed.

Labor Installing Parking Equipment

Work Element	Unit	Man-Hours Per Unit
Wood bumpers doweled in place	10 L.F.	1.0
Cast concrete bumpers doweled in place	10 L.F.	1.2
2" x 12" wood bumpers spring mounted on wall	10 L.F.	2.6
Automatic gate, 8' arm		
One way	Each	18.0
Two way	Each	20.0
Ticket dispenser	Each	15.0

Time includes handling equipment into place, installation in a prepared opening, electrical hook-up only, cleanup and repairs as needed.

15
mechanical section

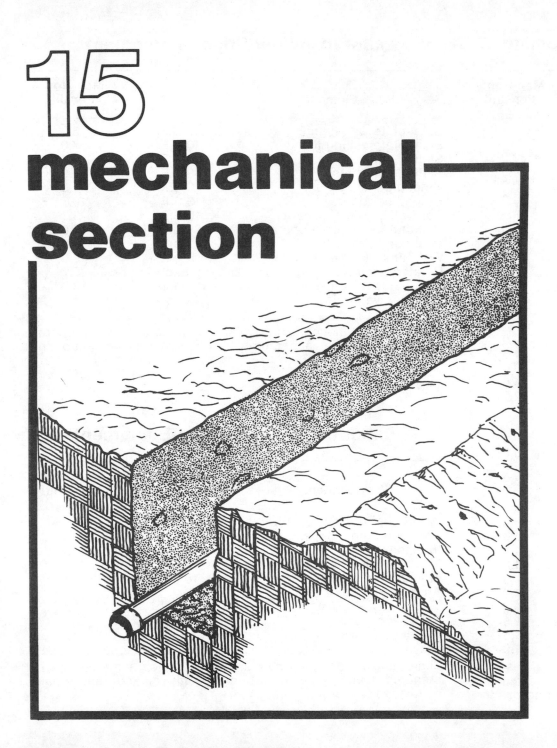

section contents

Plumbing Materials

Plumbing materials include pipe, tubing, fittings, valves, faucets, oakum, lead, solder, flashings and other items used in the construction of a plumbing system. **Roughing in** includes the plumbing work required preliminary to the setting of the fixtures. Plumbing fixtures, such as water closets and tanks, urinals, lavatories, showers, sinks and laundry tubs, drinking fountains and their connecting piping for fittings complete the plumbing system.

Most materials, such as pipe, fittings, and valves, used in plumbing systems are identified by markings indicating the type and size of the item. The metal of which the material is made can usually be determined by the color of the item or by markings on the item.

Cast-iron soil pipe used for sanitary drainage pipe and stacks in waste-disposal systems is available in two weights: service-weight and extra-heavy. The selection of service-weight or extra-heavy pipe is determined by the specification. Where sewers are to be laid under a roadway, subjecting the pipe to vibration or settling, extra-heavy pipe is used. For household drains or sewers, service-weight pipe is usually satisfactory. Extra-heavy is used where liquids are corrosive. The pipe is usually supplied in sizes varying from 2 inches to 8 inches in diameter.

Galvanized steel pipe may be used for a wide range of pressures and temperatures and is produced in a variety of wall thicknesses and different weights. It is classified by the terms "standard weight," "extra strong," and "double extra strong."

Copper tubing has many advantages over galvanized iron pipe. It is easier to transport and install and is more resistant to corrosion. Copper tubing is classified by its wall thickness and also by its hardness. It is available in three wall thicknesses (types K, L, and M). Types K and L are available in either hard (cold drawn) or soft (annealed) tubes; type M is available only in hard temper. Type K has the thickest wall and is the most durable type of tubing. Type L has a slightly thinner wall, and type M has the thinnest wall of the three types. Types K and L are used in interior water systems and type M is used in drainage lines. Hard copper tubing cannot be bent without kinking. When a change in direction is required, a fitting such as an ell must be used. Soft copper tubing may be bent readily.

Plastic pipe is extremely lightweight, can be cut with a saw or knife, and will not rust, rot, or corrode. Plastic pipe comes in two types and three forms: rigid, semirigid, and flexible. Many codes now permit PVC in most applications.

Insulation is used to prevent the loss of heat from hot-water pipes, to prevent freezing in hot-water and cold-water pipes, and to prevent the condensation of moisture on cold-water pipes. Insulating materials include asbestos, asbestos air cell, magnesia, and mineral wool.

Cast-Iron Fittings

Pipe and tubing fittings are the metal links that connect sections of pipe with or without a change in direction. Some fittings are used to connect different sizes of pipe and others to divert or divide flow, or to combine flow. Fittings are usually made of the same material.

A **tee** is used to make a right-angle line from a horizontal or vertical length of pipe. A reducing tee is a straight piece of pipe which has a side takeoff whose diameter is smaller than the diameter of the straight-through section.

A **straight tee** or tee-branch is a straight piece of pipe with a perpendicular takeoff. It is intended for venting and cleanout purpose only, and the branch opening is not to be used as a waste inlet.

A **test tee** has a large bulbous midsection plugged by a screw-in cast-iron plug. It is used in stack and waste installations where the vertical stack joins the horizontal sanitary sewer. This type of tee permits a test plug to be inserted into the system.

A **tapped tee** is a straight piece of pipe with a female thread tapped into one or two side inlets. It is generally used in a venting system as a main vent tee.

A **sanitary tee** is an extra-short-pattern 90-degree Y-branch. It is most commonly used in a main stack to allow the takeoff of a soil-pipe branch.

A **Y-branch** is used to connect two pipes end to end and a third pipe at an angle. The following types of Y-branches are most commonly used: The single **Y-branch** has one straight-through section and a side takeoff extending upward. In this type of Y-branch, all bells and spigots are for the same size pipe. Single Y-branches are used in sanitary sewer systems where a branch feeds into a main, and the incoming branch must feed into the main as nearly as possible in a line parallel to the main flow. The **reducing single Y-branch** is similar to the single type, except that the side takeoff is of a smaller size than the straight-through portion.

The **double Y-branch** is similar to the straight type, except that there is a branch takeoff on both sides of the fitting instead of on one side only. Its most important use is as an individual vent. The **inverted Y-branch** has a straight-through section and a single or double side takeoff extending downward. The **upright Y-branch** has a straight-through section and one or two side takeoffs extending upward and parallel to the straight-through section and extending beyond the straight-through section. The H-branch extends only the same length as the straight-through section.

The **combination Y-branch** and 1/8 bend has a straight-through section and one or two side takeoffs which leave the straight-through section at a 45-degree angle and then bend until they are perpendicular to the section.

A **cross fitting** has four takeoffs on the same plane and spaced 90 degrees apart. It is used to connect intersecting pipe lines. There are two types of cast-iron soil pipe crosses: regular and sanitary. Both types have the two side takeoffs extending outward at an angle of 90 degrees. The sanitary cross has a slight curve in the side takeoffs.

Bends are used to produce changes in direction in piping. Their angle is not designated in degrees, but by fractions of 360 degrees; for example, a 90-degree turn is designated as a quarter bend. Cast-iron soil pipe bends include the sixteenth, eighth, quarter, long-sweep, reducing-sweep, and closet bends. A sixteenth bend changes the direction of a soil pipeline 22½ degrees; an eighth bend changes the direction 45 degrees; and a quarter bend changes the direction 90 degrees.

A **long-sweep bend** changes the direction of a soil pipeline 90 degrees but more gradually than a quarter bend.

A **reducing-sweep bend** gradually changes the direction of the pipe 90 degrees and in the sweep portion reduces one size. For example, a 4 by 3 reducing-sweep bend would have a 4-inch spigot fitting on one end, reducing in 90 degrees to a 3-inch bell on the other end.

A **closet bend** is a specialized bend inserted into a cast-iron soil-pipe branch so that a water closet may be placed on it. It may be an untapped closet bend or it may have either one or two side tappings for additional waste or vent use. The spigot end of the closet bend, which is calked into the soil pipeline, is usually longer than the end that takes the flange to which the water closet is bolted.

Iron and Steel Fittings

Iron and steel pipe fittings are usually of two general types, recessed fittings and pressure fittings. The pressure fitting is a standard fitting used on water pipe. The recessed fitting, sometimes called a cast-iron drainage fitting, is generally required on all drainage lines. The recess creates a smooth joint, lessening the chance of grease or foreign material remaining in the joint and causing a stoppage in the line. The only pipe fittings that are made of wrought iron and steel are nipples and couplings. Other fittings, such as tees, crosses, elbows, unions, plugs, caps, and brushings, are made of cast iron or malleable iron.

There are two types of iron and steel pipe tees: straight tees and reducing tees. The **straight tee** is a fitting with a straight-through portion and a 90-degree takeoff on one side. In the straight tee, all of the openings are of the same size. the **reducing tee** is similar to the straight tee, except that one of the openings is of a different size than the others.

Iron and steel **crosses** are similar to regular cast-iron pipe crosses except that they have female threads at all four branch points.

There are four types of iron and steel pipe **elbows** in general use. These are the 90-degree, 45-degree, street, and reducing elbows. The 90- and 45-degree elbows are used to change the direction of a pipeline 90 degrees and 45 degrees, respectively. The street elbow is a 90- or 45-degree elbow which has one female and one male thread instead of two female threads. It is used to change direction of a pipeline in a close space where it would be impossible or impracticable to use an elbow and nipple and for a close bend out of a female threaded fitting. The reducing elbow is the same as a regular 90-degree elbow, except that one opening is smaller than the other.

Unions are pipe fittings which are used to connect pipe. They are also used in screwed pipelines for opening or dismantling lines. There are two types of iron and steel pipe unions: ground-joint unions and flange unions. A ground-joint union is made in three separate pieces and is used for joining two pipes. It consists of two machined pieces with female pipe threads, which are screwed on the pipes to be united, and a threaded collar which holds the two pieces of the union together. The two machined pieces of the union have a ground-ball, fitted joint to make a watertight or gastight joint.

The **flange union** is made in two parts and is used for joining two pipes. Each part has a female pipe thread and screws on one of the two pieces of pipe to be united. The two flanges are then pulled together by nuts and bolts. There may be either a cork or composition gasket between the two flanges to make a watertight or gastight union. It also may have a thin ground joint similar to that for a ground-joint union.

Couplings connect two pipes end to end where they need not be disconnected later. There are three types of iron and steel pipe couplings: straight, reducing, and eccentric reducing. A **straight coupling** is a short fitting with a female thread on either end. It is used for joining two lengths of pipe in a straight run which does not require fittings. A **reducing coupling** is used to join two pipes of different sizes. An **eccentric reducing coupling** has two female threads of different size with different centers so that, when joined, the two pieces of pipe will have one inside point in line with each other.

A **nipple** is a short length of pipe with a male thread on each end, and is used to make an extension from a fitting or to add a short length to a pipe.

Pipe plugs are iron or steel pipe fittings provided with male threads. They are screwed into other fittings to close one opening of the fittings. Pipe plugs have various type heads, such as the square head, which is most widely used; slotted head; and hexagon socket head. The slotted-head pipe plug is used mostly in close spaces where it would be impossible to turn a square-head pipe plug with a wrench. The hexagon socket-head pipe plug is used to close openings in boilers or other places where a square-head pipe plug would protrude and interfere with the installation of the insulating jacket or other parts.

A **pipe cap** is an iron and steel pipe fitting which is used similarly to a pipe plug, except that the pipe cap is used on the male thread of a piece of pipe or a nipple.

A **pipe bushing** is threaded internally and externally to make a connection from a pipe or fitting of one size to one of another size and where a reducing coupling cannot be used. Pipe bushings are provided with a hexagon head so that a machine wrench may be used for tightening them without injury to the threads.

A locknut is a hexagonal-head nut which is used for slip joints, to lock fittings subject to vibration, and to tighten nipples.

Copper Fittings

Copper-tubing fittings are of two general types, the solder joint and the flared tube, or compression joint. Each type is available in the same patterns as iron and steel pipe fittings, such as tees, elbows and couplings.

Solder-joint copper fittings are wrought copper or are cast in copper and have a machine-ground female joint into which the male end of the copper tubing is inserted. This ground joint makes a close and tight fit with the tubing. When the joint is heated and treated with a fluxing material, hot, liquid solder is drawn into the joint by capillary attraction. Solder-joint copper fittings are used more often than flare-tube fittings since they are easier to install and cost less.

Flared-tube copper fittings are often referred to as compression fittings. When the fitting is used, the nut of the fitting is slipped over the end of one pipe and the body of the fitting is slipped over the other pipe. Both ends of the pipe are flared-out with a special tool and the nut is screwed down on the body. The most important use of flare-tube fittings is in air piping. Copper pipe is joined by solder-joint pipe couplings but a flared-tube union fitting may be used when a union is required.

With adapter fittings it is possible to connect copper tubing to steel or wrought-iron pipe having threaded ends. These fittings are usually made of cast brass, but sometimes of wrought copper. Where such a connection is made, a non-conductive nipple or separator should be installed.

Valves

A valve consists of a body containing an opening and a means of closing the opening with a disc or plug which can be tightly pressed against a seating surface in the opening. Valves are made of brass, bronze, malleable iron, cast iron, cast steel, and forged steel. They are sized by the nominal inside diameter of the pipe to which they are connected.

The most widely used types of pipeline valves are the hand-operated gate, globe, and angle valves and the automatic check valves. Certain valves, such as plug valves and diaphragm valves, are special-purpose valves used mostly in particular types of piping systems. Fixture valves are used with water closets and urinals to supply water for flushing the fixtures, and with showers to control the supply of water.

Gate valves contain a sliding disc which moves vertically, perpendicular to the path of the fluid flow, and seats between and against two opposed seat races to shut off the flow. (See Figure 15-1) A threaded stem and handwheel are used to lift or lower the disc to open or close the valve. The volume of flow through the valve, however, is not in direct relation to the number of turns of the handwheel. Gate valves have either a single solid wedge-shaped disc or a double disc with a wedge-shaped spreader to make a tight closure. Gate valves may be provided with either rising stems or nonrising stems.

Gate valves are used where the valve will be used fairly seldom and where the usual position is either fully closed or open. When fully opened a gate valve permits the fluid to move through the valve in a straight line with a minimum loss of pressure at the valve. This type of valve is not suitable where the valve is kept only partly open. The flow against a partly opened valve disc may cause vibration and damage the discs and seating surfaces. Gate valves can be installed with either face on the inlet side.

Globe valves have a horizontal interior partition which shuts off the inlet from the outlet, except through a circular opening in the partition. (See Figure 15-2.) The lower end of the valve stem contains a replaceable fiber or metal disc shaped and fitted to close the circular hole in the horizontal partition. The valve is closed by turning the handwheel clockwise until the disc presses firmly on the circular opening, which is the valve seat, and closes the opening in the seat. The volume of flow through globe valves is roughly proportionate to the number of turns of the handwheel. Globe valves may be provided with either rising stems or nonrising stems. The surfaces of the seat and disc may be either flat or beveled, depending upon the type of disc.

Globe valves are used in applications requiring frequent operation and where the valve may be used to throttle and regulate flows. The design of this type of valve increases resistance to flow at the valve. The fluid flow through globe valves can be closely regulated. The discs and seats which are liable to be worn or damaged in throttling service can be reground or replaced. Globe valves can be fitted with fiber discs that are suitable for almost any type of service except steam. For steam lines a metal disc globe valve should be used. Globe valves should always be installed with the seat against the direction of flow.

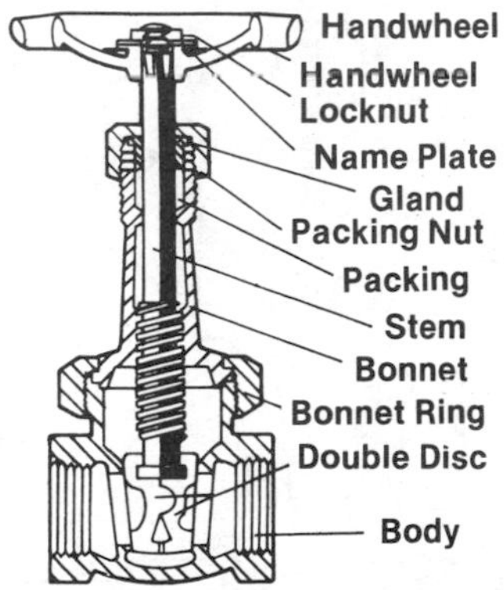

Sectional View of a Gate Valve
Figure 15-1

Handwheel
Handwheel Locknut
Name Plate
Gland
Packing Nut
Packing
Stem
Bonnet
Bonnet Ring
Double Disc
Body

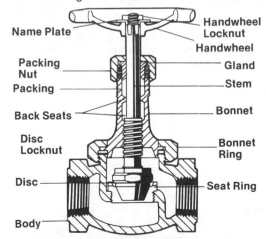

Sectional View of a Globe Valve
Figure 15-2

Name Plate
Handwheel Locknut
Handwheel
Packing Nut
Gland
Packing
Stem
Back Seats
Bonnet
Disc Locknut
Bonnet Ring
Disc
Seat Ring
Body

Angle valves are similar to globe valves except that the valve outlet is at an angle of 90 degrees to the inlet. (See Figure 15-3.)

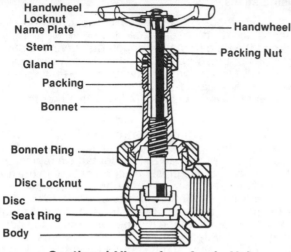

Sectional View of an Angle Valve
Figure 15-3

Check valves are used when it is necessary to control the flow in one direction only. Fluid flow in the proper direction keeps the valves open. Reversal of flow closes the valve automatically. For installation purposes, most check valves are marked to indicate the inlet opening or direction of flow. There are two basic types of check valves: swing check valves and lift check valves.

Swing check valves contain a hinged disc which seats against a machined seat in the tilted bridge wall opening of the valve body. (See Figure 15-4.) The disc swings freely on its hinge pin in an arc from a fully closed position to one parallel with the flow. The fluid or gas in the pipeline enters below the disc. Line pressure overcomes the weight of the disc and raises it, permitting a continuous flow. If the flow is reversed or back-pressure builds up, this pressure is exerted against the disc, forcing it to seat and stop the flow.

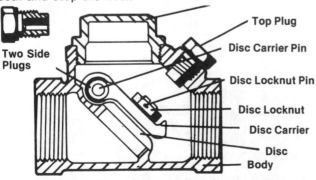

Sectional View of a Swing Check Valve
Figure 15-4

Swing check valves are used for low- and medium-pressure services. When fully opened, fluid moves through a swing check valve in a straight line, similar to the flow through a gate valve. When installed in a piping system, the valves must be positioned so that the disc is held to the seat and will close by gravity.

Lift check valves contain a disc which seats on a horizontal bridge wall in the valve body. (See Figure 15-5.) The disc is raised from its seat by the pressure of the fluid flow and moves vertically to open. Lift check valves are used for water, gas, air, steam, and general vapor services. When opened, fluid moves through a lift check valve similar to the flow in a globe valve. Lift check valves seat more positively and tightly than swing check valves.

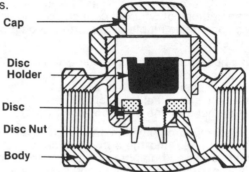

Sectional View of a Lift Check Valve
Figure 15-5

Plug valves have a circular-tapered, ground plug fitting tapered hole or seat. (See Figure 15-6.) An opening through the plug permits the unobstructed passage of fluid through the valve when the opening is aligned with the pipeline. The advantage of plug valves is that they may be completely and quickly opened by a one-quarter turn of the handle and do not have soft packing which tends to wear as valves are used. Some plug valves are lubricated.

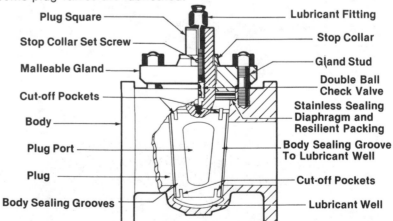

Sectional View of a Plug Valve
Figure 15-6

Ball valves, like plug valves, are quick opening, needing only a quarter turn from full open to full close. Ball valves are nonsticking and provide tight closure. They also cause a small pressure drop because of their smooth full-opening port. Major components of the ball valve are the body, spherical plug, and seats. (See Figure 15-7.)

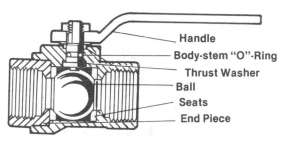

Ball Valve
Figure 15-7

Diaphragm valves contain a rubber diaphragm, that seals the bonnet from the body, and a circular, flatface disc. (See Figure 15-8.) The disc is joined to the diaphragm with a leakproof connection. Stem packing is not used with this type of valve since the diaphragm supplies the necessary sealing around the stem. The diaphragm prevents pipeline fluids or gases from contacting the operating parts of the valve.

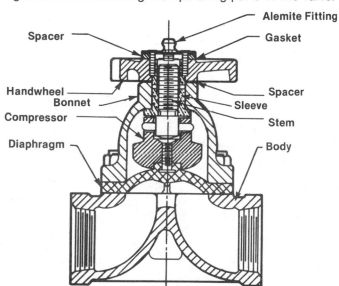

Cutaway View of a Diaphragm Valve
Figure 15-8

Butterfly valves are made for all types of pipe fittings: screwed, flanged, welded, and socket. They are made very thin for use between two flanges for close space. (See Figure 15-9.)

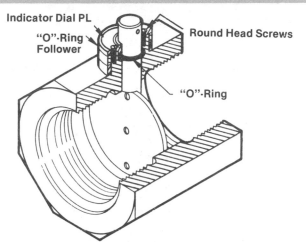

Butterfly Valve
Figure 15-9

Labor Installing Schedule 40 Galvanized Pipe

Work Element	Unit	Man-Hours Per Unit
1/4″	100 L.F.	8.9
3/8″	100 L.F.	9.7
1/2″	100 L.F.	11.9
3/4″	100 L.F.	13.7
1″	100 L.F.	16.1
1-1/4″	100 L.F.	17.7
1-1/2″	100 L.F.	19.6
2″	100 L.F.	22.2
2-1/2″	100 L.F.	23.5
3″	100 L.F.	29.4
4″	100 L.F.	44.3
5″	100 L.F.	59.0
6″	100 L.F.	70.9
8″	100 L.F.	118.3

Time includes drilling and placing of hangers, sleeves and inserts, move on, setup and all necessary cleanup.
Suggested Crew: 1 plumber and 1 laborer

Labor Installing Type K, L or M Copper Pipe

Work Element	Unit	Man-Hours Per Unit
3/8″	100 L.F.	8.5
1/2′	100 L.F.	10.3
3/4″	100 L.F.	12.5
1″	100 L.F.	13.8
1-1/4″	100 L.F.	15.3
1-1/2″	100 L.F.	17.3
2″	100 L.F.	19.3
2-1/2″	100 L.F.	19.7
3″	100 L.F.	26.3
4″	100 L.F.	38.0
5″	100 L.F.	52.9
6″	100 L.F.	59.2

Time assumes pipe is installed in building walls on supports with an average of one fitting each 8 feet. If coiled pipe is used, deduct 25%. Time includes moving on and off site, installing supports, pipe and fittings, repair and cleanup as needed.
Suggested Crew: 1 plumber and 1 laborer

Material for Soldered Joints, Type K, L, and M Copper Tubing

Tubing Size (Nominal) Inches	Tubing Outside Diameter Inches	Pounds of Solder Required per 100 Joints	
		Hard Solder Pounds	Soft Solder Pounds
3/8	1/2	0.375	0.5
1/2	5/8	0.56	0.75
3/4	7/8	0.75	1.0
1	1-1/8	1.05	1.4
1-1/4	1-3/8	1.27	1.7
1-1/2	1-5/8	1.35	1.8
2	2-1/8	1.80	2.4
2-1/2	2-5/8	2.40	3.2
3	3-1/8	2.90	3.9
3-1/2	3-5/8	3.37	4.5
4	4-1/8	4.10	5.5

Labor Installing Schedule 40 Black Steel Pipe

Work Element	Unit	Man-Hours Per Unit
1/4"	100 L.F.	8.9
3/8"	100 L.F.	9.7
1/2"	100 L.F.	11.9
3/4"	100 L.F.	13.7
1"	100 L.F.	16.1
1-1/4"	100 L.F.	17.7
1-1/2"	100 L.F.	19.6
2"	100 L.F.	22.2
2-1/2"	100 L.F.	23.5
3"	100 L.F.	29.4
4"	100 L.F.	44.3
5"	100 L.F.	59.0
6"	100 L.F.	70.9
8"	100 L.F.	118.1
10"	100 L.F.	177.3
12"	100 L.F.	196.6
14"	100 L.F.	221.7
16"	100 L.F.	295.3
18"	100 L.F.	354.1
20"	100 L.F.	392.7
24"	100 L.F.	403.1

Time includes drilling and placing of hangers sleeves, inserts, move on and off site and cleanup as necessary. 10" diameter and over 15 should be HVAC applications only.
Suggested Crew: 1 plumber and 1 laborer. Add 20% for welded joints.

Labor Installing Brass Pipe

Work Element	Unit	Man-Hours Per Unit
Threaded brass pipe		
1/8"	100 L.F.	8.9
1/4"	100 L.F.	8.9
3/8"	100 L.F.	9.8
1/2"	100 L.F.	11.7
3/4"	100 L.F.	13.7
1"	100 L.F.	16.2
1-1/4"	100 L.F.	17.7
1-1/2"	100 L.F.	19.7
2-1/2"	100 L.F.	22.2
3"	100 L.F.	29.5
4"	100 L.F.	44.2
5"	100 L.F.	59.0
6"	100 L.F.	70.0
Threadless brass pipe (silver soldered)		
1/4"	100 L.F.	8.0
3/8"	100 L.F.	8.4
1/2"	100 L.F.	10.2
3/4"	100 L.F.	11.9
1"	100 L.F.	14.1
1-1/4"	100 L.F.	15.4
1-1/2"	100 L.F.	17.1
2"	100 L.F.	19.3
3"	100 L.F.	20.3
4"	100 L.F.	25.6
5"	100 L.F.	51.3
6"	100 L.F.	61.6

Time includes move on and off site. Labor hours are based upon average of one fitting every ten feet. All cleanup and repair work included. Time includes installing supports.
Suggested Crew: 1 plumber and 1 laborer
Add 10% to labor for extra heavy brass pipe and fittings.

Labor Installing Iron Body Gate Valves

Work Element	Unit	Man-Hours Per Unit
Iron body, bronze mounted, threaded		
1½", 125 lb. rating	Each	.66
2", 125 lb. rating	Each	.82
2½", 125 lb. rating	Each	1.0
3", 125 lb. rating	Each	2.1
4", 125 lb. rating	Each	6.6
Iron body, bronze mounted, flanged		
1½", 150 lb. rating	Each	1.3
2", 150 lb. rating	Each	1.6
2½", 150 lb. rating	Each	2.1
3", 150 lb. rating	Each	3.3
4", 150 lb. rating	Each	8.9
5", 150 lb. rating	Each	13.2
6", 150 lb. rating	Each	17.5
8", 150 lb. rating	Each	20.0
10", 150 lb. rating	Each	24.3
12", 150 lb. rating	Each	26.8

Time includes move on and off site, unloading, stacking, clean-up and repairs as needed.
Suggested Crew: 1 plumber and 1 laborer

Labor Installing O.S. & Y. Gate Valves

Work Element	Unit	Man-Hours Per Unit
Threaded valves		
1½", 125 lb. rating	Each	.69
2", 125 lb. rating	Each	.86
2½", 125 lb. rating	Each	1.1
3", 125 lb. rating	Each	2.2
4", 125 lb. rating	Each	6.5
Flanged valves		
1½", 125 lb. rating	Each	1.3
2", 125 lb. rating	Each	1.6
2½", 125 lb. rating	Each	2.2
3", 125 lb. rating	Each	3.3
4", 125 lb. rating	Each	8.8
5", 125 lb. rating	Each	13.3
6", 125 lb. rating	Each	17.8
8", 125 lb. rating	Each	19.8
10", 125 lb. rating	Each	24.2
12", 125 lb. rating	Each	26.4

Use these figures for installing O.S. & Y. threaded and flanged iron body gate valves.
Time includes move on and off site, unloading, stacking, clean-up and repairs as necessary.
Suggested Crew: 1 plumber and 1 laborer

Labor Installing Angle or Globe Valves

Work Element	Unit	Man-Hours Per Unit
Bronze angle or globe valves, threaded, brazed or soldered, 125 lb. rating		
⅛" to ½"	Each	.28
¾" or 1"	Each	.33
1¼" or 1½"	Each	.55
2"	Each	.65
2½"	Each	.84
3"	Each	1.00
Iron body threaded angle or globe valves, 125 lb. rating		
2"	Each	.86
2½"	Each	1.10
3"	Each	2.20
4"	Each	6.50
5"	Each	8.80
6"	Each	13.0
8"	Each	18.00
10"	Each	22.00
12"	Each	25.00
Iron body flanged angle or globe valves, 125 lb. rating		
2"	Each	1.66
2½"	Each	2.20
3"	Each	3.20
4"	Each	8.80
5"	Each	13.00
6"	Each	18.00
8"	Each	20.00
10"	Each	24.00
12"	Each	27.00

Time includes move on and off site, handling materials into place, cleanup and repairs as needed.
Suggested Crew: 1 plumber and 1 laborer

Labor Installing Check Valves

Work Element	Unit	Man-Hours Per Unit
Swing type valves, bronze, threaded brazed or soldered installation		
1/4", 125 lb. rating	Each	.27
3/8", 125 lb. rating	Each	.27
1/2", 125 lb. rating	Each	.28
3/4", 125 lb. rating	Each	.33
1", 125 lb. rating	Each	.34
1-1/4", 125 lb. rating	Each	.55
1-1/2", 125 lb. rating	Each	.55
2", 125 lb. rating	Each	.67
2-1/2", 125 lb. rating	Each	.86
3", 125 lb. rating	Each	.82
Iron body, threaded or flanged check valves		
2", 125 lb. rating	Each	1.6
2½", 125 lb. rating	Each	2.2
3", 125 lb. rating	Each	3.3
4", 125 lb. rating	Each	8.7
5", 125 lb. rating	Each	13.6
6", 125 lb. rating	Each	18.0
8", 125 lb. rating	Each	19.8
10", 125 lb. rating	Each	24.5
12", 125 lb. rating	Each	26.4

If a welder is required, add eighteen percent to the man-hours per unit. Time includes move on and off site, unloading, stacking, clean-up and repair as necessary.
Suggested Crew: 1 plumber and 1 laborer

Labor Installing Plug Valves

Work Element	Unit	Man-Hours Per Unit
Threaded, lubricated type, wrench operated plug valves		
½", 150 lb. rating	Each	.55
¾", 150 lb. rating	Each	.55
1", 150 lb. rating	Each	.66
1¼", 150 lb. rating	Each	.84
1½", 150 lb. rating	Each	.85
2", 150 lb. rating	Each	1.1
2½", 150 lb. rating	Each	2.2
3", 150 lb. rating	Each	3.2
4", 150 lb. rating	Each	6.7
Flanged, lubricated type, wrench operated plug valves		
1", 150 lb. rating	Each	.74
1¼", 150 lb. rating	Each	.86
1½", 150 lb. rating	Each	1.0
2", 150 lb. rating	Each	2.1
2½", 150 lb. rating	Each	3.3
3", 150 lb. rating	Each	4.4
4", 150 lb. rating	Each	8.9
Flanged, lubricated type, worm gear operated		
6", 150 lb. rating	Each	17.7
8", 150 lb. rating	Each	19.7
10", 150 lb. rating	Each	23.2
12", 150 lb. rating	Each	26.3
14", 150 lb. rating	Each	34.4
16", 150 lb. rating	Each	40.9
18", 150 lb. rating	Each	45.5
20", 150 lb. rating	Each	50.2
24", 150 lb. rating	Each	57.6

Time includes move on and off site, unloading, stacking, cleanup and repair as needed. Time does not include testing.
Suggested Crew: 1 plumber, 1 laborer; for valves over 12 inches, 2 laborers, 1 plumber

Labor Installing Pipe Specialties

Work Element	Unit	Man-Hours Per Unit
Vacuum breakers		
½"	Each	.55
¾"	Each	.55
1"	Each	.83
1¼"	Each	.96
1½"	Each	1.1
Expansion joints		
½" copper bellows	Each	.82
¾" copper bellows	Each	1.1
1" copper bellows	Each	1.6
1¼" copper bellows	Each	2.2
1½" copper bellows	Each	2.2
2" copper bellows	Each	3.3
2½" copper bellows	Each	4.4
3" copper bellows	Each	6.5
4" copper bellows	Each	8.9
5" stainless steel bellows	Each	13.2
6" stainless steel bellows	Each	17.9
8" stainless steel bellows	Each	19.6
10" stainless steel bellows	Each	24.1
12" stainless steel bellows	Each	26.2
Shock absorbers		
¾" - 3" to 4" long	Each	1.1
1" - 3" to 5" long	Each	1.6
1" - 3" to 6" long	Each	1.6
1" - 3" to 7" long	Each	1.6
1" - 3" to 8" long	Each	1.7
1" - 3" to 9" long	Each	1.7
1" - 4½" to 10" long	Each	3.4

Time includes move on and off site, unloading, stacking, clean-up, and repair as required. Time does not include testing.
Suggested Crew: 1 plumber and 1 laborer

Labor Installing Special Valves

Work Element	Unit	Man-Hours Per Unit
Ball valves, 150 lb. rated		
½"	Each	.56
¾"	Each	.56
1"	Each	1.1
1½"	Each	1.6
2"	Each	2.2
2½"	Each	2.7
3"	Each	4.4
4"	Each	8.7
Butterfly valves		
2"	Each	2.2
2½"	Each	3.3
3"	Each	5.5
4"	Each	11.0
5"	Each	13.
6"	Each	17.8
8"	Each	19.8
10"	Each	26.7
12"	Each	30.8
14"	Each	37.1
16"	Each	40.0
18"	Each	44.8
20"	Each	53.1
Quick opening valves		
1½", thread	Each	.55
¾", thread	Each	.82
1" thread	Each	1.1
1¼", thread	Each	2.2
1½", thread	Each	2.8
2", thread	Each	3.0
2½", flange	Each	3.3
3", flange	Each	4.3
4", flange	Each	8.7
5", flange	Each	13.2
6", flange	Each	17.6

Labor Installing Special Valves (continued)

Work Element	Unit	Man-Hours Per Unit
Diaphragm operated valves, threaded		
½"	Each	.84
¾"	Each	.93
1"	Each	1.1
1¼"	Each	2.2
1½"	Each	2.2
2"	Each	2.2
3"	Each	4.5
4"	Each	8.8
5"	Each	13.2
6"	Each	17.6
8"	Each	19.6
10"	Each	24.6
12"	Each	26.6
14"	Each	33.8
16"	Each	40.3
Solenoid valves		
½"	Each	1.0
¾"	Each	1.1
1"	Each	1.6
1¼"	Each	2.1
1½"	Each	2.2
2"	Each	3.3
Relief valves, ASME rated		
½"	Each	.83
¾"	Each	.84
1"	Each	1.1
1¼"	Each	1.6
1½"	Each	1.8
2"	Each	2.8
Pressure reducing valves, flanged, 125-200 lb. rated		
2"	Each	3.2
2½"	Each	4.5
3"	Each	5.3
4"	Each	8.9

Labor Installing Special Valves (continued)

Work Element	Unit	Man-Hours Per Unit
Actuating control valves		
½"	Each	1.1
¾"	Each	1.1
1"	Each	1.6
1¼"	Each	2.2
1½"	Each	2.2
2"	Each	2.7
2½"	Each	3.2
3"	Each	4.3
4"	Each	9.0
5"	Each	13.4
6"	Each	17.9
8"	Each	19.9
Multiport valves, wrench operated		
3"	Each	4.4
4"	Each	8.9
5"	Each	17.4
6"	Each	19.7
8"	Each	22.7
10"	Each	25.5
Temperature valves, 90 to 200 degree (f) temp. range		
½"	Each	1.6
¾"	Each	2.1
1"	Each	2.7
1¼"	Each	3.3
1½"	Each	3.8
2"	Each	5.5
2½"	Each	7.6

Time includes move on and off site, unloading, stacking, cleanup and repair as needed. Testing not included.
Suggested Crew: Under normal conditions, 1 plumber and 1 laborer will be required. Valves greater than twelve inches will require 2 laborers and 1 plumber. Under certain conditions, rental equipment operators and additional laborers and plumber will be required.

Labor Installing Drain Valves

Work Element	Unit	Man-Hours Per Unit
Hose gate type (bronze)		
½"	Each	.28
¾"	Each	.33
1"	Each	.33
1¼"	Each	.55
1½"	Each	.62
2"	Each	.79
2½"	Each	1.1
Stop and waste valves		
1/4"	Each	.28
3/8"	Each	.28
1/2"	Each	.28
3/4"	Each	.33
1"	Each	.33
Boiler drain and sill cock type		
1/2"	Each	.28
3/4"	Each	.28
Wall and box hydrants, brass		
3/4"	Each	1.0
1"	Each	1.1

Time includes move-on and off-site, unloading, clean-up and repair as needed.
Suggested Crew: 1 plumber and 1 laborer.

Labor Installing Gas Valves

Work Element	Unit	Man-Hours Per Unit
Iron body bronze mounted, threaded and flanged gas valves, brass tee and lever handle type		
1/4", 125 lb. rating	Each	.28
3/8", 125 lb. rating	Each	.27
1/2", 125 lb. rating	Each	.27
3/4", 125 lb. rating	Each	.33
1" , 125 lb. rating	Each	.33
1-1/4", 125 lb. rating	Each	.56
1-1/2", 125 lb. rating	Each	.55
2" , 125 lb. rating	Each	.66
Iron body, threaded, flat and square head, gas valves		
½", 125 lb. rating	Each	.28
¾", 125 lb. rating	Each	.33
1", 125 lb. rating	Each	.33
1¼", 125 lb. rating	Each	.55
1½", 125 lb. rating	Each	.55
2", 125 lb. rating	Each	.66
Iron body, flanged, square head		
2", 125 lb. rating	Each	.83
2½", 125 lb. rating	Each	1.1
3", 125 lb. rating	Each	2.18
4", 125 lb. rating	Each	6.60
Brass service stops, flat, tee and square head		
1/4", 125 lb. rating	Each	.28
3/8", 125 lb. rating	Each	.29
1/2", 125 lb. rating	Each	.29
3/4", 125 lb. rating	Each	.34
1", 125 lb. rating	Each	.36
1-1/4", 125 lb. rating	Each	.55
1-1/2", 125 lb. rating	Each	.56
2", 125 lb. rating	Each	.66

Time includes move on and off site, unloading, stacking, clean-up and repairs as needed.
Suggested Crew: 1 plumber and 1 laborer

Labor Installing Duplex Water Pumps

Work Element	Unit	Man-Hours Per Unit
5 HP, 1,750 RPM, 150 GPM	Each	67.7
7½ HP, 1,750 RPM, 150 GPM	Each	70.1
10 HP, 1,750 RPM, 150 GPM	Each	83.6
15 HP, 1,750 RPM, 150 GPM	Each	87.2
20 HP, 1,750 RPM, 150 GPM	Each	91.9
25 HP, 1,750 RPM, 150 GPM	Each	98.7

Time includes installation of controls and accessories, move on and off site, unloading, cleanup and repairs as needed. Electric hookup not included.
Suggested Crew: 1 plumber and 2 laborers
Note: Under certain conditions lifting devices will be required and an operator's time will have to be added to the man-hours per unit.

Labor Installing Triplex Water Pumps

Work Element	Unit	Man-Hours Per Unit
5 HP, 1,750 RPM, 150 GPM	Each	91.6
7½ HP, 1,750 RPM, 150 GPM	Each	92.4
10 HP, 1,750 RPM, 150 GPM	Each	93.5
15 HP, 1,750 RPM, 150 GPM	Each	95.0
20 HP, 1,750 RPM, 150 GPM	Each	98.1
25 HP, 1,750 RPM, 150 GPM	Each	107.5

Time includes controls and accessories, move on and off site, unloading, clean-up and repairs as needed. Testing not included. Electric hook-up not included.
Suggested Crew: 1 plumber and 2 laborers.
Note: Under certain conditions, lifting devices will be required and an operator's time will have to be added to the man-hours per unit.

Labor Installing Hydropneumatic Simplex-Pumps With Compressors

Work Element	Unit	Man-Hours Per Unit
5 HP, 1,750 RPM, 1,000 GPM	Each	35.2
7½ HP, 1,750 RPM, 1,000 GPM	Each	44.4
10 HP, 1,750 RPM, 1,000 GPM	Each	52.4

Time includes all controls and accessories, move on and off site, unloading, cleanup and repairs as needed. Testing and electrical hookup are not included.
Suggested Crew: 1 plumber and 2 laborers
Note: Under certain conditions, lifting devices will be required and an operator's time will have to be added to the man-hours per unit.

Labor Installing Hot Water System Circulating Pumps

Work Element	Unit	Man-Hours Per Unit
1/12 HP	Each	4.4
1/6 HP	Each	4.5
1/4 HP	Each	6.1
1/3 HP	Each	6.5
1/2 HP	Each	6.5
3/4 HP	Each	13.3
1 HP	Each	16.4

Use these figures for iron body in-line pumps. Time includes move-on and off site, unloading, clean-up and repair as needed.
Suggested Crew: 1 plumber and 1 laborer.

Labor Installing Hot Water Generators

Work Element	Unit	Man-Hours Per Unit
Gas fired, commercial, cement lined		
500 gal./hr. recovery rate	Each	32.0
1,000 gal./hr. recovery rate	Each	38.6
1,500 gal./hr. recovery rate	Each	55.0
2,000 gal./hr. recovery rate	Each	60.5
2,500 gal./hr. recovery rate	Each	66.3
3,000 gal./hr. recovery rate	Each	73.9
Oil fired, commercial, cement lined		
500 gal./hr. recovery rate	Each	32.0
1,000 gal./hr. recovery rate	Each	40.8
1,500 gal./hr. recovery rate	Each	45.7
2,000 gal./hr. recovery rate	Each	62.5
2,500 gal./hr. recovery rate	Each	69.8
3,000 gal./hr. recovery rate	Each	77.4
Electric heated commercial, cement lined		
500 gal./hr. recovery rate	Each	30.4
1,000 gal./hr. recovery rate	Each	37.6
1,500 gal./hr. recovery rate	Each	53.5
2,000 gal./hr. recovery rate	Each	62.5
2,500 gal./hr. recovery rate	Each	62.8
3,000 gal./hr. recovery rate	Each	73.9
Steam, high temperature hot water heaters, cement lined		
500 gal./hr. recovery rate	Each	29.7
1,000 gal./hr. recovery rate	Each	40.0
1,500 gal./hr. recovery rate	Each	49.6
2,000 gal./hr. recovery rate	Each	61.6
2,500 gal./hr. recovery rate	Each	63.1
3,000 gal./hr. recovery rate	Each	73.6

Time includes move on and off site, unloading, lifting into position, repair and clean-up as needed. Includes typical pipe and electrical connection.
Suggested Crew: 1 plumber and 2 laborers, 1 operator

Labor Installing Water Heaters

Work Element	Unit	Man-Hours Per Unit
Electric heated, residential (glass lined)		
8 gallon	Each	8.8
10 gallon	Each	8.9
12 gallon	Each	9.0
30 gallon	Each	13.3
45 gallon	Each	13.8
60 gallon	Each	17.7
75 gallon	Each	19.8
100 gallon	Each	24.7
120 gallon	Each	26.7
Gas fired, residential (glass lined)		
8 gallon	Each	8.9
10 gallon	Each	8.9
12 gallon	Each	9.5
30 gallon	Each	13.3
45 gallon	Each	13.8
60 gallon	Each	17.7
75 gallon	Each	19.8
100 gallon	Each	26.7
120 gallon	Each	26.9

Time includes move on and off site, unloading, cleanup and repair as needed. Includes vent for gas units, water, gas or electric hookup.
Suggested Crew: 1 plumber, 1 laborer

Labor Installing Sewage Ejectors

Work Element	Unit	Man-Hours Per Unit
Simplex ejector (6 foot shaft)		
2 HP, 50 GPM, 1,750 RPM	Each	17.7
2 HP, 75 GPM, 1,750 RPM	Each	19.8
2 HP, 100 GPM, 1,750 RPM	Each	27.0
3 HP, 125 GPM, 1,750 RPM	Each	32.8
3 HP, 150 GPM, 1,750 RPM	Each	41.0
3 HP, 200 GPM, 1,750 RPM	Each	47.0
5 HP, 250 GPM, 1,750 RPM	Each	54.4
7½ HP, 300 GPM, 1,750 RPM	Each	65.7
10 HP, 350 GPM, 1,750 RPM	Each	70.4
15 HP, 400 GPM, 1,750 RPM	Each	73.9
20 HP, 500 GPM, 1,750 RPM	Each	76.1
Duplex ejectors (6 foot shaft)		
2 HP, 50 GPM, 1,750 RPM	Each	22.1
2 HP, 75 GPM, 1,750 RPM	Each	23.4
2 HP, 100 GPM, 1,750 RPM	Each	34.4
3 HP, 125 GPM, 1,750 RPM	Each	40.6
3 HP, 150 GPM, 1,750 RPM	Each	47.8
3 HP, 200 GPM, 1,750 RPM	Each	50.6
5 HP, 250 GPM, 1,750 RPM	Each	64.4
7½ HP, 300 GPM, 1,750 RPM	Each	74.2
10 HP, 350 GPM, 1,750 RPM	Each	82.8
15 HP, 400 GPM, 1,750 RPM	Each	86.9
20 HP, 500 GPM, 1,750 RPM	Each	91.5

Time includes move-on and off-site, unloading and lift into place, clean-up and repair as needed. Electrical work is not included.
Suggested Crew: 1 plumber and 2 laborers

Labor Installing Water Meters

Work Element	Unit	Man-Hours Per Unit
Water meters, disk type		
¾" diameter in and out	Each	4.4
1" diameter in and out	Each	8.8
1½" diameter in and out	Each	13.3
2" diameter in and out	Each	17.5
Compound meters with cast-iron companion flanges		
2" diameter in and out	Each	17.5
3" diameter in and out	Each	26.3
4" diameter in and out	Each	34.9
6" diameter in and out	Each	42.3
8" diameter in and out	Each	47.4
Detector fire meters with disc bypass (companion flanged)		
3" diameter in and out	Each	27.0
4" diameter in and out	Each	36.0
6" diameter in and out	Each	47.2
8" diameter in and out	Each	56.8
10" diameter in and out	Each	71.6

Time includes move-on and off-site, unloading, stack, clean-up and repairs as required. Time does not include testing or excavation.
Suggested Crew: 1 plumber and 1 laborer

Labor Installing Floor Drains

Work Element	Unit	Man-Hours Per Unit
Flat round top, cast-iron		
3" outlet	Each	2.2
4" outlet	Each	2.2
5" outlet	Each	2.8
Flat square top, cast-iron		
3" outlet	Each	2.2
4" outlet	Each	2.2
5" outlet	Each	2.8
Funnel type, rough brass top		
3" outlet	Each	2.2
4" outlet	Each	2.2
Drain with sediment bucket and cast-iron top		
3" outlet	Each	2.1
4" outlet	Each	2.1
5" outlet	Each	2.6
6" outlet	Each	2.6

Time includes move on and off site, unload, clean-up and repair as needed.
Suggested Crew: 1 plumber

Labor Installing Roof Drainage Systems

Work Element	Unit	Man-Hours Per Unit
Cast-iron roof drains (large domed top)		
3" outlet	Each	2.2
4" outlet	Each	2.2
5" outlet	Each	2.8
6" outlet	Each	2.8
Cast-iron roof drain (with promenade top)		
3" outlet	Each	2.1
4" outlet	Each	2.1
5" outlet	Each	2.6
6" outlet	Each	2.6
Cast-iron scupper drains (with cast-iron grate)		
2" outlet	Each	2.2
3" outlet	Each	2.2
4" outlet	Each	2.2
Cast-iron area drains (flat grated top)		
3" throat	Each	2.2
Area drain (round grated top)		
3" outlet	Each	2.2
4" outlet	Each	2.2
Trench drains (light duty outlet sections with extension)		
2" outlet by 9" long	Each	3.2
3" outlet by 9" long	Each	4.0
4" outlet by 9" long	Each	4.5

Time includes move on and off site, unloading, stacking, clean-up and repair as needed.
Suggested Crew: 1 plumber and 1 laborer.

Plumbing Fixtures and Equipment

A water closet tank requires two valves. One is a float-controlled valve through which water enters the tank; the other is a flush-ball valve through which water is discharged from the tank into the water closet (See Figure 15-10.)

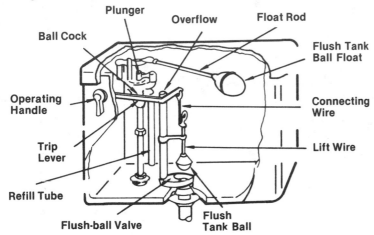

Flush Valves for Tank-type Water Closets
Figure 15-10

The **float valve** controls the flow of water into the flush tank. The quantity of water in the tank, when filled, is determined by the adjustment of the height of the float when the valve is closed. When the water is emptied from the tank, the float drops, opening the float valve which admits water into the tank. A refill tube is provided to allow a small amount of water to enter the water closet bowl through the overflow pipe. This fills the trap in the closet bowl. When the float reaches its original position, the float valve closes and the tank is ready for the next flush.

The **flush-ball valve** releases water from the closet tank into the water closet for flushing. This valve is usually operated by a trip lever which lifts the ball of the valve from the valve seat. The ball floats until the water level recedes to a point where the ball is drawn onto the valve seat by suction. This stops the flow of water from the tank. The ball is held on the valve seat by water pressure above the ball. An overflow pipe is usually provided as an integral part of the valve so the water will not overflow the closet tank.

A **flushometer valve** is a device which discharges a predetermined quantity of water directly from the supply line to fixture for flushing purposes. It is self-closing, and actuated by direct water pressure. It takes water directly from water supply pipes to flush the fixture, thereby eliminating the closet tank and its connections. These valves use less water than closet tanks and can be operated again at intervals of only a few seconds. There are several types of flushometer valves, including the

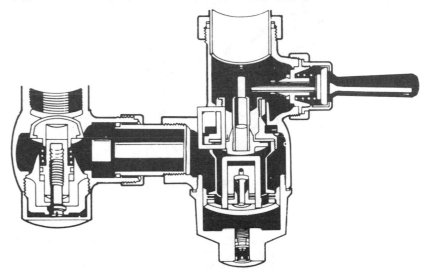

Sectional View of a Piston-type Flushometer Valves
Figure 15-11

piston and diaphragm types. A vacuum breaker is installed at the outlet of the flushometer to prevent back siphonage. Figure 15-11 shows a piston-type flushometer.

Mixing valves are used to supply wash water at a uniform temperature. The temperature of the water may be regulated between the limits of the temperatures of the cold water supply and the hot water supply.

Traps and Interceptors
Traps are installed between plumbing fixtures and waste pipes to prevent the escape of sewer gas and odors into building. They are designed to retain a small portion of the water discharged from the plumbing fixture. This water seal, which is replaced by each discharge of the fixture, prevents sewer gas from escaping back through the fixture. The deeper the seal, the more resistance there is to the passage of sewer gas but the greater the fouling area. Therefore, a minimum depth of 2 inches and a maximum depth of 4 inches is generally accepted.

A **P-trap** consists of a P-shaped piece of steel, cast-iron, or brass pipe with a vertical inlet and a horizontal outlet. (See Figure 15-12.) The U-shaped portion of the trap, which is below the overflow outlet level, constantly holds a quantity of water to form the seal. The depth of the seal is measured from the top of the dip to the crown weir or overflow. P-traps are commonly used with lavatories, sinks, urinals, drinking fountains, and in some instances, shower baths.

A **drum trap** consists of a cylindrical metal shell which is closed at the bottom and fitted with a screw-on cover. The shell is provided with a threaded inlet at the base and a threaded outlet near the top. The water seal, which usually is more than 2 inches, is the vertical distance from the top of the inlet to the bottom of the outlet. Drum traps are used with fixtures, such as bath tubs, which are installed close to the floor.

P-trap
Figure 15-12

An **integral trap** is a built-in water seal contained in the drain outlet of plumbing fixtures, such as water closets and urinals. During the flushing action, water flows from inside the bowl through the integral trap and out of the fixture. When the last portion of water passes through the inside of the fixture, a portion is retained to reseal the trap.

A **hair trap** contains baffles, partitions, and screens which collect hair, lint, and other foreign material and prevent this matter from entering drainage systems. Hair traps are usually installed in waste connections from lavatories of hair-dressing shops.

Interceptors are installed in drainage systems to remove and withhold grease, flammable wastes, sand, and other ingredients harmful to the drainage system.

A **grease interceptor** is installed in a drainage system to accumulate grease from waste water and prevent the grease from discharging into the sewer. Grease interceptors operate on the principle that grease is lighter than water and will rise to the top of water. (See Figure 15-13).

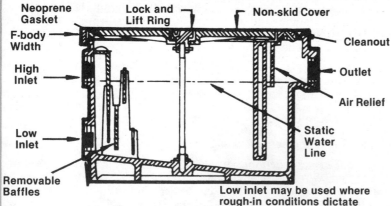

Neoprene Gasket
Lock and Lift Ring
Non-skid Cover
F-body Width
Cleanout
High Inlet
Outlet
Air Relief
Low Inlet
Static Water Line
Removable Baffles
Low inlet may be used where rough-in conditions dictate

Sectional View of a Grease Interceptor
Figure 15-13

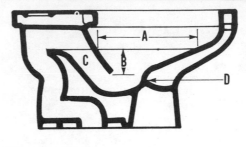

Siphon Jet

The flushing action of the siphon jet bowl is accomplished by a jet (D) of water being directed through the uplet (C) of the trapway instantaneously, the trapway fills with water, the siphonic action starts. The powerful, quick, and relatively quiet action of the siphon jet bowl combined with its large water surface (A) and deep water seal (B) contribute to its general recognition by sanitation authorities as the premier type of closet bowl. The siphon jet bowl is a logical choice for the most exacting installation.

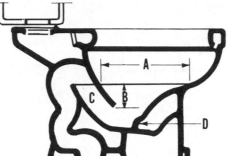

Reverse Trap

The flushing action and general appearance of the reverse trap bowl is similar to the siphon jet. However, the water surface (A) depth of seal (B) and size of trapway (C) are smaller. Consequently, less water is required for operation. Reverse trap bowls are particularly suitable for installation with flush valves or low tanks.

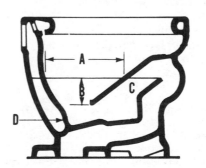

Washdown

The washdown type bowl is simple in construction and yet highly efficient within its limitations. Proper functioning of the bowl is dependent upon siphonic action in the trapway accelerated by the force of water from the jet (B) directed over the dam. Washdown bowls are widely used where low cost is a prime factor. They will operate efficiently with a flush valve or low tank.

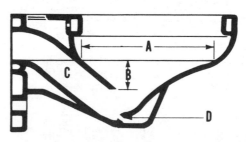

Blowout

The blowout bowl cannot be fairly compared with any other type — it depends entirely upon a driving jet (D) action for its efficiency rather than upon siphonic action in the trapway. It is economical in the use of water, yet has a large water surface (A) that reduces fouling space, a deep water seal (B) and a large unrestricted trapway (C). Blowout bowls are especially suitable for use in schools, offices, and public buildings. They are operated with flush valves only.

A — water surface
B — water seal
C — trapway
D — jet

Water Closets
Figure 15-14

A **sand interceptor** prevents the passage of sand, cinders, and other solids into the drainage system. Sand interceptors may also be used to intercept oil, gasoline, or other flammable fluids. They must be readily accessible for cleaning.

Plumbing Fixtures

Plumbing fixtures may be constructed of cast iron coated with enamel, pressed steel coated with enamel, or vitreous china. Since plumbing fixtures are both the terminals of the water supply and the beginnings of the drainage or soil and waste systems, they control both the quantity of water which must be furnished and the amount of drainage which must be handled by stacks, drains, and sewers.

Water closets used in plumbing installations may be of the following types: washdown bowl; washdown bowl with jet; reverse-trap bowl; and siphon-jet bowl. (See Figure 15-14.)

The **washdown bowl** is the simplest type of water closet. The trap is at the front of the bowl and is somewhat smaller than in other types of water closets. Proper functioning of this type of closet depends on siphon action alone. The flushing is done by small streams of water discharged from the rim of the bowl.

Some washdown bowls have a jet which provides more positive flushing action. When the bowl is flushed, a small stream of water spurts from the jet into the upper arm of the trap and starts the siphon action.

The **reverse-trap bowl** is similar to the washdown bowl except that the trapway is placed in the rear of the bowl. This type of bowl comes either with or without a jet. The reverse-trap bowl has a larger water surface, a deeper water seal, and is quieter and more efficient than the washdown type of bowl.

The **siphon-jet bowl** is similar to the reverse-trap bowl. However, the trapway is larger and the water seal is deeper. When the bowl is flushed, water flows down the sides of the bowl and a supply of flush water also enters the jet chamber near the opening into the upper arm of the trap. Siphon-jet bowls are quiet and positive in operation.

Closet tanks supply water to water closets for flushing. The tanks are made of metal or vitreous china and may be separate units or cast integrally with the water closet bowl. A separate tank may be wall-hung and connected to the bowl with an elbow or sleeve or it may be close-coupled and supported directly on the closet.

There are four types of urinals: wall hung, trough, stall, and pedestal. (See Figure 15-15). They may be flushed by a tank similar to a closet tank mounted above the fixture or, preferably, by automatic siphon valves or flushometer valves.

A **wall-hung urinal** consists of a vitreous china or porcelain bowl attached to a wall at a convenient height and with water and waste connections for flushing and discharging the waste. Most wall-hung urinals have an integral trap. The flushing action may be either washdown or siphon-jet.

A **trough urinal** consists of an enameled iron trough-shaped casting which is suspended from hangers attached to a wall. It is flushed by a perforated pipe which delivers water over the back of the fixture.

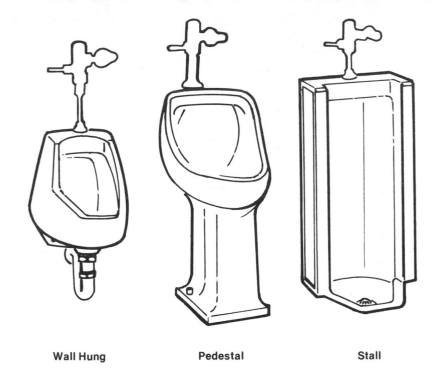

Wall Hung   Pedestal   Stall

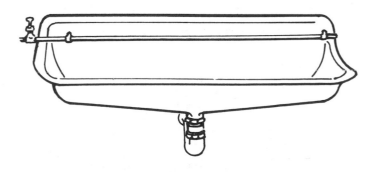

Trough

**Types of Urinals
Figure 15-15**

A **stall urinal** is an enameled iron or vitrified porcelain casting set in the floor with its back attached to a wall.

A **pedestal urinal** consists of a vitreous china bowl and stand which is bolted to the floor. The flushing action is similar to that of a water closet. Pedestal urinals are usually equipped with flushometer valves. They are used only where wall-hung urinals cannot be used.

Lavatories consist of circular, square or oval-shaped bowls and are usually made of vitreous china. Normally, lavatories are supplied with both hot and cold water, waste connections, single or combination faucets, and either a pull-out or pop-up drain. The most commonly used types of lavatories are the wall-hung or pedestal types. (See Figure 15-16.)

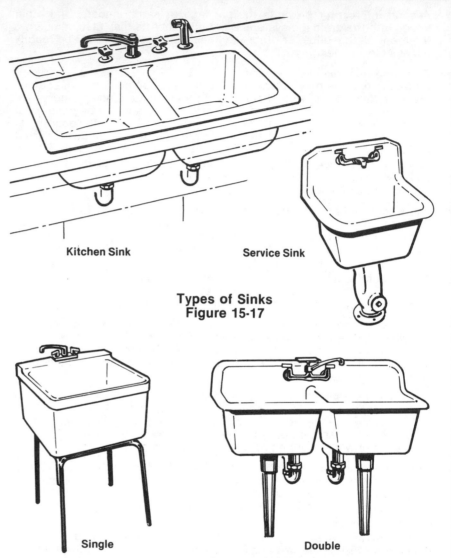

Kitchen Sink   **Service Sink**

**Types of Sinks
Figure 15-17**

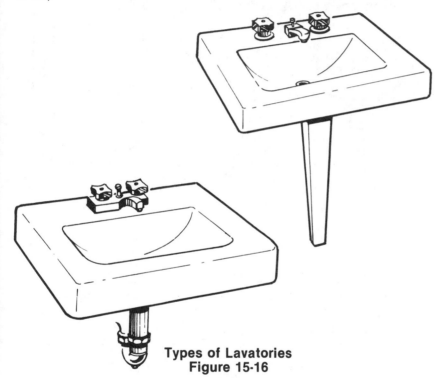

**Types of Lavatories
Figure 15-16**

Single   **Double**

**Types of Laundry Tubs
Figure 15-18**

Sinks are manufactured to serve almost any purpose. They usually are divided into two categories: residential sinks and commercial sinks such as slop sinks and scullery sinks. (See Figure 15-17.) They may have either single or swing-type combination faucets. The drains may be of the plain-strainer type or of the basket-strainer type.

Laundry tubs are plumbing fixtures used for soaking and washing clothes. They are usually made of nonabsorbent concrete or enameled iron and may contain single or double compartments. (See Figure 15-18.) Each tub is set on a metal stand which rests on the floor. Double tubs generally are provided with a swing-type combination faucet. The faucets usually are provided with a hose bibb, permitting attachment of hoses.

Drinking fountains have orifices above the rim of the fixture which project a stream of water upward so that the person drinking can not touch the orifice with his lips. Drinking fountains can be wall-hung, pedestal, or electrically cooled types. (See Figure 15-19.) Wall-hung and pedestal-type fountains are connected to cold water supply piping. Electrically cooled fountains contain individual refrigeration units and water is passed over the refrigeration coils and is cooled before it is supplied through the orifice.

Labor Installing Shower Receptors

Work Element	Unit	Man-Hours Per Unit
Terrazzo, rectangular, 32" x 48"	Each	3.3
Cast stone, rectangular, 32" x 48"	Each	2.7
Enameled steel, rectangular, 32" x 48"	Each	2.7
Plastic, rectangular, 32" x 48"	Each	2.4
Terrazzo, corner type	Each	3.8
Cast stone, corner type	Each	3.8
Enameled steel, corner type	Each	3.3
Plastic, corner type	Each	2.4
Standard lead pan type	Each	8.9
Standard copper pan type	Each	8.8

Time includes move on and off site, unload and remove from container, cleanup and repair as needed.
Suggested Crew: 1 plumber and 1 laborer. In some locations 2 laborers will be required for lifting and moving. Time must be added to the man-hours if a second laborer is required.

Labor Installing Bathtubs

Work Element	Unit	Man-Hours Per Unit
Cast-iron enamel, 4'-6", recessed	Each	6.6
Cast-iron enamel, 5', recessed	Each	6.7
Steel enamel, 5', recessed	Each	5.5
Cast-iron enamel, 5'-6", recessed	Each	7.1
Cast-iron enamel, 5', corner type	Each	7.0
Standard institutional tub, including base	Each	18.0

Time includes move on and off site, unloading, removal from crate, cleanup and repair as needed. Plumbing hookup only.
Suggested Crew: 1 plumber and 1 laborer. In some locations, 2 laborers will be required for moving and lifting cast-iron tubs. Additional time must be added to man-hours where a second laborer is required.

Wall Fountain

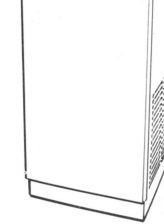

Electrically Cooled Fountain

Types of Drinking Fountains
Figure 15-19

Water Heaters

The domestic-type gas storage heater consists of a tank enclosed in an insulating jacket. A gas burner integral with the tank provides the heat which is controlled by a thermostatic element extending inside the tank. When the temperature of the water in the storage tank drops, the gas-control valve opens and the pilot light ignites the burner. The operation of the heater is completely automatic and it will maintain any water temperature from 110 degrees to 165 degrees, depending on the setting of the thermostat.

An electric water heater consists of an insulated tank which contains one or two controlled heating elements. The elements furnish the heat necessary to raise the water temperature and may consist of heating bands wrapped around the tank of immersion-type units extending into the tank. One great advantage of this type of heater is that it does not require a flue pipe since there are no products of combustion to be exhausted to the outside.

Each hot water heater is equipped with both pressure and temperature relief valves. The pressure relief valve allows the water within the tank to blow into a drain line when the pressure in the storage tank exceeds the predetermined setting. This setting is usually 10 psi above the pressure in the supply main. The temperature relief valve allows the water to blow into the drain line when the temperature in the storage tank exceeds 205 degrees. Some temperature relief valves also automatically shut off the source of heat.

Labor Installing Sinks

Work Element	Unit	Man-Hours Per Unit
Highback single compartment		
Iron enamel, 24" x 48"	Each	2.9
Enamel steel, 24" x 48"	Each	2.8
Sinks with drainboards		
Iron enamel, single, 42" x 21"	Each	4.2
Stainless steel, single, 42" x 21"	Each	4.1
Enameled steel, single, 42" x 21"	Each	4.1
Iron enamel, double, 42" x 21"	Each	6.2
Stainless steel, double	Each	6.1
Enameled steel, double	Each	6.2
Shelf back single compartment	Each	3.9
Service sink		
Cast-iron, enamel, 22" x 18"	Each	3.9
Vitreous china, 22" x 19"	Each	3.8
Vitreous china, 24" x 20"	Each	3.9
Mop receptor, 36" x 36"	Each	3.7
Laundry type, wall hung		
Cast-iron enamel, 28" x 26"	Each	3.9
Cast-iron enamel, 20" x 16"	Each	2.9
Enameled steel, 24" x 20"	Each	2.8
Plastic, 24" x 20"	Each	1.6
Cast-iron enamel, 48" x 20"	Each	5.8
Enameled steel, 48" x 20"	Each	5.6
Plastic, 48" x 20"	Each	2.2
Floor type sink, cast-iron	Each	3.9
Surgical type sink, vitreous china with foot pedals	Each	5.9
Stainless steel scrub sink with foot pedals	Each	4.7

Time includes move-on and off-site, unload, setting sink in prepared opening, plumbing hook-up only, clean-up and repair as needed.
Suggested Crew: 1 plumber and 1 laborer

Labor Installing Lavatories and Urinals

Work Element	Unit	Man-Hours Per Unit
Wall hung slab type, vitreous china, 20" x 18"	Each	3.3
Wall hung shelf type		
Vitreous china, 20" x 18"	Each	3.3
Enameled steel, 20" x 18"	Each	3.4
Iron enamel, 20" x 18"	Each	3.4
Wall mounted corner type		
Vitreous china, 17" x 17"	Each	3.3
Wall mounted prison types		
Vitreous china	Each	4.0
Stainless steel	Each	4.0
Built-in types, 18" in diameter		
Vitreous china	Each	3.4
Enameled steel	Each	3.4
Iron enamel	Each	3.5
Pedestal type urinals		
Washdown type with flush valve	Each	2.7
Washdown tank type	Each	3.8
Wall hung urinals (carrier included)		
Washdown type with flush valve	Each	3.3
Blowout type with flush valve	Each	3.3
Siphon jet type with flush valve	Each	3.4
Washdown tank type	Each	4.4

Time includes move on and off site, unloading, clean-up and repairs as needed. Time does not include wall carrier installation. Includes plumbing hook-up only.
Suggested Crew: 1 plumber

Labor Installing Water Closets

Work Element	Unit	Man-Hours Per Unit
Water closet, floor mounted (regular types)		
Washdown type	Each	3.2
Siphon jet type	Each	3.3
Flush valve type, floor mounted		
Washdown type	Each	3.3
Siphon jet type	Each	3.3
Floor mounted elongated type (tank type)		
Washdown type	Each	3.2
Siphon jet type	Each	3.3
Flush valve type, elongated		
Washdown type	Each	4.5
Siphon jet type	Each	4.5
Wall mounted regular types, including carrier		
Siphon jet type	Each	4.5
Flush valve type, wall mounted		
Siphon jet	Each	4.5
Wall mounted elongated, including carrier		
Siphon jet with tank	Each	4.5
Siphon jet with flush valve	Each	4.5

Add twenty-five percent to man-hours per unit for hospital type bed pan cleansers. Time includes move on and off site, unloading, clean-up and repair as needed. Plumbing hook-up only.
Suggested Crew: 1 plumber

Labor Installing Drinking Fountains

Work Element	Unit	Man-Hours Per Unit
Wall mounted fountains		
Vitreous china	Each	3.3
Stainless steel	Each	3.3
Stainless steel, electric	Each	6.1
Steel cabinet	Each	3.8
Steel cabinet, electric	Each	6.1
Semi-recessed type		
Vitreous china	Each	4.4
Stainless steel	Each	4.5
Stainless steel, electric	Each	7.1
Steel cabinet	Each	4.9
Steel cabinet, electric	Each	7.2
Floor mounted drinking fountain		
Vitreous china	Each	4.4
Stainless steel, electric	Each	5.5
Steel cabinet, electric	Each	5.5
Standard concrete pedestal	Each	6.5
Standard cafeteria 6 station, self-contained, stainless steel	Each	10.0

Time includes move-on and off-site, unload, remove from crate, clean-up and repair work as needed. No electrical hook-up included.
Suggested Crew: 1 plumber and 1 laborer

Labor for Rough-In of Plumbing

Work Element	Unit	Man-Hours Per Unit
Install sewer pipe		
4″ to 12″	Foot	0.45
Install cast iron drain lines and fittings		
4″	Joint[1]	0.85
6″	Joint	1.00
8″	Joint	1.30
Installing threaded steel pipe (schedule 40)		
½″ to ¾″	Joint	1.00
1″	Joint	1.16
1¼″	Joint	1.24
1½″	Joint	1.36
2″	Joint	1.60
2½″	Joint	2.00
3″	Joint	2.40
3½″	Joint	2.80
4″	Joint	3.20
Copper tubing[2]		
3/8″ to 1/2″	Joint	0.5
3/4″	Joint	0.75
1 to 1¼″	Joint	0.9
Rough in fixtures		
Lavatory	Each	5.6
Water closet	Each	6.4
Shower with stall	Each	6.4
Slop sink	Each	4.0
Urinal with stall	Each	4.8
Bathtub	Each	5.2
Kitchen sink	Each	4.8
Bathtub with shower	Each	7.6
Floor drain	Each	2.4
Grease trap	Each	3.2
Valves, faucets, etc., installed with rough plumbing		
1″ or less	Each	0.5
1 to 2″	Each	1.00
2″ and over	Each	1.25
Test plumbing system	Each	2 to 3

Notes for plumbing rough-in
[1] A joint is the connection that joins pipe with pipe, pipe with a valve, pipe with a coupling, etc.
[2] Usually less than half as many joints will be required for copper tubing than for steel pipe of equal length.

No fixture setting is included. Labor includes unloading and installing sewer and drain pipe, installing water pipe, and testing. The installation of cast iron drains includes caulking and leading joints, plumbing and grading pipe, installing pipe hangers and straps, cutting pipe, and installing fittings. The installation of galvanized steel pipe, includes cutting and threading pipe, making joints including applying joint compound, installing pipe hangers and straps, and installing fittings. The installation of copper water pipe includes cutting, cleaning and soldering copper pipe joints, plumbing and grading pipe, and installing pipe hangers and straps.
Suggested crew: one plumber and one helper.

Drain, Vent & Waste

The **drainage system** consists of all the piping, fittings, and fixtures that carry waste water, sewage, or other drainage from a building to the street sewer or other place of disposal. It also includes the piping and fittings that ventilate the system. (See Figure 15-20.) A drainage system may contain both sanitary and storm sewers. A sanitary sewer conveys liquid or waterborne waste from plumbing fixtures. A storm sewer is used to convey water, subsurface water, condensate, cooling water, or other similar discharges. Storm sewers are usually separate systems and are not connected to sanitary sewers.

Stacks, branch sewers, and fixture drains are the portions of the drainage system that connect the waste outlets of the fixtures with the building drain.

Every building that has plumbing fixtures should have a soil or waste stack or stacks extending from the building drain through the roof. These stacks should be as straight and direct as possible and free from sharp bends and turns. Do not use bends over 45 degrees. Locate stacks so that the drainage and vent lines serving fixtures will be of minimum length.

Occasionally, a stack can not extend upward in a continuous vertical line due to building construction or other interference. In this case the stack must pass around the obstruction by using an offset. Offsets are combinations of elbows, bends or special offset fittings which carry the stack to one side of the original line, then allow it to continue on a line parallel to the original direction. A return offset is a double offset which

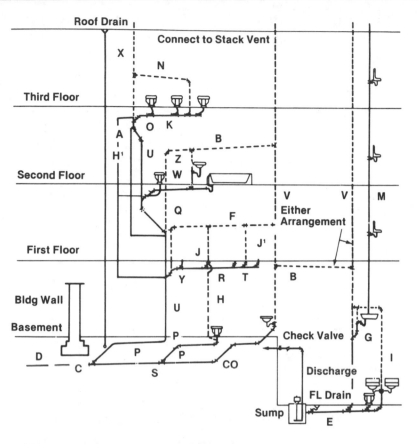

Drainage and Valve Lines of a Three Story Building
Figure 15-20

A. Branch Interval
B. Branch Vent
C. Building Drain
D. Building Sewer
E. Building Subdrain
F. Circuit Vent
G. Continuous Waste Vent
H. Double Offset
H¹. Dry Vent
I. Dual Vent (Unit Vent)
J. Fixture Drain
J¹. Group Vent
K. Horizontal Branch
M. Indirect Waste

N. Loop Vent
O. Offset
P. Primary Branch
Q. Relief Vent
R. Return Offset or Jumpover
S. Secondary Branch
T. Side Vent
U. Soil Stack
V. Vent Stack (Main Vent)
W. Wet Vent
X. Stack Vent (Main Soil and Waste Vent)
Y. Yoke Vent
Z. Back Vent (Individual Vent)
C.O. Cleanout

returns the pipe to its original direction after passing around an obstruction. Single offsets introduce less resistance to flow than return offsets.

Every stack which has soil pipe branches emptying into it must have a main vent tee. The main vent tee is a tapped tee with a 2 inch side outlet. The tee is installed in the stack at least 3 feet above the highest branch. The main vent tee forms a junction between the main vent and the main soil and waste vent.

After the main vent tee is installed, the stack is extended through the roof of the building to form the vent terminal. The stack must be carried full size or larger through the roof and project at least 6 inches above the roof. The stack opening in the roof is made watertight with roof flashing of galvanized steel, cast-iron plates, copper, or lead. Flashings are available in a number of different types and sizes. Size of flashing to be used depends on the size of the portion of the stack that passes through the roof. Use a final coat of roofing cement to insure a leakproof flashing.

In cold climates, the moisture inside of the terminal opening of the stack can freeze and clog the opening. The air within the stack and vent pipes in the building is usually very close to the moisture saturation point. When this humid air leaves the stack, it can freeze in extreme temperatures. Where possibility of frost closure exists, the terminal should be a minimum of 3 inches and preferably 4 inches in diameter. Enlarge small vent stacks with increasers which are joined to the stack approximately 12 inches below the roof. Other ways to prevent frost closure are: (1) the use of insulating material between the pipe and flashing; and (2) the installation of a high lead flashing which provides an insulating pocket of air between the flashing and the end of the main vent above the roof. The air pocket is left open to the heat of the building.

Make soil and waste pipe connections to stacks with sanitary tee. Straight tee must not be used. After the sanitary tee is installed, the branch piping is extended to the fixtures. Slope the piping to the correct grade. Cleanouts must be installed to allow cleaning of the branch in the event of stoppage.

The closet bend is a type of soil pipe branch forming a direct connection between a water closet and the soil stack or branch pipe. When the soil stack is close to the water closet, the closet bend may be caulked directly into a sanitary tee branch takeoff from the stack.

Cast-iron soil pipe stacks and branches will be properly supported so the weight of the piping does not bear on calked joints.

Vents

The vents may be made with plastic, cast-iron, galvanized steel or tube copper with plastic, malleable, cast-iron, brass, or copper fittings. Trap seal loss by direct or indirect siphon is usually caused by inadequate venting of the trap of a fixture. Trap seal loss may result from evaporation and capillary attraction. The main purpose of a vent system is to provide an opening to the atmosphere on the discharge side of a trap, preventing loss of the water seal by siphon. Any difference between the air pressure and the pressure on the discharge side of the trap will force the water seal in the direction of the least pressure. Venting the discharge side of the trap to the atmosphere equalizes these pressures.

Labor Installing Cast-Iron Soil Pipe

Work Element	Unit	Man-Hours Per Unit
Cast-iron, extra heavy (bell & spigot)		
2"	100 L.F.	25.1
3"	100 L.F.	27.1
4"	100 L.F.	35.5
5"	100 L.F.	44.2
6"	100 L.F.	58.4
8"	100 L.F.	70.6
10"	100 L.F.	88.7
12"	100 L.F.	117.0
15"	100 L.F.	179.4
Cast-iron - no hub (plain end)		
1½"	100 L.F.	23.2
2"	100 L.F.	23.6
3"	100 L.F.	27.6
4"	100 L.F.	29.6
5"	100 L.F.	35.9
6"	100 L.F.	43.8
Cast-iron silicone lined pipe		
1½"	100 L.F.	25.0
2"	100 L.F.	30.1
3"	100 L.F.	37.2
4"	100 L.F.	50.5
5"	100 L.F.	59.8
6"	100 L.F.	75.2
8"	100 L.F.	102.7

Deduct 10% from man-hours for below ground work. Time includes handling pipe into place, installation above ground on supports including placement of supports, installation of fittings an average of every 10 feet, clean-up and repairs as needed. Silicone lined cast iron pipe is assumed to be used for an acid waste and vent line with short runs and special packing.
Suggested Crew: 1 plumber and 1 laborer

Labor Installing PVC Schedule 40 or 80 DWV Plastic Pipe

Work Element	Unit	Man-Hours Per Unit
3/8"	100 L.F.	6.1
1/2"	100 L.F.	7.5
3/4"	100 L.F.	8.7
1"	100 L.F.	10.1
1-1/4"	100 L.F.	11.7
1-1/2"	100 L.F.	12.9
2"	100 L.F.	14.4
2-1/2"	100 L.F.	15.1
3"	100 L.F.	18.7
4"	100 L.F.	29.1
6"	100 L.F.	46.2

Time is based on average of one fitting every 10 feet. Time includes unloading, stacking, move on and off site, cleanup and repair as needed. This table assumes pipe is installed in a building on hangers. Deduct 10% for pipe laid in a trench.
Suggested Crew: 1 plumber and 1 laborer

Labor Intalling DWV Copper Pipe

Work Element	Unit	Man-Hours Per Unit
1¼"	100 L.F.	13.9
1½"	100 L.F.	15.4
2"	100 L.F.	18.3
3"	100 L.F.	21.6
4"	100 L.F.	33.0
5"	100 L.F.	44.0
6"	100 L.F.	54.0

Hours include cutting, soldering, fitting, and handling for runs of 100' or more in trenches. Add 20% for runs in buildings including hangers.
Suggested Crew: 1 plumber and 1 helper

Compressed Air Systems

Compressed air systems are used in hospitals, automotive service stations, machine shops, or laundries. The air is compressed in an electric-, gasoline-, or diesel-driven compressor and stored in a tank until needed. Air is drawn into the compressor, where it is reduced in volume and passed through a check valve into the storage tank. A combination pressure-control and safety valve regulates the operation of the compressor and power unit and, when the amount of air stored in the tank reaches the desired pressure, the compressor automatically shuts off. If the pressure-control valve fails to operate, the safety valve functions and relieves the pressure on the tank preventing its exploding. In some cases, compressed air is drawn from the tank, when needed, through a reducing valve. The compressor equipment is commonly housed in one centralized location. If the piping costs for the distribution of air to the required locations and the pressure line loss from one source are excessive, two or more compressor units located near the specific load centers may be more economical and efficient.

Air compressors usually operate automatically, the operation being controlled by a diaphragm-type pressure switch. Install a pressure gage adjacent to the pressure switch. Safety valves should be installed on the compressor auxiliaries, such as intercoolers, aftercoolers, and air receivers, and on the compressor discharge line between the compressor and any shutoff valve in the line.

Compressed air piping for laboratories and clinics in hospital buildings is copper tubing, type L, hard-drawn, with wrought-copper or cast-red-brass sweat fittings; for shops, garages, laundries, and similar installations, the piping is normally plain steel or wrought iron with fittings of malleable iron. Installation standards for compressed air systems are similar to gas systems. Because compressed air for hospital, laboratory, and dental use requires nonpulsating, clean, dry, oil-free air, it lends itself to a common system. This system uses rotary or centrifugal watersealed compressors (75 to 100 psi), galvanized receivers, and copper tube piping to deliver the required quality of air.

Water in compressed air will often cause operating difficulties. The water causes water hammer in pipe lines, reduces the capacity of the lines, and washes away lubricants in pneumatic tools. Oil also is carried along in the air stream, is condensed along with moisture in the aftercooler or in other areas, and must be removed from the system. Provide drains in the low points of lines for the removal of the water and oil. This will also reduce internal corrosion.

Labor Installing Air Compressors and Pumps

Work Element	Unit	Man-Hours Per Unit
Supply piping		
1½" and smaller	1000 L.F.	100
2" to 4"	1000 L.F.	200
Pickling pipe for oxygen and acetylene system	1000 L.F.	12
Valves and quick connectors		
1½" and smaller	Each	.8
2" to 4"	Each	1.2
Hose reels	Each	2.0
Install electric driven compressor tank units		
5 cfm to 20 cfm	Each	6.0
25 cfm to 50 cfm	Each	8.0
Install electric motor and compressor on anchor bolts		
100 cfm to 250 cfm	Each	16.0
275 cfm to 500 cfm	Each	32.0
550 cfm to 750 cfm	Each	40.0
Install gasoline or diesel motor and compressor		
100 cfm to 250 cfm	Each	18.0
275 cfm to 550 cfm	Each	36.0
550 cfm to 750 cfm	Each	48.0
Install electric motor and pump sets on anchor bolts		
50 to 200 gpm @ 100' tdh	Each	5.2
250 to 800 gpm @ 100' tdh	Each	8.4
850 to 1250 gpm @ 100' tdh	Each	12.8
Install gasoline or diesel motor & pump		
50 to 200 gpm @ 100' tdh	Each	8.4
250 to 800 gpm @ 100' tdh	Each	14.0
850 to 1250 gpm @ 100' tdh	Each	20.0

Suggested crew: two to six men, depending on job scope.

Labor Installing Medical Gas Systems

Work Element	Unit	Man-Hours Per Unit
Wall type outlets, for nitrous oxide, oxygen, air		
Single outlet	Each	2.4
Double outlet	Each	3.2
Triple outlet	Each	3.9
Quadruple outlet	Each	4.9
Ceiling type outlets for nitrous oxide, oxygen, air or vacuum		
Single outlet	Each	2.7
Double outlet	Each	3.4
Triple outlet	Each	4.2
Quadruple outlet	Each	5.2
Gas system alarm		
Single, with pressure gauge, self contained	Each	3.1
Double, with pressure gauge, self contained	Each	3.9
Single audio/visual legend	Each	3.0
Double audio/visual legend	Each	3.7
Triple audio/visual legend	Each	4.2
Quadruple audio/visual legend	Each	6.1
Zone valve and box assemblies for concealed piping		
½" valve	Each	2.4
¾" valve	Each	2.4
1" valve	Each	2.8
1¼" valve	Each	3.4
1½" valve	Each	3.7
2" valve	Each	4.3
Zone valve and box, double type for concealed piping		
½" to ½" valve	Each	3.1
½" to ¾" valve	Each	3.3
½" to 1" valve	Each	3.7
¾" to ¾" valve	Each	3.7
¾" to 1" valve	Each	3.7

Time includes move on and off site, cleanup and repairs.

Automatic Fire Sprinkler Systems

Automatic sprinkler systems are the most reliable and effective fire protection systems available. A sprinkler system can detect fire, transmit alarms, and most important, extinguish or control a fire before major loss of property or life can occur. Sprinkler systems are normally installed to protect high valued occupancies, special hazards, or to provide adequate life safety.

There are four basic types of sprinkler systems:

Wet-pipe sprinkler systems are the most common type and the simplest. This system's piping has closed automatic sprinkler heads and is filled with water under pressure. Immediate discharge occurs from individual sprinkler heads when fused by heat. Water continues to flow until manually shut off.

Dry-pipe systems are normally used in locations where pipes are subject to freezing. They are similar to the wet-pipe system except piping is filled with air under pressure, which holds closed a "dry-pipe valve." The actuation of a sprinkler head releases air pressure, permitting water pressure to open the "dry-pipe valve." Water then flows into piping and out of sprinkler heads which have opened.

Water and foam-water **deluge systems** are most commonly used in aircraft hangers or special hazard areas, where it is essential to deliver water through all sprinkler heads simultaneously when actuated by a heat detection system installed throughout the protected area. When the valve opens, water flows into piping and discharges through all sprinkler heads. Supplemental foam concentrate deluge valves which supply foam concentrate into the sprinkler system are required in foam-water systems.

Pre-action systems employ closed sprinkler heads attached to piping containing air which may or may not be under pressure. Water is held back by a valve activated by a fire detection system generally more sensitive than the sprinklers. Actuation of the detection system opens the valve, permitting water to flow into piping and discharge from any sprinkler heads which have opened.

Pipe used is normally black steel, designed to withstand a working pressure of not less than 175 psi. Some common terms used to describe components of piping include:

Risers are the vertical pipes supplying the sprinkler system. Feed mains supply cross mains. Cross mains directly supply the lines on which the sprinkler heads are placed. Branch lines are pipes to which the sprinkler heads are directly attached.

Fittings should be of a type specifically approved for sprinkler systems and designed for a minimum working pressure of 175 psi. Screwed or flanged fittings are normally used. Welded joints are occasionally used in large risers and large feed mains.

All **valves** should be of an approved type. All valves on connections to water supplies and in supply pipes to sprinkler systems should be outside screw and yoke (O.S. & Y.) or approved indicator type. Check valves are a straightway type installed in a vertical or horizontal position.

Hangers are used to attach sprinkler piping to structural elements of the building. The adequate support of sprinkler piping is an important consideration because of the water damage possibilities which might result from broken piping.

Sprinkler heads generally have a 1/2" outlet, a releasing mechanism (most commonly a fusible solder link melted by heat), and a deflector. There are two distinct sprinkler head designs, one approved for installation in the upright position and one to be installed in the pendant position. Heads are designed to produce a relatively uniform distribution of water at all levels below the sprinkler. Foam-water sprinkler heads are available for installation in foam-water deluge systems.

A sprinkler alarm system normally consists of components constructed and installed so that any flow of water through the system equal to or greater than that from a single sprinkler head will result in an alarm signal. The sprinkler alarm is normally arranged to activate the building evacuation alarm system, transmit a signal to the local fire department, and sound a water motor gong.

Some common components used in alarm assemblies are:

The **water motor gong** is an alarm bell mechanically operated by the flow of water through it. Usually located on a building wall as near the sprinkler riser as possible.

An **alarm check valve** is used in wet-pipe sprinkler systems. Located in the sprinkler system riser, it is arranged to activate the necessary attachments required to give an alarm by opening upon flow of water.

Water pressure surges can inadvertently open alarm check valves. To prevent needless alarms, **retard chambers** are located in the alarm assembly and must fill with water before the alarm signal is activated.

A **pressure switch** is normally located on top of the retard chamber and arranged to activate all electrically operated alarms.

The **excess pressure pump** is a small-capacity pump which maintains higher pressure on the sprinkler system side of the alarm check valve to prevent needless alarms due to surges on the water system.

The **paddle alarm** is a device located directly on the sprinkler riser piping with a paddle extending into the piping. It is arranged to activate electrically operated alarm systems upon the flow of water. This type alarm is used only with wet-pipe systems in lieu of an alarm check valve and should never be used with dry-pipe, pre-action or deluge systems.

Labor Installing Automatic Sprinkler Systems

Work Element	Unit	Man-Hours Per Unit
Sprinkler heads and accessories		
Pendant heads	Each	1.1
Upright heads	Each	1.1
4" alarm valve, wet system	Each	35.9
4" alarm valve, dry system	Each	39.6
6" alarm valve, wet system	Each	39.6
6" alarm valve, dry system	Each	43.3
Concealed pipe dry system (heads only)	Each	4.8
Exposed piping dry system (heads only)	Each	4.4
Wet system, single story building		
Concealed pipe (head only)	Each	4.0
Exposed pipe (head only)	Each	3.6

Time is based on jobs with 500 or more sprinkler heads. Man-hours per unit include move-on and off-site, clean-up and repairs as needed. Time does not include scaffold set-up or moving.
Suggested Crew: 1 plumber and 1 laborer.

Labor Installing Carbon Dioxide Fire Fighting Systems

Work Element	Unit	Man-Hours Per Unit
Carbon dioxide cylinders		
Simplex cylinders	Each	4.4
Duplex cylinders	Each	8.7
Heat or ion detector with brackets	Each	1.8
Dispersion nozzle	Each	.62
½" rubber tubing	100 L.F.	2.4

Time includes move-on and off-site, unloading, clean-up and repairs as needed.
Suggested Crew: 1 plumber

Labor Installing Standpipe and Firehose Stations

Work Element	Unit	Man-Hours Per Unit
Simplex fire pumps		
20 HP, 500 GPM, 1,750 RPM	Each	19.0
30 HP, 500 GPM, 1,750 RPM	Each	20.7
40 HP, 750 GPM, 1,750 RPM	Each	28.3
50 HP, 1,000 GPM, 1,750 RPM	Each	33.5
75 HP, 1,000 GPM, 1,750 RPM	Each	43.7
Firehose cabinets, primed steel		
125 L.F. hose, fully recessed	Each	6.6
125 L.F. hose, semi-recessed	Each	5.4
Firehose racks with 2½'' brass angle valves		
125 L.F. hose on open rack	Each	4.2
Siamese connections, brass		
2½'' x 2½'' x 4''	Each	8.7
Roof manifold, brass (vertical)		
2½'' x 2½'' x 2½'' x 4''	Each	8.9
Fire department valve		
2½''	Each	2.2
Fire extinguishers, stainless steel case		
Water, hand pumped	Each	2.3
Carbon dioxide (cast-iron)	Each	2.1
Carbon dioxide extinguisher with cabinet	Each	3.3
5 lb. dry chemical extinguisher	Each	2.1
1½'' fog nozzles for firehoses, chrome plated	Each	2.1

Time includes move-on and off-site, unload, clean-up and repairs as needed.

Suggested Crew: 1 plumber

Steam Heating Systems

Water heated to the boiling point evaporates and produces steam as long as heat is added. If the heat is removed or reduced, evaporation will stop or decrease. The quantity of heat contained in each pound of steam depends on its pressure and temperature. Steam can be generated and used as either saturated (dry or wet) or superheated steam.

For each steam pressure there is a specific temperature which will produce **saturated steam.** When one of these specific combinations of temperature and pressure exists, it is called saturated steam. When steam is saturated, a drop in temperature or an increase in pressure will cause part of the steam to revert to water. There are two types of saturated steam: dry, i.e., without moisture, and wet, which is intermingled with moisture, mist or spray. Saturated steam is commonly used for space heating and process heat utilization.

When steam has a temperature higher than its corresponding saturation pressure, it is called **superheated steam.** The difference between the temperature of superheated steam and its saturation temperature is called the superheat. Usually, superheated steam is generated in central heating plants when necessary to avoid condensation in the steam lines of the plant and the distribution system.

A steam heating system is known as a **one-pipe system** when a single main serves the dual purpose of supplying steam to the heating unit and conveying condensate from it. Ordinarily, to each heating unit there is but one connection which must serve as both the supply and the return, although separate supply and return connection may be used.

A steam heating system is known as a **two-pipe system** when each heating unit is provided with two piping connections, and when steam and condensate flow in separate mains and branches.

Heating systems may also be described as **up-flow or down-flow,** depending on the direction of steam flow in the risers, and as a **dry-return** or a **wet-return** depending on whether the condensate mains are above or below the water line of the boiler or condensate receiver.

Steam heating systems may also be classified as high pressure, low pressure, vapor, and vacuum systems, depending on the pressure conditions under which the system is designed to operate.

When condensate is returned to the boiler by gravity, the system is known as gravity return system. In this system all heating units must be elevated above the water line of the boiler, so that the condensate can flow freely to the boiler. Elevation of the heating units above the water line must be enough to overcome pressure drops due to flow, as well as pressure differences due to operation.

Note in Figure 1 that the boiler and wet-return form a U-shaped container, with the boiler steam pressure on the top of the water at one end, and the steam main pressure on the top of the water at the other end. The difference between these two pressures is the pressure drop in the system. The water in the far end will rise to overcome this difference to balance the pressures. It will rise far enough to produce a flow through the return pipe and overcome the resistance of check valves, if installed.

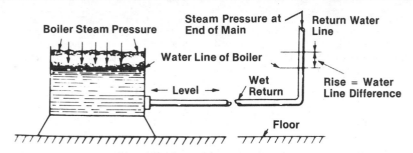

Difference in Steam Pressure on Water in Boiler and at End of Steam Main
Figure 1

When condensate cannot be returned to the boiler by the action of gravity, and either traps or pumps must be used, the system is known as a **mechanical return system.** There are three general types of mechanical condensate return devices in common use: (a) the alternating return trap, (b) the condensate return pump, and (c) the vacuum return pump.

In systems where pressure conditions in the system vary between that of a gravity return and forced return system, a boiler return trap or alternating receiver is used and the system may be known as an **alternating return system.**

When condensate is pumped to the boiler under pressures of the atmosphere or above, the system is known as a **condensate pump return system.**

When condensate is pumped to the boiler under vacuum conditions, the system is known as a **vacuum pump return system.**

Radiators and other heating units, in general, have only one piping connection from main to unit, although it is possible to use two connections to the same main, as indicated in Figure 2. Unit heaters in one-pipe systems may also have separate connections to the wet-return.

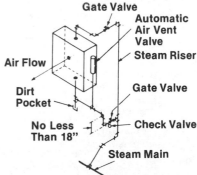

Typical Two-pipe Connections to Unit Heaters in One-pipe Air Vent Systems
Figure 2

There are several variations in the piping arrangement of a one-pipe system:

Up-feed one-pipe systems where the radiators and other heating units are located above the supply mains. The mains in this instance convey both steam and condensate. This system is illustrated in Figure 3. Typical connections to radiator or risers are illustrated in Figure 4, and method of changing sizes of mains in Figure 5.

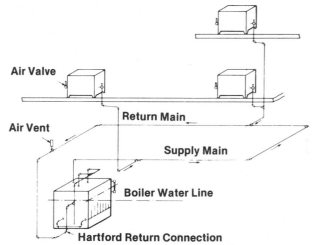

Typical Up-feed Gravity One-pipe Air Vent System
Figure 3

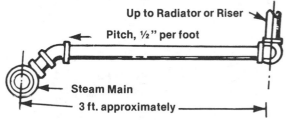

Typical Steam Runout Where Risers are Not Dripped
Figure 4

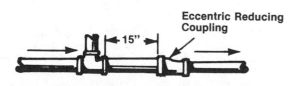

Method of Changing Sizes of Steam Main When Runouts are Taken From Top
Figure 5

Up-feed one-pipe systems where the radiators and other heating units are located above the mains, and the mains are dripped at each radiator connection to a wet-return. This way the steam main carries a minimum of the condensate. This system is illustrated in Figure 6. Typical connections to radiators and risers are illustrated in Figure 7. Up-feed systems are not recommended for use higher than four stories.

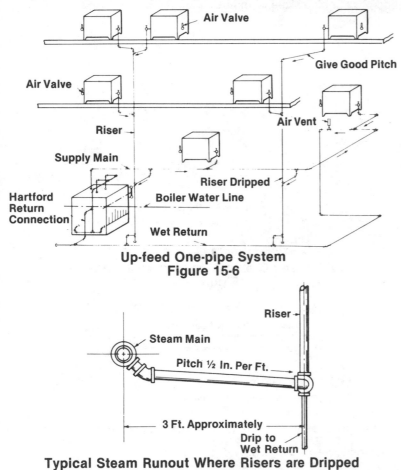

Up-feed One-pipe System
Figure 15-6

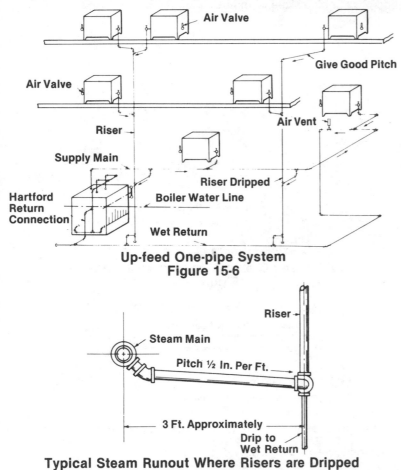

Typical Steam Runout Where Risers are Dripped
Figure 15-7

In down-feed one-pipe systems, the radiators and other heating units are located below the supply main. In this arrangement only risers and connections to heating units convey both steam and condensate, and both are flowing in the same direction. The steam main is kept relatively free of condensate by dripping through the drop risers.

Each radiator or heating unit in a one-pipe system must be supplied with a thermostatic air valve which relieves air from the heating unit under pressure, and closes when steam heats the thermostatic element of the valve.

To improve steam circulation in one-pipe systems, quick-vent air valves are provided at the ends and at intermediate points where the steam main is brought to a higher elevation, or where dropped below the water line. It is best to install the air-vent valves about a foot ahead of the drips, as indicated in Figure 6.

Air valves are of two general types, the pressure and the vacuum types. The pressure type permits the inflow of atmospheric air to the system when the steam pressure in the system falls below atmospheric pressure. The vacuum type, which contains a small check valve, prevents the air from flowing back to the system and thereby maintains vacuum conditions in the system. Systems which use vacuum valves are known as vapor or vacuum one-pipe systems. The vapor or vacuum systems will maintain a more uniform temperature condition than the pressure systems.

Each heating unit in a one-pipe system may also be provided with a valve on the connection to the unit. This is not essential except to shut the unit off when it is not needed for heating. Valves on one-pipe systems must be either fully opened or fully closed.

Two-pipe high-pressure systems operate at pressures above 15 psi. usually from 30 to 150 psi. They are usually used in large industrial type buildings, which are equipped with unit heaters or large built-up fan units, or in which high pressure steam is required for process work.

Figure 8 illustrates a typical high-pressure system. Because of the high pressures and the great differential between steam and return mains, it is possible to locate returns above the heating units and lift the condensate to these returns.

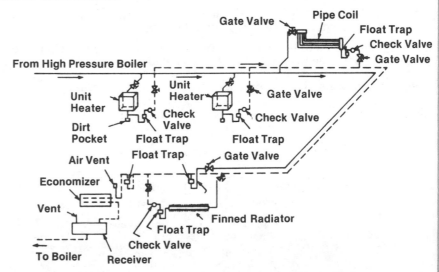

Typical High Pressure Heating System
Figure 8

The condensate can be flashed into steam in low-pressure mains if any are available, or passed through an economizer heater before being discharged to a vented receiver. It is, of course, necessary to provide for the elimination of air from high-pressure systems, the same as in low-pressure systems.

Return traps used on high-pressure systems are usually of the bucket, inverted bucket, float or impulse type.

Low-pressure systems operate at pressures of 0 to 15 psi. The piping arrangement of both up-feed and down-feed low-pressure systems is identical with those of two-pipe vapor systems. The only difference between the two systems is in the type of air valve used. The air valves used in low-pressure systems usually do not contain the check discs and hence the system cannot operate under a vacuum. The low-pressure systems are not as popular as the vapor systems, because they have the disadvantage of not holding heat when the rate of steam generation is falling. They also have the disadvantage of corroding to a greater extent than vapor systems, due to the continued presence of new air in the system.

Low-pressure systems have the advantage, however, of returning condensate to the boiler readily and not retaining it in the piping. Figure 9 illustrates a typical low-pressure system with condensate pump.

Two-pipe vapor systems operate at pressures varying from 20 in. vacuum or more to 15 psi. without the use of a vacuum pump. A typical two-pipe up-feed vapor system is shown in Figure 10. A typical two-pipe down-feed system is illustrated in Figure 11. The method of dripping drop risers in a down-feed system is illustrated in Figure 12. Radiators discharge their condensate and air through thermostatic traps to the dry-return main. Air is eliminated, when the system is under pressure, at the ends of the supply and return mains just before they drop to the wet return.

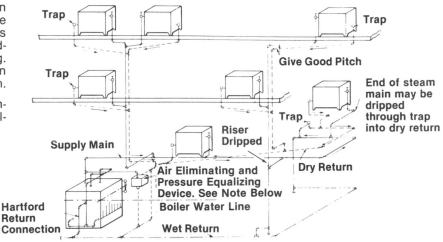

Proper piping connections are essential with special appliances for pressure equalizing and air elimination.

Typical Up-feed Two-pipe System with Automatic Return Trap
Figure 10

Vapor systems may also be provided with an automatic return trap or alternating receiver which automatically returns condensate to the boiler when the boiler is steaming under pressure conditions which would prevent the return of condensate by gravity. The typical connections for an automatic return trap are illustrated in Figure 13.

Each heating unit in a vapor system, as in all two-pipe systems, is provided with a graduated or modulating valve which permits the control of heat in the radiator by varying the opening of the valve.

Vacuum systems operate under conditions of both low pressure and vacuum, but use a vacuum pump to insure a vacuum in the return piping during all operating conditions.

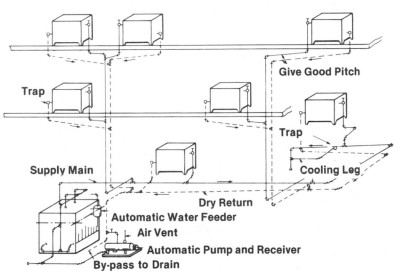

Typical Installation Using Condensate Pump
Figure 9

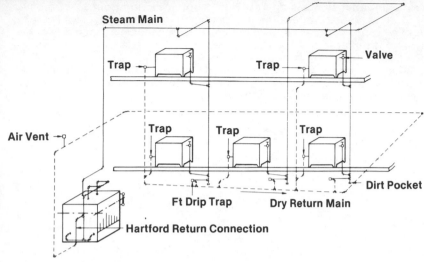

Typical Down-feed Two-pipe System
Figure 11

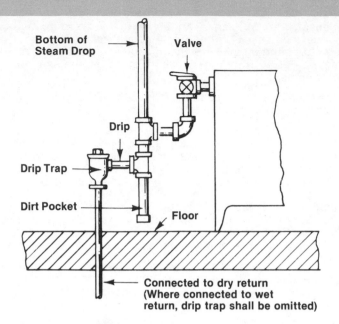

Detail of Drip Connections at Bottom of Down-feed Steam Drop
Figure 12

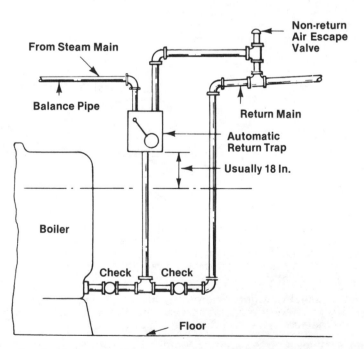

Typical Connections for Automatic Return Trap
Figure 13

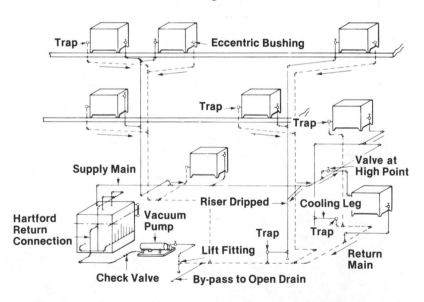

Typical Up-feed Vacuum Pump System
Figure 14

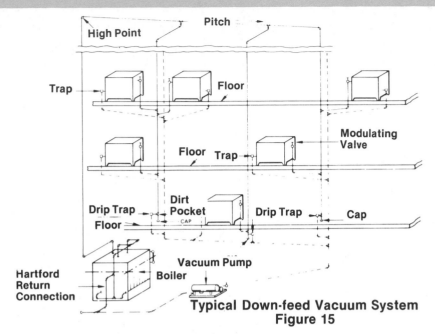

**Typical Down-feed Vacuum System
Figure 15**

A typical two-pipe up-feed vacuum system is illustrated in Figure 14. A down-feed arrangement is shown in Figure 15.

The return risers are connected in the basement into a common return main which slopes downward toward the vacuum pump. The vacuum pump withdraws the air and water from the system, separates the air from the water, expels air to atmosphere, and pumps the water back to the boiler or other receiver. It is essential that no connection be made from the supply side to the return side at any point except through a trap.

Sub-atmospheric systems are similar to vacuum systems but provide control of building temperature by varying the heat output from the radiators. Radiator heat emission is controlled by varying the pressure, temperature and specific volume of steam in circulation. These systems differ from the ordinary vacuum system in that they maintain a controllable partial vacuum on both the supply and return sides of the system, instead of only on the return side. In the vacuum system, steam pressure above that of the atmosphere exists in the supply mains and radiator practically at all times. In the sub-atmospheric system, atmospheric pressure or higher exists in the steam supply piping and radiators only during severe weather. Under average winter temperature, the steam is under partial vacuum. Further reduction in heat output is obtained by restricting the quantity of steam.

The rate of steam supply is controlled by a valve in the steam main or by thermostatically controlling the rate of steam production in the boiler.

Cast-iron sectional heating boilers usually have several outlets in the top. Two or more outlets should be used whenever possible to reduce the velocity of the system in the vertical uptakes from the boiler, and thus to prevent carrying of water into the steam main.

Cast-iron boilers are generally provided with return tappings on both sides. Steel boilers are generally equipped with only one return tapping. Where two tappings are provided, both should be used to guarantee proper circulation through the boiler. The return connection should include either a Hartford return connection or a check valve to prevent the accidental loss of boiler water to the returns. The Hartford return connection is preferred to the check valve, because the check valve is apt to stick slightly open. The check valve also offers additional resistance to the condensate coming back to the boiler.

To prevent the boiler from losing its water under any circumstances, the Hartford return connection is usually recommended. This connection for a one or two-boiler installation is shown in Figure 16. A Hartford return connection includes: (1) a direct connection (made without valves) between the steam side of the boiler and the return side of the boiler, and (2) a close nipple, or preferably an inverted Y-fitting connection about 2 inches below the normal boiler water line.

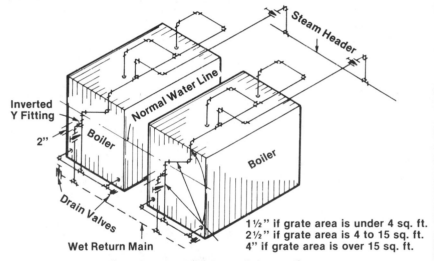

1½" if grate area is under 4 sq. ft.
2½" if grate area is 4 to 15 sq. ft.
4" if grate area is over 15 sq. ft.

**The Hartford Return Connection
Figure 16**

Condensate return pumps are used in gravity systems when the local conditions do not permit condensate to return to the boiler under the existing static head.

A common condensate pump unit for low-pressure heating systems consists of a motor-driven centrifugal pump with receiver and automatic float control. Other types in use include rotary, screw, turbine and reciprocating pumps with steam turbine or motor drive, and direct-acting steam reciprocating pumps.

On vacuum systems, where the returns are under a vacuum, and sub-atmospheric systems, where the supply piping, radiation and the returns are under a vacuum, a vacuum pump is needed to discharge the air and non-condensable gases to atmosphere and to dispose of the condensate. Direct-acting steam-driven reciprocating vacuum pumps were once used where high-pressure steam was available, or where the exhaust steam from the pump could be used. These have been replaced by the automatic motor-driven return line heating pumps developed for this service. The usual vacuum pump unit consists of a compact assembly of an exhausting unit for withdrawing the air-vapor mixture and discharging the air to atmosphere, and a water removal unit which discharges the condensate to the boiler. It is furnished complete with receiver, separating tank and automatic controls mounted as an integrated unit on one base. There are also special steam turbine-driven units which are operated by passing the steam to be used in heating the building through the turbine with only a 2 to 3 psi drop across the turbine required for its operation.

In the ordinary vacuum system, the vacuum pump is controlled by a vacuum regulator which cuts in when the vacuum drops to the lowest point desired, and cuts out when it has been increased to the highest point. These points can be varied to suit the particular system or operating conditions. In addition to this vacuum control, a float control is included which will start the pump whenever sufficient condensate accumulates in the receiver, regardless of the vacuum on the system.

Steam traps are automatic devices used to trap or hold steam until it has given up its latent heat, and to allow condensate and air to pass as soon as it accumulates. In general, traps consist of a vessel to accumulate the condensate, an orifice through which the condensate is discharged, a valve to close the orifice port, mechanisms to operate the valve, and inlet and outlet openings for the entrance and discharge of the condensate from the trap vessel.

Riser, radiator and convector connections must be properly pitched when installed and must be arranged so that the pitch will be maintained under the strain of expansion and contraction. These connections may be made by swing joints which permit the expansion or contraction to occur under heating and cooling without bending the pipe. To take care of expansion in long risers, either expansion joints or pipe swing joints are used. Anchoring of pipes between expansion joints is desirable.

Labor Installing Cast Iron Sectional Boilers

Work Element	Unit	Man-Hours Per Unit
Light oil or combination oil & gas boilers		
40 H.P.	Each	78
50 H.P.	Each	78
60 H.P.	Each	80
70 H.P.	Each	83
80 H.P.	Each	86
90 H.P.	Each	88
100 H.P.	Each	91
110 H.P.	Each	97
120 H.P.	Each	101
130 H.P.	Each	103
140 H.P.	Each	105
150 H.P.	Each	106
Gas fired boilers		
40 H.P.	Each	79
50 H.P.	Each	79
60 H.P.	Each	80
70 H.P.	Each	81
80 H.P.	Each	83
90 H.P.	Each	85
100 H.P.	Each	86
120 H.P.	Each	92
130 H.P.	Each	93
140 H.P.	Each	95
150 H.P.	Each	97
Heavy oil fired L-burner boilers		
60 H.P.	Each	75
80 H.P.	Each	76
100 H.P.	Each	79
120 H.P.	Each	86
140 H.P.	Each	90
160 H.P.	Each	94

Use these figures for installing cast iron sectional boilers and controls. Figures assume boiler sections are factory assembled but jackets, burners and trim are job site installed. No piping or valve work included in these times.
Suggested Crew: 2 steamfitters and 1 helper.

Labor Installing Oil Fired Packaged Boilers

Work Element	Unit	Man-Hours Per Unit
Light oil fired boilers		
40 H.P.	Each	63
50 H.P.	Each	63
60 H.P.	Each	64
70 H.P.	Each	70
80 H.P.	Each	71
90 H.P.	Each	73
100 H.P.	Each	74
125 H.P.	Each	85
150 H.P.	Each	92
200 H.P.	Each	97
250 H.P.	Each	98
300 H.P.	Each	99
350 H.P.	Each	99
400 H.P.	Each	113
500 H.P.	Each	114
No. 2 oil fired or gas boiler		
40 H.P.	Each	63
50 H.P.	Each	64
60 H.P.	Each	64
70 H.P.	Each	71
80 H.P.	Each	74
100 H.P.	Each	81
No. 6 oil (heavy) fired boilers		
80 H.P.	Each	64
100 H.P.	Each	64
125 H.P.	Each	74
150 H.P.	Each	92
200 H.P.	Each	94
250 H.P.	Each	100
300 H.P.	Each	106
350 H.P.	Each	108
400 H.P.	Each	111

Use these figures for installing steel box boilers or marine type packaged boilers. Time includes installation of the boiler complete with burners and basic controls but no piping or valves.
Suggested Crew: 2 steamfitters and 1 helper.

Hot-Water Heating Systems

Hot water is a very useful carrier of heat. Circulating in a closed system, the water absorbs heat in a boiler or heat exchanger and releases it to the heat-using equipment. Hot-water systems can be classified as high temperature, medium temperature, and low temperature.

High-temperature hot water is usually generated in central heating plants and then delivered to the consumers by a distribution system.

The design supply-water temperature for **medium-temperature water systems** ranges from 220 degrees to 300 degrees for outside distribution, large space heaters, absorption refrigeration, and industrial purposes.

The design supply-water temperature for **low-temperature hot-water systems** ranges from 100 degrees to 220 degrees for space heating.

In gravity systems, flow is produced by the difference between the weights of the column of hot water in the supply risers and the column of cooler water in the return risers of the system's circuits. The density (weight) of water decreases as its temperature rises. Therefore, the size of the head available for circulation depends on the temperature difference of the water in the supply and return risers. The force-producing circulation is exactly balanced by the friction loss caused by the water flow. The circulation automatically reaches the velocity at which this condition occurs. Velocity or flow through the piping is low. Therefore, the pipe size must be relatively large to prevent the friction loss at the desired flow rate from exceeding the available head.

In forced-circulation systems, pumps supply the head required to circulate the water. Therefore, forced-circulation systems are especially adaptable for commercial and industrial uses, since the necessary head can be obtained independently of the temperature difference between supply and return. The velocity of water flow in forced systems is higher than in gravity systems. Therefore, smaller pipes and space-heating equipment can be used for the same heating requirements and temperature drop. For the same water flow in pounds per hour (or gallons per minute), the higher the speed of flow, the smaller the section of the pipe can be.

Every hot-water heating system should have an **expansion tank** to handle the expansion and contraction of water that occurs as its temperature changes. The expansion tank should be large enough to permit the water volume change without causing undue strains on the equipment. When water is heated from 40 degrees F to 200 degrees, its volume expands approximately 4 percent. This expansion is handled by an expansion tank or a relief valve which discharges the excess water. The excess water is stored in the expansion tank until the water temperature lowers. Then it is returned to the system.

In **open-tank systems** the expansion tank is freely vented to the atmosphere. Normally, these systems are limited to installations with operating temperatures of 180 degrees or less.

Closed-tank systems use an airtight tank, sealed to prevent free venting to the atmosphere. A closed-tank system can be operated over a wide range of temperatures and pressures.

A **converter** is a shell-and-tube heat exchanger which transfers heat from steam to water. The heated water is used as the heating medium for a hot-water system. The converter shell (of cast iron or steel) must be thick enough, and the tubes adequate to carry the required pressures. Tubes or coils are usually installed in a horizontal, multipass arrangement that gives an equal flow velocity through all the tubes. Normally, flanged openings permit easy removal of tubes for cleaning and repair. In steam converters, the shell or tank holds the steam, and the water to be heated flows through the tubes. In high-temperature converters, the water usually flows through the tubes, and the heated water flows through the shell or tank.

Low-pressure heating boilers are used most often as the heat source for hot-water heating systems. As a rule, when these boilers are used in hot-water systems, the pressure rating is 30 psi. However, pressures as high as 160 psi. can be carried if the water temperature is kept at or below 250 degrees. These relatively high pressures are needed to serve large areas or tall buildings. The boiler operating pressure and the size and design of the expansion tank determine the maximum possible height from the boiler to the top of the highest element of the system.

The various accessories of low-pressure heating boilers include the following:

Every hot-water boiler must have a **relief valve** with a capacity adequate for the gross output of the boiler. The valve, which is connected to the top of the boiler, is necessary to prevent the operating pressure from exceeding the maximum permissible level. A discharge pipe, bottom-mitered to prevent capping and arranged so that there is no danger of scalding anyone, is also essential. Do not place valves in relief or discharge lines, or run discharge piping outside the building in areas where freezing may occur.

A **thermometer** must be placed in the upper part of the boiler to indicate the water temperature.

The boiler also needs a **pressure or altitude gauge,** with a range of at least twice the maximum operating boiler pressure or permissible altitude, to indicate the operating pressure.

The following connections are necessary on low-pressure water-heating boilers:

The **outlet connection** should be fully as large as the manufacturer's tappings. Some installations use a dip tube to keep air out of the supply main. The dip tube, an extension of the outlet pipe, projects into the boiler and causes a dead space at the highest part (where the temperature is higher). The air that separates from the boiler water into the dead space flows through proper connections into the expansion tank.

Properly used, **return connections** prevent thermal shock. If water enters the boiler at a high velocity, it disturbs the thermal circulation of the boiler water and may cause sludge or sediment to pile up in the mud ring. To keep entering water at a low velocity, use a return header connection the same size as the boiler inlet. If a boiler has two return connections, use both. When more than one circulation pump is to be used, connect their discharges to a common pump discharge header before making the boiler connection. Be sure that each pump has a check valve on its discharge side (between the pump and common discharge header) to prevent backflow through the pump circuit. Return header connections must be readily accessible for cleaning, draining, and rodding, and must have plugged openings. Use plugged tees instead of elbows at boiler return connections to assure ample clean-out facilities.

Forced-circulation hot-water systems must have single-flow **control valves** in the supply main. Although the circulating pumps of these systems are turned off when heat is not required for the space-heating equipment, the density difference between the hot and cold water could produce gravity circulation. The single-flow control valve, by remaining closed until the thermostat control starts the circulation pump, blocks gravity circulation to the space heater.

Place separate **blow-off or drain connections** near each boiler, so that the entire system can be drained through the drain valve.

Water-heating boilers must have **water service connections** in the return main, near the boiler, through which to fill the boiler and supply make-up water.

When connected to the chimney, the **breeching** should not project beyond the chimney lining. If a number of boilers are connected to a common breeching, each should have an outlet gas damper. The design of breeching and gas connections should promote unrestricted gas flow.

Because the water circulates in a closed system, the water requirements of hot-water heating systems are low.

Normally the feedwater system of a hot-water installation consists of the following:

A **pressure-reducing valve** is usually installed in the make-up or cold-water line to the boiler. This valve automatically keeps the closed system supplied with water at a predetermined safe pressure. The maximum permissible pressure for standard hot-water heating systems is 30 psi. Therefore, the reducing valve setting should be as low as possible. Valves are usually factory set at 12 psi. This equals a static head of 27.6 feet of water (suitable for buildings with 1, 2 or 3 stories). However, if the static head of the system is high, boilers with higher operating pressure may be required, and the reducing valve must be set higher. Reducing valves have either "inlet" or "outlet" markings or directional arrows to indicate the proper flow direction.

Usually a connection for manual feeding is made in the return main near the boiler. Feeding is regulated by a manually operated globe valve.

A certified approved **backflow preventer** is used to prevent cross connection and is usually installed in the make-up, cold-water line, ahead of manual feed by-pass.

All hot-water heating systems must provide for pressure relief. Otherwise, water expansion can subject equipment to excessive pressures. Each system has a conventional hot-water, pressure-relief valve.

Circulators are circulating water pumps, usually centrifugal. They circulate the hot water through the system and inject it into the boiler against boiler pressure. They vary in size from small, 1/6 hp booster pumps with a capacity of 5 gpm, at 6 or 7 feet head, to large units which pump hundreds of gallons a minute against high heads. Most hot-water heating pumps use mechanical seals. Usually, booster pumps are installed by setting the unit in the pipe line as if it were a flanged fitting and providing the proper wiring. They can be installed in either the supply or the return header adjacent to the boiler, but the supply header is more desirable.

Figure 1 shows a **one-pipe circulation system.** In this system, a single main is carried entirely around the building from the pump discharge to the boiler inlet (return line) as shown. The water flow through the main is constant, except when some of it is by-passed through the heat-distributing units. Therefore, size of the main circuit remains constant from the first to the last heating unit. Supply risers are taken off the main to feed the heat-transmitting units and the returns from the units are reconnected to the main, a short distance along the line. This can be done without interfering with normal operations because differential pressures of heating unit inlets and outlets cause water flow. Differential pressure is produced by the constant drop, from pump discharge to suction, of the pressure head in the main. Therefore, if a supply riser is removed at any point and the return is reconnected into the main at a proper distance along the line, there will be enough pressure difference between the two points to produce circulation through radiation. Usually, 8 or 10 feet between connections will produce desired circulation.

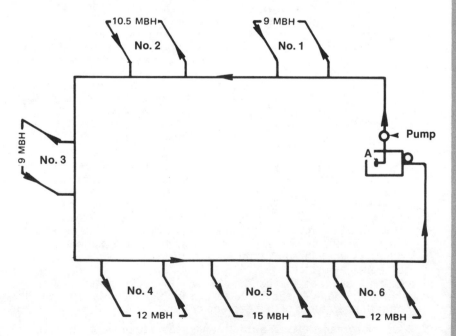

**One-pipe Water Circulation System
Figure 1**

If special flow fittings are installed at the inlet of the supply riser, or at both the supply and the return branches, connections can be placed within one foot of each other. Some designers increase the size of radiation to compensate for the gradual temperature drop in the main. However, if a system is designed for a temperature of 20 degrees, the normal temperature decrease will reduce the boiler capacity only slightly.

The **two-pipe system** has two mains: the supply main feeds water to the risers that serve the heating units; the return main collects the water returned from those units. The two mains run side by side. The supply main decreases and the return main increases in size where the branches connect. Since the heating units of a two-pipe system are connected parallel, it requires a minimum pumping head. Also, if throttle valves or restricting orifices are used in the risers, the flow through individual units can be adjusted easily over a wide range. However, the two-pipe system requires more pipe and pipe fittings than the one-pipe system. Two-pipe systems are classified as direct-return and reverse-return. See Figure 2.

The heating units of the two-pipe, **direct-return system** are parallel. Nevertheless, the water taken from the main to feed the first radiator is returned first, that removed for the second radiator is returned second, and so successively throughout the heating units. Since this procedure causes a progressively greater friction loss in each additional circuit, the flow circuits become hydraulically unbalanced. This condition may cause the first radiator to have a greater flow than is required to develop its full capacity. In a large system the flow through the last unit may be so small that practically no heat is delivered.

In the two-pipe, **reversed-return system**, the water taken from the main to feed the first radiator is returned last to the return main. The water supplied to the last radiator is returned first. As a result, all unit circuits are of approximately equal length, a condition conducive to system balance. The reversed-return system may require more pipe than the direct-return. However, its inherently better flow distribution and natural balance without the aid of additional valves or orifices compensates for the additional cost.

A **series-loop system** may have one or more loops or circuits. All the heat transmitters in a circuit are installed in succession (in series) and the same amount of water flows through each and through the connecting main. For a given available head, the length of the circuit and the number and type of heat transmitters determine the water flow rate and temperature drop. The water temperature decreases progressively as the water flows through each successive heating unit. To compensate for the temperature drop, some designers increase each unit heating surface in proportion to the distance from the source of heat supply. Series systems are frequently used with baseboard radiation units. But neither the flow nor the temperature of the water supplied to individual heating units can be regulated. Therefore, unit heat delivery is usually controlled by air dampers in the baseboard cabinets.

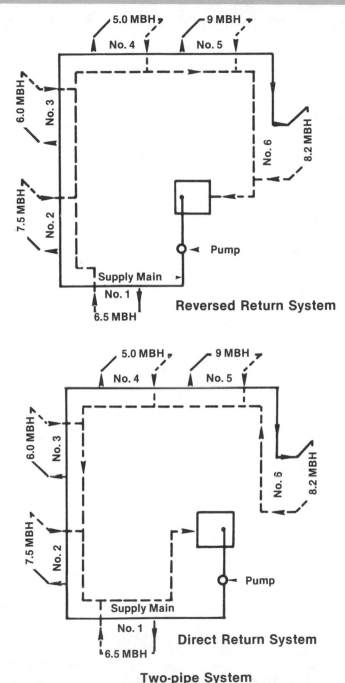

Reversed Return System

Direct Return System

**Two-pipe System
Figure 2**

Labor Installing Hot Water Heating System Pumps

Work Element	Unit	Man-Hours Per Unit
Horizontal split case pumps[1]		
3" x 2½", 1½ H.P.	Each	31
5" x 4", 3 H.P.	Each	41
8" x 6", 7½ H.P.	Each	57
4" x 3", 15 H.P.	Each	37
6" x 5", 25 H.P.	Each	48
8" x 6", 40 H.P.	Each	57
10" x 8", 60 H.P.	Each	63
10" x 8", 100 H.P.	Each	65
Centrifugal type pumps[1]		
1½" x 1¼", ¾ H.P.	Each	15
1½" x 1¼", 1¼ H.P.	Each	15
2" x 1½", 1 H.P.	Each	16
2" x 1½", 2 H.P.	Each	17
2½" x 2", 1 H.P.	Each	31
2½" x 2", 2 H.P.	Each	32
3" x 2½", 1 H.P.	Each	31
3" x 2½", 2 H.P.	Each	33
4" x 3", 1 H.P.	Each	35
4" x 3", 2 H.P.	Each	36
4" x 3", 5 H.P.	Each	37
5" x 4", 3 H.P.	Each	42
5" x 4", 7½ H.P.	Each	43
Inline type pumps[2]		
¾", 1/12 H.P., low velocity	Each	2.1
1¼", 1/12 H.P., low velocity	Each	2.2
1½", 1/6 H.P., low velocity	Each	3.2
2½", ¼ H.P., low velocity	Each	4.3
3", 1/3 H.P., low velocity	Each	4.4
1¼", 1/6 H.P., low velocity	Each	2.5
3", ½ H.P., medium velocity	Each	4.5
3", ¾ H.P., medium velocity	Each	4.6
3", 1 H.P., medium velocity	Each	4.6

[1] Pumps are listed by suction diameter, discharge diameter and horsepower.
[2] Pumps are listed by flange size, horsepower and velocity.
Time includes handling materials into place, installation in a prepared location, cleanup and repairs as needed. Includes electrical hook-up only and no piping.
Suggested Crew: 1 plumber and 1 helper

Labor Installing Compression Tanks & Fittings

Work Element	Unit	Man-Hours Per Unit
Compression tanks		
15 gallons	Each	3.9
24	Each	4.0
30	Each	4.0
40	Each	4.2
60	Each	4.9
80	Each	6.5
100	Each	7.6
120	Each	14.0
144	Each	14.6
180	Each	14.7
230	Each	15.0
280	Each	15.4
Airtrol tank fittings	Each	1.0
Airtrol boiler fittings		
1½" boiler connection	Each	3.9
3" boiler connection	Each	3.9
4" boiler connection	Each	7.8
6" boiler connection	Each	7.8
8" boiler connection	Each	7.9
Air separators		
56 GPM, 2", screwed	Each	3.8
90 GPM, 2½", screwed	Each	5.9
170 GPM, 3", screwed	Each	7.8
300 GPM, 4", flanged	Each	7.9
500 GPM, 5", flanged	Each	9.1
700 GPM, 6", flanged	Each	11.3
1300 GPM, 8", flanged	Each	11.5
2000 GPM, 10", flanged	Each	15.0
2950 GPM, 12", flanged	Each	16.8

Labor hours include handling materials into place, setting in prepared locations, cleanup and repairs as needed.
Suggested Crew: 1 pipefitter and 1 helper

Labor Installing Hot Water Convection Units

Work Element	Unit	Man-Hours Per Unit
Convector type baseboard panels, non-ferrous type		
½" tube, 7½" high enclosure	10 L.F.	1.9
¾" tube, 7½" high enclosure	10 L.F.	2.2
1" tube, 7½" high enclosure	10 L.F.	2.6
¾" tube, 9⁹⁄₁₆" high enclosure	10 L.F.	2.2
1" tube, 9⁹⁄₁₆" high enclosure	10 L.F.	2.6
Radiant cast iron baseboard panels		
7¼" high	10 L.F.	3.9
9¾" high	10 L.F.	4.0
Steel or copper fin tube with 1¼" pipe, commercial type		
Single tier fin tubes	10 L.F.	3.9
Double tier fin tubes	10 L.F.	5.2
Triple tier fin tubes	10 L.F.	7.8
Cast iron free-standing radiators, per S.F. of face		
6 tube, 19" to 32" high	10 S.F.	.8
5 tube, 22" to 25" high	10 S.F.	.7
4 tube, 19" to 25" high	10 S.F.	.6
3 tube, 25" high	10 S.F.	.6
Convector enclosures, 24" high		
4" to 8" deep, 24" long	Each	1.9
4" to 8" deep, 36" long	Each	2.2
4" to 8" deep, 48" long	Each	2.6
10" deep, 24" long	Each	2.2
10" deep, 36" long	Each	2.6
10" deep, 48" long	Each	3.0

Labor includes handling materials into place, installation in prepared locations, connection, cleanup and repairs as needed.
Suggested Crew: 1 pipefitter and 1 helper

Labor Installing Heat Pumps

Work Element	Unit	Man-Hours Per Unit
Single packaged units, air to air		
24 MBTU	Each	3.4
36 MBTU	Each	5.3
48 MBTU	Each	7.6
60 MBTU	Each	8.0
Split system outdoor sections		
36 MBTU	Each	3.7
48 MBTU	Each	3.9
60 MBTU	Each	7.3
Split system indoor sections		
36 MBTU	Each	3.6
48 MBTU	Each	3.8
60 MBTU	Each	7.2

Time includes handling materials into place, setting unit on prepared pad, cleanup and repairs as needed. No electrical run or ducting included.
Suggested Crew: 1 mechanic and 1 helper

Labor Installing Pipe Insulation

Work Element	Unit	Man-Hours Per Unit
Up to ¾" diameter	100 L.F.	13.0
1"	100 L.F.	15.0
1¼"	100 L.F.	17.0
1½"	100 L.F.	17.0
2"	100 L.F.	19.0
2½"	100 L.F.	21.0
3"	100 L.F.	25.0
4"	100 L.F.	26.0
5"	100 L.F.	32.0
6"	100 L.F.	38.0
8"	100 L.F.	48.0
10"	100 L.F.	56.0
12"	100 L.F.	65.0

Use these figures for installing ¾" insulation to pipe, fittings and valves. Hours assume one fitting or valve every 10 feet. Includes handling materials into place, wrapping insulation, repairs and cleanup as needed.

Warm-Air Heating Systems

Warm-air heating systems provide space heating for buildings by circulating warm air. The air is warmed in a furnace and is carried through ducts by gravity flow or forced circulation. A warm-air heating system consists of the following main elements: a source of warm air (usually a furnace or heat exchanger), a supply duct system (including diffusers or registers to carry the warm air from the source to the heated areas), a return air system to carry the cool air back to the source, and a control system.

In **gravity warm-air systems** the warm air is circulated through the air distribution system by natural convection. The difference in weight between the heated air leaving the furnace and the cooler air entering the furnace casing provides the circulation. Since this force is small, the application of gravity warm-air systems is limited.

The quantity of heat that can be supplied to a heated area depends on the temperature and weight of the air discharged. Therefore, three interrelated factors determine the satisfactory operation of a warm-air heating system: (1) Size and location of air ducts, and available circulating head. (2) Heat loss of the building. (3) Heat available from the furnace.

Figure 1 illustrates a simple gravity warm-air system. Warm air is conveyed from the furnace bonnet (top section of furnace casing), through metal ducts designed for minimum flow resistance, to the rooms to be heated. The horizontal runs are called "leaders," the vertical ducts, "stacks" or "risers." Stacks connect the registers, usually installed in room baseboards, floors, or sidewalls just above the baseboard. Usually the stacks are located in inside partitions to prevent chilling (and consequent reduction of the available circulating head). The cooled air returns to the furnace through return registers and ducts, usually located in the floor. Return registers may be placed at either cold or warm wall locations, but cold walls are preferable unless a long, high-frictional-loss duct design must be used. Often, gravity warm-air systems have only one or two centrally located return registers, all on the first floor.

Before connecting with the furnace casing, return ducts usually join a single larger duct which enters the casing near the floor or furnace foundation. This connection must be made below all parts of the furnace which radiate enough heat to cause a countercurrent of warm air. Such a current would oppose the flow of cooled air into the furnace.

A **forced warm-air heating system** is similar to a gravity system. The main difference is that the forced system uses a centrifugal fan or blower instead of natural convection to attain positive air circulation. Figure 2 illustrates a typical forced-air furnace unit which provides the required air circulation head with a motor-driven blower.

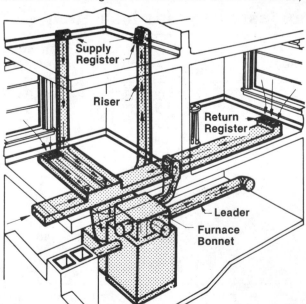

Simple Gravity Warm Air Heating System
Figure 1

Typical Forced-Air Furnace Unit
Figure 2

Since outlet air velocities and temperatures determine the amount of heat supply, they are essential factors in room comfort. Varied distribution of air in rooms can be obtained by specific location of registers.

Locate supply registers near the floor and set them to direct air upward along the wall. You can also place them high in a side wall with horizontal vanes which give a downward air deflection. In either case, locate the supply registers so that warm air from the registers blankets a cold wall and mixes with the cold air which descends from the exposed wall and glass areas.

Locate return registers near the greatest outside exposure. The cold air is then taken directly to the furnace for heating, and the cold draft conditions across the floor are eliminated.

Locate supply openings near the source of the greatest heat loss. Equip them with registers designed to blanket the cold area. This type of location delivers the warm air so that it mixes with cool air from the heat-loss area and infiltration points, and prevents or reduces drafts.

Usually, **dampers** installed in the various warm air ducts control warm air distribution. Locate them either at the branch take-off or at the warm air outlet. Be sure that damper position is indicated by suitable outside arrows, and that labels indicate the proper summer and winter settings.

Ducts can be either rectangular or round. They are usually made of sheet steel and vary in thickness with the cross-sectional area. The larger the area, the thicker the sheet. Warm-air ducts which pass through cold spaces or are located in exposed walls are normally insulated with at least 2-inch insulation. Special connections to and from the furnace casing and fan casings use strips of fire-resistant fabric to reduce or eliminate noise.

Warm-air furnaces can be made of either cast iron or steel. Generally, the minimum thickness of **cast iron furnace** sections is 1/2 inch. These furnaces resist corrosion and high temperature and, because of their relatively large mass, have a large heat-storage capacity. This characteristic gives them a flywheel heating effect. However, they are slow to respond to heat changes.

The metal parts of **steel furnaces** are joined by riveting, welding, or both. Because of their relatively small mass, they can deliver heat rapidly on demand, and can adapt to fast changes in heat requirements. However, their heat-storage capacity is rather small.

Coal-fired furnaces are made of cast iron or steel. Cast iron furnaces are constructed in sections, made gastight by liberal use of furnace cement, asbestos rope, or both. The radiator (secondary heating surface) is usually located on top of the combustion chamber (primary heating surface.) Steel furnaces are made of heavy-gage steel and are riveted and calked or welded at the joint to make them gastight. The fronts, which include the fire, ash pit and draft doors, are usually cast iron. Small steel furnaces usually have a single radiator attached to the rear of the combustion chamber. In larger sizes two radiators may be installed on the furnace sides.

Oil-fired furnace designs are conducive to maximum oil combustion because of their longer gas travel, which means adequate combustion

chamber volume, and their large heating surfaces. These furnaces are usually of the blow-through type, with an air-space pressure higher than the gas-space (combustion chamber or flue) pressure. Compact fan-furnace-burner units designed to install in basements, closets or attics are available.

Gas-fired, warm-air furnaces are direct-fired; that is, the heat is transferred directly from the hot combustion gases to the air which circulates around the furnace and radiator. The two types of gas-fired furnaces, horizontal and vertical, are described below. Each consists of a gas burner, gas controls, fan (on forced-air types), filters, heat exchanger, casing, and sometimes a humidifier. Each furnace has a draft diverter, built into the furnace on some models, and installed separately on the smoke outlet to the breeching on others. The majority of gas burners are Bunsen types and operate with a nonluminous flame. Figure 3 illustrates this type of furnace.

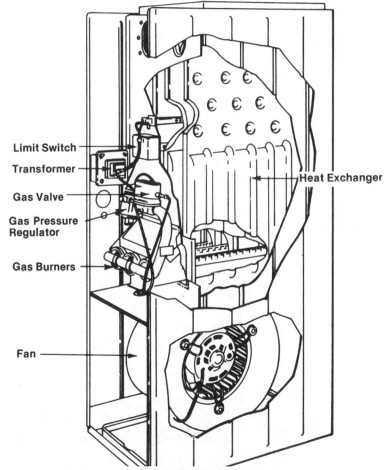

Vertical Gas-fired Warm Air Furnace
Figure 3

In the **horizontal furnace,** the fan, filters, and heat exchanger are aligned in a horizontal position. Since this furnace can be suspended from the ceiling, or installed in attics or crawl spaces, it takes up little or no floor space.

The **vertical furnace** has the same components as the horizontal, but is designed for floor installation. There are two types: the duct type for basement installation and the counterflow which discharges air downward for installation above basement level.

The **duct-type furnace** is designed especially for installation in a duct system with a central air supply. It is built exactly like a gas-fired unit heater except that it has no fan, and flanges are added to facilitate mounting in the duct system.

To assure safety and facilitate maintenance, install the gas-fired duct furnace with the clearances specified by the manufacturer. Include a limit control to prevent overheating in case of an air-supply failure. If the duct in which the furnace is installed also carries refrigerated air for an air-conditioning system, install a bypass duct with dampers to shut off and isolate the furnace unit when the duct is used for cooling.

Size the duct furnace to handle the required amount of air in cubic feet per minute. Do not allow the air to fall below the minimum amount specified by the manufacturer.

Gas furnace manufacturers use a variety of gas furnace manifold controls. Basically, they consist of the following items, in the order of gas flow:

A **main valve** or stop cock at the upstream side of manifold controls. The pilot takeoff valve is installed at the upstream side of the stop cock, so that pilot burners can be lighted without turning on main gas supply.

A low-pressure **gas regulator** to reduce gas pressure from the system, or from the building gas pressure regulator. The low-pressure gas regulator delivers gas at uniform pressure to the furnace burners, regardless of fluctuation in upstream pressure.

An automatic **shut-off safety valve** which stops the gas flow to the burner if the flame fails or pilot goes out.

Gas burner **primary controls** are valves which control the gas flow to gas burners by opening or closing in response to heat demand from the thermostat.
When diaphragm-operated valves are used, a bleeder tube is installed from the chamber on top of the diaphragm to a point directly under the pilot or into the burner chamber.

Observe the following rules when installing the gas line:
Use as short a gas line, and as few fittings as possible. Use pipe dope sparingly and on male threads only; place piping where it will be accessible, and support it adequately. Avoid installing gas pipe through air ducts, supporting beams, or under floor slabs. Install a ground joint union close to the control assembly of the unit, and a drip leg in the vertical riser which supplies the unit.

The diaphragm-operated **gas regulator** is installed in the line, upstream from the operating valve, to provide constant gas pressure at the burners. The bleeder pipe from the chamber above this diaphragm must be vented into the burner chamber, or into an opening in the flue gas vent above the downdraft diverter. A constant escape of gas from the bleeder pipe indicates that the diaphragm is ruptured and must be replaced.

The following requirements must be met when venting a gas furnace: Flue pipe diameter must be at least as large as the flue connection and never less than 3 inches. Maintain a minimum upward slope of 1 inch per linear foot for horizontal runs; a pitch of at least 45 degrees is advisable in cold attics. The flue pipe must run as directly as possible, with a minimum number of turns. Chimneys and flues running through roofs must be extended at least two feet above the roof peak or other object within a 15-foot radius. All flue pipe that extends through the roof must be equipped with a hood.

Be sure that furnace rooms are ventilated to provide adequate air for fuel combustion. Provide an air inlet opening of at least one square inch of free air area for each 1000 BTU per hour of furnace input rating, but never less than 200 square inches. Locate the opening at or near the floor line whenever possible, and be sure that drapes or other furnishings do not block the inlet register. Also install, at or near the ceiling, a louvered outlet-opening with a free area of at least one square inch per 1000 BTU per hour of furnace input rating, but never less than 200 square inches.

In large buildings, a central fan or **hot-blast system** (Figure 4) sometimes performs both heating and ventilation functions. In this system, the air heaters are heat exchangers consisting of pipe coils, finned tubes, or cast iron sections connected into stacks or units by nipples. The intermediate heat carrier is either hot water or steam (from boilers, converters, etc) circulated through the heating elements of the heat exchanger. A fan draws air through the air heater and distributes it to the heated areas through ducts. Because the amount of air required for

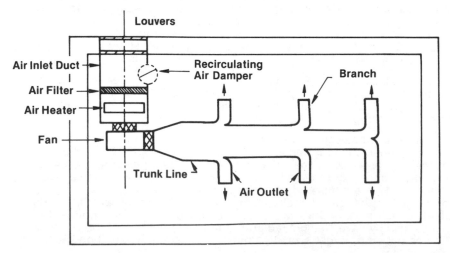

Central Fan System of Heating
Figure 4

heating purposes usually exceeds that required for ventilation, economy is improved by recirculating a portion of the heated air. A common hot-blast arrangement includes the following:

Cold-air inlet duct has louvers to control the inflow of outside air to the heated area.

An **air filter** is located in the inlet air duct just before the air heater.

A **recirculating air damper,** installed before the air filter, permits a regulated recirculation of heated air into the cold-air inlet duct. The combined operation of this damper and the louvers (which control the inflow of cold air) permits the outside air introduced into the system to be tempered.

The **air heater** (a heat exchanger heated by steam or hot water) is located in the air duct, after the air filter.

A **motor-operated fan,** located after the air heater draws the tempered air through the heat exchanger and discharges it to a trunk line.

Labor Installing Gas Furnaces

Work Element	Unit	Man-Hours Per Unit
Upflow heating furnaces		
50 MBTU	Each	3.6
65 MBTU	Each	3.8
80 MBTU	Each	4.0
105 MBTU	Each	5.0
120 MBTU	Each	6.0
140 MBTU	Each	6.7
160 MBTU	Each	7.1
180 MBTU	Each	7.3
Counterflow heating furnaces		
80 MBTU	Each	3.5
105 MBTU	Each	4.9
120 MBTU	Each	6.0
140 MBTU	Each	7.3
Horizontal heating furnaces		
80 MBTU	Each	14.0
105 MBTU	Each	14.0
125 MBTU	Each	14.4
140 MBTU	Each	14.6

Hours include handling materials into place, installation in a prepared location, supply hook-up, cleanup and repairs as needed.
Suggested Crew: 1 mechanic and 1 apprentice

Labor Installing Oil & Electric Furnaces

Work Element	Unit	Man-Hours Per Unit
Electric furnaces, 240 volt, 8 to 27 KW, upflow, counterflow or horizontal	Each	3.7
Oil fired upflow furnaces		
85 MBTU	Each	3.8
100 MBTU	Each	5.1
125 MBTU	Each	7.3
Oil fired counterflow furnaces		
100 MBTU	Each	4.9
125 MBTU	Each	7.1
Low-boy pattern oil furnaces		
100 MBTU	Each	4.8
125 MBTU	Each	7.3
150 MBTU	Each	8.2
200 MBTU	Each	10.9
250 MBTU	Each	12.3
Oil fired horizontal furnaces		
85 MBTU	Each	4.1
100 MBTU	Each	5.0
125 MBTU	Each	7.3
150 MBTU	Each	8.7
200 MBTU	Each	11.0
250 MBTU	Each	12.4
335 MBTU	Each	14.8
Fuel oil storage tanks set on floor		
275 to 500 gallons	Each	10.0
1,000 to 2,000 gallons	Each	17.0
5,000 gallons	Each	35.0

Man-hour figures include the positioning and connecting of units complete with fans, filters, safety controls and burners.
Man-hour figures for installation of fuel storage tanks set on floors include supports, saddles, coatings and fittings.
Suggested crew: two to four men depending on size of equipment and scope of job.

Cooling Systems

The four basic components of every mechanical refrigeration unit are the evaporator, the expansion valve, the condenser and the compressor.

An **evaporator** is that part of the system where heat is removed from the air or water that is to be cooled. Refrigerant in the system evaporates inside this coil and, in so doing, absorbs heat from the air or water being cooled. The refrigerant enters the evaporator in a liquid state through an expansion valve, capillary, or other metering device, then absorbs heat as it boils or evaporates into a vapor. The refrigerant finally leaves the evaporator in the form of a gas.

The **direct expansion evaporator** is a fin-and-tube type coil which cools the air directly. The air that is cooled is in direct contact with the cold exterior surface of the coil. Ductwork may be used to deliver the conditioned air to the rooms to be cooled, or, an evaporator coil can be installed within the conditioned space. The direct expansion evaporators are used in the window-type air-conditioning unit.

The direct expansion **shell-and-tube type** has a heavy shell which contains the water which is passed over the outside of the refrigerant tubes. Because the water circulating through the system is under pressure, this outer shell must be strong enough to withstand the pressure in it. The liquid refrigerant enters the direct expansion shell-and-tube evaporator and passes back and forth through the inside of the tubes. By the time the refrigerant reaches the discharge point, it has absorbed enough heat to evaporate. This heat has been given up by the water being chilled. The water to be chilled enters and gradually proceeds to the other end of the shell, flowing up and down over the refrigerant-carrying tubes. Baffles direct the flow. Having transmitted or lost considerable heat to the colder refrigerant in the tubes, the water leaves the shell at the discharge point chilled to a predetermined temperature. Thus, it goes into the system to pick up more heat and cool the areas being conditioned. Refrigerant temperatures naturally are lower than the water entering or leaving the shell, but not so low as to cause it to freeze.

In the **flooded shell-and-tube type** evaporator the refrigerant is vaporized on the outside of finned or plain tubes which are within a closed shell. The tube bundle has water pumped through it to be chilled, and heat is absorbed by the refrigerant. Space is usually provided above the tubes submerged in the boiling liquid refrigerant for the separation of liquid droplets from the leaving vapor.

The **expansion valve** provides a means for controlling the flow of the liquid refrigerant into the evaporator. While flowing through the expansion valve, the liquid cools to the temperature in the evaporator. This expansion valve also serves to maintain the high pressure in the condenser which the compressor made possible.

The **condenser** is able to condense hot refrigerant gas into liquid because the refrigerant is at a high pressure and, therefore, will condense at a high temperature.

There are three types of condensers, air-cooled, water-cooled, and evaporative.

The **air-cooled condenser** usually is a fin-and-tube coil, and therefore looks very much like a direct expansion evaporator coil. The refrigerant vapor is condensed inside its tubes by giving up heat to air which is circulated across the condenser's outer surface. This air usually is discharged outdoors.

Air-cooled condensers are most popular in areas where use of water for air conditioning is restricted. These condensers are also used in mobile applications, such as railway and bus air conditioning, where the use of water is impractical. In addition, air-cooled condensers are commonly used in small window-sill type units because they eliminate the cost and inconvenience of installing water piping.

Centrifugal fan air-cooled condensers operate on the same basic principle as propeller fan units. Use of a centrifugal fan allows indoor, as well as outdoor, condenser installation.

Air-cooled condensers have two principal disadvantages: they require power to operate their fan or fans and the condenser's capacity is lowest on hot days when maximum capacity is required. However, the air-cooled condenser is gaining steadily in popularity for both residential and commercial applications. In many areas it is the most economical and convenient condensing method.

If water supply is adequate and economical, a **water-cooled condenser** is usually selected, especially for large commercial applications. Most water-cooled condensers are of the shell-and-tube type. And they are more compact than either the air-cooled condenser or the evaporative condenser. In addition, the water-cooled condenser does not require a fan, as the other two do. At first glance, the water-cooled condenser reminds you of the direct expansion water chiller. There's one big difference. Most water-cooled condensers have the cooling water flowing inside the shell but outside the tubes. This is exactly opposite of the chilling water evaporator where the water is on the outside of the tubes.

In the shell-and-tube type water-cooled condenser, the water is brought in one end, conducted back and forth inside the tubes (water movement from one end to the other is referred to as single pass), and is finally removed. Most shell-and-tube type water-cooled condensers have an even number of passes so the outlet water connection can be at the same end of the shell as the inlet connection. Because water picks up heat as it flows through the condenser, it will leave the condenser several degrees warmer than it was when it entered.

The compressed refrigerant, in the form of a hot gas, is usually forced into the top of the water-cooler condenser. This hot gas spreads to both ends of the shell. As the cooled water removes heat from this hot gaseous refrigerant, the refrigerant condenses on the outside of the tubes and drips down to the bottom of the shell. The liquid refrigerant then collects in a small well, or "reservoir," at the bottom of the shell and leaves the condenser from this point. Liquid in this reservoir is maintained even when there is very little of it in the shell itself. This accumulation of high-pressure liquid refrigerant at the bottom of the shell not only serves as a small reserve supply, but also as a liquid seal.

A major disadvantage of the water-cooled condenser lies in the fact that it uses a large amount of water. After this water has picked up heat in the process of condensing the refrigerant, it must either be disposed of, or cooled so that it can be reused.

An **evaporative condenser** is a different type of condenser than either the air-cooled or water-cooled types because it uses an entirely different principle. As the name implies it uses evaporation to cool and thus condense the refrigerant.

The refrigerant which is to be condensed in an evaporative condenser is passed through the coil tubes. Water is sprayed downward over the outside of these finless tubes. At the same time, air from either outdoors or indoors is blown or drawn upward through the water spray and over the outside of the same tubes. Some of the water which is sprayed evaporates as the air moves over it. The remainder of the water is collected in a tank beneath the tubes and is pumped back to the spray nozzles over and over again. Evaporation is a cooling process because it is accompanied by the absorption of heat. As the water evaporates, it absorbs heat from the hot gaseous refrigerant inside the tubes and condenses the refrigerant. This liquid refrigerant then is drained into a tank called the liquid receiver.

The **compressor** takes the cool gas from the evaporator to compress it to a hot gas at a high pressure. The two most popular types of compressors used on commercial mechanical refrigeration systems are reciprocating and centrifugal.

There are two classifications of **reciprocating compressors:** open compressors, and hermetic compressors.

In the reciprocating-type compressor each piston is forced up and down within a cylinder by means of a connecting rod. The connecting rod is driven by a crankshaft which is propelled by some source of mechanical power. Practically all reciprocating compressors used in air conditioning are driven by electric motor. The open-type compressors are so designated because one end of the crankshaft extends outside the crankcase.

In the **hermetic type** the motor and compressor are sealed inside the same enclosure. All hermetic-type compressors are driven directly by electric motors. There are two types of hermetic compressor which are commonly classified as serviceable hermetic and welded hermetic or non-serviceable hermetic. The motor and compressor of this serviceable hermetic compressor are enclosed in a common housing. The cylinder head, end bell, and crankcase cover plates can be removed for servicing the internal mechanism. The compressor section is basically the same as the open-type compressor. However, the compressor and motor are connected by a common shaft inside the compressor housing. Suction gas is drawn thru the motor section and is used to cool the motor. The welded hermetic compressor and motor are sealed within a welded steel shell. The shell can not normally be removed for service in the field. Usually, the motor and compressor are mounted in a vertical plane. The only basic difference between this compressor and the serviceable hermetic compressor is the means of access to the compressor and motor assembly.

The compressors are oil lubricated. The only purpose of oil in a refrigeration system is the lubrication of the compressor. In the operation of the compressor, some oil will be pumped out to the discharge line, and must be returned to the crankcase.

If you place water in a bucket and then swing the bucket in a circular plane fast enough, all the water will remain in the bucket. Centrifugal compression is a result of centrifugal force. The pressure is developed by whirling the gas at a high speed. The pressure in a rotating impeller is lowest at the eye or in the center. Here is where the vapor enters. The gas is thrown to the tip of the wheel by centrifugal force and develops high gas velocities at the periphery of the wheel. This velocity or kinetic energy is converted into pressure and supplies a continuous gas flow at an increased pressure.

The **centrifugal compressor** inherently runs at high speeds and pumps a large volume of gas at relatively low compression ratios. Some centrifugal compressors run at speeds as high as 18,000 rpm. Centrifugal compressors can be single or multi-stage depending on the application and refrigerant used. In order to achieve good efficiency, normal practice has been to use compression ratios up to 2.5 for multi-stage units and up to 4.5 for single-stage units. The majority of centrifugal compressors use one or two stages of compression. For large refrigeration loads, there is a need for single refrigeration units of large capacity. The centrifugal refrigeration unit satisfies this need. For example, a single centrifugal refrigeration unit can be used in place of many reciprocating units. It has also been proven through years of actual field operation that centrifugal refrigeration equipment is very reliable and dependable. There are relatively few moving parts compared to reciprocating equipment.

The refrigeration cycle or system is divided in half with respect to pressure. From the expansion valve through the evaporator to the compressor is the low-pressure section. From the compressor through the condenser and back to the expansion valve constitutes the high-pressure section. These divisions are commonly referred to as the "high side" and the "low side."

In general, the maximum size of a single reciprocating compressor is about 100 tons. Systems having loads of more than 100 tons may have either a multiple reciprocating compressor or a single centrifugal compressor, depending on economic considerations.

For loads over 100 tons, the best choice is a centrifugal compressor, or an absorption refrigeration system. An absorption system may be considered for smaller loads. But for all absorption systems a heat source such as steam, high-temperature water, or gas, must be available.

Cooling towers cool the water that is circulated through the condensers. A cooling tower will cool water below the dry-bulb temperature of ambient air. The cooling effect by evaporation is completely dependent upon the ability of the surrounding air to absorb moisture; this is directly related to the wet-bulb temperature.

Cooling towers are classified according to the method of moving air through the tower: natural draft, induced draft, or forced draft.

The **natural-draft cooling tower** is designed to cool water by means of air moving through the tower at the low velocities common in open spaces during the summer.

An **induced-draft cooling tower** is provided with a top mounted fan that induces atmospheric air to flow up through the tower, as warm water falls downward.

A **forced-draft cooling tower** uses a fan to force air into the tower.

Air Handlers

Fans are divided into two general classifications: (1) centrifugal or radial flow, in which the air flows radially through the impeller within a scroll-type housing, and (2) axial flow, in which the air flows axially through the impeller within a cylinder or ring.

Centrifugal fans are further subdivided into types denoted by the curvature or slope of the impeller blades, the angle of which largely determines the operating characteristics. For a given output, a forward inclination of blade indicates a relatively low speed of operation, and a backward inclination indicates a relatively high speed of operation. Many intermediate forms are also found.

For comfort air conditioning the following components are included in the air distribution system:

A **supply fan or blower** of a capacity which will produce the air changes in the occupied area as required for the type of occupancy. The air supply to this fan is usually made up of fresh air and return air and the total passed through an air filter. The filter may be any one of the several types. The fan discharges into a discharge plenum for the purpose of supplying air to either a heating coil or cooling coil or both simultaneously. After passing the air through these coils, the air enters the respective hot and cold plenums, sometimes called hot and cold decks, or ducts. The various zones are then connected to these plenums. This may be done in pairs (in case of a double-duct system) or the zones may consist of a single duct. Duct systems are also classified as "high pressure" or "high velocity" ductwork and "low pressure" or "low velocity."

The term **"high pressure"** or **"high velocity"** ductwork includes ductwork systems and plenums from the fan discharge to the final high-velocity mixing boxes with a static pressure range of 3" through 7" W.C. (Water Column).

High-velocity or high-pressure systems with fan static pressures of 3 inches or greater are defined as high pressure. Usually the static pressure will be limited to a maximum of 7 inches and duct velocities limited to 4000 feet per minute.

A high-velocity double duct system begins with a high-pressure fan of Class II or III design and conveys air through sound-treated high-velocity ductwork connected to sound and pressure-attenuating mixing units. Connections to the outlets of the reduction units are treated as low velocity.

Smaller-sized ductwork utilizing higher velocities permits conveyance of air to areas limited by construction and reduces floor-to-floor heights in most, but not all cases. Ductwork is generally round for greatest strength, tightness, and economy, but oval and rectangular ducts may be used when necessary, particularly for large risers.

A necessary component of the high-pressure system is the **mixing box**, the function of which is to blend air at two different temperatures for proper delivery to the rooms. This requires special pressure-reducing air valves at both hot and cold inlets, mixing baffles to prevent stratification of air and sound attenuation treatment to absorb noise generated by the air valves.

The term **"low pressure"** or **"low velocity"** ductwork applies to systems with fan static pressures less than 3 inches. Generally duct velocities are less than 2,000 feet per minute.

Low-velocity double duct systems have been in use for many years. But it was not until the 1950's that their use became common. Space for the installation of the double ducts is a main consideration for this system and must be provided during the initial planning. Difficulties in providing for this space in modern structures with low floor-to-floor heights and flush ceilings, together with the need for developing a compact distribution system for existing buildings, has brought about the development of high-velocity double duct systems. High velocity saves ceiling space and duct shaft space, but requires greater attention in the selection of fans and equipment. Also, higher duct velocities require increased fan static pressures, hence, increased operating costs. On the other hand, high-velocity systems have been found to be easy to balance and control and have much greater flexibility for partition changes.

Generally, high-velocity systems are applicable to large multi-story buildings, where the advantage of saving in duct shafts and floor to floor heights is more substantial. Small two and three-story buildings normally would use low velocity.

A **double duct system** generally consists of a blowthrough fan unit discharging filtered air through stacked or adjacent heating and cooling coils into separate plenums and ductwork with thermostatically controlled mixing dampers at room locations.

The inherent advantage of a double duct system is that individual room conditions may be maintained from a central system, within the limitations of supply air temperatures, by blending of hot and cold air through automatically controlled mixing devices, without the need of separate systems. Another important plus is flexibility in that additional individually controlled rooms can be easily incorporated, at modest costs, after the building is completed.

In modern buildings of multiple exposures designed for variable functions and changing occupancy, individual room control is essential. The double duct systems will be the usual choice.

Double duct systems for low pressure are usually tiered hot and cold ducts within the furred space, generally above corridors. The manner of distributing proper temperature air to the room is through right-angle interlinked mixing dampers operated by motors controlled through thermostats. This type of system also generally uses the same corridor plenum area around the ducts for conveyance of return or exhaust air.

Labor Installing Electric Chillers

Work Element	Unit	Man-Hours Per Unit
Centrifugal electric chillers		
70 ton capacity	Each	67
80	Each	69
100	Each	70
120	Each	76
140	Each	82
170	Each	86
200	Each	95
240	Each	102
280	Each	108
320	Each	115
370	Each	136
500	Each	166
700	Each	204
900	Each	228
Reciprocating electric chillers		
19 ton capacity	Each	45
28	Each	47
45	Each	49
69	Each	62
99	Each	68
158	Each	80
Steam absorption chillers		
50 ton capacity	Each	79
75	Each	83
100	Each	91
150	Each	101
200	Each	109
270	Each	130
350	Each	150
480	Each	168
580	Each	205
710	Each	225
890	Each	260

These figures include handling materials into place, installation of chiller, starter and trim in prepared locations, supply hook-up, cleanup and repairs as needed. No electrical run included.
Suggested Crew: 1 mechanic and 1 helper

Labor Installing Cooling Towers

Work Element	Unit	Man-Hours Per Unit
Centrifugal blow-through towers		
50 ton capacity	Each	46
75	Each	47
100	Each	59
150	Each	63
250	Each	75
500	Each	90
1000	Each	133
Axial-flow cooling towers		
100 ton capacity	Each	45
200	Each	73
400	Each	76
700	Each	98
Vertical draw-through towers		
10	Each	30
20	Each	31
40	Each	32
75	Each	45
125	Each	64
Air-cooled condensers		
10 ton capacity	Each	31
25	Each	33
50	Each	51
75	Each	52
100	Each	55
150	Each	71
250	Each	82
Water-cooled condensers		
1 ton capacity	Each	16
3	Each	46
5	Each	47
10	Each	48
20	Each	58
30	Each	63

Time includes handling materials into place, installation in prepared locations, supply hook-up, cleanup and repairs as needed.
Suggested Crew: 1 mechanic and 1 helper

Labor Installing Air Handlers

Work Element	Unit	Man-Hours Per Unit
1 or 4 zone standard units with coils, filter, damper and vibration isolators		
1,750 to 2,750 CFM	Each	43
1,800 to 3,750 CFM	Each	44
2,250 to 6,600 CFM	Each	44
2,700 to 8,300 CFM	Each	45
3,350 to 8,700 CFM	Each	46
4,250 to 12,500 CFM	Each	47
5,050 to 16,000 CFM	Each	58
6,500 to 19,000 CFM	Each	59
7,700 to 23,500 CFM	Each	60
10,200 to 27,000 CFM	Each	61
13,200 to 35,000 CFM	Each	70
18,600 to 47,000 CFM	Each	71
55,000 CFM	Each	74
Add for drip pans & miscellaneous structural components	S.F.	.2

Time includes handling materials into place, securing in prepared positions, supply hook-up, cleanup and repairs as needed.
Suggested Crew: 1 or 2 mechanics and 1 helper.

Labor Installing Split System Air Conditioning Units

Work Element	Unit	Man-Hours Per Unit
Outdoor condenser sections		
18 MBTU	Each	7.3
24 MBTU	Each	7.7
30 MBTU	Each	8.1
36 MBTU	Each	10.2
42 MBTU	Each	11.2
48 MBTU	Each	13.2
60 MBTU	Each	15.0
90 MBTU	Each	20.5
120 MBTU	Each	30.0
180 MBTU	Each	33.0
240 MBTU	Each	35.5
360 MBTU	Each	44.0
480 MBTU	Each	55.0
Indoor section with evaporative coil and casing		
18 MBTU	Each	3.7
24 MBTU	Each	4.1
30 MBTU	Each	5.1
36 MBTU	Each	6.1
42 MBTU	Each	7.2
48 MBTU	Each	8.3
60 MBTU	Each	9.3
90 MBTU	Each	11.0
120 MBTU	Each	11.5

Includes handling materials into place, setting units in prepared locations, electrical hook-up, cleanup and repairs as needed. Indoor section assumes installation as part of a packaged heating system and includes coil, capillary feed, drip pan, drain connection, liquid and suction lines, insulated casing and motor relays in a upflow, counterflow or horizontal evaporator unit.
Suggested Crew: 1 mechanic and 1 helper

Labor Installing Air Terminal Units

Work Element	Unit	Man-Hours Per Unit
Induction units with cabinets		
Primary 50 to 250, secondary 90 to 510 CFM	Each	3.5
Primary 200 to 450, secondary 350 to 860 CFM	Each	3.6
Primary 300 to 550, secondary 515 to 985 CFM	Each	3.7
Fan coil units with cabinets		
155 to 215 CFM range	Each	4.0
210 to 300 CFM range	Each	4.1
305 to 410 CFM range	Each	4.3
395 to 600 CFM range	Each	4.5
530 to 805 CFM range	Each	4.7
670 to 1015 CFM range	Each	4.8
1005 to 1200 CFM range	Each	5.0
Mixing boxes, under window type		
200 to 400 CFM range	Each	2.9
300 to 500 CFM range	Each	3.2
Double ducted ceiling type mixing boxes		
100 to 200 CFM range	Each	2.9
300 to 550 CFM range	Each	3.6
800 to 1400 CFM range	Each	4.9
2200 to 3200 CFM range	Each	5.0
Single duct terminal reheat unit for hot water		
1500 to 2000 CFM range	Each	3.7
2000 to 3000 CFM range	Each	3.8
3000 to 4000 CFM range	Each	5.1
4000 to 5000 CFM range	Each	5.3
5000 to 6000 CFM range	Each	7.0
6000 to 7000 CFM range	Each	7.3

Labor includes handling materials into place, securing unit in a prepared location, cleanup and repairs as needed.
Suggested Crew: 1 sheet metal worker and 1 laborer

Labor Installing Diffusers

Work Element	Unit	Man-Hours Per Unit
Diffusers, round step-down type, fixed pattern		
6" to 10" neck	Each	1.0
12" to 16" neck	Each	1.6
18" to 24" neck	Each	2.3
30" to 36" neck	Each	3.0

Labor includes handling materials into place, securing in a prepared location, cleanup and repairs as needed.

Labor Installing Expansion Type Air Conditioners

Work Element	Unit	Man-Hours Per Unit
Incremental self-contained A/C units with direct cooling coils		
Expansion cooling with hot water heating coils, 9,000 to 15,000 CFM	Each	15.0
Expansion cooling with steam heating coils, 9,000 to 19,000 CFM	Each	15.0
Expansion cooling with electric heating coils, 9,000 to 15,000 CFM	Each	7.8
Wall mount room air conditioners, 6,000 to 32,00 BTU	Each	4.0

Labor includes handling materials into place, installations in prepared locations, electrical hook-up, cleanup and repairs as needed.
Suggested Crew: 1 mechanic and 1 helper

Labor Installing Sheet Metal and Fiber Duct Work

Work Element	Unit	Man-Hours Per Unit
Fabricate sheet metal duct		
20" to 94" perimeter	100 L.F.	21
96" to 126" perimeter	100 L.F.	41
128" to 190" perimeter	100 L.F.	61
192" to 240" perimeter	100 L.F.	76
242" to 360" perimeter	100 L.F.	93
Install sheet metal duct		
20" to 94" perimeter	100 L.F.	26
96" to 126" perimeter	100 L.F.	44
128" to 190" perimeter	100 L.F.	90
192" to 240" perimeter	100 L.F.	125
242" to 360" perimeter	100 L.F.	165
Insulate sheet metal duct (asbestos paper or aluminum foil)	100 L.F.	4
Fiber duct for slab heating or cooling system		
6" inside diameter	100 L.F.	1.1
8" inside diameter	100 L.F.	1.4
10" inside diameter	100 L.F.	1.4
12" inside diameter	100 L.F.	1.6
16" inside diameter	100 L.F.	2.0
20" inside diameter	100 L.F.	2.6
24" inside diameter	100 L.F.	3.9

Installation includes hangers but does not include grills and registers

Installation of fiberglass duct will vary with the manufacturer. Fabrication is usually performed in the sheet metal shop and includes making patterns, cutting, forming, seaming, soldering, attaching stiffeners, and hauling to the site. Installation includes unloading, storing on site, handling into place, hanging, fastening, and soldering.
Suggested Crew: 2 to 3 sheet metal workers depending on duct size and scope of job.

Labor Installing Grills and Registers

Work Element	Unit	Man-Hours Per Unit
4" x 8"	Each	0.5
6" x 12"	Each	0.7
16" x 16"	Each	0.8
20" x 36"	Each	1.0

Time includes handling materials into place, installation, cleanup and repairs as needed.

Size to Weight Conversions for Galvanized Iron Duct

Largest side dimension and weights, 3.5 lb. static pressure
To 12", 26 gauge minimum, 1.3 lbs. per S.F.
13" to 30", 24 gauge minimum, 1.5 lbs. per S.F.
31" to 54", 22 gauge minimum, 1.8 lbs. per S.F.
55" to 84", 20 gauge minimum, 2.1 lbs. per S.F.
85" and over, 18 gauge minimum, 2.8 lbs per S.F.
16 gauge, 3.5 lbs per S.F.

Labor Installing Dampers

Work Element	Unit	Man-Hours Per Unit
6" to 8" diameter	Each	.7
12" to 14" diameter	Each	.8
16" to 18" diameter	Each	.9
20" to 24" diameter	Each	1.0
30" to 36" diameter	Each	1.2

Labor includes fitting fabricated opposed blade dampers in ductwork, cleanup and repairs as needed.
Suggested Crew: 1 sheet metal worker

Labor Installing Exhaust Fans

Work Element	Unit	Man-Hours Per Unit
10" to 15" diameter fans, 150 to 4000 CFM		
¼ H.P.	Each	2.2
⅓ H.P.	Each	2.2
½ H.P.	Each	2.9
¾ H.P.	Each	3.7
1 H.P.	Each	3.9
18" to 20" diameter fans, 1500 to 6000 LFM		
⅓ H.P.	Each	2.3
½ H.P.	Each	3.2
¾ H.P.	Each	3.9
1 H.P.	Each	4.1
1½ H.P.	Each	4.3
2 H.P.	Each	4.6
3 H.P.	Each	5.8
25" to 36" diameter fans, 2700 to 21,000 CFM		
½ H.P.	Each	3.2
¾ H.P.	Each	4.0
1 H.P.	Each	4.2
1½ H.P.	Each	4.5
2 H.P.	Each	4.6
3 H.P.	Each	5.8
5 H.P.	Each	6.0
7 H.P.	Each	6.4

Labor includes handling materials into place, securing in a prepared location, electrical hook-up, cleanup and repairs as needed.

Labor Installing Heating & A/C Controls

Work Element	Unit	Man-Hours Per Unit
Aquastats	Each	1.0
Pressure control or limiting devices	Each	1.0
Remote bulb temperature controls	Each	1.7
Modulating motorized valves		
½" valve	Each	1.0
¾" to 2" valve	Each	1.7
2½" valve	Each	2.1
3" valve	Each	2.7
Humidity controls	Each	1.3
Room thermostats	Each	.5
Modulating damper motors	Each	1.0
Modulating valve motors	Each	1.0
Universal relays	Each	1.0

Labor includes handling materials into place, securing, connecting, cleanup and repairs as needed.

16 electrical section

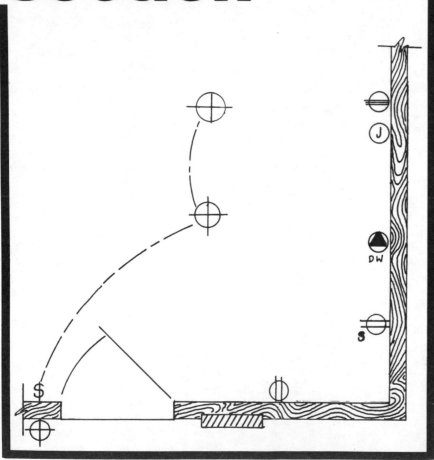

Interior Wire Systems

Electrical wiring in buildings is encased in conduit to reduce the potential hazard of electrical current.

Rigid steel conduit (RSC) is approved for both exposed and concealed work. For ordinary conditions rigid metal conduit is the most expensive. The advantages of metal conduit are (1) it is fireproof, (2) it is moistureproof, (3) it is strong mechanically and can withstand physical abuse, and (4) it successfully resists the normal action of cement when embedded in concrete.

The two primary disadvantages are the high material costs and the high labor costs for installation. Both the inside and outside surfaces of rigid steel conduit require protection against corrosion. The most common protection is a zinc coating. Other protections are enamel or polyvinyl chloride coatings. To help distinguish zinc coated rigid steel conduit from ordinary commercial pipe, the interior is coated with black lacquer or a clear enamel or lacquer. The coating gives a glossy finish to the interior.

Polyvinyl chloride coated conduit is used only in severe corrosion areas. In these areas all conduit and fittings require treatment and all damage to the coating during installation should be repaired.

Rigid steel conduit which is not composed of approved corrosion-resistant material cannot be installed in or beneath cinder fill where it may be subject to permanent moisture unless the conduit is buried 18 inches beneath fill or is protected on all sides by two inches of non-cinder concrete.

Rigid aluminum conduit can be used in the same manner as rigid steel conduit except that it is not suitable for embedding in concrete, installing in the ground or in the fill beneath floor slabs, or for coating with polyvinyl chloride. The major advantage of aluminum conduit is its light weight which results in lower handling and labor costs.

Electrical metallic tubing is commonly called "EMT" or "Thin Wall Conduit." It is similar in many respects to rigid metal conduit and is suitable for circuits rated at 600 volts or less. The advantages of electrical metallic tubing are (1) lower material cost, (2) light weight and (3) lower handling and installation costs. It may be used for concealed as well as exposed work. However, it may not be used (1) where subject to severe physical damage during or after construction, (2) if protected from corrosion solely by enamel, or (3) in cinder concrete or fill, unless located and protected as required for rigid metal conduit.

Rigid nonmetallic conduit includes products formed from nonmetallic compositions such as polyvinyl chloride, fiber or asbestos cement. When properly formed, treated and installed, nonmetallic conduit offers resistance to crushing, impact, corrosion, electrolysis, rust, rot, high and low temperatures, chemicals and sunlight reactions.

Where use is permitted by the National Electrical Code, nonmetallic conduit offers several advantages such as light weight, ease of cutting and the use of slip-on fittings with welding solvents. Connect nonmetallic conduit to metallic conduit or fittings with appropriate adapters. These are threaded for the metal termination and solvent-welded to the nonmetallic conduit.

Rigid nonmetallic conduit and fittings can be used under the following conditions to carry 600 volts or less:

1. Direct earth burial not less than 18 inches below the surface. If less than 18 inches, it should be encased in not less than 2 inches of concrete.

2. In concrete walls, floors and ceilings.

3. In locations subject to severe corrosive influences and where subject to chemicals for which the materials are specifically approved.

4. Under cinder fill.

5. In wet locations. In laundries or other wet locations and in locations where walls are frequently washed, the entire conduit system including boxes and fittings must be installed to prevent water from entering the conduit. All supports, bolts, straps and screws must be of corrosion-resistant materials or protected with corrosion-resistant materials.

Flexible metal conduit is used where flexibility is required due to movement or vibration of equipment, or where bends and offsets would be difficult with rigid conduit. To meet various physical or chemical requirements, "flex" is made of galvanized steel, aluminum, brass or bronze. The most common type is galvanized steel. Practically the same National Electrical Code rules apply to the flexible as to the rigid conduit. The rules governing the insulation on the wires are the same as for rigid conduit; outlet or switch boxes must be installed at all outlets or switches; the conduit must be continuous from outlet to outlet, must be securely fastened to the boxes, and must be provided with proper bushings.

Flexible metal conduit can not be used (1) in wet locations unless conductors are of the lead-covered type or of other type specially approved for the conditions; (2) in hoistways except as provided by the National Electrical Code; (3) in storage-battery rooms; (4) in any hazardous location except as permitted by the National Electrical Code; or (5) where rubber-covered conductors are exposed to oil, gasoline, or other materials having a deteriorating effect on rubber. Flexible metal conduit may not be used as a grounding means unless both the conduit and fittings are approved by Underwriter's Laboratories for that use.

Liquid-tight flexible metal conduit is similar to regular flexible metal except that it is covered with a liquid-tight nonmetallic sunlight-resistant jacket. Liquid-tight flexible metal conduit may be used in exposed or concealed locations where conditions of installation, operation or maintenance require flexibility or protection from liquids, vapor or solids.

Surface metal raceway and multi-outlet assembly are wiring methods intended for exposed installation on walls, ceilings, and floors in dry locations. The raceway consists of base and cover sections that, when assembled complete with fittings, form a complete raceway system. A common name for this type of system is Wiremold which is a trade name of a surface metal raceway manufacturer. The trade name for surface raceway complete with integral mounted outlets is Plugmold.

The number and trade sizes of conductors installed in any raceway should not be more than those for which the raceway was designed. This data is marked on the raceway or on the package in which it is shipped. Power and telephone service can be carried through combination raceways with separate channels to segregate high and low potential wiring.

Outlet boxes and fittings are constructed of sheet-steel, sheet-aluminum, cast metal, malleable-iron, or molded-composition material. Metals not inherently resistant to corrosion must be protected against corrosion. Terms for boxes and fittings include:

An **outlet box** is a box of metal as insulating means for connection to a wiring system. It is intended primarily to enclose splices and wiring devices or to support a fixture or other equipment intended for similar installation. This box may or may not be provided with studs or a bar hanger or with clamps for securing cable, tubing, or conduit.

A **conduit box** is a cast metal box provided with threaded hubs or threaded holes for connection to rigid metallic conduit or electrical metallic tubing.

A **flush-device box** is a box provided with ears or flanges with tapped holes spaced to accept the mounting yokes of a wiring device or devices. This box is provided with a mounting means (which may be the raceway system on cast boxes with threaded hubs), and it may or may not include clamps for connection of cable, tubing, or conduit.

All conductors are made of either copper or electrical-conductor (EC) grade aluminum. The insulator used for conductor wire is either thermoplastic or rubber.

Thermoplastic insulated wire that does not carry a suffix number is rated 600 volts and is identified as follows:

T has a single conductor with flame-retardant thermoplastic insulation.

TW indicates a single conductor with flame-retardant, moisture-resistant thermoplastic insulation.

THHN has a single conductor with flame-retardant and heat-resistant thermoplastic insulation with a jacket of extruded nylon.

THW wire has a single conductor and flame-retardant moisture and heat-resistant thermoplastic insulation.

THWN indicates a single conductor with flame-retardant, moisture-, and heat-resistant thermoplastic insulation with a jacket of extruded nylon.

MTW wire has a single copper conductor and flame-retardant, moisture, heat-, and oil-resistant thermoplastic insulation; or a single copper conductor with flame-retardant, moisture-and heat-resistant thermoplastic insulation and an oil-resistant jacket of extruded nylon.

TA is a single conductor wire with thermoplastic and felted asbestos insulation and a flame-retardant non-metallic covering.

TBS wire has a single conductor, thermoplastic insulation and a flame-retardant nonmetallic covering.

Power and control cable is a multi-conductor cable. The individual strands are either type TFN, TFFN, THNN or THWN conductors. Cable with two conductors is flat or round. Cable with three or more conductors is round.

Rubber-insulated wire is rated 600 volts except when designated with a numerical suffix and is identified by the following letters:

R indicates a single conductor with rubber insulation and a moisture-resistant, flame-retardant, nonmetallic covering.

RW identifies single conductor wire with moisture-resistant rubber insulation and a moisture-resistant, flame-retardant, nonmetallic covering.

RH wire is single conductor with rubber insulation of the heat-resistant grade and a moisture-resistant, flame-retardant, nonmetallic covering.

RHW has a single conductor with moisture- and heat-resistant rubber insulation and a moisture-resistant, flame-retardant, nonmetallic covering.

RHH is single conductor wire with rubber insulation of the heat-resistant grade and a moisture-resistant, flame-retardant nonmetallic covering.

RU wire has a single conductor with latex insulation and moisture-retardant, nonmetallic covering.

RUW indicates wire with a single conductor, moisture-resistant latex insulation and a moisture-resistant, flame-retardant nonmetallic covering.

RUH indicates single conductor wire with heat-resistant latex insulation and a moisture-resistant, flame-retardant, nonmetallic covering.

XHHW wire has a single conductor and cross-linked thermosetting polyethylene moisture-resistant and heat-resistant insulation.

SA wire has a single conductor with silicone rubber insulation and a moisture-resistant, flame-retardant, nonmetallic covering.

SIS wire has a single conductor with synthetic thermosetting insulation which is heat-resistant, moisture-resistant and flame-retardant. No overall covering is provided as it is used for switchboard wiring only.

D used as a suffix indicates wiring with two rubber-insulated nonmetallic-covered conductors laid parallel under an outer nonmetallic covering.

M used as a suffix indicates a cable with two or more rubber-insulated nonmetallic-covered conductors twisted together under an outer nonmetallic covering.

L used as a suffix indicates wire or cable with a lead sheath.

AL used as a suffix indicates wire or cable with an aluminum sheath. AL as a prefix indicates aluminum conductor.

A suffix number in combination with R, RH, RW, or RHW indicates working voltage greater than 600 volts. The suffix number is the maximum working voltage in hundreds of volts.

Types R, RH, RHW, RW and SA are made in sizes No. 14 AWG and larger with copper conductor, and No. 12 AWG and larger with aluminum conductor. Types RU, RUH, and RUW are made in sizes No. 14 to No. 2 AWG with copper conductor and No. 12 to No. 2 AWG with aluminum conductor. Type SIS is made in sizes No. 14 to No. 4/0 AWG with copper conductor and No. 12 to No. 4/0 AWG with aluminum conductor.

Power and control cable conductors may be any size from No. 14 AWG through No. 2 AWG for copper conductor and from No. 12 AWG through No. 2 AWG for aluminum conductors. An extruded thermoplastic sheath provides the overall covering. Power cable is intended for installation in continuous rigid cable supports.

Wiring Devices

Manufacturers have established an informal quality designation for switches and receptacles. The superior quality device of each manufacturer is usually designated "Specification Grade." The common quality device which conforms with the minimum requirements of Underwriters Laboratories Inc. and National Electrical Manufacturers Association is usually designated "Standard Grade." Some manufacturers have an "Intermediate Grade" which is usually between the two other grades. No standards have been established for each grade. Therefore one manufacturer's "Specification Grade" will not necessarily be equal in quality to any other manufacturers' "Specification Grade" devices.

A "Specification Grade" device generally has a heavy duty molded body and is both back and side wired.

Receptacles are intended to be installed in surface or flush-mounted outlet boxes with integral mounting yokes. Those listed as "grounding type" have a single voltage and single current rating and are provided with a contact member intended for equipment grounding purposes only. Where a terminal for the connection to the grounding conductor is provided, it is indicated by a green-colored finish or the word "Green."

Grounding type receptacles have a special grounding circuit between the device yoke and the metallic flush type boxes. They can be installed in walls without a bonding jumper as permitted by the National Electrical Code. The terminal for connection of a grounded circuit (neutral conductor) is white and is marked label "WH" or "WHITE." All other terminals are a different color. Receptacles and their matching plugs have ratings which indicate the design amperes, volts, poles, wires and whether the device is intended for grounding.

Switches

General-use snap switches can be installed in flush device boxes or on outlet box covers. Snap switches are classified in two categories: AC-DC General Use and AC current circuits. AC general-use switches are marked "AC" to limit their use to alternating-current circuits. AC-DC general-use switches are not limited to "AC" circuits.

A **cabinet** is an enclosure designed either for flush or surface mounting, and provided with a frame, matt, or trim in which a swinging door or doors may be hung. A **cutout box** is an enclosure designed for surface mounting and with a swinging cover secured directly to and telescoping with the walls of the box proper. A **junction** or **pull box** is an enclosure designed for either flush or surface mounting with a solid unhinged cover secured directly to the box proper.

Cabinets and cutout boxes are made of sheet steel, sheet aluminum, cast iron or copper-free aluminum. Unless the metal is inherently resistant to corrosion, both the inside and outside surfaces are protected by enameling, galvanizing or plating. A cabinet or box which is designated as a NEMA Type I is reasonably tight and does not have any open holes or slots unless they are to be closed by subsequently installed equipment.

An enclosure intended for outdoor use is designed to exclude beating rain. The enclosure usually has some external means for mounting.

Installation

Exposed conduit must be installed parallel with or at right angles to building walls and must be supported as required by the National Electrical Code. The code requires supports at different intervals depending on the size of conduit and its type. Supports can be one- or two-hole pipe straps, one-hole malleable iron pipe straps or conduit hangers. Fasten conduit supports and boxes as follows: to wood with screws or screw type nails; to masonry with threaded metal inserts, metal expansion screws, toggle bolts or powder-actuated fasteners; and to steel with machine screws, bolts, or powder-actuated fasteners.

Conduit larger than one inch in reinforced concrete slabs should be parallel with or at right angles to the main reinforcement. When at right angles to the reinforcement, the conduit should be close to one of the supports of the slab. Conduit in concrete should be located so that it does not affect the strength of the slab. It should be surrounded by a minimum of one inch of concrete. Where embedded conduit crosses expansion joints, concrete-tight expansion fittings and bonding jumpers must be provided. Conduit installed in the ground or in fill beneath floor slabs may be encased in concrete not less than three inches thick. The top of the concrete envelope must be directly under the floor slab.

Changes in directions of runs are made with symmetrical bends or metal fittings. Field-made bends and offsets should be made with a conduit-bending machine or "hickey." Crushed or deformed raceway must not be installed.

In suspended ceilings only lighting system branch circuit raceways should be fastened to ceiling supports.

Conduit should be fastened to all sheet-metal boxes and cabinets with two locknuts where insulating bushings are used and where bushings cannot be brought into firm contact with the box. Otherwise a single locknut and bushing may be used.

A pull wire is inserted in each empty raceway in which wiring is to be installed if the raceway is more than 50 feet long and has more than two 90-degree bends or in any raceway that is more than 150 feet long.

Boxes are provided wherever required for pulling of wires, making connections or mounting of devices or fixtures. Boxes for metallic raceways should be the cast-metal hub type when located in normally wet locations and when mounted on an exterior surface. Boxes in other locations can be sheet steel except that aluminum boxes must be used with aluminum conduit and nonmetallic boxes may be used with nonmetallic wiring systems.

16 Electrical

Boxes in brick, block or tile walls should be square-cornered tile type or standard boxes with square-cornered tile-type covers. Cast-metal boxes installed in wet locations and boxes installed flush with the outside of exterior surfaces should be gasketed. Separate boxes must be provided for flush or recessed fixtures when required by the fixture terminal operating temperature. Fixtures must be readily removable for access to the boxes unless ceiling access panels are provided. Boxes and supports should be fastened to wood with wood screws or screw-type nails of equal holding strength, with bolts and expansion shields on concrete or brick, with toggle bolts on hollow masonry units and with machine screws or welded studs on steel work.

When pulling wire, avoid the cutting or abrasion of insulation by using a lubricant on the wire. But the lubricating compound must not damage the insulating materials. All wires in a conduit are bundled and pulled at one time. Pulling lines are attached by direct connection to the conductors or with a cable grip.

Nonmetallic-sheathed cable (Romex) may be installed exposed on walls and ceilings in protected areas or concealed in hollow walls under floors or above ceilings. Outlets and switches are connected by running the cable into the outlet box. All splices must be enclosed in outlet or junction boxes. This requirement applies to both exposed and concealed installation.

Wiring devices such as receptacles and switches are located at standard heights. The location of wall outlets is measured from the finished floor to the center of the outlet or switch box. Standard heights in inches are as follows:

Telephone outlets..12
Duplex receptacles..12
Light switches...48
Receptacle for wall fan..78
Receptacle over counter for bench.......................46
Clock outlets...84 to 96

In concrete, tile or other noncombustible walls, boxes and fittings should be installed so that the front edge of the box or fitting will not set back of the finished surface more than ¼ inch. In walls constructed of wood or other combustible materials, outlet boxes and fittings are set flush with the finished surface.

Labor Installing Conduit In Buildings

Work Element	Unit	Man-Hours Per Unit
Rigid galvanized conduit		
½"	100 L.F.	6.9
¾"	100 L.F.	9.3
1"	100 L.F.	11.5
1¼"	100 L.F.	16.2
1½"	100 L.F.	18.5
2"	100 L.F.	22.0
2½"	100 L.F.	29.0
3"	100 L.F.	30.0
3½"	100 L.F.	40.0
4"	100 L.F.	50.0
5"	100 L.F.	95.0
6"	100 L.F.	122.0
Aluminum conduit		
½"	100 L.F.	6.8
¾"	100 L.F.	8.1
1"	100 L.F.	9.3
1¼"	100 L.F.	11.5
1½"	100 L.F.	13.8
2"	100 L.F.	16.0
2½"	100 L.F.	18.5
3"	100 L.F.	25.0
3½"	100 L.F.	28.0
4"	100 L.F.	35.0
5"	100 L.F.	58.0
6"	100 L.F.	80.0
Electric metallic tubing		
½"	100 L.F.	6.8
¾"	100 L.F.	8.1
1"	100 L.F.	9.3
1¼"	100 L.F.	11.6
1½"	100 L.F.	13.8
2"	100 L.F.	16.0
2½"	100 L.F.	18.0
3"	100 L.F.	21.0
4"	100 L.F.	31.0

Labor for handling, securing, cleanup and repairs as needed for conduit installed under an exposed ceiling up to 15' high and including installation of hangars at 8 foot intervals, fittings, and supports. Conduit over 1" assumes installation of 3 elbows and 3 couplings each 100 feet. Aluminum conduit over 2" assumes rigid steel fittings.
Suggested Crew: 1 electrician and 1 helper

Labor Installing Conduit Under Slabs

Work Element	Unit	Man-Hours Per Unit
Rigid galvanized conduit		
½"	100 L.F.	4.6
¾"	100 L.F.	5.7
1"	100 L.F.	8.0
1¼"	100 L.F.	8.6
Electric metallic tubing		
½"	100 L.F.	3.5
¾"	100 L.F.	4.9
1"	100 L.F.	6.3
1¼"	100 L.F.	9.3

Labor includes handling, setting supports, laying conduit in an open trench, 13 fittings for each 100 feet, cleanup and repairs as needed.
Suggested Crew: 1 electrician and 1 helper

Conduit and Wire Conversion Factors

Wire Ampacity	Wire Size		Conduit Size	
	AWG	Millimeters	Inches	Millimeters
20	12	2.5	¾	20
25	10	4	1	25
30	8	6	1¼	30
45	6	10	1½	40
60	4	16	2	50
80	3	25	2½	60
100	2	35	3	75
130	1/0	50	3½	90
165	3/0	70	4	100
200	4/0	95	4½	125
235	250 MCM	120		
280	280	150		
325	400	185		
360	500	240		
415	600 MCM	300		

Labor Pulling Wire In Conduit

Work Element	Unit	Man-Hours Per Unit
No. 14 solid wire	100 L.F.	.7
No. 12 solid wire	100 L.F.	.8
No. 10 solid wire	100 L.F.	1.1
No. 8 solid wire	100 L.F.	1.4
No. 6 stranded wire	100 L.F.	1.5
No. 4 stranded wire	100 L.F.	1.7
No. 3 stranded wire	100 L.F.	1.8
No. 2 stranded wire	100 L.F.	1.9
No. 1 stranded wire	100 L.F.	2.2
No. 1/0 stranded wire	100 L.F.	2.8
No. 2/0 stranded wire	100 L.F.	3.0
No. 3/0 stranded wire	100 L.F.	3.6
No. 4/0 stranded wire	100 L.F.	4.1
No. 250 MCM stranded wire	100 L.F.	4.3
No. 300 MCM stranded wire	100 L.F.	4.5
No. 350 MCM stranded wire	100 L.F.	4.8
No. 400 MCM stranded wire	100 L.F.	5.7
No. 500 MCM stranded wire	100 L.F.	5.8
No. 600 MCM stranded wire	100 L.F.	5.9
No. 750 MCM stranded wire	100 L.F.	7.6
No. 1,000 MCM stranded wire	100 L.F.	9.1

Labor assumes three strands of type TW copper wire are pulled in conduit in a 100 foot run with a simple layout. Deduct 5% for aluminum wire. No connecting is included.
Suggested Crew: 1 electrician and 1 helper

Cable Trays

A cable tray is a structure of metal units forming a rigid assembly for carrying electrical cables. In its simplest form, the cable tray can be a continuous "U" shaped section, although there are different types according to function. Cables rest on the bottom of the tray or supports and are held in by two side rails.

It is not the intent of this section to cover the types of cables that can be installed in cable trays. The cable types are in accordance with the plans and specifications and Article 318 of the National Electrical Code.

Cables trays are designed to meet the requirements of the National Electrical Manufacturers Association. NEMA has established four classes of cable trays. Class I can span 12 feet and support 35 pounds per linear foot. Class II can also span 12 feet but can support 50 pounds per linear foot. Class III can span up to 20 feet and support 45 pounds per linear foot. Finally, Class IV cable tray can span 20 feet and carry 75 pounds.

Four types of trays are available to meet differing service requirements. They are the ladder, trough, channel and solid bottom types.

A **ladder tray** is a prefabricated metal structure with two side rails connected by crosswire members, uniformly spaced, generally referred to as rungs.

Trough tray has a ventilated bottom and closely spaced supports supported by side rails.

A **channel cable tray** is prefabricated metal structure consisting of a one-piece ventilated channel section not exceeding 4 inches in width.

The fourth type of tray is the **solid bottom tray.** It is engineered to meet the applicable NEMA Standards for the ventilated type of cable tray and load specifications for the class II tray on 12-foot spans.

The term **ventilated** means that the tray has at least 40 percent open area in the surface supporting the cables.

Trays manufactured to NEMA Standards are available with the following dimensions for the ladder and trough types. Nominal lengths of straight sections are 12 and 24 feet; widths are 6, 12, 18 and 24 inches; inside depths are 3, 4 and 6 inches (outside depth dimensions can not exceed inside depths by more than $1\frac{1}{4}$ inches); the radii of elbows are 12, 24 and 36 inches; elbows are 30, 45, 60 and 90 degrees. Rung spacing for ladder-type trays are 6, 9, 12 and 18 inches on centers. For trough-type trays the maximum open spacing between transverse elements is 4 inches, measured in a direction parallel to the tray side rails.

Channel-type trays are also made in 12- and 24-foot lengths; widths are 3 and 4 inches nominal outside dimensions; depths are $1\frac{1}{4}$ to $1\frac{3}{4}$ inches nominal outside dimensions; the radii are 12, 24 and 36 inches; degree of arc of elbows, 30, 45, 60 and 90 degrees.

Fittings are available to allow change in horizontal and vertical alignment, branch runs or transition from one width to another.

Accessories include covers, cover clamps and clips, conduit adapters, expansion joint splice plates, drop outs, blind ends, tray to box connectors, divider strips, cable clamps and structural supports.

Covers are available as solid, louvered or ventilated types. Solid covers provide protection from sunlight and accumulation of dirt. They also guard against unauthorized contact with the cables, mechanical damage, help isolate cables from fire and reduce radio frequency interference. Louvered covers provide shelter from sunlight and accumulation of dirt. They guard against unauthorized contact or mechanical damage to the cables and permit some circulation of air around the cables. Ventilated covers normally have rungs that are spaced approximately 1/2 inch apart. They guard the cables against unauthorized contact or mechanical damage and provide full air flow ventilation.

Cover clamps are generally a pair of flat bars that are placed on top of the cover and bottom of the tray and clamped by a pair of bolts on the exterior of both side rails. Cover clips are an economical device for holding down horizontal or slowly rising runs. Clips of the locking screw type should not be used on trays that are positioned at an angle more than 30 degrees from the horizontal. Clips without the locking screw are used on horizontal runs that are free from up-lift forces (generally created by short circuit currents in power cables) and vibration.

Conduit **adapter dropouts** are used to run cables from trough-type cable tray to conduit. They consist of a sized plate with a hole for specified conduit size, nuts, bolts and washers. Installation of the plate requires cutting the tray bottom to clear the conduit. Dropouts are used for vertical drops out of the tray without use of support for the suspended cable.

Blind ends provide a cover for the "dead-end" termination of runs. Tray to box connectors are used to form a neat connection of cable trays to junction boxes, switchgear, or any flat surface perpendicular to the trays. **Divider strips** provide a barrier, inside the cable trays, when you have to separate cables of different voltages or cables serving different functions.

Cable tray supports come in varied shapes, but can be either cantilever bracket, trapeze or individual rod suspension. Bracket types are used to support cable trays from columns, framing, concrete walls or other vertical supports. Trapeze or individual rod suspension types are used to support trays from overhead structural members or in combination with cantilever brackets.

Application

Supports should be located so that connectors between horizontal straight sections of cable tray runs fall between the support point and the quarter point of the span. Unspliced straight sections should be used on all simple spans. Vertical straight lengths should be supported at intervals not more than 8 feet on centers. The figures show support requirements for the various types of tray fittings. Sloping trays are supported at the same intervals as horizontal trays.

Article 318 of the National Electrical Code establishes installation requirements:

1. Each run of cable trays must be installed before installations of cables.

2. Cable trays must be mechanically connected to any enclosure or raceway into which the cables contained in the cable tray extend or terminate.

3. When cable trays are installed in tiers, the minimum vertical clearance must be 12 inches.

4. Cable trays may extend through partitions or walls, other than fire walls, provided the section of tray is continuous and unventilated.

5. Cable trays may extend vertically through dry floors and platforms if the tray is totally enclosed where it passes through the floor or platform and for a distance of six feet above the floor to provide protection from physical injury.

6. Cable trays can go vertically through floors and platforms in wet locations where there are curbs or other suitable means to prevent water flow through the openings if the cable tray is totally enclosed for a distance of 6 feet above the openings to provide protection from physical damage.

7. A minimum working vertical clearance of 6 inches must be maintained from the top of the cable tray to the ceilings, beams and other obstructions.

Cable trays with splice plates, used for joining sections, should have all contact surfaces cleaned. Nuts should be on the outside of the side rails. Metal-to-metal contact is undependable in joints using adjustable splice plates. Flexible bonding jumpers must be used across all adjustable splice plate connections. Trays that are vinyl or epoxy coated generally do not have the coating at the splice sections. After the splice plates have been secured and tested for electrical continuity, all uncoated areas should be coated with the proper materials furnished by the tray manufacturer.

Depending upon the spacing between supports, a fully loaded tray can show a deflection of 2 inches at midspan. This does not necessarily mean that the trays are overstressed, provided they were properly designed and installed in accord with plans and specifications. Where appearance is important, provide additional supports.

Cable trays and fittings must be bonded and grounded. Grounding must be to the building's grounding system. The number of grounding points should be as indicated on the plans, but the minimum is two grounding points.

Cables are placed in the trays either by laying over the side or by being pulled along the tray. Avoid damaging the cables and trays during installation. Trays supported on brackets make it quite convenient to place cables by laying over the side. Trays supported by trapeze or rods will generally require pulling the cable into place. Short lengths of cable may be "pulled" by a basket grip, providing the strain does not damage the cable sheath. Continuous lengths of cable up to 1,000 feet with as many as 12 bends can be installed with the tray manufacturer's installation tools. They consist of rollers, single and triple pulleys. Location and spacing of rollers and pulleys should be indicated on a drawing prepared by the cable tray manufacturer. Drawings should take into consideration the number and size of cables that will be installed.

The following points should be observed during installation:

The specified tray (or trough) sections, fittings and accessories are as indicated on the contract plans.

When you have to field-cut the trays to meet job requirements, remove all burrs and sharp edges from cut pieces. Rung spacing between adjacent sections should not exceed that specified for the tray.

Splice places must be properly secured to form a tight connection. Bolt heads must be located on the inside of the tray.

Expansion splice connector fasteners should be securely locked but must leave the splice plate free to move. The splice assembly must be positioned to permit expansion and contraction.

Adjustable or expansion-type splices must be equipped with a bonding jumper cable of the size specified.

Vertical and horizontal spacing between trays must allow for placement and removal of cable and cable rollers, but in no case should it be less than the requirements of the NEC.

No holes can be made in the side rails, except for splicing side rails of field-cut sections and attachment of vertical cover shear connectors.

Cable rollers used during the installation of cable should be located as closely as possible to a support.

Labor Installing Wireway & Cable Tray

Work Element	Unit	Man-Hours Per Unit
Square type wireway		
2½" x 2½"	10 L.F.	2.3
4" x 4"	10 L.F.	2.9
6" x 6"	10 L.F.	3.9
8" x 8"	10 L.F.	4.9
Galvanized cable tray		
6"	10 L.F.	1.9
12"	10 L.F.	2.2
18"	10 L.F.	2.4
24"	10 L.F.	2.5
Aluminum cable tray		
6"	10 L.F.	1.8
12"	10 L.F.	2.0
18"	10 L.F.	2.2
24"	10 L.F.	2.4

Labor includes handling materials into place, installation of supports, fittings and tray, cleanup and repairs as needed.
Suggested Crew: 1 electrician and 1 helper

Busways

A busway is a prefabricated assembly of bus bars mounted in a protective sheetmetal enclosure or trough. Busway wiring systems provide a flexible means of distributing power in industrial and commercial buildings.

Conductors in busways are made of either copper or aluminum. These conductors are in the form of flat bars with a rectangular cross section, except for busway of the lighting type and trolley busway where the conductors may be round, square or odd shapes. The conductors are electroplated at all joints to reduce the contact resistance. For grounded-neutral systems, an insulated ground bus is provided, in addition to the phase conductors.

Insulation used between phases and from phase to ground includes such materials as mylar, varnished glass cloth, glass tape, mica tape, molded phenolic and molded polyester. Insulation supports at plug-in openings, at joints, and at intervals along the bus provide mechanical support and determine the short-circuit rating of feeder and plug-in busways.

Enclosures for feeder and plug-in busways are of two types: ventilated and totally enclosed or unventilated. Enclosures for busways with plug-in provision have access openings at regular intervals on all standard straight lengths except for lighting busways where the plug-in access hole is continuous along the bottom of the busway.

Busway systems are widely used as risers and feeders in large commercial and apartment buildings. **Prefabricated busway** usually allows a substantial saving in installation cost where long runs of feeder and branch circuits are involved. Each type of busway offers special features which make it particularly well suited to certain types of installations. Standard sections are available in lengths from one to ten feet.

Feeder busway is used primarily in industrial buildings for feeder circuits between the service entrance and the main switchboard, or for high capacity feeders to load centers. It is ideally suited for inserts and feeders to resistance welders and high frequency equipment where the impedance of the feeder circuit must be minimized to reduce voltage regulation. It is available in ratings of from 600 to 4000 amperes and in two-pole or three-pole assemblies.

Plug-in busway is used in industrial plants where the feeder circuits must provide high current capacity, low-voltage-drop and frequent power tap-offs for branch circuits. It is available in ratings from 600 to 4000 amperes and in two-pole or three-pole assemblies.

Power plug-in busway is used in industrial plants, commercial buildings, manual training shops, laboratories and garages for feeding individual power loads when flexibility of power plug-in receptacles is necessary. This permits machines to be relocated and plugged-in to the busway at a new location with a minimum of re-wiring. It is available in ratings from 100 to 1500 amperes in two-pole and three-pole assemblies. Three-pole assemblies are available for three-wire and four-wire systems.

Lighting plug-in busway is used in industrial plants, office buildings, department stores, garages, warehouses, truck terminals, shipping docks and railway freight terminals where the loads are primarily lighting fixtures, small power tools and small machines. It provides maximum flexibility for these low-capacity loads, since loads can be tapped into the busway at any point along the busway. Only one rating, 50 amperes, is available for two, three, and four-wire systems.

Current limiting busway is used in commercial, industrial and institutional buildings for connection between the service entrance and the main switchboard, or wherever it is necessary to reduce the available short-circuit current. It is available in ratings from 1000 to 4000 amperes for three-phase, three-wire or four-wire systems.

Trolley busway provides a flexible power distribution system for supplying individual loads wherever a mobile power takeoff is required, such as for cranes and portable tools.

Busway Fittings

Hardware for all electrical connections (nuts, bolts, washers, etc.) is furnished by the manufacturer of the busway. This provides a spring-type connection at busway joints and low contact resistance. Only connection hardware furnished with the busway should be used.

Transposition fittings are used on feeder and plug-in busways to change the bus bar arrangement from a phase rotation of ABC to CBA, or ABCN to NCBA.

Power takeoffs are sometimes used where feeder busway connects to a vertical riser or other branch circuit. Use of power takeoffs including overcurrent protection is required by the National Electrical Code at points where the busway is reduced in size under the following conditions:

If the length of the smaller busway exceeds 50 feet or

If the smaller busway has a current rating less than one-third the current setting of the overcurrent device next back on the line.

The power takeoff consists of a power takeoff box, factory assembled as an integral part of the feeder busway length, plus a circuit breaker or fusible takeoff device which is also normally factory assembled to the takeoff box. Panelboards can be used as power takeoff devices, but must be assembled on the site.

Reducers may be used in lieu of power takeoffs where the size of busway is reduced to a smaller ampere rated busway and overcurrent protection is not required by the National Electrical Code; i.e., the length of the smaller busway does not exceed 50 feet and has a current rating at least one-third that of the overcurrent device next back on the line.

Cable tap boxes are used for connection of feeder busway or plug-in busway to cable and are similar to power takeoff boxes but are fitted with lugs only.

Flanges are used to encircle the busway where it passes through a wall or floor, or enters electrical equipment enclosures.

Elbows, offsets and tees are furnished with feeder and plug-in busway for changes in direction, offsets or branches in the busway run. Elbows are also available for lighting busways.

Busway plugs are used for power takeoffs from plug-in busway and are available in fusible cover-operated plugs, switch-operated fusible plugs and circuit breaker plugs.

Expansion lengths for feeder or plug-in power busway allow for expansion and contraction of the busway due to changes in temperature, or where differential expansion exists between the building and busway. In general horizontal runs do not require expansion lengths. However, if both ends of the busway are fixed and expansion cannot be tolerated, expansion lengths are necessary.

Feed-in boxes are used with lighting busway and trolley busway and serve as a junction box for power connections to the busway from the power source. These are available for center feed or end feed.

Tap boxes for lighting busway are used for connections from the busway to loads and are available in three types: unfused, fusible and circuit breaker type. These are used when the lighting bus tap is to a branch circuit consisting of wire in conduit. The box provides means for terminating the conduit.

Plugs for lighting busway are used for cord connections to lighting fixtures.

Trolleys are used for movable power taps from trolley busway and can be unfused, fused, or receptacle type. Either standard-duty or heavy-duty trolleys are available.

Fire stops are used where busway passes through fire-rated walls or floors.

Standard hangers for flatwise or edgewise mounting of feeder busway or plug-in busway are furnished by the busway manufacturer. However, hangers for "trapeze mounting" can be substituted for the standard hangers for most types of feeder and plug-in busway. Spring hangers are recommended for vertical riser applications. Rigid hangers are recommended for special mounting conditions.

Installation

Busway designed for indoor use may be installed in exposed dry locations for circuits not exceeding 600 volts. It can't be installed when subject to severe physical damage or corrosive vapors, in hoistways, in hazardous locations, or outdoors unless specifically approved for the purpose.

Busway designed for outdoor use is also limited to 600 volts maximum. Outdoor busway systems are not be confused with isolated phase bus and high-voltage bus structures which are designed for above 600 volts.

A typical installation procedure for busway would be as follows:

Make up and install support rods or brackets on five-foot centers which are to support standard plates or channels.

Raise busway to the final level and put the hanger plates or channels in place to carry the weight.

The standard sections butt together. The housing is joined by fastening the splice plate, which is already attached to one end of the section of busway. After the splice plate is fastened, the bus bars are ready to be joined.

The bus bars are easily bolted together by use of carriage bolts and lock washers. Most bolts are accessible from one side of the housing. This is particularly important when the run is mounted against a wall or other surface. Generous spacings between bolts provide real "elbow room." Any type of flat, opened wrench can be used. However, a torque wrench should be used. The conductors are bare at the joints to simplify initial installation and subsequent maintenance. Ample air clearance eliminates the need for taping. Solid covers provide protection from the entry of foreign material. Two carriage bolts are used to join each pair of bus bars. Since the bolt holds itself from turning, only one wrench is required. Use two bolts per joint to guarantee a large contact area. Bolts have tapered shoulders so that they pull the bars into line as they are tightened. Lock washers prevent loosening after the bars have been pulled up tight. The joint cover is put into place and fastened to complete the joint.

After the entire run is in final alignment, install the clip-type hangers, which fasten the busway securely to the hanger plates, or channels. Refer to the manufacturer's installation procedure for the busway actually furnished.

Plug-in busway can be mounted on walls, or suspended from ceilings, I-beams or messenger cables. These methods of installation are shown in the illustration.

Plug-in busway of the lighting type can be suspended below a ceiling on hanger rods or straps, surface mounted against the ceiling, suspended from a messenger cable support or, in case of a hung-type ceiling, recessed into the ceiling. It may also be mounted on a wall or up a wall as a riser. Field cut lighting bus with a hack-saw to the exact length required. Individual sections of busway are joined either with a built-in spring-type joint or standard hardware (alignment pins, spring-loaded connectors, coupling plates, etc.), furnished with the busway.

Trolley busway is available from the factory in either curved or straight sections complete with support hardware, feed-in boxes and trolleys. The trolley connectors are inserted or removed at slide-out sections provided at intervals along the busway housing.

The busway enclosure is grounded to the service entrance equipment ground if there is continuity throughout the length of the busway enclosure. Bonding jumpers must be installed at all expansion joints or telescoping sections to provide the required electrical continuity.

Field cutting of lighting busway to the required length is permissible. However, field cutting of other types of busway is not allowed. When special lengths are required, they must be ordered from the busway manufacturer.

Busways must be securely supported at intervals not over 5 feet, unless specifically approved for supports at greater intervals. But spacing of supports can never exceed 10 feet. Where a busway is installed in a vertical position, the supports for the bus bars must be designed for vertical installation.

Labor Installing Busway

Work Element	Unit	Man-Hours Per Unit
Copper conductance busway in steel case		
225 AMP	10 L.F.	4.7
400 AMP	10 L.F.	5.7
600 AMP	10 L.F.	6.8
800 AMP	10 L.F.	8.1
1000 AMP	10 L.F.	9.0
1350 AMP	10 L.F.	11.5
1600 AMP	10 L.F.	16.9
2000 AMP	10 L.F.	24.0
2500 AMP	10 L.F.	28.0
3000 AMP	10 L.F.	34.0
4000 AMP	10 L.F.	46.0
5000 AMP	10 L.F.	58.0
Deduct for aluminum bus	%	15.0
Cable tap boxes		
225 AMP	Each	6.5
400 AMP	Each	9.1
600 AMP	Each	13.2
800 AMP	Each	21.0
1000 AMP	Each	25.0
1350 AMP	Each	30.0
1600 AMP	Each	32.0
2000 AMP	Each	36.0
2500 AMP	Each	41.0
3000 AMP	Each	44.0
4000 AMP	Each	48.0
5000 AMP	Each	53.0
Add for 4 wire boxes	%	10.0
Bus duct plug-in circuit breakers		
60	Each	2.9
100	Each	3.6
225	Each	4.6
400	Each	6.8
600	Each	9.7
800	Each	15.0

Busway labor includes handling materials into place, surface installation of 3 or 4 wire busway including supports and fittings to 15' height, cleanup and repairs as needed.
Suggested Crew: 1 electrician and 1 helper

Labor Installing Bus Duct Switching

Work Element	Unit	Man-Hours Per Unit
3 pole, 3 fuse switches		
30	Each	2.6
60	Each	2.6
100	Each	3.5
200	Each	4.7
400	Each	6.4
600	Each	9.3
800	Each	14.0
4 pole, 3 fuse switches		
30	Each	2.8
60	Each	2.8
100	Each	3.5
200	Each	5.2
400	Each	7.0
600	Each	9.8
800	Each	14.8

Labor includes handling materials into place, installation, cleanup and repairs as needed.
Suggested Crew: 1 electrician

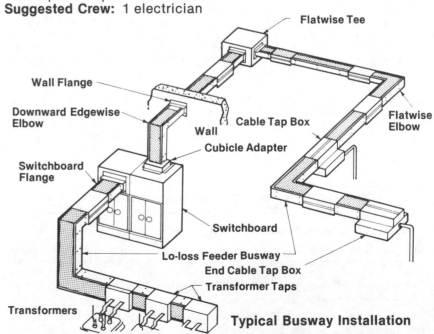

Typical Busway Installation

Labor Installing Cable

Work Element	Unit	Man-Hours Per Unit
Cross linked poly cable, 5 KV, un-shielded		
No. 8	100 L.F.	2.3
No. 6	100 L.F.	2.5
No. 4	100 L.F.	2.7
No. 2	100 L.F.	3.4
No. 1/0	100 L.F.	3.6
No. 2/0	100 L.F.	4.0
No. 3/0	100 L.F.	4.2
No. 4/0	100 L.F.	4.9
No. 250 MCM	100 L.F.	5.4
No. 350 MCM	100 L.F.	5.7
No. 500 MCM	100 L.F.	6.3
No. 750 MCM	100 L.F.	7.0
No. 1000 MCM	100 L.F.	8.8
Add for shielded 15 KV cable with grounded neutral	%	10.0
Type XHHW 3 wire 600 volt (without ground) laid in cable tray		
No. 1/0	100 L.F.	5.8
No. 2/0	100 L.F.	6.4
No. 3/0	100 L.F.	7.3
No. 4/0	100 L.F.	8.2
No. 250 MCM	100 L.F.	9.3
No. 350 MCM	100 L.F.	10.5
No. 500 MCM	100 L.F.	12.2
No. 750 MCM	100 L.F.	12.8
Add for 15,000 volt poly cable	%	5.0
Low voltage alarm & communications cable		
1 pair cable	100 L.F.	.6
2 pair cable	100 L.F.	.7
3 pair cable	100 L.F.	.9
4 pair cable	100 L.F.	1.1
5 pair cable	100 L.F.	1.5

Time includes handling materials into place, installation, cleanup and repairs as needed but no connecting.
Suggested Crew: 1 electrician and 1 helper

Labor Installing Cable In Buildings

Work Element	Unit	Man-Hours Per Unit
Non-metallic sheathed cable		
No. 14, 2 conductor	100 L.F.	2.9
No. 12, 2 conductor	100 L.F.	3.0
No. 10, 2 conductor	100 L.F.	3.2
No. 14, 3 conductor	100 L.F.	3.3
No. 12, 3 conductor	100 L.F.	3.4
No. 10, 3 conductor	100 L.F.	3.5
No. 8, 3 conductor	100 L.F.	3.7
No. 6, 3 conductor	100 L.F.	3.8
BX armored cable		
No. 14, 2 conductor	100 L.F.	3.0
No. 12, 2 conductor	100 L.F.	3.1
No. 10, 2 conductor	100 L.F.	3.3
No. 8, 2 conductor	100 L.F.	3.4
No. 14, 3 conductor	100 L.F.	3.4
No. 12, 3 conductor	100 L.F.	3.6
No. 10, 3 conductor	100 L.F.	4.0
No. 8, 3 conductor	100 L.F.	4.1
No. 14, 4 conductor	100 L.F.	3.6
No. 12, 4 conductor	100 L.F.	4.0
No. 10, 4 conductor	100 L.F.	4.3
No. 8, 4 conductor	100 L.F.	4.5

Time includes handling materials into place, drilling for concealed wall installation, running wire into place, clean-up and repair as needed.
Suggested Crew: 1 electrician and 1 helper

Labor Installing Direct Burial Cable

Work Element	Unit	Man-Hours Per Unit
No. 12	100 L.F.	.35
No. 10	100 L.F.	.37
No. 8	100 L.F.	.46
No. 6	100 L.F.	.58
No. 4	100 L.F.	.81
No. 2	100 L.F.	.90
No. 1	100 L.F.	1.2
No. 1/0	100 L.F.	1.3
No. 2/0	100 L.F.	1.7
No. 3/0	100 L.F.	1.8
No. 4/0	100 L.F.	2.2
No. 250 MCM	100 L.F.	2.5
No. 300 MCM	100 L.F.	2.9
No. 350 MCM	100 L.F.	3.2
No. 400 MCM	100 L.F.	3.7
No. 500 MCM	100 L.F.	4.1
No. 600 MCM	100 L.F.	4.8
No. 750 MCM	100 L.F.	5.8
No. 1000 MCM	100 L.F.	6.2

Labor assumes 3 cables of single conductor type RR copper wire is laid in trench. Deduct 10% for aluminum wire. No trenching or backfill included.
Suggested Crew: 1 electrician and 1 helper

Labor Installing Grounding Systems

Work Element	Unit	Man-Hours Per Unit
Grounding rods, 10' long		
½"	Each	1.1
⅝"	Each	1.2
¾"	Each	1.3
Ground rod clamps		
½", 2 strand	Each	.4
⅝", 1 strand	Each	.4
¾", 2 strand	Each	.4
Cold water main clamps		
2"	Each	.6
3"	Each	.9
4"	Each	1.2
5"	Each	1.6
6"	Each	1.8
Copper ground bar, ¼"		
2" wide	10 L.F.	1.0
3" wide	10 L.F.	1.6
4" wide	10 L.F.	2.1
Thermoweld coupling	Each	1.0

Labor includes handling and placing materials, connection, cleanup and repairs as needed.
Suggested Crew: 1 electrician

Labor Installing Motor Generator Sets

Work Element	Unit	Man-Hours Per Unit
30 KW	Each	32
40 KW	Each	41
50 KW	Each	47
75 KW	Each	50
100 KW	Each	65
125 KW	Each	74
150 KW	Each	82
175 KW	Each	95
200 KW	Each	117
250 KW	Each	135
300 KW	Each	147
Connect motor up to 5 H.P.	Each	1.0
Connect 7½ to 10 H.P. motor	Each	2.0
Connect 15 to 30 H.P. motor	Each	3.2
Connect 40 to 50 H.P. motor	Each	8.4
Connect 60 to 100 H.P. motor	Each	9.5

Labor assumes a 3 phase, 4 wire, 1800 RPM 120/208 volt motor generator set is installed on a prepared pad one or two levels below ground. No fresh air exhaust duct, power wiring or conduit is included.

Labor Installing Motor Control Stations

Work Element	Unit	Man-Hours Per Unit
Control stations, including NEMA type I enclosure		
1 button	Each	1.0
2 button	Each	1.5
3 button	Each	2.0
4 button	Each	2.1

Labor includes handling materials into place, mounting in a prepared position, connecting, cleanup and repairs as needed.

Labor Installing Motor Control Center Starters

Work Element	Unit	Man-Hours Per Unit
Class I type starter		
Size 1	Each	5.3
Size 2	Each	5.3
Size 3	Each	6.4
Size 4	Each	6.7
Size 5	Each	8.0
Size 6	Each	10.3
Class II type starter		
Size 1	Each	8.1
Size 2	Each	8.5
Size 3	Each	11.0
Size 4	Each	11.8
Size 5	Each	15.0
Size 6	Each	15.5

Hours assume a full voltage non-reversing type control center with combination circuit breaker is being installed in a prepared location, including handling into position, cleanup and repairs as needed.

Labor Installing 600 Volt 3 Pole Starters

Work Element	Unit	Man-Hours Per Unit
Push button or switch type starter, including NEMA type I enclosure		
Size 0	Each	3.1
Size 1	Each	4.2
Size 2	Each	5.3
Size 3	Each	6.3
Size 4	Each	8.5
Size 5	Each	10.2
Size 6	Each	18.2
Size 7	Each	26.0
Size 8	Each	32.0

Labor includes handling materials into place, mounting in a prepared position, connecting, cleanup and repairs as needed.

Open Type Substations

Substations are used to transform the high voltage used in distribution systems to working voltages. Substations are usually located as near the electrical load center as practical.

Two types of substations used:

The exterior type: requires outdoor equipment with enclosures that will withstand weather. They are installed outdoors on transformer pads or on overhead pole structures.

The interior type: requires only conventional NEMA I Switch enclosures and is installed indoors in transformer rooms or vaults.

A substation consists of a high-voltage section, a transformation section and a low-voltage section. The high-voltage sections consist of cable terminations or potheads, lightning arresters, primary porcelain fuse cutouts, oil fuse cutouts, fused or non-fused load break switches, switches that only interrupt exciting current and primary oil or air circuit breakers. This equipment is used to provide overload protection to the transformers.

The transformation section will normally consist of a two winding (conventional) or single winding (auto transformer) type. A ferrous metal enclosure usually encases the copper or aluminum windings.

The low-voltage section consists of secondary devices such as disconnect switches (safety switches) switchboards, air circuit breakers or secondary terminal chambers. This equipment is used to provide protection to the secondary feeders.

The substation bus structure also includes necessary copper or aluminum bus bars, cables, insulators and structural supports for the equipment.

Substations are used in various voltage and amperage ratings to meet the system characteristics specified in the contract documents.

Transformers may be either liquid-filled or the dry type. Transformers will be either pole type, pad mounted, subway, network type, or unit substation. Oil-filled transformers are used outdoors or indoors in transformer vaults, if constructed to NEC requirements which requires curbs, fire doors etc. Dry transformers are used indoors where the room need not be a fire resistant vault as defined by the NEC. Voltages of transformers installed indoors will rarely exceed 15KV. Dry transformers are also used outdoors where located closer to the building than the minimum distances specified for oil filled transformers. Transformer KVA ratings can be increased by the addition of forced air or forced oil cooling.

Outdoor transformer stations must be located so as not to interfere with access to the building and, if possible, clear of any other underground utilities. Where possible, they should be protected from the weather. Indoor transformers should be located as near as possible to large electrical loads. Transformer rooms must not contain through pipes, ducts or

other foreign systems but must have adequate ventilation to dissipate heat given off by transformers. Gravity ventilation is usually not adequate; mechanical ventilation may be required.

Labor Installing Transformers

Work Element	Unit	Man-Hours Per Unit
Dry transformers, 5 KV		
100 KVA	Each	30
150 KVA	Each	31
225 KVA	Each	32
300 KVA	Each	34
400 KVA	Each	35
500 KVA	Each	38
750 KVA	Each	40
Oil filled transformers, 15 KVA		
100 KVA	Each	32
150 KVA	Each	33
225 KVA	Each	33
300 KVA	Each	35
500 KVA	Each	38
750 KVA	Each	44
1000 KVA	Each	47
1500 KVA	Each	52
2000 KVA	Each	59
Primary disconnect switch, indoor, 600 AMP		
5 KV	Each	39
15 KV	Each	44

Labor includes handling the materials into place, floor mounting in a prepared compartment, connecting, cleanup and repairs as needed.

Load Centers and Switchboards

National Electrical Manufacturers Association classifies switchboard enclosures and construction according to classes.

There are three common classes. All three classes have the following common features: 600 volts or less, a nominal height of 90 inches excluding floor sills, vertical sections which are electrically interconnected, totally enclosed except the bottom, with spaces provided at the top or bottom of the vertical sections for conduit and cable entry.

Class I switchboard is a supported structure with the following features.

Ampere rating of 22,000 amperes or less.

Vertical sections designed to be front accessible only, directly mounted against a wall or partition for support.

Circuit breakers or fusible devices for branch circuits group mounted with a bus in an integrated assembly.

Depth of each vertical section may vary but must all line up in the rear.

Class II switchboard is a self-supporting structure with the following features:
Ampere rating of 4,000 amperes or less.

Vertical sections designed to be front or rear accessible, self-supporting, free standing.

Circuit breakers or fusible devices for branch circuits group mounted with a bus in an integrated assembly.

Depth of each vertical section may vary but must line up in front or rear.

Class III switchboard is a self-supporting structure with the following features:
Ampere rating of 4,000 amperes or less.

Vertical sections are rear accessible only, designed to be self-supporting, free standing.

Circuit breaker or fusible devices for branch circuits are individually mounted complete with bus and device (line and load) connections made in the rear.

Depth of each vertical section must be same and line-up in front and rear.

Class II and Class III switchboards may be combined into a single switchboard line-up.

The **framework** of switchboards consists of cold gauge formed steel channels and angles welded and bolted together. The floor mounting base consists of a single die-formed channel-shaped member or commercial channel steel. Base mounting holes are provided in each section frame. Holes are furnished in the sides of each frame assembly so that any number of sections can be bolted together to form multiple section

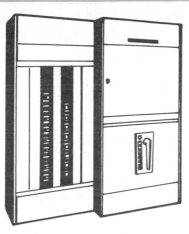

NEMA Class I   NEMA Class II

NEMA Class III

switchboards. The multiple section switchboard is factory constructed as one complete unit or taken apart for shipment.

Closure plates on the front of all switchboards are screw removable and formed with rolled edges for increased rigidity. Where plates contain meters or front-mounted devices or require opening for access to fuses, they are hinged on one side and held closed with knurled screws. These doors are interlocked where required by code.

The spacing and positioning of the **bus bars** depends on the rating of the switchboard and the devices it contains. The bus bars are classed as either main bus or distribution panel bus. The distribution panel bus is the vertical bus which directly feeds the overcurrent protective devices in group mounted construction. Main bus is all other bus which would include cross-over bus between distribution bus and all bus bars ahead of a group mounted distribution panel.

Main devices for overcurrent protection or disconnect purposes are available as fusible or non-fusible switches, molded case circuit breakers, and power circuit breakers. Molded case circuit breaker frames and power circuit breakers should be sized so that the available short-circuit current of the system is within the interrupting capability of the device.

Metering can be provided in all switchboards for either power company or user purposes. This generally consists of a current transformer compartment which is a front accessible, front connected portion of a switchboard section. Where front accessibility is not required, instrument transformers can be located elsewhere in the switchboard and front connected space conserved.

Load center means the low-voltage outgoing feeder section of a unit substation. In this case the load center is an integral part of the substation and the entire substation is a single compact unit. The low-voltage section or load center is essentially the same as the switchboards described previously. The load center is a part of the unit whereas switchboards are usually free standing, individual structures. The secondary devices are essentially the same as those used in a switchboard. All application and inspection procedures would be the same as those for switchboards.

The low-voltage outgoing feeder section receives low-voltage power from the transformer and controls its distribution to the loads. It consists of a metal enclosure with devices to provide disconnecting means, overcurrent protection and instrumentation for the main and feeder circuits. The low-voltage section may contain:

Low-voltage power circuit breakers-drawout or stationary mounted.

Molded case circuit breakers, group or individually mounted.

Fusible switches and motor starting equipment.

Instruments, meters and auxiliary equipment.

Provisions are made in the equipment to accommodate feeder cable and conduit and connections to bus duct.

Switchboards are used to subdivide electric power. They may be located where the power service enters a building as "service entrance switchboards" or centrally located through an installation as "distribution switchboards." The National Electrical Code requires that overcurrent protection be provided at every point in a distribution system where conductor sizes are reduced. Switchboards house these overcurrent protective devices. Screw-on cover plates or hinged doors limit all inadvertent unintentional access to any current-carrying parts. Switchboards are designed to be free standing, totally enclosed "dead front" enclosures for safety to operating personnel.

Switchboards are generally built-in NEMA Type 1 general purpose enclosures for indoor installation where atmospheric conditions are normal. NEMA Type 2 industrial enclosures are required to keep out dust. All screw removable plates and hinged doors are gasketed with neoprene sponge stripping. NEMA Type 3, weather-resistant construction for outdoor installations, is also available. This construction permits exterior exposure to rain or moisture and includes a welded base, gasketed closure plates, front bulk-head doors and special paint finish.

Switchboards may consist of single section containing a few devices or multiple sections containing many devices. Frame structures are standard and line up one with another. Switchboards can be applied in a system to serve any one of three basic functions such as service entrance switchboards, distribution switchboards and combination switchboards.

Service entrance switchboards are located as near as possible to places where the service feeder enters the building carrying the full current load of the installation. This major feeder conductor should be as short as possible. The probability of a fault or mechanical damage occurring is less in shorter conductors. Service entrance switchboards should be arranged to serve this function. The NEC requires that each set of service entrance conductors be provided with a readily accessible means of disconnecting all conductors from the source of supply. The means of disconnect is often housed in the service entrance switchboard as a main switch or main circuit breaker feeding the branch circuit devices. Exception is made when the service can be disconnected by means of not more than six switches or six circuit breakers, manually operable, located at a readily accessible point. This is commonly known as the "six-subdivision rule."

The **service** or **system ground** is the intentional grounding of one conductor in the system. In the conventional 3-phase 4-wire wye connected system, the neutral point of the distribution transformer wye secondary windings establishes the service or system ground. The **equipment ground** consists of a low impedance electrical bonding of all subsequent bus-way enclosures, switchboard frames, conduit, motor frames, receptacles, etc. Both the service or system ground and the equipment ground of a wye system should be connected in common and grounded at the service entrance switchboard. Throughout the remainder of the system, the two should remain separate and independent of one another.

Distribution switchboards are used in systems large enough to warrant their need. A distribution switchboard would be fed by a branch device located in the service entrance switchboard and its branch devices would feed subsequent panelboards or large loads directly.

Combination switchboards serve the function of both the service entrance switchboard and the distribution switchboard. They tend to be used in medium-sized installations where one multiple section switchboard can feed all the loads.

Switchboard location depends primarily on code requirements, safety, and economy. Economy would best be served by positioning a switchboard as close as possible to the load it feeds. However, the National Electrical Code requires that the disconnecting means be located at a readily accessible point nearest to the entrance of the conductors, either inside or outside the building wall. The NEC adds that conductors buried in 2 inches of concrete or brick are considered outside the building.

Metal-Clad Switchgear

The term **metal-clad switchgear** indicates a type of design in which all the equipment required to control an individual circuit, including bus, circuit breaker, disconnecting devices, current and potential transformers, controls, instruments and relays, is assembled in one metal cubicle. The circuit breaker is removable from the cubicle. Circuit breakers can be of the oil or air type, although the trend is strongly toward use of air circuit breakers.

Circuit-breaker disconnection can be by vertical-lift or horizontal-drawout designs. Interlocks are provided in metal-clad assemblies to prevent disconnecting or connecting the circuit breaker if it is closed and to keep anyone from coming in contact with the high-voltage circuits when the circuit breaker is removed from the cubicle.

In metal-clad switchgear the basic incoming or feeder unit consists of a stationary, self-supporting cubicle which houses the controls, instrumentation and the removable circuit breaker. The cubicle is fabricated from structural steel channels and sheets welded together to form a rigid, completely enclosed unit. The assembly of the cubicle is done in jigs which assures complete uniformity of all cubicles. Uniformity of cubicles is inportant because breakers of the same size and rating must be physically interchangeable between cubicles.

The design of metal-clad switchgear provides maximum safety for the operator. During normal service, there is no danger of his accidentally coming in contact with high-voltage equipment because the equipment and connections are enclosed in grounded metal compartments. To gain access to these compartments, the bolted-on covers must be removed. The operator can have access to control wiring and secondary connection compartments while the unit is in service because steel barriers isolate these sections from the high-voltage circuits. Access to these compartments is provided by hinged access doors on panels. The front of the switchgear is the side on which the instrument panel is inserted. On indoor units, the breaker is located on the panel side. In outdoor units, the breaker is inserted in the rear (side opposite instrument panel).

Wiring, including the front panel wiring is stranded and flexible and should have an extremely long life. The panel wire is carried across the hinge in a bundle without the use of terminal blocks, eliminating a possible source of poor connections. This bundle is not subjected to sharp bends, thus eliminating the possibility of wire breakage. The inherent durability of stranded wire over solid wire further reduces the possibility of wire breakage from vibration during shipment. Terminal blocks for connection of external or off-panel wires are located in the control voltage compartment in a vertical column.

Metal-clad switchgear assemblies are used for indoor and outdoor installations with a voltage range of 2.4 KV. through 15 KV., continuous current rating. They provide centralized circuit control for medium voltage systems and are used for control and protection of motors, transformers, generators, distribution lines, rectifiers, and similar power equipment. Metal-clad switchgear can be furnished as single units or in special arrangements such as double bus units, back to back construction, transfer bus arrangements and double bus arrangements.

Labor Installing Secondary Distribution Sections

Work Element	Unit	Man-Hours Per Unit
Secondary distribution switch with main circuit breaker		
225 AMP	Each	18
600 AMP	Each	20
1600 AMP	Each	37
2000 AMP	Each	46
3000 AMP	Each	59
4000 AMP	Each	69
Secondary distribution fusible switch		
800 AMP	Each	20
1200 AMP	Each	23
1600 AMP	Each	30
2000 AMP	Each	41
2500 AMP	Each	49
3000 AMP	Each	58
4000 AMP	Each	67
Additional secondary feeder circuit breakers		
100 AMP	Each	3.0
225 AMP	Each	4.3
400 AMP	Each	6.2
600 AMP	Each	8.2
800 AMP	Each	9.2
1000 AMP	Each	11.6
1200 AMP	Each	17.0
Additional secondary feeder switching, NEMA class I enclosure		
30 to 100 AMP, 2 branch	Each	3.0
60 AMP, 2 branch	Each	3.0
100 AMP, 2 branch	Each	3.1
200 AMP, 1 branch	Each	4.0
400 AMP, 1 branch	Each	5.3
600 AMP, 1 branch	Each	7.0
800 AMP, 1 branch	Each	8.8
1200 AMP, 1 branch	Each	17.0
Complete metering section	Each	21.0

Includes handling material into place, installation of 3 pole, 240 volt capacity units in a prepared position, connecting, cleanup and repairs as needed.

Panelboards

The primary function of a panelboard is to provide protection for light, heat or power circuits. Panelboards are designed to be placed in or against a wall or partition and must be accessible from the front. There are an infinite number of panelboard combinations possible and each of these has its own peculiarities. The panelboard is the distribution center of the circuits supplied from it.

A lighting and appliance branch circuit panelboard is one having 10% of its overcurrent devices rated 30 amperes or less, for which neutral connections are provided.

A distribution panelboard or power panelboard has branch circuit switching and overcurrent protective devices primarily used as distribution circuits to lighting or appliance branch circuit panelboards. Branch circuit switching may also be provided to other distribution or power panels or to feed a group of branch circuits other than lighting or appliance type.

A combination service entrance distribution panel has sections for lighting and power and is suitable for underground or overhead service and has provisions for load cables leaving the bottom as well as the top of the section. The panelboard must meet the requirements of the N.E.C. for each particular installation.

Panelboards consist of the following main components: bussing and breaker assembly, main terminal lugs, neutral terminal connections, steel cabinet and front or trim. (See Figure 16-1).

The main assembly consists of a bus structure and main lugs or main breaker as specified. The current rating (maximum capacity in amperes) of a panelboard must not exceed the current-carrying capacity of the main bus bars, or the main circuit breaker if included as a part of the panelboard.

Circuit breakers must be of the ampere rating, number of poles, frame size and interrupting capacity specified. Breakers are usually the bolt-on types unless otherwise specified. Circuit breaker arrangement is usually in two vertical rows. Circuit breakers are in the "on" position when moved horizontally towards the center of the board. Odd number breakers are typically on the left side, even numbers on the right. These conventions do not apply to column type panelboards in which the over-current devices are numbered consecutively vertically.

The cabinet is an enclosure designed either for surface or flush mounting and has a frame or trim in which swinging doors are hung. A branch circuit directory should be provided inside the door. Door locks may be provided.

Solid neutral assemblies are usually mounted on the panelboard interior end opposite the mains. All neutrals should be listed for aluminum or copper conductors for branch circuits or mains.

Labor Installing Panelboards

Work Element	Unit	Man-Hours Per Unit
Light and power panels		
8 circuit	Each	10.5
10 circuit	Each	10.6
12 circuit	Each	10.8
14 circuit	Each	10.9
16 circuit	Each	11.1
18 circuit	Each	11.3
20 circuit	Each	11.6
22 circuit	Each	15.8
24 circuit	Each	16.3
26 circuit	Each	17.4
28 circuit	Each	18.4
30 circuit	Each	19.7
32 circuit	Each	21.0
34 circuit	Each	23.0
36 circuit	Each	25.6
38 circuit	Each	28.0
40 circuit	Each	29.0
42 circuit	Each	31.6
Power panel, 600 volts, with main lugs only		
225 AMPS, 45" high	Each	4.0
400 AMPS, 63" high	Each	6.3
600 AMPS, 81" high	Each	8.3
1200 AMPS, 45" high, 400 AMP branch	Each	12.4
1200 AMPS, 99" high, 800 AMP branch	Each	15.0
Power panels with main circuit breakers only		
225 AMPS, 18" high	Each	4.3
400 AMPS, 27" high	Each	4.6
600 AMPS, 36" high	Each	4.9
800 AMPS, 72" high	Each	6.4
600 AMPS, 72" high with 400 AMP branch	Each	7.6
800 AMPS, 72" high with 400 AMP branch	Each	7.6

These figures assume 3 phase, 4 wire type NQO, NHIB or NQHB panelboards with main lugs only and 3 pole circuit breakers are installed in a prepared location, including handling, cleanup and repairs as needed.
Suggested Crew: 2 or 3 electricians

Labor Installing Branch Circuit Breakers

Work Element	Unit	Man-Hours Per Unit
1 pole, 60 AMP	Each	.8
1 pole, 100 AMP	Each	1.1
2 pole, 60 AMP	Each	1.2
2 pole, 100 AMP	Each	1.3
3 pole, 60 AMP	Each	2.1
3 pole, 100 AMP	Each	2.3
3 pole, 225 AMP	Each	2.6
3 pole, 400 AMP	Each	3.9
3 pole, 600 AMP	Each	4.2

Use the figures for installing branch breakers in light and power panels.

Suggested Crew: 1 electrician

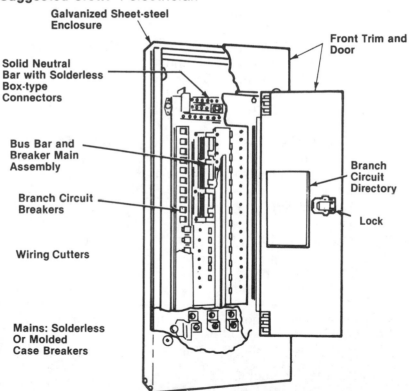

Galvanized Sheet-steel Enclosure

Solid Neutral Bar with Solderless Box-type Connectors

Bus Bar and Breaker Main Assembly

Branch Circuit Breakers

Wiring Cutters

Mains: Solderless Or Molded Case Breakers

Front Trim and Door

Branch Circuit Directory

Lock

Typical Panelboard and Breakers
Figure 16-1

Underfloor Electrical Systems

The six most common underfloor electrical distribution systems are: (1) ducts in fill; (2) ducts in structural slabs; (3) cellular metal floors; (4) cellular concrete floors; (5) ducts between the bottom of floor and a hung ceiling; (6) ducts running between or through bar joists.

An underfloor raceway system consists of ducts laid below the surface of the floor and interconnected by means of special cast iron junction boxes. The ducts are made of either fiber or steel. Fiber ducts are made in two types, the open bottom type and the completely enclosing type. Steel ducts are always of the completely enclosing type, usually with a rectangular cross-section.

Provision is made for outlets by means of specially designed floor-outlet fittings which are screwed into the walls of the ducts. Junction boxes are located at each end of a run of raceway. The tops are placed flush with the floor covering.

Underfloor raceway systems are used principally in fire resistant construction. The raceways provide space in the floor slab for installation of wiring for telephone and signal systems and for outlets for office machinery.

The NEC allows their use when embedded in concrete or in concrete fill of floors. They are not allowed in locations which are exposed to corrosive or hazardous conditions. Generally, the ducts or raceways are laid out in the floor to form a network. This type of construction is particularly suitable for large office areas or where outlet locations are subject to change.

Steel-duct raceways are recommended for use where expansion of facilities is anticipated. The wires and cables are pulled through the ducts after the duct system is completely installed.

Cellular-metal-floor raceways are formed in the floor construction by structural members made of corrugated steel. Hollow spaces or cells in the concrete form raceways for the wires. The cells are cross connected by headers formed over the beams in the space left between adjacent lengths of flooring. The headers are also wire raceways, providing electrical access from distribution points.

Cellular-metal-floor raceways are used principally for office buildings where the needs of the tenants for wiring outlets are not known very definitely in advance and where plenty of outlets and good interior appearance must be maintained.

The cellular steel floor is considered most adaptable for future rearrangement or additions to the building's electrical or signal systems. The cost may be lower than if a standard floor with a large number of underfloor raceways were installed. The position of the raceways may be varied in 6 inch increments, and a very flexible system of outlets for telephones, business machines, and desk lamps may be provided with all outlets located under the desks and out of sight.

16 Electrical

Connections to the cells from a distribution center are made with metal header ducts which are run horizontally across the precast slab and embedded in the concrete fill over the slab. Connections from the headers to the cells are made through handholes or metal junction boxes. Outlets can be located at any point along a cell.

Labor Installing Underfloor Electric Systems

Work Element	Unit	Man-Hours Per Unit
Underfloor ducting		
3.3 sq. inch cross section	10 L.F.	.8
8.6 sq. inch cross section	10 L.F.	1.0
Header duct, 8.6 sq. inch	10 L.F.	1.0
Single duct junction box	Each	2.0
Double duct junction box	Each	3.0
Triple duct junction box	Each	4.2
Power or phone outlet	Each	2.0
Trench duct		
12" wide	10 L.F.	2.1
18" wide	10 L.F.	2.7
24" wide	10 L.F.	3.0
36" wide	10 L.F.	4.4
Cross fittings		
12"	Each	2.0
18"	Each	2.7
24"	Each	3.3
36"	Each	3.7
Tee fittings		
12"	Each	1.6
18"	Each	2.0
24"	Each	2.6
36"	Each	3.1
Ell fittings		
12"	Each	1.1
18"	Each	1.6
24"	Each	2.1
36"	Each	2.6
Panel connections		
12"	Each	1.1
18"	Each	1.6
24"	Each	2.1
36"	Each	2.6

Labor includes handling materials into place, installing in prepared locations, cleanup and repairs as needed.
Suggested Crew: 1 electrician and 1 helper

Labor Installing Residential and Commercial Electrical Rough-In

Work Element	Unit	Man-Hours Per Unit
Install service main, four wire conductor[1]		
60 amp	Each	11
100 amp	Each	13
200 amp	Each	14
400 amp	Each	16
Install 8' ground rod and ground wire	Each	2.4
Install type NM cable[2]		
No. 10/3 with ground and smaller	250 L.F.	8.0
No. 8/2 with ground and larger	100 L.F.	2.4
Install boxes for type NM cable[3]	10 each	1.6
Install pull boxes[4]		
12" x 12" x 6", NEMA type 1	Each	2.4
16" x 20" x 8", NEMA type 1	Each	3.3
24" x 36" x 8", NEMA type 1	Each	4.0
6" x 6" x 6", NEMA type 3R and 4	Each	3.2
10" x 6" x 6", NEMA type 3R and 4	Each	4.0
16" x 16" x 6", NEMA type 3R and 4	Each	6.5
24" x 18" x 8", NEMA type 3R and 4	Each	10.4
Pull and splice wire in conduit		
No. 10 and smaller	100 L.F.	0.9
No. 8 and larger	100 L.F.	1.4
Install nonmetallic cable (romex), fittings and outlet boxes		
No. 10 and smaller	100 L.F.	3.0
No. 8 and larger	100 L.F.	4.0
Install armored cable (BX), fittings and outlet boxes		
No. 10 and smaller	100 L.F.	3.6
No. 8 and larger	100 L.F.	4.1

[1] Increase man-hour figures by 25% for underground service.
[2] Surface mounted on wood or behind wall or ceiling.
[3] Plastic or metal boxes on wood surface or behind wall or in ceiling.
[4] Includes attaching conduit to pull box.
Suggested Crew: 1 or 2 electricians.

Labor Installing Residential Power Outlets

Work Element	Unit	Man-Hours Per Unit
Duplex (standard)	Each	.6
Duplex (weatherproof)	Each	.7
Outlets 30 amp	Each	.9
Outlets, 40, 50, 60 amp	Each	.9
Motor connection, 3 phase	Each	3.0
J-box	Each	.4
Floor box, flush	Each	1.0
Floor box, with special connectors	Each	1.7
Clock outlets	Each	.4
Fan connections	Each	.4
Outlet testing	Each	.02
Intercom wall outlets	Each	.5
4" octagon steel	10 Each	.4
4" square	10 Each	.4
Floor box	10 Each	.4
Connect small appliances		
Water heater (80 gallon capacity)	Each	1.0
Space heater (200 watts)	Each	0.8
Air conditioning units (up to 18,000 Btu)	Each	1.5

Time does not include wiring but does include move on and off site, unloading, placing, wire hookup, cleanup and repairs as needed.
Suggested Crew: 1 electrician

Labor Installing Industrial Fixtures

Work Element	Unit	Man-Hours Per Unit
Suspended striplights		
1 or 2 tube, 40 watt	Each	.5
2 75 watt tubes	Each	.7
3 75 watt tubes	Each	.8
4 75 watt tubes	Each	1.0

Labor includes handling, hanging and connecting fixture, cleanup and repairs as needed.
Suggested Crew: 1 electrician and 1 helper

Labor Installing Fluorescent Fixtures

Work Element	Unit	Man-Hours Per Unit
Recess mounted in grid ceiling		
1' x 2', 2 20 watt tubes	Each	1.0
1' x 4', 2 40 watt tubes	Each	1.0
2' x 4', 4 40 watt tubes	Each	2.6
3' x 3', 4 30 watt tubes	Each	3.3
3' x 3', 6 30 watt tubes	Each	3.0
4' x 4', 6 40 watt tubes	Each	6.7
4' x 4', 8 40 watt tubes	Each	8.0
Surface mounted on grid or ceiling		
1' x 4', 2 40 watt tubes	Each	1.6
1' x 8', 4 40 watt tubes	Each	3.1
16" x 4', 4 40 watt tubes	Each	1.6
16" x 8', 4 40 watt tubes	Each	2.1
2' x 4', 4 40 watt tubes	Each	1.7
2' x 2', 4 20 watt tubes	Each	3.5
4' x 4', 6 20 watt tubes	Each	3.5
Add for installation in a concealed grid ceiling	%	15.0
Pendant mounted commercial fixtures		
1' x 4', 2 40 watt tubes	Each	1.6
16" x 4', 4 40 watt tubes	Each	2.0
Air handling with boot for grid mounting		
1' x 4', 2 40 watt tubes	Each	1.0
2' x 2', 4 20 watt tubes	Each	1.0
2' x 4', 4 40 watt tubes	Each	1.5
3' x 3', 6 30 watt tubes	Each	3.2
4' x 4', 8 40 watt tubes	Each	4.3
Explosion proof fixtures		
1' x 4', 2 40 watt tubes	Each	4.2
1' x 4', 2 100 watt tubes	Each	6.5
Dust and vapor proof fixtures		
1' x 4', 2 40 watt tubes	Each	2.0
1' x 8', 4 40 watt tubes	Each	3.0

Labor includes handling fixture into position, installing and connecting in a prepared location, cleanup and repairs as needed. Tube installation not included.
Suggested Crew: 1 electrician and 1 helper

Labor Installing Incandescent Fixtures

Work Element	Unit	Man-Hours Per Unit
Ceiling mounted fixtures		
Recessed	Each	1.0
Surface mounted	Each	.7
Pendant mounted	Each	1.0
Wall mounted fixture	Each	.7
Specialty fixtures		
Vapor tight	Each	2.0
Explosion proof	Each	3.1
Mercury vapor ceiling fixtures		
250 watt single lamp	Each	3.2
400 watt single lamp	Each	3.2
400 watt twin lamp	Each	4.2
1000 watt single unit	Each	4.2
Mercury vapor wall fixtures		
175 watt	Each	2.0
Exit lights		
Ceiling mounted	Each	1.0
End mounted	Each	1.0
Pendant mounted	Each	1.6
Emergency 2 lamp battery fixture	Each	4.3

Includes handling materials into place, installing fixtures and trim in prepared locations, connection, cleanup and repairs as needed.
Suggested Crew: 1 electrician

Labor Installing Exhaust Fans

Work Element	Unit	Man-Hours Per Unit
Exhaust fans:		
12" to 24"	1	2
26" to 42"	1	6
48" to 60"	1	6
Power ventilators, industrial & commercial		
6" to 20"	1	8
24" to 42"	1	21

Labor Installing Special Electrical Systems

Work Element	Unit	Man-Hours Per Unit
Electrical snow melting equipment		
Control cabinet	Each	17.0
Sensing bulb	Each	2.0
Junction box	Each	2.0
Wire mesh mats	100 S.F.	2.1
MI cable	100 L.F.	4.0
Lightning protection gear		
24" copper point	Each	1.0
10' ground rod	Each	2.2
Thermal connection	Each	1.0
Cable clamps	Each	.5
Nurses call system		
Main equipment panel	Each	3.6
Nurses control station	Each	16.4
Duty station	Each	3.0
Staff station	Each	1.0
Single bedside station	Each	1.0
Double bedside station	Each	2.0
Emergency call button	Each	1.0
Pull cord button	Each	1.0
Single dome light	Each	1.0
Double dome light	Each	1.0
Pillow speaker	Each	.5
Cord set	Each	.5
Public address system, conventional		
Office	Each	1.5
Industrial	Each	3.0
Sound system		
Speakers, ceiling or wall mounted	Each	1.0
Speakers, trumpet	Each	2.5
Volume control	Each	0.8
Amplifier, 250 watt	Each	8.0
Cabinets	Each	8.0
Intercom, master		
Up to 25 station capacity	Each	8.0
Remote station	Each	0.9

Labor includes handling materials into place, installation and connection, cleanup and repairs as needed.
Suggested Crew: 1 electrician and 1 helper

Fire Alarm Systems

Fire alarm systems are of two basic types: manual, which requires the person discovering a fire to activate the system; and automatic, which detects fire and activates the system without human assistance.

These systems can be designed to perform one or combinations of the following basic functions:

Notify occupants of a fire. This is normally done with fire alarm horns or bells located throughout the building. In locations where they may cause panic, such as hospitals, chimes or other special signaling systems to alert trained personnel are normally specified.

Notify the local fire department. The method normally depends on the type fire alarm system existing at each station. The most common methods are to trip a transmitter which sends a coded signal to the fire department, or trip an open or closed relay tied directly into the fire department.

Supervise fire extinguishing systems. The most common example is activating sprinkler or carbon dioxide systems in the building. Alarm systems can also be arranged to give a trouble signal to warn when the system is not working.

Actuate fire control equipment. Alarm systems can be arranged to close fire doors, shut down air conditioning equipment, or operate other fire controls.

The manual system uses fire alarm stations located near exits. Although the alarm station may be a telephone or other signaling devices, factory assembled fire alarm boxes are most common.

The automatic systems use detection elements to activate the alarm system. Some common detecting elements are: (a) thermostats that operate at a fixed temperature or set rate of rise. (b) heat-thermopile detectors; (c) ionization type smoke detectors; (d) resistant bridge type smoke detectors; (e) photo electric type smoke detectors; (f) heat-detecting wire or pneumatic tube types. These systems are also normally equipped with manual activating stations.

Fire alarm systems consist of several factory assembled components. The most commonly used are described below:

The **Control panel** is the heart of the fire alarm system. It normally contains all switches, relays, visual and audible indicating devices and other parts required for complete control, supervision and testing of the alarm system.

The **annunciator panel** is normally used with noncoded systems in larger facilities to indicate the area or zone in the building from which the alarm signal was activated. Its purpose is to aid fire fighters in finding the fire or problem as quickly as possible.

The **coded transmitter** is used with some fire alarm systems as the means of notifying the fire department. The transmitter houses electrically or mechanically driven mechanisms which turn a code wheel to open or close an electric circuit to transmit a telegraphic signal.

The **master alarm box** is a special type coded transmitter. It is normally located outside the facility near the main entrance and is part of the municipal fire alarm system. The building alarm system is normally connected to trip this transmitter when operated. Shunt trip or local energy trip will be specified. In the shunt trip system, the interior building alarm initiating devices are electrically part of the municipal system circuit and are connected in series in a shunt around the transmitter trip coil. When this shunt circuit is broken, power is applied to the trip coil and the transmitter is activated. The local energy trip type system does not depend on electrical current. It is a separate system which supplies the power to the trip coil to activate the transmitter. The master box also can be tripped manually by use of a pull lever located on the face of the box.

The **master alarm box supervisory panel** is a special panel used in conjunction with a master alarm box. The supervisory panel gives an audible and visual signal when the master box has been activated by an auxiliary device. Its purpose is to assure that the mechanical tripping mechanism is restored to normal position after activation.

Manual fire alarm boxes are provided for manual operation of the fire alarm system. These boxes will be specified as coded or noncoded type. The noncoded type is a simple electric switch. The coded box houses electrically or mechanically driven mechanisms which turn a code wheel to open and close an electric circuit to transmit a telegraphic signal.

Heat detectors are used to activate the alarm system automatically. Heat detectors fall in two general categories: fixed-temperature type, which respond when the detection element reaches a predetermined temperature, and rate-of-rise type, those which respond to an increase in heat at a rate greater than some predetermined value. Some detectors combine both the rate-of-rise and fixed temperature principles.

Two basic types of **smoke detectors** are commonly used: photoelectric type and ionization type. Photoelectric type detectors use a light beam source and a current generating light receiver. The change in current resulting from partial obscuration or reflection of this beam by smoke is measured. When a critical point is reached, an alarm relay is tripped. The ionization detector's sensing element is an ionization chamber which detects the conductivity of air. A voltage applied across the chamber causes a very small electrical current to flow as the ions travel to the electrode of opposite polarity. When either visible or invisible particles of combustion enter the chamber they attach themselves to the ions causing an increase in voltage between electrodes. A predetermined voltage increase trips an alarm relay.

Installation of a fire alarm system includes locating approved prefabricated alarm equipment and interconnecting these components with required wiring. Most alarm system components are located on the design drawings. Automatic detector units are usually located throughout areas requiring protection. Heat detectors are spaced according to the UL listed spacing requirements for that specific detector. Wiring is installed in conduit as required to interconnect these components. Wire sizes may vary from No. 19 AWG to No. 10 AWG, depending on voltage and energy requirements.

Labor Installing Fire Alarm and Signal Systems

Work Element	Unit	Man-Hours Per Unit
Burglar alarm, mechanical or electrical	Each	2.5
Card reader, flush, standard or multi-channel	Each	3.5
Door switch, hinge or magnetic	Each	1.5
Exit control lock, horn or flashing light	Each	2.4
Indicating panels		
1 channel	Each	3.0
10 channel	Each	5.5
70 channel	Each	8.0
Ultrasonic unit with horn; 12, 24 or 120 V.	Each	4.0
Control panel, fire, sprinkler and stand pipe		
4 zone	Each	4.0
8 zone	Each	8.0
12 zone	Each	12.0
Battery rack	Each	2.5
Automatic charger	Each	2.4
Signal bell, trouble buzzer or manual station	Each	0.9
Detector, rate of rise or fixed temperature	Each	0.8
Smoke detector,		
ceiling type	Each	1.5
duct type	Each	2.5
Light and horn	Each	2.5
Fire alarm horn	Each	2.5
Master box	Each	3.5
Break glass station	Each	0.8
Remote annunciator		
8 zone drop	Each	4.0
12 zone drop	Each	5.6
16 zone drop	Each	6.5
Standpipe or sprinkler alarm, alarm device	Each	0.8
Actuation device	Each	0.9

Suggested Crew: 2 electricians

Labor Installing Air-Conditioning, Dehumidifiers, and Refrigerators

Work Element	Unit	Man-Hours Per Unit
Install window air conditioning units		
½ ton to ¾ ton capacity	Each	2.0
1 ton to 1½ ton capacity	Each	3.5
2 ton capacity	Each	4.0
Install self contained water cooled units		
3 ton to 5 ton capacity	Each	18.0
5 ton to 8 ton capacity	Each	24.0
10 ton to 15 ton capacity	Each	36.0
Install air conditioning equipment for:		
25 ton system	Each	240.0
26 ton to 50 ton system	Each	360.0
51 ton to 75 ton system	Each	560.0
76 ton to 100 ton system	Each	840.0
Evaporative condenser, chilled water	Per ton	80.0
Evaporative condenser, direct expansion	Per ton	81.0
Water cooled condenser direct expansion	Per ton	56.0
Set and connect dehumidifiers	Per 1000 C.F. space	.32
Erect and connect walk-in refrigerators		
1,000 C.F. and smaller	Each	40.0
1,050 to 3,900 C.F.	Each	80.0
4,000 to 5,000 C.F.	Each	120.0

Man-hour figures include setting and connecting all equipment except remote coils or units.

Man-hour figures do not include the installation of piping or electrical work between various pieces of equipment, nor installation of ductwork or diffusers.

Suggested crew: two to seven men depending on equipment size and job scope.

Index

Practical References for Builders

National Construction Estimator

Current building costs in dollars and cents for residential, commercial and industrial construction. Prices for every commonly used building material, and the proper labor cost associated with installation of the material. Everything figured out to give you the "in place" cost in seconds. Many time-saving rules of thumb, waste and coverage factors and estimating tables are included. **544 pages, 8½ x 11, $22.50. Revised annually**

Estimating Home Building Costs

Estimate every phase of residential construction from site costs to the profit margin you should include in your bid. Shows how to keep track of man-hours and make accurate labor cost estimates for footings, foundations, framing and sheathing finishes, electrical, plumbing and more. Explains the work being estimated and provides sample cost estimate worksheets with complete instructions for each job phase. **320 pages, 5½ x 8½, $17.00**

Building Cost Manual

Square foot costs for residential, commercial, industrial, and farm buildings. In a few minutes you works up a reliable budget estimate based on the actual materials and design features, area, shape, wall height, number of floors and support requirements. Most important, you include all the important variables that can make any building unique from a cost standpoint. **240 pages, 8½ x 11, $14.00. Revised annually**

Masonry Estimating

Step-by-step instructions for estimating nearly any type of masonry work. Shows how to prepare material take-offs, how to figure labor and material costs, add a realistic allowance for contingency, calculate overhead correctly, and build competitive profit into your bids. **352 pages, 8½ x 11, $26.50**

Cost Records for Construction Estimating

How to organize and use cost information from jobs just completed to make more accurate estimates in the future. Explains how to keep the cost records you need to reflect the time spent on each part of the job. Shows the best way to track costs for sitework, footing, foundations, framing, interior finish, siding and trim, masonry, and subcontract expense. Provides sample forms. **208 pages, 8½ x 11, $15.75**

Estimating Tables for Home Building

Produce accurate estimates in minutes for nearly any home or multi-family dwelling. This handy manual has the tables you need to find the quantity of materials and labor for most residential construction. Includes overhead and profit, how to develop unit costs for labor and materials, and how to be sure you've considered every cost in the job. **336 pages, 8-1/2 x 11, $21.50**

Berger Building Cost File

Labor and material costs needed to estimate major projects: shopping centers and stores, hospitals, educational facilities, office complexes, industrial and institutional buildings, and housing projects. All cost estimates show both the man-hours required and the typical crew needed so you can figure the price and schedule the work quickly and easily. **288 pages, 8½ x 11, $30.00. Revised annually**

Contractor's Guide to the Building Code Revised

This completely revised edition explains in plain English exactly what the Uniform Code requires and shows how to design and construct residential and light commercial buildings that will pass inspection the first time. Suggests how to work with the inspector to minimize construction costs, what common building shortcuts are likely to be cited, and where exceptions are granted. **544 pages, 5½ x 8½, $24.25**

Bookkeeping for Builders

This book will show you simple, practical instructions for setting up and keeping accurate records — with a minimum of effort and frustration. Shows how to set up the essentials of a record-keeping system: the payment journal, income journal, general journal, records for fixed assets, accounts receivable, payable and purchases, petty cash, and job costs. You'll be able to keep the records required by the I.R.S., as well as accurate and organized business records for your own use. **208 pages, 8½ x 11, $19.75**

Carpentry Estimating

Simple, clear instructions show you how to take off quantities and figure costs for all rough and finish carpentry. Shows how much overhead and profit to include, how to convert piece prices to MBF prices or linear foot prices, and how to use the tables included to quickly estimate man-hours. All carpentry is covered; floor joists, exterior and interior walls and finishes, ceiling joists and rafters, stairs, trim, windows, doors, and much more. Includes sample forms, checklists, and the author's factor worksheets to save you time and help prevent errors. **320 pages, 8½ x 11, $25.50**

Estimating Electrical Construction

A practical approach to estimating materials and labor for residential and commercial electrical construction. Written by the A.S.P.E. National Estimator of the Year, it explains how to use labor units, the plan take-off and the bid summary to establish an accurate estimate. Covers dealing with suppliers, pricing sheets, and how to modify labor units. Provides extensive labor unit tables, and blank forms for use in estimating your next electrical job. **272 pages, 8½ x 11, $19.00**

Electrical Construction Estimator

If you estimate electrical jobs, this is your guide to current material costs, reliable manhour estimates per unit, and the total installed cost for all common electrical work: conduit, wire, boxes, fixtures, switches, outlets, loadcenters, panelboards, raceway, duct, signal systems, and more. Explains what every estimator should know before estimating each part of an electrical system. **416 pages, 8½ x 11, $25.00. Revised annually**

Wood-Frame House Construction

From the layout of the outer walls, excavation and formwork, to finish carpentry and painting, every step of construction is covered in detail, with clear illustrations and explanations. Everything the builder needs to know about framing, roofing, siding, insulation and vapor barrier, interior finishing, floor coverings, and stairs — complete step-by-step "how to" information on what goes into building a frame house. **240 pages, 8½ x 11, $14.25. Revised edition**

Contractor's Survival Manual

How to survive hard times in construction and take full advantage of the profitable cycles. Shows what to do when the bills can't be paid, finding money and buying time, transferring debt, and all the alternatives to bankruptcy. Explains how to build profits, avoid problems in zoning and permits, taxes, time-keeping, and payroll. Unconventional advice includes how to invest in inflation, get high appraisals, trade and postpone income, and how to stay hip-deep in profitable work. **160 pages, 8½ x 11, $16.75**

Planning and Designing Plumbing Systems

Explains in clear language, with detailed illustrations, basic drafting principles for plumbing construction needs. Covers basic drafting fundamentals: isometric pipe drawing, sectional drawings and details, how to use a plot plan, and how to convert it into a working drawing. Gives instructions and examples for water supply systems, drainage and venting, pipe systems, refrigeration, gas, oil, and compressed air piping, storm, roof and building drains, fire hydrants, and more. **224 pages, 8½ x 11, $13.00**

Audio: Estimating Remodeling

Listen to the "hands-on" estimating instruction in this popular remodeling seminar. Make your own unit price estimate based on the prints enclosed. Then check your completed estimate with those prepared in the actual seminar. After listening to these tapes you will know how to establish an operating budget for your business, determine indirect costs and profit, and estimate remodeling with the unit cost method. **Includes seminar workbook, project survey and unit price estimating form, and six 20-minute cassettes, $65.00**

Construction Surveying & Layout

A practical guide to simplified construction surveying: How land is divided, how to use a transit and tape to find a known point, how to draw an accurate survey map from your field notes, how to use topographic surveys, and the right way to level and set grade. You'll learn how to make a survey for any residential or commercial lot, driveway, road, or bridge — including how to figure cuts and fills and calculate excavation quantities. If you've been wanting to make your own surveys, or just read and verify the accuracy of surveys made by others, you should have this guide. **256 pages, 5½ x 8½, $19.25**

Builder's Guide to Accounting Revised

Step-by-step, easy-to-follow guidelines for setting up and maintaining an efficient record keeping system for your building business. Not a book of theory, this practical, newly-revised guide to all accounting methods shows how to meet state and federal accounting requirements, including new depreciation rules, and explains what the tax reform act of 1986 can mean to your business. Full of charts, diagrams, blank forms, simple directions and examples. **304 pages, 8½ x 11, $20.00**

Estimating Plumbing Costs

Offers a basic procedure for estimating materials, labor, and direct and indirect costs for residential and commercial plumbing jobs. Explains how to interpret and understand plot plans, design drainage, waste, and vent systems, meet code requirements, and make an accurate take-off for materials and labor. Includes sample cost sheets, manhour production tables, complete illustrations, and all the practical information you need to accurately estimate plumbing costs. **224 pages, 8½ x 11, $17.25**

Estimating Painting Costs

Here is an accurate step-by-step estimating system, based on a set of easy-to-use manhour tables that anyone can use for estimating painting costs: from simple residential repaints to complicated commercial jobs — even heavy industrial and government work. Explains taking field measurements, doing take-offs from plans and specs, predicting productivity, figuring labor, material costs, overhead and profit. Includes manhour and material tables, plus samples, forms, and checklists for your use. **448 pages, 8½ x 11, $28.00**

Contractor's Growth and Profit Guide

Step-by-step instructions for planning growth and prosperity in a construction contracting or subcontracting company. Explains how to prepare a business plan: selecting reasonable goals, drafting a market expansion plan, making income forecasts and expense budgets, and projecting cash flow. Here you will learn everything required by most lenders and investors, as well as solid knowledge for better organizing your business. **336 pages, 5½ x 8½, $19.00**
